Multivariable CALCULUS

from Graphical, Numerical, and Symbolic Points of View

Revised Preliminary Edition

Arnold Ostebee
Paul Zorn
St. Olaf College

Saunders College Publishing
Harcourt Brace College Publishers
Fort Worth Philadelphia San Diego New York Orlando Austin
San Antonio Toronto Montreal London Sydney Tokyo

This material is based upon work supported by the National Science Foundation under Grant No. DUE- 9450765.

Any opinions, findings, and conclusions or recommendations expressed in this material are those of the authors and do not necessarily reflect the views of the National Science Foundation.

Printed in the United States of America

Ostebee & Zorn: _Multivariable Calculus from Graphical, Numerical and Symbolic Points of View._ Revised Preliminary Edition.

0-03-023786-6

890123 202 987654

Contents

Notes for instructors

This book aims to do what its title suggests: present multivariable calculus from graphical, numerical, and symbolic points of view. In doing so, we continue the philosophy and viewpoints embodied in our two volumes on single-variable calculus, *Calculus from Graphical, Numerical, and Symbolic Points of View*, also published by Saunders College Publishing. For many more details on philosophy, strategy, use of technology, and other issues, see either those volumes or our World Wide Web site: `http://www.stolaf.edu/people/zorn/ozcalc/mvcindex.html`

What's new? This edition embodies several changes from the previous (1996) version. They include: (i) **reorganization** and reordering of topics to achieve a more "linear" coverage (e.g., multivariable integrals appear in one chapter, not two); (ii) **new problems** have been added to most sections; (iii) **Appendix C**, a brief introduction to matrices and determinants, has been added; (iv) several student **Projects** have been appended to various sections (see the table of contents); (v) **Chapter 11**, on infinite series, has been reprinted from Volume 2 of our single-variable text to serve instructors who teach this material in Calculus III.

Audience and prerequisites. The text addresses a general mathematical audience: mathematics majors, science and engineering majors, and non-science majors. We assume a little more mathematical maturity than for single-variable calculus, but the presentation is not rigorous in the sense of mathematical analysis. We want students to encounter, understand, and use the main concepts and methods of multivariable calculus, and to see how they extend the simpler objects and ideas of elementary calculus. A fully rigorous logical development belongs later in a student's mathematical education.

We assume that students have the "usual" one-year, single-variable calculus preparation, but little or nothing more than that. A basic familiarity with numerical integration techniques (such as the midpoint rule) is helpful, but it could be developed enroute if necessary. (We do not assume that students have studied single-variable calculus from our own text!)

Although we stress linear functions and linear approximation, we do not assume that students have formal experience with linear algebra. Vectors are used often, but are introduced "from scratch." Matrices appear only occasionally (but are important when they do appear). For students unfamiliar with matrices, or who need basic review, Appendix C should be useful. (It can be either covered "officially" or left to student reading.)

Technology. Technology is an important tool for illustrating and comparing graphical, numerical, and symbolic viewpoints in calculus—especially in multivariable calculus, where calculations can be messy, and where geometric intuition is harder to come by. Although we refer occasionally to computations done with *Maple*, other programs (*Mathematica*, *Derive*, the TI-92, etc.) would do just as well. In any event, we strongly recommend that students have access to (and use!) *some* capable and flexible technology, especially for graphical representations. In particular, some exercises effectively require technology. Less tangibly, but just as important, technology helps foster an experiment-oriented, hands-on, concrete approach to the subject.

Exercises, solutions. Exercises in multivariable calculus are often more involved, use more steps, and are more open-ended than those in single-variable calcuclus. With this in mind, the *Solutions Manual* provides relatively complete solutions, rather than just answers, to many of the exercises. Some of the exercises and solutions could be used to augment and extend the formal Examples in the text.

Thanks. We owe thanks to many people for useful suggestions, criticism, and advice. We took some—but not all—of their good advice; all errors of omission and commission are ours alone. It's impossible to list all our creditors and our specific debts to them, but Reg Laursen (Luther College), Richard Mercer (Wright State University), Ruth Dover (Illinois Mathematics and Science Academy), Allen Holmes (Union College), and Mark Omodt (Anoka-Ramsey Community College) deserve special thanks, as do our students and theirs. Participants in two summer workshops on multivariable calculus also taught us at least as much as we taught them.

We also thank the following respondents to a useful survey, which partly guided this revision: George Ashline (St. Michael's College), Peter Atwood (Cornerstone College), Richard Calhoun (Concordia University), Stephan C. Carlson (Rose-Hulman Institute of Technology), Caren Diefenderfer (Hollins College), Ervin Eltze (Fort Hays State University), Jackie Hall (Longwood College), Allen Hibbard (Central College), Joan Hundhausen (Colorado School of Mines), Dan Kemp (South Dakota State University), John Kurtzke (University of Portland), Sergio Loch (Grand View College), James Marshall (Illinois College), Michael May (Saint Louis University), Mike Panahi (University of Texas at Dallas), Sharon Robbert (Trinity Christian College), Randy Ross (Morehead State University), Joanne Snow (Saint Mary's College), Richard Speers (Skidmore College), Douglas Swan (Morningside College), Scott Wright (Loyola Marymount University).

Staying in touch. Our Web site (address above) offers various resources and information (*Maple* worksheets, information on obtaining review copies, etc.) that instructors may find useful. We also appreciate hearing your suggestions, comments, and advice on this revised preliminary edition. Our physical and e-mail addresses are below.

Arnold Ostebee and Paul Zorn
Department of Mathematics
St. Olaf College
1520 St. Olaf Avenue
Northfield, Minnesota 55057-1098

e-mail: `ostebee@stolaf.edu` `zorn@stolaf.edu`

Acknowledgment. This text was prepared with support from the National Science Foundation (Grant DUE-9450765).

October, 1997

How to use this book: notes for students

All authors want their books to be *used*: read, studied, thought about, puzzled over, reread, underlined, disputed, understood, and, ultimately, enjoyed. So do we.

That might go without saying for *some* books—beach novels, user manuals, field guides, etc.—but it may need repeating for a calculus textbook. We know as teachers (and remember as students) that mathematics textbooks are too often read *backwards*: faced with Exercise 231(b) on page 1638, we've all shuffled backwards through the pages in search of something similar. (Very often, moreover, our searches were rewarded.)

A textbook isn't a novel. It's a peculiar hybrid of encyclopedia, dictionary, atlas, anthology, daily newspaper, shop manual, *and* novel—not exactly light reading, but essential reading nevertheless. Ideally, a calculus book should be read in *all* directions: left to right, top to bottom, back to front, and even front to back. That's a tall order. Here are some suggestions for coping with it.

Read the narrative. Each section's narrative is designed to be read from beginning to end. The examples, in particular, are supposed to illustrate ideas and make them concrete—not just serve as templates for homework exercises.

Read the examples. Examples are, if anything, more important than theorems, remarks, and other "talk." We use examples both to show already-familiar calculus ideas "in action," and to set the stage for new ideas.

Read the pictures. We're serious about the "graphical points of view" mentioned in our title. The pictures in this book are not "illustrations" or "decorations." The are an important part of the language of calculus. An ability to think "pictorially"—as well as symbolically and numerically—about mathematical ideas may be the most important benefit calculus can offer.

Read the language. Mathematics is not a "natural language" like English or French, but it has its own vocabulary and usage rules. Calculus, especially, relies on careful use of technical language. Words like **rate**, **amount**, **concave**, **stationary point**, and **root** have precise, agreed-upon mathematical meanings. Understanding such words goes a long way toward understanding the mathematics they convey; misunderstanding the words leads inevitably to confusion. Whenever in doubt, consult the index.

Read the instructors' preface (if you like). Get a jump on your teacher.

In short: *read the book.* Read it actively, with paper and pencil at hand, and, if possible, with technology at your elbow. Do the calculations for yourself. Plot some curves and surfaces for yourself. You bought the book—do whatever you can to make it your own.

A last note

Why study calculus at all? There are plenty of good practical and "educational" reasons: because it's good for applications; because higher mathematics requires it; because it's good mental training; because other majors require it; because jobs require it. The ideas and methods of multivariable calculus, in particular, are even more powerful and flexible than those of single-variable calculus in modeling the physical and human worlds, in all their higher-dimensional richness.

But there's another, different, reason to study the subject: calculus is among our species' deepest, richest, farthest-reaching, and most beautiful intellectual achievements. We hope this book will help you see it in that spirit.

A last request

Last, a request. We sincerely appreciate—and take very seriously—students' opinions, suggestions, and advice on this book. We invite you to offer your advice, either through your teacher or by writing us directly. Our addresses appear below.

Arnold Ostebee and Paul Zorn
Department of Mathematics
St. Olaf College
1520 St. Olaf Avenue
Northfield, Minnesota 55057-1098

Chapter 1

Curves and Vectors

1.1 Three-dimensional space

Single-variable calculus is done mainly in the two-dimensional xy-plane. The Euclidean plane, also known as $\mathbb{R}^2$, is the natural home of such familiar calculus objects as the graph $y = f(x)$, tangent lines to this graph at various points, and regions whose areas we might measure by integration.

To do *multivariable* calculus we'll need more room. The graph of $z = f(x, y)$, where f is a function of two input variables, lives in *three*-dimensional xyz-space. With three dimensions to work in we'll be able to "see" not only these graphs but also a variety of multivariable analogues of derivatives and integrals. This section explores three-dimensional Euclidean space, also known as $\mathbb{R}^3$, where we'll be spending most of our time.

In another sense, of course, we spend *all* of our time in $\mathbb{R}^3$. In everyday use, the word "space" connotes three physical dimensions. The intuition we gain from living in three spatial dimensions is often useful in mentally picturing and manipulating the objects of multivariable calculus. Familiar as three-dimensional space is, however, it poses special problems for visualization. Two-dimensional pictures (on paper or on a computer screen) of three-dimensional objects are *always* more or less distorted or incomplete. Minimizing such problems is an active science (and an art) in its own right; doing so means carefully controlling viewpoint, perspective, shading, lighting, and other factors.

This book's main subject is multivariable calculus, not computer graphics or technical drawing, so we'll draw pictures to illustrate ideas as simply as possible—not necessarily to look as "lifelike" as possible. It's worth remarking, though, that many of the basic tools and methods of computer graphics draw directly on the very ideas we'll develop in this book.

Cartesian coordinates in three dimensions

The *idea* of Cartesian coordinates is the same in both two and three dimensions, but the pictures look a little different. Compare these:

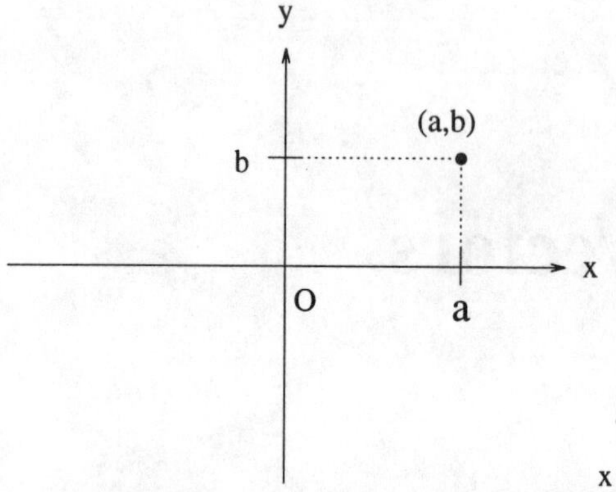

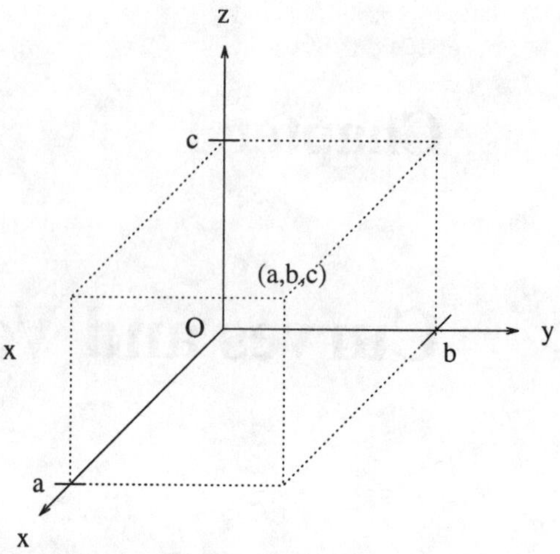

Recall the formalities in the xy-plane. A **Cartesian coordinate system** consists of an **origin**, labeled O, and horizontal and vertical **coordinate axes**, labeled x and y, passing through O. On each axis we choose a positive direction ("east" and "north", usually) and a unit of measurement (not necessarily the same on both axes).

We'll occasionally use other axis labels.

Given such a coordinate system, every point P in the plane corresponds to one and only one ordered pair (a, b) of real numbers, called the **Cartesian coordinates** of P. The pair (a, b) can be thought of as P's "Cartesian address": To reach P from the origin, move a units in the positive x-direction and b units in the positive y-direction.

If a or b is negative, go the other way.

Coordinates in three-dimensional xyz-space work the same way—but with *three* coordinate axes, labeled x, y, and z. Each axis is perpendicular to the other two. To reach the point $P = (a, b, c)$ from the origin, go a units in the positive x-direction, b units in the positive y-direction, and c units in the positive z-direction. As the figure above illustrates, the resulting point $P = (a, b, c)$ can also be thought of as a corner (the one opposite to the origin) of a rectangular solid with dimensions $|a|$, $|b|$, and $|c|$.

There's "room" in $\mathbb{R}^3$ for three mutually perpendicular axes. $\mathbb{R}^2$ has room for only two.

Quadrants and octants. The two axes divide the xy-plane into four **quadrants**, defined by the pattern of positive or negative x- and y-coordinates. The analogous regions in xyz-space are called **octants.** The **first octant**, for instance, consists of all points (x, y, z) with all three coordinates positive. In the picture above, only the first octant is visible. The picture below gives another view. There are *eight* octants in all in xyz-space—one for each of the possible patterns $(+, +, +)$, $(-, +, +)$, ..., $(-, -, -)$ of signs of the three coordinates.

As the name suggests.

Coordinate planes. One can think of the first octant as a room, with the origin at the lower left corner of the front wall. Three walls meet at this point, all at right angles. These "walls" are known as the **coordinate planes**: the yz-plane (the front wall), the xy-plane (the floor), and the xz-plane (the left wall). These coordinate planes correspond to simple equations in the variables x, y, and z. The yz-plane, for

Be sure you agree: look carefully at the standard picture.

example, is the graph of the equation $x = 0$, i.e., the set of all points (x, y, z) that satisfy this equation. The xy- and xz-planes, similarly, are graphs of the equations $z = 0$ and $y = 0$, respectively.

Many possible views. The xy-plane, being "flat," is relatively easy to draw. Simulating three-dimensional space on a flat page or computer screen is, of course, much harder, and there is always some price to be paid in distortion. For instance, the x-, y-, and z-axes are, in 3-d reality, all perpendicular to each other, but no flat picture can really show this. The axes shown below, for instance, don't make right angles on the page:

Study this picture carefully; it's worth the effort.

Three-dimensional coordinates

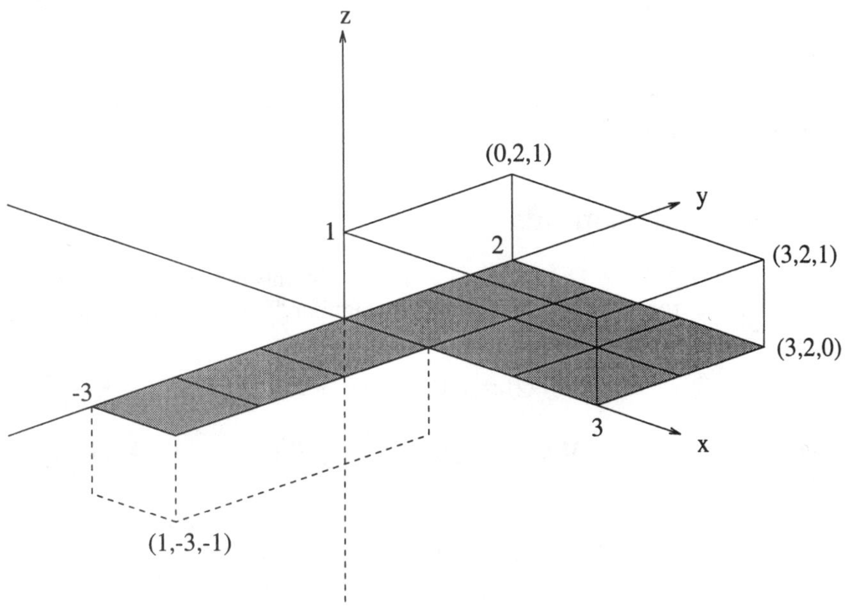

This view of xyz-space is somewhat different from that in the preceding picture. Observe:

Horizontal and vertical. The xy-plane (in which the shaded "floor tiles" lie) is drawn to appear *horizontal*. The z-axis is *vertical*; the positive direction is up. This is a standard convention; we'll usually follow it.

Hidden lines. The dashed lines in the picture lie "below" the xy-plane. They would be hidden from view if the xy-plane—the "floor"—were opaque. How much of xyz-space is considered visible is a matter of choice. Sometimes, only the first octant is shown.

Positive directions. An arrow on each axis indicates the positive direction. In particular, the 3×2 block of shaded squares lies in the first quadrant of the xy-plane. The other shaded squares lie in the plane's fourth quadrant.

Plotting points: positive and negative coordinates. Every point $P = (a, b, c)$ is plotted in the same way: starting from the origin, move a, b, and c units in the positive x-, y-, and z-directions, respectively. Negative coordinates cause no special problem—just move the other way.

Where's the viewer? The picture is drawn as though the viewer were floating somewhere above the *fourth quadrant* of the xy-plane. In the previous 3-d picture, by contrast, the viewer floats somewhere above the *first* quadrant.

Think about this. How can you tell?

There's nothing sacred about either viewing angle; we'll use various viewpoints as we go along. So, for that matter, do the various computer plotting packages readers may have at hand.

No perspective. To a human viewer, rectangular boxes like those above would appear "in perspective"; the sides would seem to taper toward a vanishing point. The closer the viewer, the more pronounced these effects would be. For the sake of simplicity, perspective effects are not shown in the picture above (and elsewhere in this book). In effect, the viewer is assumed to be very, very far from the origin.

There is no single "best" picture of a 3-d object; choosing a good or convenient view may depend on properties of the object, what needs emphasis, or even on the drawing technology at hand.

Computer, calculator, pencil, sharp stick, ...

Distance and midpoints

Let $P = (x_1, y_1)$ and $Q = (x_2, y_2)$ be any two points in the xy-plane. Recall that the **distance** from P to Q is given by the familiar "Pythagorean" formula

Or from Q to P—it doesn't matter.

$$d(P, Q) = \sqrt{(x_2 - x_1)^2 + (y_2 - y_1)^2},$$

and that the **midpoint** M of the segment joining P to Q has the following coordinates:

Each coordinate of M is the average of the corresponding coordinates of P and Q.

$$M = \left(\frac{x_1 + x_2}{2}, \frac{y_1 + y_2}{2}\right).$$

The formulas in three dimensions aren't much different:

> **Definition:** The **distance** between $P = (x_1, y_1, z_1)$ and $Q = (x_2, y_2, z_2)$ is
>
> $$d(P, Q) = \sqrt{(x_2 - x_1)^2 + (y_2 - y_1)^2 + (z_2 - z_1)^2}.$$
>
> The **midpoint** of the segment joining P and Q has coordinates
>
> $$M\left(\frac{x_1 + x_2}{2}, \frac{y_1 + y_2}{2}, \frac{z_1 + z_2}{2}\right).$$

Both definitions are simply three-dimensional variations of the corresponding formulas in the xy-plane. In both two and three dimensions, for example, distance is computed as the square root of the sum of the squared differences in coordinates.

In either two or three dimensions, the distance formula reflects the Pythagorean rule.

■ **Example 1.** Consider the points $P = (0, 0, 0)$ and $Q = (2, 4, 6)$. Find the distance from P to Q and the midpoint M of the segment joining them. How far is M from P and from Q?

Solution: By the distance formula,

$$d(P,Q) = \sqrt{(2-0)^2 + (4-0)^2 + (6-0)^2} = \sqrt{56} \approx 7.483.$$

The midpoint, according to the formula, is $M(1,2,3)$—each coordinate of M splits the difference between the corresponding coordinates of P and Q. To see why M deserves the name "midpoint," notice that

$$\begin{aligned} d(P,M) &= \sqrt{(1-0)^2 + (2-0)^2 + (3-0)^2} \\ &= \sqrt{(2-1)^2 + (4-2)^2 + (6-3)^2} = \sqrt{14} = \frac{\sqrt{56}}{2} =\approx 3.742. \end{aligned}$$

Thus M lies—as a midpoint should—halfway between P and Q. □

Equations and their graphs

The graph of an equation in x and y is the set of all points (x, y) that satisfy the equation. The graph of $x^2 + y^2 = 1$, for instance, is a circle of radius one, centered at the origin. The graph of the equation $x = 0$ is the y-axis.➽

The graph of an equation may or may not be the graph of a function. The unit circle is not a function graph.

The same idea applies for three variables: the graph of an equation in x, y, and z is the set of points (x, y, z) that satisfy the equation. Here, graphically, are three simple examples (only the first octant is shown):

Three simple graphs

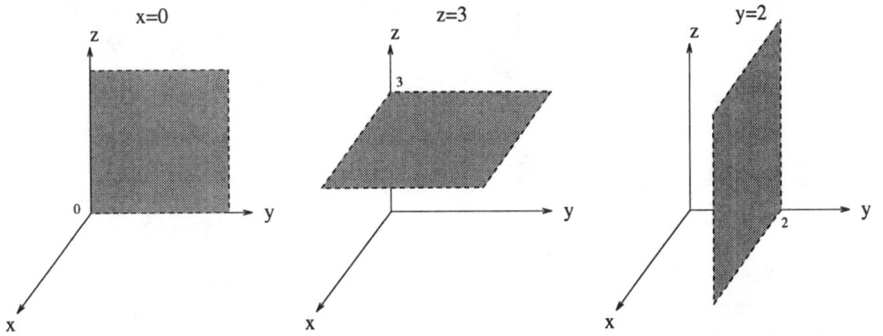

Solutions of the equation $x = 0$ are points of the form $(0, y, z)$, so the graph is the yz-plane. Similarly, solutions of $z = 3$ are all points of the form $(x, y, 3)$, so the graph is a horizontal plane floating 3 units above the xy-plane. The graph of $y = 2$ is parallel to the xz-plane, but moved two units in the positive y-direction.

Notice that in each case above the graph of an equation in x, y, and z is a plane—a *two-dimensional* object. By comparison, the graph of one equation in x and y is usually a curve or a line—a *one-dimensional* object. The pattern is the same in both cases: the graph of an equation has dimension *one less than* the number of variables.

A few special types of graphs in xyz-space deserve special mention.

Planes

A **linear equation** is one of the form $ax + by + cz = d$, where a, b, c, and d are constants, with at least one of a, b, and c nonzero.➽ All three equations plotted above are linear; they illustrate an important general fact:

What goes wrong if $a = b = c = 0$?

The graph of any linear equation is a plane.

(We'll explain this fact carefully in a later section.) To draw planes in xyz-space, we'll use the the fact that a plane is uniquely determined by three points (unless the points happen to be on a straight line).

■ **Example 2.** Plot the linear equation $x + 2y + 3z = 3$ in the first octant.

Solution: First we'll find some points (x, y, z) that satisfy $x + 2y + 3z = 3$. This is, if anything, too easy—there are infinitely many possibilities. Given *any* values for x and y, the equation determines a corresponding value for z. If, say, $x = 1$ and $y = 1$, then $x + 2y + 3z = 3$ can hold only if $z = 0$. Similarly, setting $y = 2$ and $z = 3$ forces $x = -10$.◂◂ Among all possible solutions, here are three:

Check this calculation.

$$P = (3, 0, 0); \quad Q = (0, 3/2, 0); \quad R = (0, 0, 1).$$

These solutions are both easy to find (set two coordinates to zero and solve for the third) and easy to plot (they lie on the coordinate axes). Using these three points, we can plot our plane:

A plane in the first octant: x+2y+3z=3

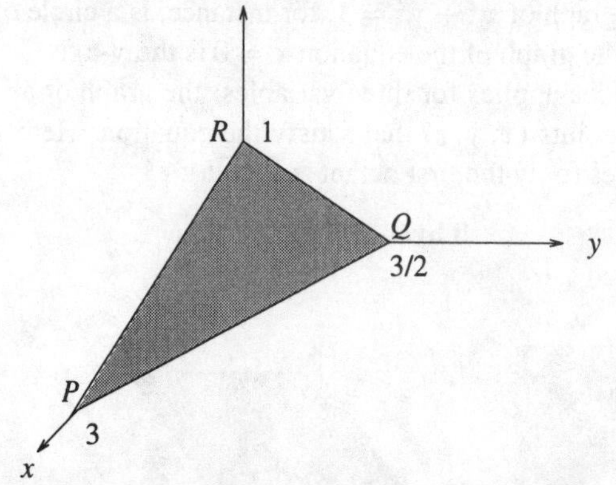

Notice:

Intercepts. A typical line in the xy-plane has x- and **y-intercepts**, where the line intersects the coordinate axes. In a similar sense, a typical plane in xyz-space has x-, y-, and z-intercepts. In the picture, the intercepts are P, Q, and R.◂◂

Not every line in the xy-plane intercepts both axes; not every plane in space intercepts all three axes. See the exercises for more on this.

Traces. If we "slice" a surface in xyz-space with a plane, the intersection of the surface with the plane is called the **trace** of the surface in that plane. The plane p shown above meets each of the three coordinate planes in a straight line. Specifically, p meets the xy-plane in the line through P and Q; p meets the xz-plane in the line through P and R, and p meets the yz-plane in the line through Q and R. These three lines, therefore, are the traces of the surface $x + 2y + 3z = 3$ in the three coordinate planes.◂◂

See the picture.

To find the equation for any trace, we combine the equation for the surface S and the equation for the slicing plane. For example, the xy-plane has equation $z = 0$, so a point (x, y, z) lies in *both* the plane p *and* in the xy-plane if and only if (x, y, z) satisfies both $x + 2y + 3z = 3$ *and* $z = 0$. Setting $z = 0$ in the first equation gives $x + 2y = 3$—this is, as expected, the equation of a line in the xy-plane, namely, the trace of p in the xy-plane. □

Spheres

In the plane, a circle of radius $r > 0$, with center $C = (a, b)$, is the set of points $P = (x, y)$ at distance r from (a, b). Translating this description into symbolic language produces the familiar formula for a circle in the plane:

$$d(P, C) = \sqrt{(x-a)^2 + (y-b)^2} = r, \quad \text{or} \quad (x-a)^2 + (y-b)^2 = r^2.$$

(Squaring both sides does no harm, and simplifies the equation's appearance.)

The object in space analogous to a circle in the plane is a **sphere** of radius r. Like a circle, a sphere is "hollow," similar to the skin of an orange. Adding the interior (the edible part of the orange) produces a **ball**. Like a circle, a sphere is the set of points at some fixed distance—the radius—from a fixed center point. Given a radius $r > 0$ and a center point $C = (a, b, c)$, the sphere of radius r, centered at C, is the set of points (x, y, z) such that

$$d(P, C) = \sqrt{(x-a)^2 + (y-b)^2 + (z-c)^2} = r,$$
$$\text{or} \quad (x-a)^2 + (y-b)^2 + (z-c)^2 = r^2.$$

The simplest example, the **unit sphere**, has center (0, 0, 0) and radius 1; its equation reduces to this simple form:

$$x^2 + y^2 + z^2 = 1.$$

Drawing a circle in the xy-plane is easy, even by hand. Drawing a sphere (or *any* "curved" object, for that matter) convincingly by hand is much harder.➻ Fortunately, rough sketches usually suffice.

For starters, circles in space don't always look circular. Depending on the viewing angle, they may look like ellipses.

The algebraic technique of completing the square may reveal an equation's spherical form.

■ **Example 3.** Is the graph of $x^2 - 2x + y^2 - 4y + z^2 - 6z = 0$ a sphere? Which one?

Solution: Completing the square➻ in each variable separately gives

Do you recall this useful technique? If not, check each step of the calculation carefully—it contains three examples of the method.

$$\begin{aligned} x^2 - 2x + y^2 - 4y + z^2 - 6z &= 0 \iff \\ (x^2 - 2x + 1) + (y^2 - 4y + 4) + (z^2 - 6z + 9) &= 1 + 4 + 9 \iff \\ (x-1)^2 + (y-2)^2 + (z-3)^2 &= 14. \end{aligned}$$

The last form shows that our equation describes a sphere of radius $\sqrt{14}$, centered at $(1, 2, 3)$. As the equation shows, this sphere passes through the origin. The picture shows this, too:

Graph of x^2-2*x+y^2-4*y+z^2-6*z=0

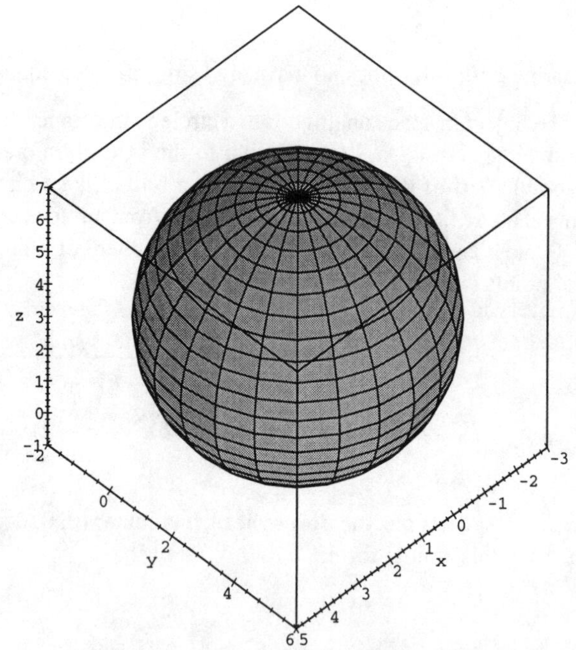

Cylinders

What is the graph of the equation $y = x^2$? The answer depends on where we're working. In the xy-plane, the graph is the familiar parabola—all points of the form (x, x^2). Here's the graph of the same equation in xyz-space:

Graph of y=x^2 in xyz-space

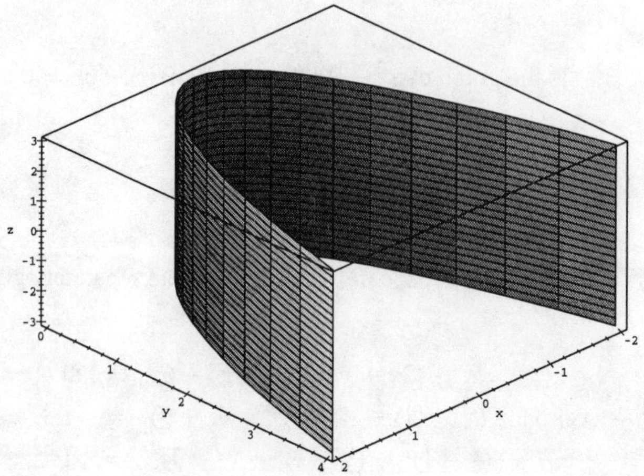

Notice:

In other words, the graph has "vertical walls."

A missing variable. The graph is *unrestricted in the z-direction*—it contains all points that lie directly above or below the graph of $y = x^2$ in the xy-plane.

The graph has this property, of course, because z is "missing" in the equation $y = x^2$. This means that if (x, y) satisfies the equation, then so does *every* point (x, y, z)—regardless of the value of z.

What's a cylinder? Graphs like this one, in which one (or more) of the variables is unrestricted, are called **cylinders**. Any equation that omits one or more variables—$y = z$, say—has a cylindrical graph. Plotting cylinders is comparatively easy.➽ If the equation involves only y and z, for instance, we first plot the equation in the yz-plane, and then "extend" the graph in the x-direction.

"Easy" is a relative thing—drawing anything in xyz-space poses certain challenges.

In everyday speech, "cylinder" usually means a circular tube. As the next example illustrates, the mathematical idea of a cylinder is much more general.

■ **Example 4.** Discuss the graph in xyz-space of the equation $z = 2 + \sin y$. Is the graph a cylinder?

Solution: There's no variable x in the equation, so the graph is unrestricted in the x-direction—i.e., the graph is a cylinder in x. Here's a representative view:➽

The surface, like many graphs, continues forever; a picture shows only part of the graph.

Graph of z=2+sin(y)

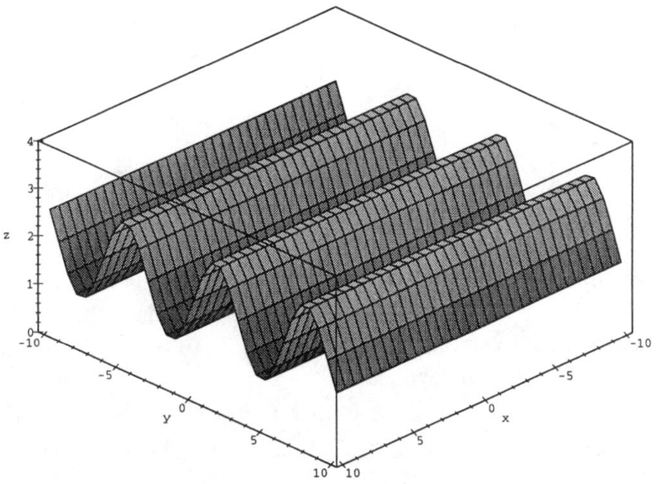

The graph looks something like the surface of an idealized ocean, with regular waves moving parallel to the y-axis. Each wave is infinitely long, with its trough and crest parallel to the x-axis. Notice especially how the surface meets the yz-plane.➽ The curve of intersection between the surface and the yz-plane (what we've called the trace of the surface in the plane) is the sinusoidal curve $z = 2 + \sin y$. In fact, the entire surface can be thought of as infinitely many identical copies of this curve, one for each value of x. □

The yz-plane isn't explicitly shown in the picture, but the y- and z-directions are shown.

■ **Example 5.** Consider the following surface, defined by the equation $z = x^2 + y^2$:

Graph of $z = x^2 + y^2$: a paraboloid

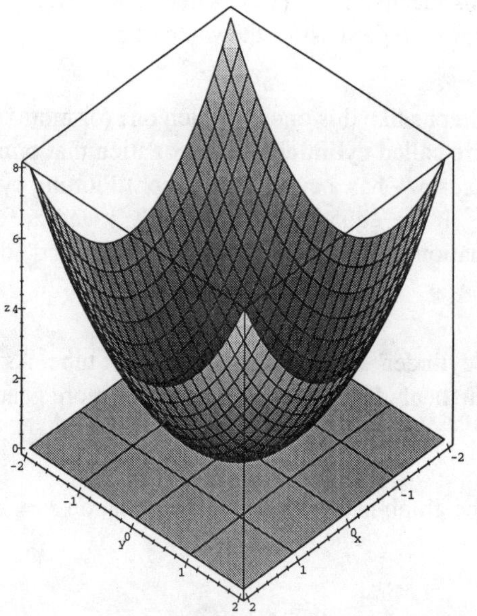

This surface—we'll call it S—is a **paraboloid**. Find the traces of S in several planes P.

Solution: The trace of S in any plane P is the intersection of S and P. Therefore, points on the trace of S in P are those that satisfy *two* equations—those of S and of P. Points on the trace of S in the xy-plane, for instance, satisfy both

$$z = 0 \quad \text{and} \quad z = x^2 + y^2.$$

Plugging the first equation into the second gives $0 = x^2 + y^2$; clearly, the only solution is $x = y = 0$. Thus the trace of S in the xy-plane is a single point—the origin. (This is evident from the picture.)

Now suppose that P is the xz-plane, whose equation is $y = 0$. Plugging this equation into $z = x^2 + y^2$ gives $z = x^2$—i.e., a parabola in the xz-plane. Similarly, the trace of P in the yz-plane (with equation $y = 0$) is the parabola $z = y^2$. These traces, too, can be seen in the picture.

All the grid lines on the surface shown are traces of S with planes of the form $x = a$ or $y = b$.

Now let P be any plane of the form $x = a$, where a is a constant. Substituting $x = a$ into $z = x^2 + y^2$ gives $z = a^2 + y^2$—once again the equation of a parabola, opening upward. Similarly, the trace of S in any plane $y = b$ is the parabola with equation $z = x^2 + b^2$. (All these parabolas suggest the reason for the surface's name.)

Finally, let P be any "horizontal" plane, with equation $z = c$; then the trace equation is $c = x^2 + y^2$. If $c > 0$, this equation describes a circle, with radius $\sqrt{c}$ and center $(0, 0)$; if $c < 0$, the equation $c = x^2 + y^2$ has *no* solutions so the trace is empty. The picture illustrates these facts, too. Notice, for example, that if $c < 0$ then the plane $z = c$ misses S entirely. □

Orientation (optional)

Any three mutually perpendicular lines can serve as coordinate axes. Here are several possible choices:

Three systems of axes

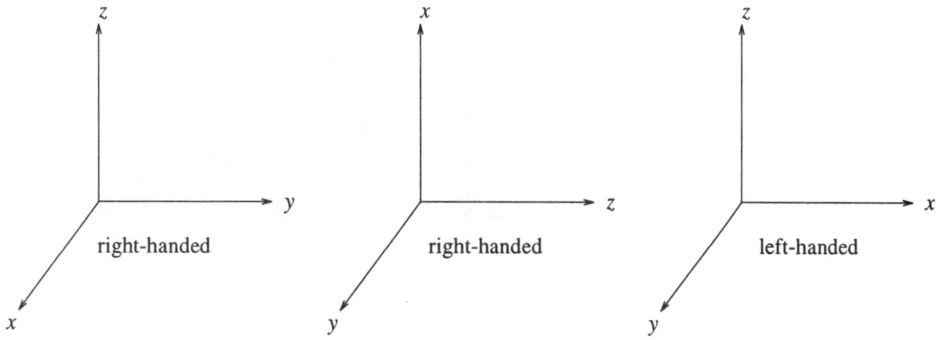

A close look reveals a subtle difference, called **orientation**, between the first two systems and the third. The first two systems are called **right-handed systems**, for the following reason:

> If the fingers of the *right* hand are swept from the positive x-axis to the positive y-axis, then the thumb points in the positive z-direction.

For a similar reason, the third system of axes is called **left-handed**. (We'll always use right-handed systems of axes in this book. The question of orientation will arise again, however, when we discuss the cross product of vectors in space.)

Exercises

1. Throughout this problem, let S be the surface with equation $z = x^2 + y$. (Use technology, if available, to draw S.) Using the ideas in Example 5, page 9, describe the trace of S in each plane below. (Give the equation of the trace and describe its shape.)

   (a) the xy-plane

   (b) the yz-plane

   (c) the xz-plane

   (d) the plane $x = 1$

   (e) the plane $y = 1$

2. Let S be the unit sphere, with equation $x^2 + y^2 + z^2 = 1$. In each part below, describe the trace of S in the given plane P.

   (a) $z = 0$

   (b) $z = 1/2$

   (c) $z = 1$

   (d) $z = 2$

3. Let S be the surface (it's a plane; see the picture on page 6) with equation $x + 2y + 3z = 3$. In each part below, describe the trace of S in the given plane P.

(a) $x = 0$

(b) $y = 0$

(c) $z = 1$

(d) $z = 2$

(e) $z = 3$

4. We said in this section that the graph of $y = x^2$ is a cylinder in xyz-space, unrestricted in the z-direction. (See the picture on page 8.)

   (a) Plot the equation $z = y^2$ in xyz-space. What is the unrestricted direction? [HINT: Start in yz-space.]

   (b) Plot the equation $z = x^2$ in xyz-space. What is the unrestricted direction?

5. Plot the equation $x^2+y^2 = 1$, first in xy-space, then in xyz-space. (The second graph should resemble a vertical pipe—a cylinder in the everyday sense of the word.)

6. In each part, plot the given equation in xyz-space (a rough sketch is fine), then describe the graph in words.

   (a) $y^2 - z^2 = 0$

   (b) $y^2 + z^2 = 0$

   (c) $y^2 + z = -1$

   (d) $y^2 + z^2 = -1$ [HINT: Does the graph contain any points?]

7. Find an equation in x, y, and z for the graph in xyz-space of

   (a) a sphere of radius 2, centered at the origin.

   (b) a sphere of radius 1, centered at (1, 1, 1).

   (c) a circular cylinder of radius 1, centered along the y-axis.

   (d) a circular cylinder of radius 2, centered along the z-axis.

   (e) a cylindrical surface that resembles an ocean, with waves rolling in the x-direction (see Example 4, page 9).

8. The equation $z = 3$ omits *two* variables. Therefore, its graph in xyz-space should be a cylinder in both the x-direction and the y-direction. Is it? What is the trace of the graph in each of the coordinate planes?

9. The graph of $x^2 + y^2 - 6y + z^2 - 4z = 0$ is a sphere. Find the center and radius; then draw the sphere.

10. Consider the unit sphere S, with equation $x^2+y^2+z^2 = 1$. If we set $z = 0$ in this equation, we get $x^2+y^2+z^2 = x^2+y^2 = 1$. This means, geometrically, that S intersects the xy-plane in the unit circle $x^2+y^2 = 1$. In mathematical language, the unit circle is the trace of S in the xy-plane. (To put it another way, the unit circle is the "equator" of the unit sphere.)

    (a) Set $x = 0$ in the original equation to find the equation of the intersection of S and the yz-plane.

    (b) What is the intersection of S and the xz-plane? Describe the answer geometrically.

(c) Use the results in parts (a) and (b) to sketch the part of S that lies in the first octant. [HINTS: First draw a set of coordinate axes. Then draw the traces of S in each of the three coordinate planes.]

(d) Set $z = 1/2$ in the original equation to show that S intersects the plane $z = 1/2$ in the circle of radius $\sqrt{3}/2 \approx 0.87$, centered at $(0, 0)$.

(e) What is the trace of S in the plane $z = 0.9$? In the plane $z = 1$? In the plane $z = 2$? Explain your answers.

11. The distance formula in xyz-space can be thought of as just another instance of the Pythagorean rule for right triangles. (The square of the hypotenuse is the sum of the squares of the sides.) This exercise illustrates why.

(a) Plot and label the points $O(0, 0, 0)$, $P(1, 0, 0)$, $Q(1, 2, 0)$, and $R(1, 2, 3)$ in an xyz-coordinate system. Observe that the triangles $\triangle OPQ$ and $\triangle OQR$ are both right triangles. Mark the sides OP, PQ, and QR with their lengths. (The lengths should be obvious from the picture.)

(b) Use the Pythagorean rule (not the distance formula) on the triangle $\triangle OPQ$ to find the length of OQ.

(c) Use the Pythagorean rule (not the distance formula) on the triangle $\triangle OQR$ to find the length of OR.

(d) For comparison, use the distance formula to compute the lengths of OQ and OR.

12. Any reasonable formula for distance should satisfy some commonsense requirements. For example, the distance $d(P, P)$ from any point P to *itself* should certainly be zero. So it is. If $P(x, y, z)$ is any point, then the distance formula says

$$d(P, P) = \sqrt{(x - x)^2 + (y - y)^2 + (z - z)^2} = 0.$$

In the same spirit, use the distance formula to verify the following commonsense properties. Throughout, use the points $P(x, y, z)$ and $Q(a, b, c)$.

(a) If $P \neq Q$, then $d(P, Q) > 0$.

(b) $d(P, Q) = d(Q, P)$.

(c) If M is the midpoint of P and Q, then $d(P, M) = d(M, Q) = d(P, Q)/2$.

13. In this section we said that xyz-space contains 8 different octants. List 8 points, all with coordinates ± 1, one in each octant. Draw a picture showing all 8 points.

14. Any line in the xy-plane (even a vertical line) has an equation of the form $Ax + By = C$ for some constants A, B, and C. For each of the following lines, state an equation in this form.

(a) The line $y = 3x + 5$

(b) The line $y = mx + b$, where m and b are any constants

(c) The horizontal line through $(2, 3)$

(d) The vertical line through $(2, 3)$

15. Consider the linear equation $Ax + By = C$ and its graph (a line) in the xy-plane. Here A, B, and C are constants, and we assume that A and B aren't both zero.

    (a) We assumed that A and B aren't both zero. What goes wrong if $A = B = 0$?

    (b) Find the slope of the line $Ax + By = C$. Which lines have undefined slope?

    (c) Find the y-intercept of the line $Ax + By = C$. Which lines have no y-intercept?

    (d) Find the x-intercept of the line $Ax + By = C$. Which lines have no x-intercept?

16. Consider the linear equation $Ax + By + Cz = D$ and its graph (a plane) in xyz-space. Here A, B, C, and D are all constants, and we assume that A, B, and C aren't all zero.

    (a) We assumed that A, B, and C aren't all zero. What goes wrong if $A = B = C = 0$?

    (b) Find (if possible) an x-intercept of the plane $Ax + By + Cz = D$. (Set $y = 0$ and $z = 0$, and then solve for x.) Give an example of a plane with no x-intercept.

    (c) Find (if possible) a z-intercept of the plane $Ax + By + Cz = D$. Give an example of a plane with no z-intercept.

17. The linear equation $x + 2y + 3z = 3$ defines a plane p in xyz-space. (See the picture on page 6.)

    (a) Find the equation of the trace of p in the xz-plane. Where does the trace intercept the x- and z-axes?

    (b) Find the equation of the trace of p in the yz-plane. Where does the trace intercept the y- and z-axes?

    (c) Find the equation of the trace of p in the plane $x = 1$.

18. Consider the plane p with equation $4x + 2y + z = 4$.

    (a) Find the x-, y-, and z-intercepts of p. Use them to draw p in the first octant.

    (b) Find the traces of p in each of the three coordinate planes. How do your answers appear in the picture in part (a).

19. We said in this section that, as a rule, a line in the xy-plane intercepts both coordinate axes, and a plane in xyz-space intercepts all three coordinate axes. But exceptions are possible, as this exercise explores.

    (a) Give an example of a line in the xy-plane that intercepts the x-axis but not the y-axis. Write an equation for your line in the form $ax + by = c$.

    (b) Consider the plane $x = 1$ in xyz-space. Find all possible intercepts with the three coordinate axes.

    (c) Consider the plane $x + 2y = 1$ in xyz-space. Find all possible intercepts with the three coordinate axes.

(d) Give the equation of a plane in xyz-space that intersects the the y-axis and the z-axis but not the x-axis.

20. Suppose that $P_1(x_1, y_1, z_1)$ and $P_2(x_2, y_2, z_2)$ both lie on the plane with equation $Ax + By + Cz = D$. Show that the midpoint of P_1 and P_2 also lies on this plane.

21. Here's another way of defining orientation of a system of axes. The system is right-handed if it's possible to point (all at once!) the index finger, the middle finger, and the thumb of the right hand along the positive x-, y-, and z-axes. Otherwise the system is left-handed. Use this definition to draw two more systems (different from those on page 11), one with each orientation.

22. Two systems of positive x-, y-, and z-axes have the same orientation if one can be "rotated into the other," i.e., if one system of axes can be picked up and turned around to "fit onto" the other. Explain why this can be done with the left and middle systems on page 11, but not with the left and right systems.

23. Is the graph of $y = 2$ a cylinder in xyz-space? Justify your answer.

24. Find an equation for the trace of the surface $y = x^2 - z^3$ in the yz-plane.

25. Let $P = (1, 2, 3)$ and $Q = (4, 6, 8)$ be points in 3-space.

    (a) Find the distance from P to Q.

    (b) Write an equation for the sphere with center at P that passes through Q.

    (c) The graph of $f(x, y) = x^2 - y + 4$ is a surface in 3-space. Does P lie on the graph? Justify your answer.

26. Find an equation for the smallest ball centered at the point $(4, -8, 1)$ that contains the origin.

1.2 Curves and parametric equations

Introduction: all kinds of curves

Curves in the xy-plane come in enormous variety. (Curves in xyz-space are equally plentiful; we'll see them at the end of this section.) Here are four basic examples.

C_1: A sine curve

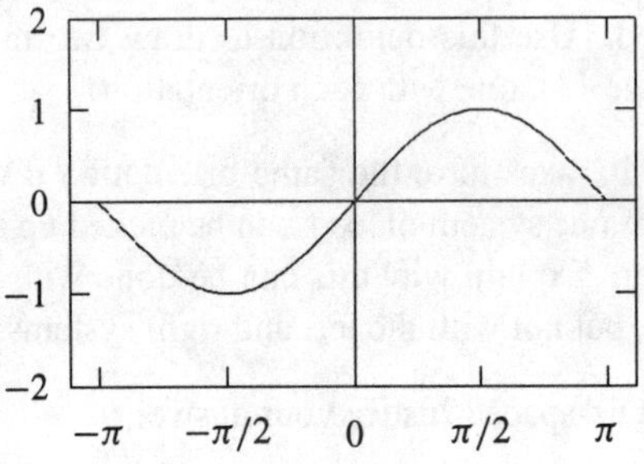

C_2: A parabolic curve

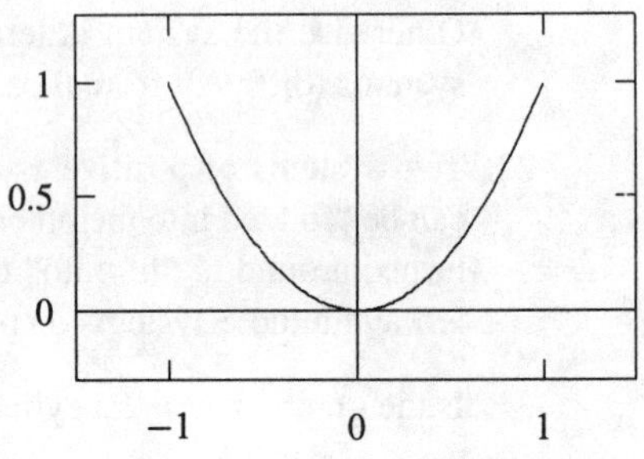

C_3: An ellipse

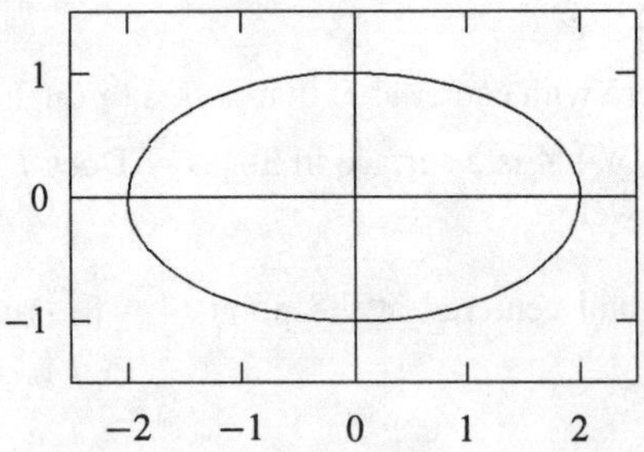

C_4: A spiral

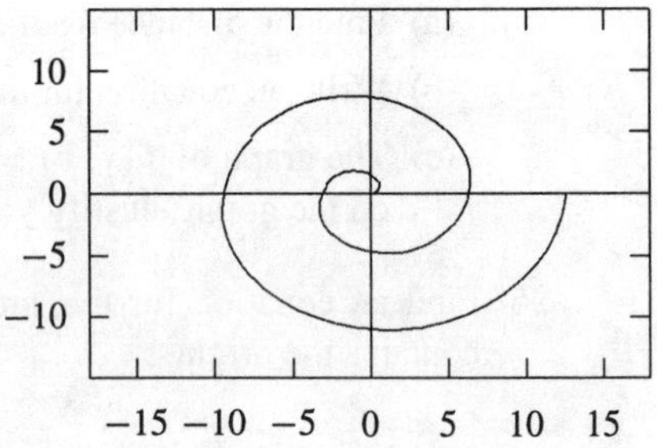

Let's take some closer looks:

Pieces of familiar graphs. The curves C_1 and C_2 are pieces of familiar function graphs. Specifically, C_1 is that piece of the graph of $y = \sin x$ for which $-\pi \le x \le \pi$; C_2 is that piece of the graph of $y = x^2$ for which $-1 \le x \le 1$. Using basic calculus properties of these functions and their derivatives, we can describe these curves in detail: where they rise and fall, their slopes at various points, their concavity, etc.

Graph vs. piece-of-a-graph. The difference between "graph" and "piece-of-a-graph" will sometimes matter to us. When it does matter we'll need to say carefully *which* piece we're interested in, usually by specifying (as we did above) a particular domain interval.

Not graphs (but still important). Curves C_3 and C_4 are *not* ordinary function graphs (or even pieces of such graphs), because some x-values on these curves correspond to more than one y-value.◂ Nevertheless, curves like C_3 and C_4 are indispensable in real-life applications. A moving object, for instance, might well follow such a curve. Modeling physical motion—for centuries among the most important applications of calculus, and still so today—would be impossible without mathematical tools for handling general curves.

C_3 and C_4 fail the "vertical line test".

400 years of modeling motion. Modeling physical motion has occupied mathematicians for centuries. Around 1600, the German astronomer Johannes Kepler asserted (among other things) that the planets follow *elliptical* orbits. In the late 1600's Isaac Newton used his new calculus to verify and extend Kepler's models of planetary motion.

Modeling motion mathematically is still important. Choreographing the twists, turns, and extensions of an industrial robot arm, for instance, would be all but impossible without ideas and tools from multivariable calculus.

Before we can apply calculus ideas and methods to general curves, we'll need to describe them in a concrete, mathematical manner. Parametric equations are the key. For the moment, we'll restrict our attention to curves in the xy-plane.

Parametric equations

The idea. Imagine a point P wandering about in the xy-plane during a time interval $a \le t \le b$. The two coordinates of P, $x = f(t)$ and $y = g(t)$, are both real-valued functions of t, defined for t in $[a, b]$.

Over the time interval $a \le t \le b$, the point $P = (f(t), g(t))$ traces out some image figure, say C, in the xy-plane. Depending on f and g, the image figure C can have almost any shape: a line segment, a circular arc, a single point, a spiral, a sine curve, a jagged mess of line segments, or worse. Even if f and g are continuous functions, it's possible for the figure C to be surprisingly bizarre.➽

C could be a filled-in square, for instance, or a spaghetti-like tangle.

Very often, however, f and g are sufficiently "well-behaved" that the image figure C turns out to be a "smooth curve," like C_1–C_4 above. (After some examples, we'll say more precisely what "well-behaved" and "smooth curve" mean; here, we use these phrases informally.)

Parametric vocabulary. In sorting out similarities and differences between parametric equations and ordinary functions, it will help to agree on some standard terminology. Here's a quick summary:

Equations of the form

$$x = f(t) \quad \text{and} \quad y = g(t) \quad \text{for} \quad a \le t \le b$$

are called **parametric equations**. The functions f and g are **coordinate functions**. The interval $a \le t \le b$ is the **parameter interval**; the variable t is called the **parameter**.➽ The figure C traced out in the xy-plane by $P(t) = (x(t), y(t))$, for t in $[a, b]$, is called a **parametric curve**; the functions f and g and the parameter interval $[a, b]$ are said to **parametrize** the curve C.

In this setting, "parameter" refers to a variable. In other mathematical situations "parameters" may be constants.

A **parametric curve in space** is similar, except that there are three coordinate functions: $x = f(t)$, $y = g(t)$, $z = h(t)$.

■ **Example 1.** At each time t with $0 \le t \le 10$, the coordinates of P are given by the parametric equations

$$x = t - 2\sin t; \qquad y = 2 - 2\cos t.$$

What curve does P trace out over the time interval? Where is P at $t = 1$, and in which direction is P headed?

Solution: The simplest way to draw a curve is to calculate many points (x, y), plot each one, and "connect the dots." The first step is to calculate, for many inputs t, corresponding values of x and y. Here's a sampler of results, rounded to two decimals:

Check several entries.

Parametric plot points for $x = t - 2\sin t$, $y = 2 - 2\cos t$

t	0	0.1	0.2	0.3	0.7	0.8	0.9	1.0	...	9.8	9.9	10
x	0	−0.10	−0.20	−0.29	−0.59	−0.63	−0.67	−0.68	...	10.53	10.82	11.09
y	0	0.01	0.04	0.09	0.47	0.61	0.76	0.92	...	3.86	3.78	3.68

Here's the result, a curve C:

Computers and calculators plot parametric curves in various ways. Try yours.

The parametric curve $x = t - 2\sin t$, $y = 2 - 2\cos t$, $0 \le t \le 10$

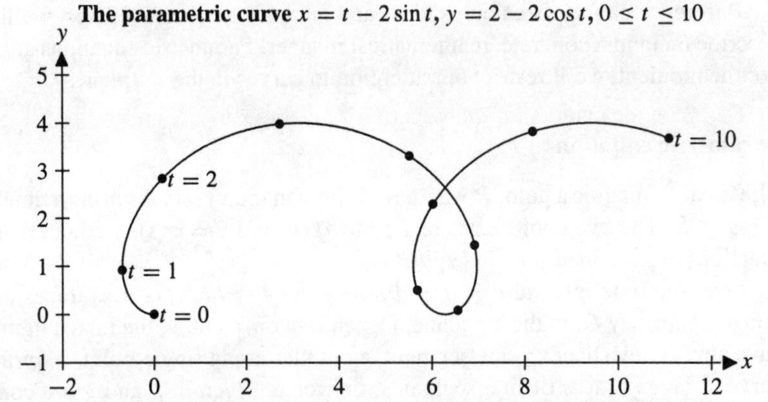

Notice:

Bullets. Points corresponding to *integer* values of t are shown bulleted. At $t = 1$, P has coordinates $(-0.68, 0.92)$; at this instant, P is heading almost due north.

Not a graph. The curve C is *not* the graph of a function $y = f(x)$, because some x-values correspond to more than one y-value.

$x = 6$, for instance.

No t-axis. The picture shows the x- and y-axes, *but no t-axis*: t-values are indicated only by the bulleted points. Bullets appear above only for convenience; most curves don't have them. Often, therefore, t-values can be found only from the coordinate functions, not directly from the picture.

Equal-time bullets. The bullets on the graph appear at equal *time* intervals, but not at equal *distances* from each other, because P speeds up and slows down as it moves. We'll see soon how to use derivatives of the coordinate functions to *calculate* the speed of P at any point along C.

In this example, t represents time.

When is P moving fastest? Slowest?

Loops, vertical tangents, and other oddities. Parametric curves can have many features that don't appear on ordinary function graphs. The curve above, for instance, has both loops and points with vertical tangent lines—even though both coordinates are given by familiar, well-behaved functions. We'll encounter other oddities (e.g., sharp corners) as we go along.

A sampler of plane parametric curves

Plane curves come in mind-boggling variety. *Any* choice of two equations $x = f(t)$ and $y = g(t)$ and a t-interval produces one. Surprisingly often the result is beautiful, useful, or interesting. The following examples hint at some of the possibilities, and at connections between parametric curves and ordinary function graphs.

Ordinary function graphs. Pieces of ordinary function graphs are easy to write in parametric form. The next example shows how.

■ **Example 2.** Parametrize the curves C_1 and C_2 shown on page 16.

Solution: One idea—letting x serve as the parameter t—works for both curves. For C_1:

$$x = t; \qquad y = \sin t; \qquad -2\pi \le t \le 2\pi.$$

For C_2:

$$x = t; \qquad y = t^2; \qquad -1 \le t \le 1.$$

Plotting these data produces the desired curves C_1 and C_2. □

Use technology to see for your self.

The same idea works for *any* function f, defined on any interval $[a, b]$. Setting

$$x = t; \qquad y = f(t); \qquad a \le t \le b$$

gives a parametrization of the graph $y = f(x)$ from $x = a$ to $x = b$.

Line segments. Line segments are among the simplest and most useful "curves." In computer graphics, for instance, complicated curves are drawn by laying many line segments end to end.

Hundreds or thousands, sometimes.

$m = \frac{b-d}{a-c}$

■ **Example 3.** **(The line segment joining two points.)** Let $P = (a, b)$ and $Q = (c, d)$ be any two points in the xy-plane. Parametrize the segment from P to Q.

Solution: Here's one possibility. Let

Other possibilites are outlined in the exercises.

$$x(t) = a(1-t) + ct; \quad y(t) = b(1-t) + dt; \quad 0 \le t \le 1.$$

$y_1 - y_2 = m(x_1 - x_2)$

Is the resulting curve, C, really the segment we want? Notice first that

$$(x(0), y(0)) = (a, b) = P \quad \text{and} \quad (x(1), y(1)) = (c, d) = Q.$$

In other words, C starts and ends at the right places. It can also be shown that for all t in $[0, 1]$, $(x(t), y(t))$ lies on the *line* through P and Q. Thus C is, indeed, the line segment from P to Q. □

See the exercises for details.

Circles and circular arcs. Circles and circular arcs, like line segments, are important modeling and drawing tools. Fortunately, circles are easy to parametrize. The **unit circle**, with radius 1 and center $(0, 0)$, illustrates the main ideas.

■ **Example 4.** Parametrize the unit circle $x^2 + y^2 = 1$.

Solution: The simplest parametrization uses trigonometric ingredients. Let

$$x = \cos t; \qquad y = \sin t; \qquad 0 \le t \le 2\pi.$$

Here's the picture; some important t-values are shown bulleted:

Recall: the cosine and sine functions are sometimes called ***circular functions****. The picture illustrates why.*

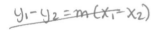

The parametric curve $x = \cos t$, $y = \sin t$, $0 \le t \le 2\pi$

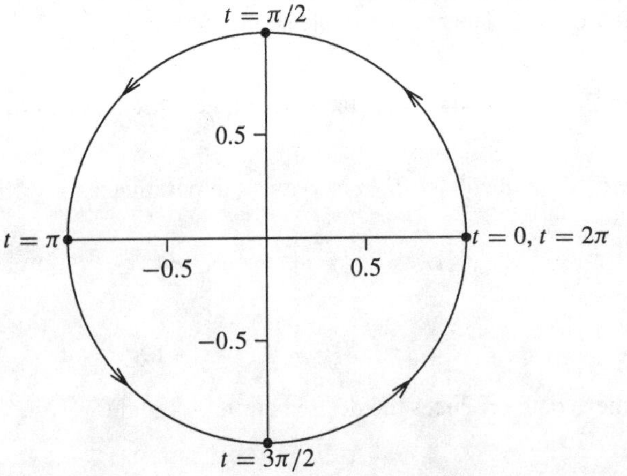

This very important example deserves a very close look.

Time or angle or arclength? On any parametric curve, the parameter t can be understood as *time*—the time at which a moving point $P(t) = (x(t), y(t))$ arrives at a given point along the curve.

Here, however, there are other useful ways to interpret the parameter t. One possibility is to think of t as the *angle* determined by the positive x-axis and the segment from the origin to $P(t)$. Another possibility is to interpret t as *arclength*, i.e., as the distance (measured along the circle) from the starting point $(1, 0)$ to the point $P(t)$. Which of these views is "best" depends on the situation.

Measured counterclockwise, in radians.

Eliminating the parameter. Because $x = \cos t$ and $y = \sin t$, it follows that

$$x^2 + y^2 = (\cos t)^2 + (\sin t)^2 = 1$$

for all t. This means that every point on our parametric curve satisfies the Cartesian equation $x^2 + y^2 = 1$ and therefore (as we planned) lies on the unit circle.

"Reducing" two parametric equations to xy-form, as we just did, is called **eliminating the parameter**. (Doing so is sometimes possible, sometimes not.)

Varying the t-interval. The full circle is traced *once*, counterclockwise, as t runs from $t = 0$ to $t = 2\pi$. (The curve starts and ends at the same point.) With larger or smaller t-intervals the circle would be traced more or less often. If, say, $0 \le t \le 4\pi$, then the circle is traversed *twice*. If, instead, we restrict t to the interval $[\pi/2, 3\pi/2]$, then only the "left" semicircle is covered. □

Other circles and arcs. The idea in the previous example extends to other circles. If (a, b) is any point in the plane and r is any positive number, then the parametrization

$$x = a + r\cos t; \qquad y = b + r\sin t; \qquad 0 \le t \le 2\pi$$

produces the circle with center (a, b) and radius r.

Curves in space

Parametric curves in three-dimensional space are described exactly like curves in the plane, except that a third coordinate function is present, and the resulting curve wanders through space rather than through the plane. Adding a third dimension to the picture allows, in principle, very wild curves—imagine, say, a knotty, tangled mess of string. (We'll stick to relatively tame space curves in this book.) On the other hand, essentially the same ideas and calculus methods apply in three dimensions as in two, so we'll content ourselves here with some basic examples.

■ **Example 5.** The following picture shows two curves in space: a **helix** (or **spiral**) making two turns, and an ordinary unit circle (assume that both curves start at the point $(1, 0, 0)$ and turn counterclockwise).

The xy-plane is shaded.

A helix and a circle: two space curves

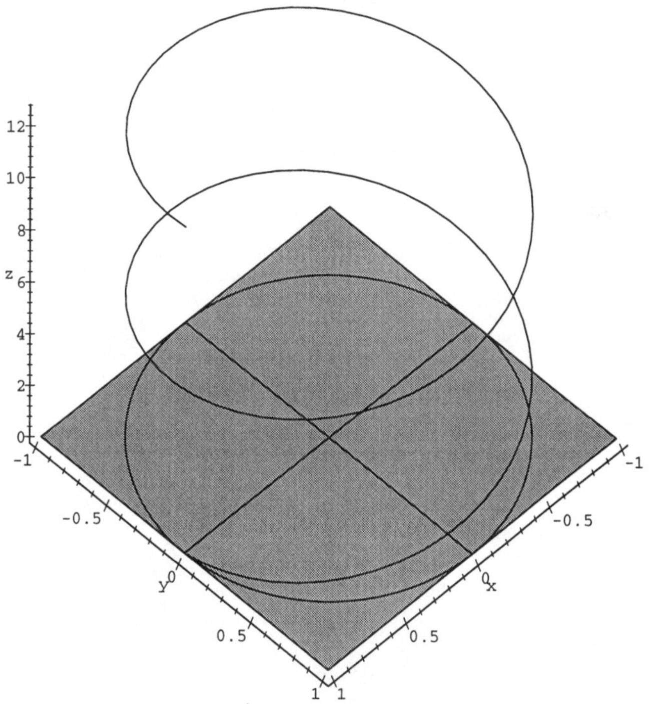

Find a parametrization for each curve.

Solution: We've just seen how to parametrize the unit circle in the xy-plane; to locate this curve in three dimensions we simply assign z a constant value:

$$x = \cos t; \qquad y = \sin t; \qquad z = 0; \qquad 0 \le t \le 2\pi.$$

The helix is a close cousin to the circle; the difference is that the helix *rises* as it turns. This suggests a parametrization almost like the one just seen, except that we'll set $z = t$ to effect the desired "rising." Since we want *two* turns, we also double the parameter interval:

$$x = \cos t; \qquad y = \sin t; \qquad z = t; \qquad 0 \le t \le 4\pi.$$

The result is indeed what's shown in the picture—notice, in particular, that z takes values from 0 to $4\pi \approx 12.6$.

As with all curves, there are many possible parametrizations. For instance, we could traverse the same helix in "half the time" by using the parametrization

$$x = \cos(2t); \qquad y = \sin(2t); \qquad z = 2t;. \qquad 0 \le t \le 2\pi. \qquad \square$$

Plane curves to order

By combining such "elements" as line segments and circular arcs, we can draw various composite curves "to order".

■ **Example 6.** A dog track is designed as shown, with three straight runs and three circular arcs.

A simple racetrack

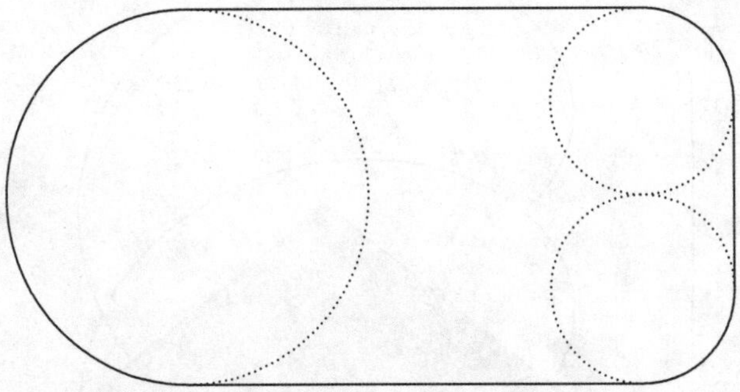

"Construct" such a track mathematically by parametrizing each of its six elements, running counterclockwise.

Solution: Imagine the track as lying in the xy-plane; the coordinates shown are chosen for convenience:

There's nothing special about these particular coordinates or units of size.

A racetrack in the xy-plane

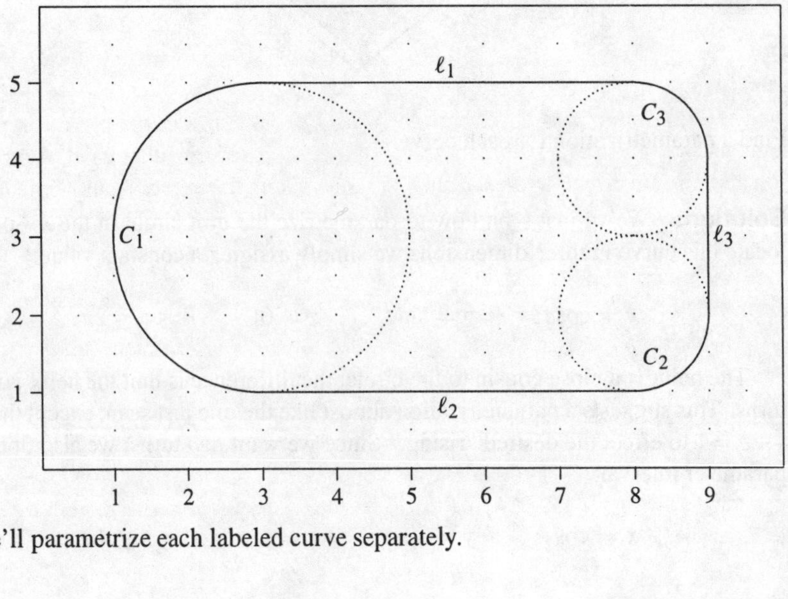

We'll parametrize each labeled curve separately.

The line segments. As we said above, setting

$$x(t) = a(1-t) + ct; \quad y(t) = b(1-t) + dt; \quad 0 \le t \le 1$$

parametrizes the segment from (a, b) to (c, d). Now ℓ_1 runs from $(8, 5)$ to $(3, 5)$, so substituting appropriate values for a, b, c, and d gives

$$x(t) = 8(1-t) + 3t = 8 - 5t; \quad y(t) = 5(1-t) + 5t = 5; \quad 0 \le t \le 1$$

parametrizes ℓ_1. Similarly,

$$x(t) = 3(1-t) + 8t = 3 + 5t; \quad y(t) = 1(1-t) + 1t = 1; \quad 0 \le t \le 1$$

parametrizes ℓ_2, and

$$x(t) = 9(1-t) + 9t = 9; \quad y(t) = 2(1-t) + 4t = 2 + 2t; \quad 0 \le t \le 1$$

parametrizes ℓ_3.

The circular arcs. Each circular arc shown has a parametrization of the general form

$$x = a + r\cos t; \qquad y = b + r\sin t; \qquad \alpha \le t \le \beta.$$

Choosing a, b, and r is easy, since each circle's center and radius appear clearly in the picture. Choosing appropriate parameter intervals $[\alpha, \beta]$ takes a little more care. In each case, the parameter interval $[0, 2\pi]$ would produce a *full* circle. Since we want only part of each circle, we'll need shorter parameter intervals.

Thinking of t in *angular* terms helps. Points on the arc C_1, correspond to angles with radian measure between $\pi/2$ and $3\pi/2$. Thus the prescription

$$x(t) = 3 + 2\cos t; \quad y(t) = 3 + 2\sin t; \quad \pi/2 \le t \le 3\pi/2$$

parametrizes C_1. Similarly,

$$x(t) = 8 + \cos t; \quad y(t) = 2 + \sin t; \quad 3\pi/2 \le t \le 2\pi$$

parametrizes C_2, and

$$x(t) = 8 + \cos t; \quad y(t) = 4 + \sin t; \quad 0 \le t \le \pi/2$$

parametrizes C_3. □

One road, many journeys

Many cars journey between Mission, South Dakota, and Valentine, Nebraska, but there's only one road: US Highway 83.

The same important difference holds between a curve in the plane, on the one hand, and any particular parametrization of that curve, on the other hand. A parametrization amounts, in effect, to prescribing a journey along the curve: At any time in some interval $a \le t \le b$, the coordinate functions $x = f(t)$ and $y = g(t)$ place a moving point (a car, say) somewhere along the curve (US 83, for instance). Many journeys—in either direction, at various speeds, over various time intervals, etc.—are possible along that stretch of US 83.

The moral is that any particular curve has many possible parametrizations. Thus we'll often emphasize parametrizations as much as the curves they determine. A parametrization depends on all three of its ingredients: (i) a coordinate function $x = x(t)$; (ii) a coordinate function $y = x(t)$; (iii) a parameter interval $a \le t \le b$.

We'll return often to the theme of finding and comparing various parametrizations of a given curve or other object. The next example is a simple introduction to the idea.

■ **Example 7.** Give two *different* parametrizations of the parabola C_2 on page 16. How do the parametrizations differ? How are they the same?

Solution: Different pairs of parametric equations may produce exactly the same geometric curve; labeling t-values can help show differences between the parametrizations. For example, both parametrizations

$$x = t; \qquad y = t^2; \qquad -1 \le t \le 1$$

and

$$x = t^3; \qquad y = t^6; \qquad -1 \le t \le 1$$

produce exactly the same parabolic curve:

The curve $x = t,\ y = t^2,\ -1 \le t \le 1$

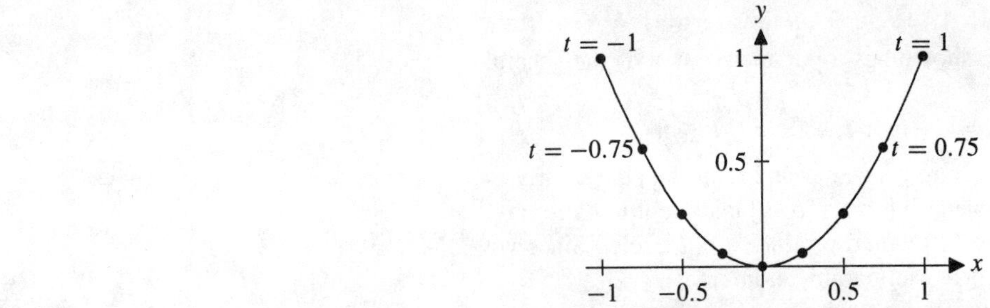

The curve $x = t^3,\ y = t^6,\ -1 \le t \le 1$

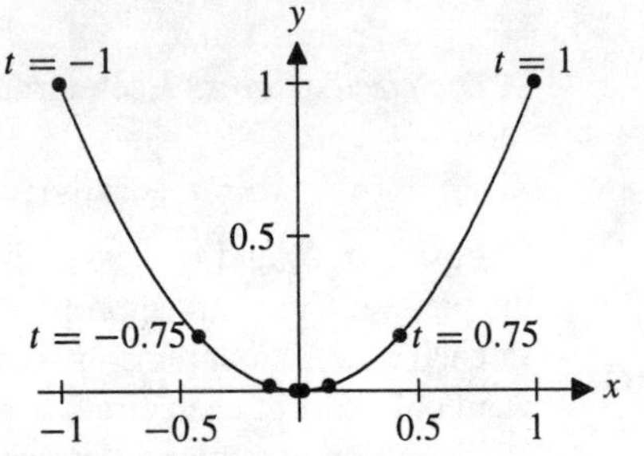

The bullets (they appear at 0.25-second time intervals) show, however, that the two parametrizations represent quite different "trips" along the same "road." Most important, the trips differ in speed. We'll return to this idea later in this chapter.

Tricks of the trade

There are various standard devices for trading one sort of parametrization for another. We mention some briefly here; more appear in the exercises.

Reversing direction. US Highway 83 runs both ways between Mission and Valentine. Similarly, any parametric curve can be "reversed." If C is parametrized by

$$x = f(t); \qquad y = g(t); \qquad a \le t \le b,$$

then the reversed curve—we'll denote it by $-C$—can be parametrized by

$$x = f(a + b - t); \qquad y = g(a + b - t); \qquad a \le t \le b.$$

(As t runs from a to b, $a + b - t$ runs from b to a.)↞

Use the formulas to convince yourself of this.

For the curve C_2 on page 16, for instance, the reverse curve $-C_2$ can be parametrized as follows:↞

Convince yourself of this by plotting.

$$x = -t; \qquad y = (-t)^2 = t^2; \qquad -1 \le t \le 1.$$

Different parameter intervals. US 83 takes the same route at all times of the day or night, regardless of how fast you drive it. For similar reasons, a parametric curve can be parametrized using *any* parameter interval.

For instance, if C is parametrized by

$$x = f(t); \qquad y = g(t); \qquad a \le t \le b,$$

then C can also be parametrized using the time interval $0 \le t \le 1$, using

$$x = f(a + (b - a)t); \qquad y = g(a + (b - a)t); \qquad 0 \le t \le 1.$$

Conversely, if C is parametrized by

$$x = f(t); \qquad y = g(t); \qquad 0 \le t \le 1,$$

then C can also be parametrized using the time interval $a \le t \le b$, via

$$x = f\left(\frac{t-a}{b-a}\right); \qquad y = g\left(\frac{t-a}{b-a}\right); \qquad a \le t \le b.$$

Polar curves as parametric equations. Curves are often described in polar coordinates by equations of the form $r = f(\theta)$, say for $\alpha \le \theta \le \beta$. Now recall that in polar coordinates, $x = r\cos\theta$ and $y = r\sin\theta$. Points on such a polar curve, therefore, have a natural parametric form, with θ as the parameter:

$$x = r\cos\theta = f(\theta)\cos\theta; \qquad y = r\sin\theta = f(\theta)\sin\theta; \qquad \alpha \le \theta \le \beta.$$

The polar curve $r = 3$, for instance, has the parametric form

$$x = 3\cos\theta; \qquad y = 3\sin\theta; \qquad 0 \le \theta \le 2\pi.$$

The result is the familiar parametrization for the circle of radius 3.

Exercises

1. Plot each parametric curve below. (Using technology is fine, but also draw or copy your own curve onto paper.) In each case, mark the direction of travel and label the points corresponding to $t = -1$, $t = 0$, and $t = 1$.

   (a) $x = t$, $y = \sqrt{1 - t^2}$, $-1 \le t \le 1$

   (b) $x = t$, $y = -\sqrt{1 - t^2}$, $-1 \le t \le 1$

   (c) $x = \sqrt{1 - t^2}$, $y = t$, $-1 \le t \le 1$

   (d) $x = -\sqrt{1 - t^2}$, $y = t$, $-1 \le t \le 1$

   (e) $x = \sin(\pi t)$, $y = \cos(\pi t)$, $-1 \le t \le 1$

2. Write each polar curve below in parametric form; then plot the result, using technology, as a parametric curve. Are the answers what you expect?

   (a) The cardioid $r = 1 + \cos\theta$, for $0 \le \theta \le 2\pi$.

   (b) The circle $r = 2$, for $0 \le \theta \le 2\pi$.

   (c) The polar curve $r = \sin\theta$, for $0 \le \theta \le \pi$.

3. Find a parametrization (there's more than one possibility!) for each curve described below. Check answers with technology.

   (a) The straight line segment from $(0, 0)$ to $(1, 2)$ (moving from left to right).

   (b) The straight line segment from $(1, 2)$ to $(0, 0)$ (moving from right to left).

(c) The entire unit circle, starting and ending at the east pole, but moving clockwise.

(d) The left half of the unit circle, moving counterclockwise, from the north pole to the south pole.

(e) The entire unit circle, starting and ending at the east pole, but using the parameter interval $0 \le t \le 1$.

4. Each parametric "curve" below is actually a line segment. In each part, state the beginning point ($t = 0$) and ending point ($t = 1$) of the segment. Then state an equation in x and y for the line each segment determines.

   (a) $x = 2 + 3t$, $y = 1 + 2t$, $0 \le t \le 1$
   (b) $x = 2 + 3(1 - t)$, $y = 1 + 2(1 - t)$, $0 \le t \le 1$
   (c) $x = t$, $y = mt + b$, $0 \le t \le 1$
   (d) $x = a + bt$, $y = c + dt$, $0 \le t \le 1$
   (e) $x = x_0 + (x_1 - x_0)t$, $y = y_0 + (y_1 - y_0)t$, $0 \le t \le 1$

5. Consider the curve C shown in Example 1; suppose that t tells time in seconds.

   (a) At which bulleted points would you expect P to be moving quickly? Slowly? Why?
   (b) *Estimate* the speed of P at $t = 3$. (One approach: Estimate how far P travels—i.e., the length of the curve—over the one-second interval from $t = 2.5$ to $t = 3.5$. If d is this distance, then d distance units per second is a reasonable speed estimate at $t = 3$.)
   (c) Use the curve to estimate the speed of P at $t = 6$.

6. Plot the parametric curve

$$x = t^3; \qquad y = \sin t^3; \qquad -2 \le t \le 2.$$

   What familiar curve is produced? Why does the result happen?

7. Let (a, b) be any point in the plane, and $r > 0$ any positive number. Consider the parametric equations

$$x = a + r\cos t; \qquad y = b + r\sin t; \qquad 0 \le t \le 2\pi.$$

   (a) Plot the parametric curve defined above for $(a, b) = (2, 1)$ and $r = 2$. Describe your result in words.
   (b) Show by calculation that if x and y are as above, then $(x - a)^2 + (y - b)^2 = r^2$. Conclude that the curve defined above is the circle with center (a, b) and radius r.
   (c) Write parametric equations for the circle of radius $\sqrt{13}$, centered at $(2, 3)$.
   (d) What "curve" results from the equations above if $r = 0$?

8. Let a and b be any positive numbers, and let a parametric curve C be defined by $x = a\cos t$; $y = b\sin t$; $0 \le t \le 2\pi$. The resulting curve is called an **ellipse**.

(a) Plot the curve defined above for $a = 2$ and $b = 1$. Describe C in words. Where is the "center" of C? Why do you think the quantities $2a$ and $2b$ are called the **major** and **minor axes** of C.

(b) What curve results if $0 \leq t \leq 4\pi$? Why?

(c) Write parametric equations for an ellipse with major axis 10 and minor axis 6.

(d) Write parametric equations for *another* ellipse with major axis 10 and minor axis 6.

(e) Show that for all t, $\frac{x^2}{a^2} + \frac{y^2}{b^2} = 1$.

(f) How does the "ellipse" look if $a = b$? How does its xy equation look?

(g) How does an ellipse look if $a = 1000$ and $b = 1$?

(h) How does an ellipse look if $a = 1$ and $b = 1000$?

9. Give two different parametrizations of the upper half of the ellipse C_3 on page 16. (One can be based on the graph of an ordinary function.)

10. If a parametric curve C has *linear* coefficient functions $x = f(t) = at + b$ and $y = g(t) = ct + b$, then C is a straight line (or part of a line).

(a) Plot the parametric curve $x = 2t$, $y = 3t + 4$, $0 \leq t \leq 1$. Where does the curve start? Where does it end? What is its shape?

(b) Eliminate the variable t in the two equations above to find a single equation in x and y for the line of the previous part.

11. Look at the racetrack in Example 6, page 22. It resembles a belt drawn snugly around three circles. Consider another similar racetrack, this time in the shape of a belt drawn snugly around two circles, the first of radius 2, centered at $(3, 3)$, and the second of radius 2, centered at $(3, 7)$.

(a) Draw the racetrack by hand; label its four elements. (Two are straight lines and two are circular arcs.)

(b) Parametrize each of the four arcs; all should run counterclockwise. (More than one method is possible.)

(c) Use technology and your results from the previous part to draw the new racetrack.

12. Suppose that a particle moves clockwise once around a circle of radius 3 centered at the point $(4, 5)$ in the xy-plane, that it moves with constant speed 2, and that it starts at the position $(1, 5)$. Give a position function that models the motion of the particle.

13. Give a parameterization for the *left* half of the circle of radius 3 centered at the point $(1, 2)$.

14. Give a clockwise-oriented parameterization for the bottom half of a circle of radius 4 centered at $(2, -3)$.

15. Give a parametric representation of the circular helix (i.e., a spring) with radius R that makes N complete turns between $(R, 0, 0)$ and $(R, 0, H)$.

1.3 Vectors

A **vector** is a quantity that has both magnitude and direction. Vectors can "live" in any dimension—in the real line, in the xy-plane, in xyz-space, or in higher-dimensional spaces. This section is about vectors in the plane and in space: what they are, why they're important, how to describe them mathematically, and how to calculate with them. Vectors are so important in both pure and applied mathematics that a general mathematical area, **linear algebra**, is devoted to studying their theory, behavior, and generalizations.

One generalization, for example, is to infinite-dimensional spaces.

Vectors in the plane

We'll think of two-dimensional vectors either geometrically, as arrows in the xy-plane, or algebraically, as 2-tuples (a, b). (In three dimensions we use arrows in xyz-space, and 3-tuples (a, b, c).) For the moment, we'll concentrate mainly on 2-dimensional vectors. This is mainly for convenience, since pictures are easier to draw in the plane than in space. Almost all of the main ideas, however, translate readily to vectors in 3-dimensional space.

Concrete examples rather than abstract theory will drive our study of vectors. The next example, in particular, is worth special attention; it illustrates some of the ways we'll see vectors in action, and it introduces some important vocabulary. Formal definitions will come later.

■ **Example 1.** The curve below is a path in the xy-plane; an ant walks along the path. At each marked point on the curve, an arrow—or vector—describes the ant's instantaneous velocity as it passes that point. What does the picture mean?

Velocity vectors on an ant's path

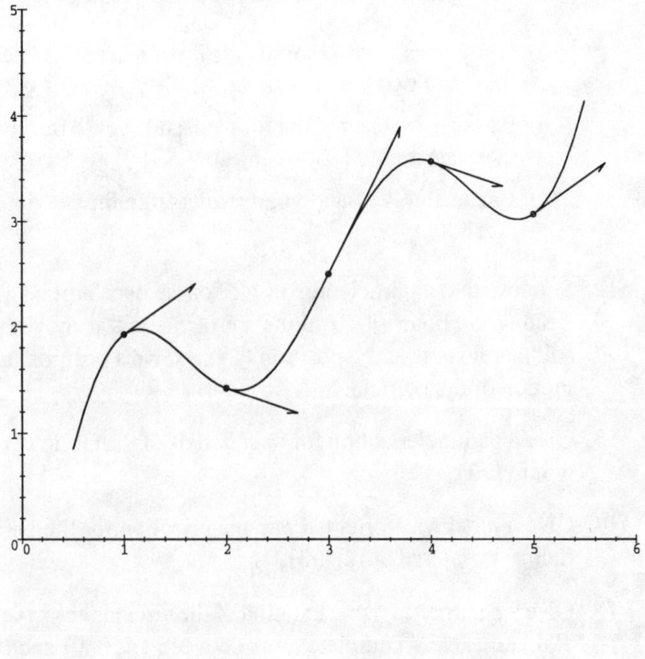

Solution: We'll explain later how the picture was drawn. First, let's discuss what it means.

Velocity: a vector. Velocity is among the simplest and most familiar vector quantities. We'll return to it again and again in this book. At any instant, the velocity of a moving object involves both a **magnitude** (the object's speed at that instant) and a **direction** (the direction of movement at that instant).

The arrows along the curve describe the ant's velocity in a convenient, compact form. Each arrow's direction is tangent to the curve (since the ant moves along the curve), and each arrow's length tells the ant's speed as the ant passes the point in question. The arrows show (among other things) that the ant moves from left to right along the curve, and that it moves faster in the middle of the curve than at either end. At the leftmost marked point the arrow has length about one, so the ant's speed is about one distance unit per second.

Speed: a scalar. Speed and velocity are sometimes equated in everyday language, but in mathematics and physics there's an important difference. Speed is an ordinary number—what a car's speedometer shows at any given instant. When vectors and numbers arise together, numbers are often called **scalars**. The name is appropriate because scalars describe the size, or "scale," of vectors. Velocity, by contrast, is a vector quantity, incorporating *both* speed and direction.➡

In a car, a "velocity-meter" would combine a speedometer and a compass.

Telling time. Both axes in the picture correspond to spatial directions; neither axis tells time. The velocity vectors, however, *are* related to time, because they indicate *speed.* To draw such a picture, therefore, we had to know not only where the ant walked but also *when* it reached any given point—even though this information doesn't appear directly in the picture. (We'll explain soon exactly *how* we used this information.)

Movable arrows. The arrows at $x = 2$ and at $x = 4$ are *identical*: both have the same length and the same direction. Entomologically, this means that at these two points on its journey the ant had exactly the *same* velocity—even though the points themselves are different.

The mathematical lesson is important:

> *A vector is determined only by its length and direction—not by where in the plane the vector happens to be drawn.*

The arrows at $x = 2$ and $x = 4$, in other words, are really two pictures of the same vector. We'll see that it's often useful to move vectors (without twisting or stretching!) from place to place to suit the purpose at hand. □

What is a vector?

Vectors can be defined mathematically in more or less abstract ways, depending on the setting and the purpose at hand. However, any reasonable definition of vector must capture the key idea of a quantity that has both magnitude and direction. Vectors in the xy-plane can be described either as arrows or as 2-tuples; both descriptions will be useful.

Vectors as arrows. A vector $\boldsymbol{v}$ can be described by an arrow in the xy-plane. Two arrows with the same length and the same direction describe the same vector—regardless of where the arrows are placed in the plane.

Vectors as 2-tuples. A vector $\boldsymbol{v}$ can be written as a 2-tuple of real numbers, as in $\boldsymbol{v} = (a, b)$. The numbers a and b are called the **components** (or, alternatively, the **coordinates**) of the vector $\boldsymbol{v}$.

The following pictures illustrate these two different views of vectors, and the connections between them:

Vectors as arrows and as 2-tuples

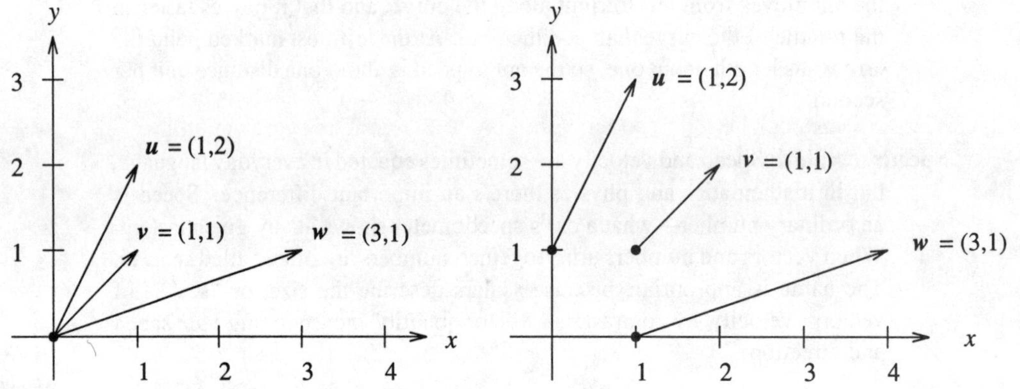

Here are some of the things the pictures show; watch, too, for some important vocabulary and notation.

Different tail points, same vectors. The two pictures show exactly the same vectors $\boldsymbol{u}$, $\boldsymbol{v}$, and $\boldsymbol{w}$. The only difference concerns where in the plane the vectors are placed; in the pictures, the vectors' tail points are shown as dark dots.

Vectors based at the origin. In the left-hand picture all three vectors have their tails at the origin, so each vector is of the form $\overrightarrow{OP}$, where O is the origin and P is the point at the tip. In this special case, the *components* of each vector are the same as the Cartesian *coordinates* of the tip point P. For this reason, a vector of the form $\boldsymbol{v} = \overrightarrow{OP}$ is called the **position vector** of the point P.

There is a technical difference between a point P with coordinates (x, y) and the vector from the origin O to P, with components (x, y). In practice, however, the difference seldom matters much, so we will occasionally fuzz the distinction by thinking of (x, y) sometimes as a point and sometimes as a position vector.

This is a little sloppy, but it's standard practice.

What the components say. In the right-hand picture, each vector's components are *not* the Cartesian coordinates of its tip point. Instead, they describe the **displacement** between the vector's head and its tail. The vector $\boldsymbol{w} = (3, 1)$, for instance, involves moving 3 units to the right and 1 unit up.

The vector from P to Q. If $P = (a, b)$ and $Q = (c, d)$ are any points in the plane, then the vector from P to Q, denoted by $\overrightarrow{PQ}$, has components $(c - a, d - b)$. In the right-hand picture, for instance, the vector $\boldsymbol{w}$ starts at $(1, 0)$ and ends at $(4, 1)$, so its coordinates are indeed $(3, 1)$.

Finding lengths. The **length** (or **magnitude**) of a vector is the distance between its head and its tail. The components make this easy to calculate, using the distance formula in the xy-plane. If $\boldsymbol{v} = (a, b)$, then

$$\text{length of } \boldsymbol{v} = |\,\boldsymbol{v}\,| = \sqrt{a^2 + b^2}.$$

(Notice that we use the same notation (| |) to denote both the length of a vector and the absolute value of a number. It should be clear from context which is which. Using the same symbol makes sense because we're measuring size in both cases.)

Bold vectors, faint scalars. Vectors and scalars are often seen together. To keep clear which is which, it's common to denote vectors by using either bold type, as in $\boldsymbol{v}$, or an arrow on top, as in $\overrightarrow{PQ}$. (The latter is convenient when describing a vector with tail at P and head at Q, or when writing by hand.) Scalars appear in ordinary type. For example, in the expression

$$a\,\boldsymbol{u} + b\,\boldsymbol{v} + c\,\boldsymbol{w},$$

a, b, and c denote scalars, while $\boldsymbol{u}$, $\boldsymbol{v}$, and $\boldsymbol{w}$ denote vectors.

Algebra with vectors

One reason vectors are so useful is that they allow algebraic operations very like those with ordinary numbers. We begin with some easy samples:

$$(1, 2) + (2, 1) = (3, 3); \quad -3\,(1, 2) = (-3, -6); \quad 3\,(1, 2) - (3, 4) = (0, 2).$$

The first operation illustrates the **sum of two vectors**; the result is another vector. The second operation illustrates **scalar multiplication**; a scalar is multiplied by a vector to produce a new vector. The third operation involves both a scalar product and the **difference of two vectors**.

None of these computations is either very surprising or, for that matter, very interesting. What *are* interesting and useful are the geometric meanings of addition and scalar multiplication. Consider addition first:

Adding vectors

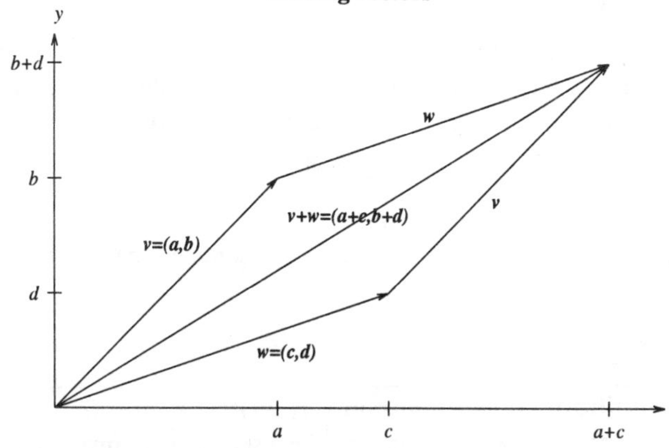

The picture shows what it means geometrically to add two vectors $\boldsymbol{v} = (a, b)$ and $\boldsymbol{w} = (c, d)$. The sum $\boldsymbol{v} + \boldsymbol{w}$ has components $(a + c, b + d)$. Geometrically, the sum $\boldsymbol{v} + \boldsymbol{w}$ is the vector obtained by putting the tail of one vector at the head of the other. (The order doesn't matter.) The picture also illustrates the **parallelogram rule**: the sum $\boldsymbol{v} + \boldsymbol{w}$ is the diagonal of the parallelogram with sides $\boldsymbol{v}$ and $\boldsymbol{w}$.

Now consider scalar multiplication:

Scalar-multiplying vectors

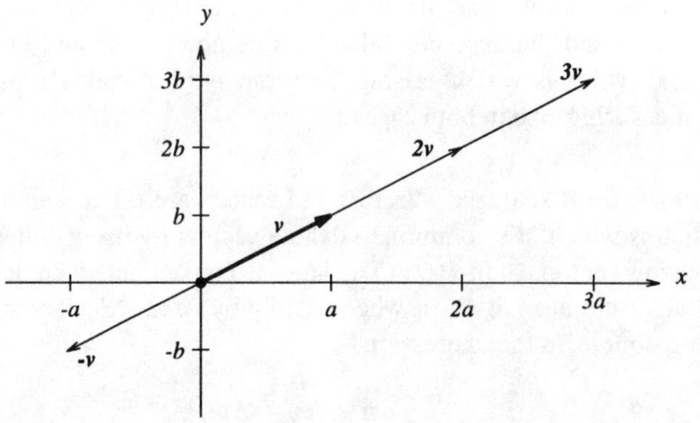

(All vectors shown have tails at the origin.) This picture shows what it means geometrically to multiply the vector $\boldsymbol{v}$ by a scalar a. The result, written $a\boldsymbol{v}$, is a vector with direction either the same as $\boldsymbol{v}$ (if $a > 0$) or opposite to $\boldsymbol{v}$ (if $a \leq 0$).

Length, addition, and scalar multiplication. The preceding pictures show how vector operations affect the lengths of the vectors involved. For scalar multiplication, we see that the length of $a\boldsymbol{v}$ is $|a|$ times the length of $\boldsymbol{v}$.⬅ In math-speak, scalar multiplication by a **dilates** $\boldsymbol{v}$ by the factor $|a|$. In symbols: $|a\,\boldsymbol{v}| = |a|\;|\boldsymbol{v}|$.

Why is the absolute value needed?

The picture "Adding vectors" illustrates the so-called **triangle inequality**⬅ for vectors. In symbols:

In a triangle, the length of each side is less than the sum of the other two lengths—hence the name.

$$|\,\boldsymbol{v} + \boldsymbol{w}\,| \leq |\boldsymbol{v}| + |\,\boldsymbol{w}|\,.$$

In words: *The length of a sum of vectors is less than or equal to the sum of the vectors' lengths.*

The triangle inequality certainly looks believable, if not downright obvious. But an algebraic proof based purely on the definitions is slightly messy; so we'll defer giving one until we've developed a few more vector tools.

Formal definitions. The following definitions make precise the insights suggested by the pictures:

> **Definition: (Operations on vectors)** Let $\boldsymbol{v} = (a, b)$ and $\boldsymbol{w} = (c, d)$ be plane vectors, and let r be a scalar. The **sum** of $\boldsymbol{v}$ and $\boldsymbol{w}$ is
>
> $$\boldsymbol{v} + \boldsymbol{w} = (a, b) + (c, d) = (a + c, b + d).$$
>
> **Scalar multiplication** of $\boldsymbol{v}$ by r is defined by
>
> $$r\,\boldsymbol{v} = r\,(a, b) = (ra, rb).$$

The following picture shows some "hybrids" of these vector operations:

Algebra with vectors: samples

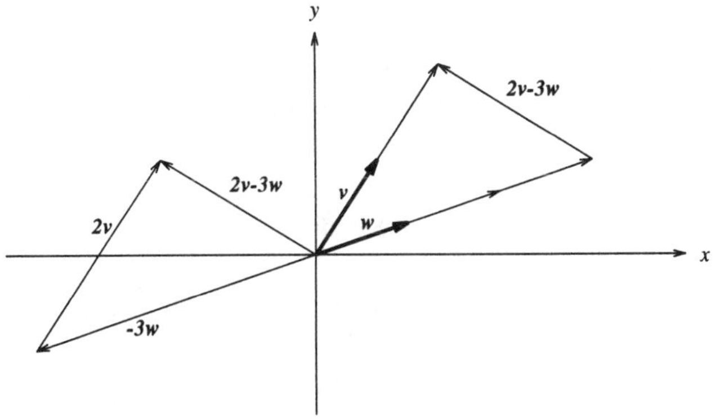

Notice, in particular, how the **difference of two vectors** appears: If $\boldsymbol{v}$ and $\boldsymbol{w}$ have the same tail point, then the difference $\boldsymbol{v} - \boldsymbol{w}$ points from the tip of $\boldsymbol{w}$ to the tip of $\boldsymbol{v}$. (Alternatively, $\boldsymbol{v} - \boldsymbol{w}$ is the vector sum of $\boldsymbol{v}$ and $-\boldsymbol{w}$.)

The picture shows the difference $2\boldsymbol{v} - 3\boldsymbol{w}$.

Properties of vector operations

The vector operations enjoy many of the pleasant properties that we've come to expect of operations with ordinary numbers. We collect a slew of such properties, together with their names, in one big theorem:

Theorem 1. (Properties of vector operations) Let $\boldsymbol{u}$, $\boldsymbol{v}$, and $\boldsymbol{w}$ be vectors, and let r and s be scalars. The following algebraic properties hold:

(a) $\boldsymbol{u} + \boldsymbol{v} = \boldsymbol{v} + \boldsymbol{u}$ (commutativity);

(b) $\boldsymbol{u} + (\boldsymbol{v} + \boldsymbol{w}) = (\boldsymbol{u} + \boldsymbol{v}) + \boldsymbol{w}$ (associativity of addition);

(c) $r(s\boldsymbol{v}) = (rs)\boldsymbol{v}$ (associativity of scalar multiplication);

(d) $(r + s)\boldsymbol{v} = r\boldsymbol{v} + s\boldsymbol{v}$ (distributivity);

(e) $r(\boldsymbol{v} + \boldsymbol{w}) = r\boldsymbol{v} + r\boldsymbol{w}$ (distributivity again).

All parts of the theorem can be shown to hold using similar properties of ordinary addition and multiplication.

The zero vector. Among numbers, 0 has many special properties. The vector $\boldsymbol{O} = (0, 0)$, called the **zero vector**, is special in similar ways. It has length zero and, therefore, no meaningful direction. Algebraically, it behaves "as a zero should." If $\boldsymbol{v}$ is any vector and r is any scalar, we have

Note carefully the different uses of similar symbols.

$$\boldsymbol{O} + \boldsymbol{v} = \boldsymbol{v}, \quad r\,\boldsymbol{O} = \boldsymbol{O} \quad \text{and} \quad 0\,\boldsymbol{v} = \boldsymbol{O}.$$

Unit vectors. A vector $\boldsymbol{v} = (a, b)$ is called a **unit vector** if $\boldsymbol{v}$ has length 1, i.e., if

$$|\boldsymbol{v}| = \sqrt{a^2 + b^2} = 1.$$

Any non-zero vector $\boldsymbol{v}$ can be multiplied (or divided) by an appropriate positive scalar to produce a unit vector with the same direction as $\boldsymbol{v}$. Suppose, for instance, that $\boldsymbol{v} = (3, 4)$. Then

$$|\boldsymbol{v}| = \sqrt{3^2 + 4^2} = 5,$$

Or, equivalently, multiply by 1/5.

so to scale $\boldsymbol{v}$ down to unit length, we can divide by 5. This produces the unit vector $\boldsymbol{u} = (3/5, 4/5)$, called the **unit vector in the direction of $\boldsymbol{v}$**. The same process works more generally:

Fact: Let $\boldsymbol{v} = (a, b)$ be any non-zero vector. Then the new vector

$$\boldsymbol{u} = \frac{\boldsymbol{v}}{\sqrt{a^2 + b^2}} = \left(\frac{a}{\sqrt{a^2 + b^2}}, \frac{b}{\sqrt{a^2 + b^2}} \right)$$

is a unit vector in the direction of $\boldsymbol{v}$.

Standard basis vectors. The two-dimensional vectors

$$\boldsymbol{i} = (1, 0) \quad \text{and} \quad \boldsymbol{j} = (0, 1)$$

are especially simple—but especially useful. Both $\boldsymbol{i}$ and $\boldsymbol{j}$ have length 1, and they point in perpendicular directions—along the x- and y-axes, respectively. The vectors $\boldsymbol{i}$ and $\boldsymbol{j}$ are known as the **standard basis vectors** for $\mathbb{R}^2$. They deserve this important-sounding name for a good reason:

> *Every vector $\boldsymbol{v} = (a, b)$ in $\mathbb{R}^2$ can be written as a sum of scalar multiples of $\boldsymbol{i}$ and $\boldsymbol{j}$.*

If, say, $\boldsymbol{v} = (2, 3)$, then

$$\boldsymbol{v} = (2, 3) = (2, 0) + (0, 3) = 2(1, 0) + 3(0, 1) = 2\boldsymbol{i} + 3\boldsymbol{j}.$$

Students who have taken linear algebra should recognize the words "basis" and "linear combination."

A sum of the form $r\boldsymbol{v} + s\boldsymbol{w}$, where r and s are any scalar constants, is called a **linear combination** of $\boldsymbol{v}$ and $\boldsymbol{w}$. The calculation with (2, 3) illustrates that *every* vector (a, b) in $\mathbb{R}^2$ can be written, if we wish, as a linear combination of $\boldsymbol{i}$ and $\boldsymbol{j}$. That is,

$$(a, b) = a\boldsymbol{i} + b\boldsymbol{j}.$$

Indeed, some authors use the $\boldsymbol{i}\,\boldsymbol{j}$-notation for almost all 2-vectors.

Vectors in space

Or even more than three.

So far we have considered only vectors in the xy-plane—those that "fit" naturally into two dimensions. Sometimes, however, a third dimension is essential. Physical motion, for instance, may be inescapably three-dimensional, as when a fly buzzes aimlessly around the room, or a maple seed spirals from a tree to the ground.

Vectors in space vs. vectors in the plane In most respects, vectors behave almost identically in two and three dimensions. In both settings, for instance, vectors can be thought of geometrically as arrows, with both magnitude and direction. In the following paragraphs we collect some similarities and differences between two- and three-dimensional vectors.

Two-tuples and three-tuples. A plane vector $\boldsymbol{v}$ corresponds to a 2-tuple (a, b). Similarly, a space vector $\boldsymbol{v}$ corresponds to a 3-tuple (a, b, c), and we sometimes simply write $\boldsymbol{v} = (a, b, c)$. The vector $\boldsymbol{v}$ can be thought of as the arrow (or **position vector**) from the origin $(0, 0, 0)$ to the point (a, b, c). The numbers a, b, and c are called the **coordinates** of $\boldsymbol{v}$.

Algebra with 3-vectors. Three-dimensional vectors are added and scalar multiplied "term-by-term," exactly like two-dimensional vectors. If $\boldsymbol{v} = (v_1, v_2, v_3)$ and $\boldsymbol{w} = (w_1, w_2, w_3)$ are vectors, and a is any scalar, then

$$\begin{aligned} \boldsymbol{v} + \boldsymbol{w} &= (v_1, v_2, v_3) + (w_1, w_2, w_3) = (v_1 + w_1, v_2 + w_2, v_3 + w_3); \\ a\boldsymbol{v} &= (av_1, av_2, av_3). \end{aligned}$$

Geometric meanings. Vector addition and scalar multiplication have exactly the same geometric meaning for vectors in space as for vectors in the plane. If $\boldsymbol{v}$ and $\boldsymbol{w}$ are space vectors, and k is a scalar,➡ then $\boldsymbol{v} + \boldsymbol{w}$ is the vector formed by laying $\boldsymbol{v}$ and $\boldsymbol{w}$ head to tail; the scalar multiple $k\boldsymbol{v}$ is formed by stretching $\boldsymbol{v}$ by the factor $|k|$. (If $k < 0$, we reverse the direction of $\boldsymbol{v}$.)

I.e., a number.

Length and the triangle inequality. The **length** of a 3-vector is defined using the distance formula in xyz-space: if $\boldsymbol{v} = (v_1, v_2, v_3)$, then

$$|\boldsymbol{v}| = \text{length} = \sqrt{v_1^2 + v_2^2 + v_3^2}.$$

The same algebraic properties of length hold in three dimensions as in two:

$$|a\boldsymbol{v}| = |a|\ |\boldsymbol{v}|; \qquad |\boldsymbol{v} + \boldsymbol{w}| \le |\boldsymbol{v}| + |\boldsymbol{w}|.$$

The right-hand expression above is the **triangle inequality** again—this time for 3-vectors.

Standard basis vectors. In three-dimensional space, there are *three* standard basis vectors:

$$\boldsymbol{i} = (1, 0, 0), \quad \boldsymbol{j} = (0, 1, 0), \quad \text{and} \quad \boldsymbol{k} = (0, 0, 1).$$

(Note that the symbols $\boldsymbol{i}$ and $\boldsymbol{j}$ denote slightly different objects in the 2-d and 3-d cases.) Each standard basis vector has length 1, and points in the positive direction along one of the coordinate axes. In space, as in the plane, *every* vector can be written as a linear combination of $\boldsymbol{i}$, $\boldsymbol{j}$, and $\boldsymbol{k}$:

$$(a, b, c) = a\,(1, 0, 0) + b\,(0, 1, 0) + c\,(0, 0, 1) = a\,\boldsymbol{i} + b\,\boldsymbol{j} + c\,\boldsymbol{k}.$$

Exercises

1. In all parts, use the vectors $\boldsymbol{u} = (1, 2)$, $\boldsymbol{v} = (2, 3)$, and $\boldsymbol{w} = (-2, 1)$. Find each of the following quantities:

   (a) $\boldsymbol{u} + \boldsymbol{v}$

   (b) $\boldsymbol{u} - \boldsymbol{v}$

   (c) $2\boldsymbol{u} - 3\boldsymbol{v}$

(d) $\boldsymbol{u} + \boldsymbol{v}/2 + \boldsymbol{w}/3$

(e) the unit vector in the direction of $\boldsymbol{u}$

(f) the unit vector in the direction of $\boldsymbol{v}$

(g) the unit vector in the direction of $\boldsymbol{w}$

2. Consider the three vectors $\boldsymbol{u} = (a, b)$, $\boldsymbol{v} = (c, d)$, and $\boldsymbol{w} = (e, f)$, and let r be a scalar. By the definition of vector addition,

$$\boldsymbol{u}+\boldsymbol{v} = (a, b)+(c, d) = (a+c, b+d) = (c+a, d+b) = (c, d)+(a, b) = \boldsymbol{v}+\boldsymbol{u}.$$

(The middle step holds because ordinary addition of numbers is commutative.) This argument shows that vector addition is commutative, as asserted in Theorem 1, page 33.

Use a similar argument to show each claim below.

(a) $\boldsymbol{u} + (\boldsymbol{v} + \boldsymbol{w}) = (\boldsymbol{u} + \boldsymbol{v}) + \boldsymbol{w}$

(b) $r(\boldsymbol{v} + \boldsymbol{w}) = r\boldsymbol{v} + r\boldsymbol{w}$

3. In this section we mentioned the parallelogram rule for vector addition: The sum $\boldsymbol{v} + \boldsymbol{w}$ is the diagonal of the parallelogram with adjacent sides $\boldsymbol{v}$ and $\boldsymbol{w}$.

(a) Draw and label a picture to illustrate this fact for the vectors $\boldsymbol{v} = (2, 1)$ and $\boldsymbol{w} = (1, 3)$. (Base both vectors at the origin.) What are the components of the diagonal vector?

(b) The parallelogram you drew actually has *two* diagonals. You drew a northeast-pointing diagonal in the previous part. Now draw the northwest-pointing diagonal vector. What are its components? How is it related to $\boldsymbol{v}$ and $\boldsymbol{w}$?

(c) Draw the southeast-pointing diagonal vector. What are its components? How is it related to $\boldsymbol{v}$ and $\boldsymbol{w}$?

4. For each vector $\boldsymbol{v}$ below, find the unit vector $\boldsymbol{u}$ in the direction of $\boldsymbol{v}$. Then sketch all 8 vectors on the same axes.

(a) $\boldsymbol{v} = (1, 1)$

(b) $\boldsymbol{v} = (-1, 1)$

(c) $\boldsymbol{v} = (1, 2)$

(d) $\boldsymbol{v} = (-1, -2)$

5. Let $\mathbf{v} = (3, 4)$. Find the vector of length 7 that points in the *opposite* direction as $\mathbf{v}$.

6. Let $\boldsymbol{v}$ and $\boldsymbol{w}$ be nonparallel vectors. What basic fact from Euclidean geometry is expressed by the relation $|\boldsymbol{v} + \boldsymbol{w}| \le |\boldsymbol{v}| + |\boldsymbol{w}|$?

7. Let $\boldsymbol{v} = (1, 2, 3)$.

(a) Find the length $|\boldsymbol{v}|$.

(b) Find a unit vector $\boldsymbol{u}$ in the direction of $\boldsymbol{v}$. By hand, draw $\boldsymbol{u}$, $\boldsymbol{v}$, and $2\boldsymbol{u}$ on the same axes.

8. Repeat the preceding exercise, but use $\boldsymbol{v} = (1, -2, 3)$.

9. Suppose that **u** and **v** are the vectors in the xz-plane pictured below.

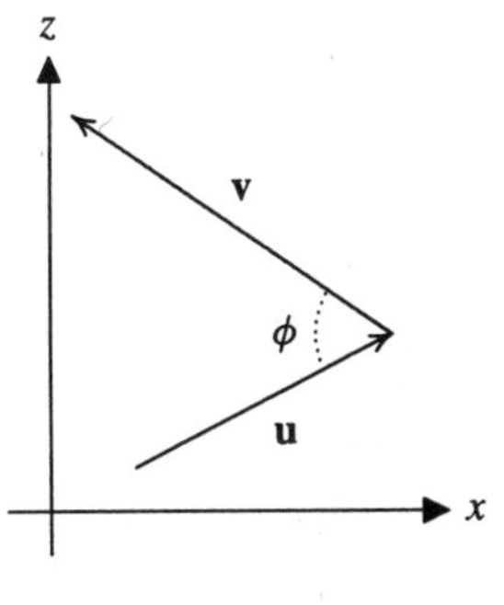

(a) Draw the vector $\boldsymbol{w} = \boldsymbol{u} + \boldsymbol{v}$.

(b) Draw the vector $\boldsymbol{a} = \boldsymbol{u} - \boldsymbol{v}$.

(c) Draw the vector $\boldsymbol{b} = \boldsymbol{v} - \boldsymbol{u}$.

10. Suppose that **u** and **v** are the vectors in the yz-plane pictured below.

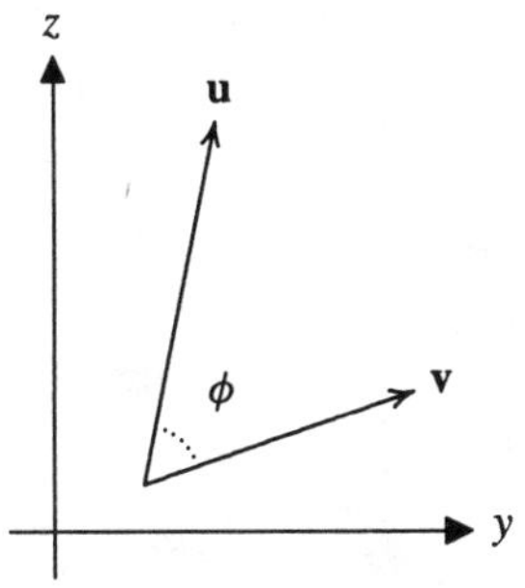

(a) Draw the vector $\boldsymbol{w} = \boldsymbol{u} + \boldsymbol{v}$.

(b) Draw the vector $\boldsymbol{a} = \boldsymbol{u} - \boldsymbol{v}$.

(c) Draw the vector $\boldsymbol{b} = \boldsymbol{v} - \boldsymbol{u}$.

1.4   Vector-valued functions, derivatives, and integrals

Scalars, vectors, and functions

The standard functions of single-variable calculus are *real-valued* functions of a *real* variable. The sine function, for instance, takes a real number, x, as input, and produces another real number, $\sin x$, as output.

In multivariable calculus, with vector quantities available, the possibilities for functions are considerably more diverse. A function might accept, say, real numbers as inputs and produce 2-vectors as outputs. Another function might accept 3-vectors as inputs and produce real numbers as outputs. The first function is called **vector-valued**; the second, **scalar-valued**. This added variety of function types accounts for much of the remarkable power of multivariable calculus to model real-world phenomena.

In this book we'll encounter functions that use all possible combinations of vectors and scalars as inputs and outputs. The good news, however, is that most of the basic ideas of single-variable calculus—graph, derivative, linear approximation, integral, etc.—extend in reasonably natural ways to functions with vector inputs or outputs (or both). But due care is needed, both to keep track of what type of object is under consideration at a given time, and to find the "right" extension of a given single-variable idea to the multivariable setting.

Vector-valued functions

A **vector-valued function** is one that produces vectors as outputs. Consider, for example, the function defined by the rule

$$f(t) = (\cos t, \sin t).$$

For this function, the notation $f : \mathbb{R} \to \mathbb{R}^2$ makes sense, because f accepts a single number t as input and produces the vector $(\cos t, \sin t)$ as output. Notice too that, because the function is vector-valued, we use the boldface symbol f rather than an ordinary f to denote it. The two functions inside the parentheses are called **component functions**, or **coordinate functions**. (If f has 3-vectors as values, there are three component functions.)

Parametric curves and vector-valued functions.   In Section 1.2 we studied curves in the plane, parametrized by two coordinate functions $x(t)$ and $y(t)$. (A curve in space has three coordinate functions, $x(t)$, $y(t)$, and $z(t)$.) As the next example shows, such curves are close cousins to vector-valued functions.

■ **Example 1.**   Consider the vector-valued function $f : \mathbb{R} \to \mathbb{R}^2$ defined by $f(t) = (\cos t, \sin t)$. How is f related to a parametric curve? Which curve?

Solution: We have seen that setting

$$x = \cos t; \qquad y = \sin t; \qquad 0 \le t \le 2\pi$$

gives a parametrization for the unit circle $x^2 + y^2 = 1$, traversed once counterclockwise, starting from the east pole. Thus, for any input t, the vector $f(t) = (\cos t, \sin t)$ can be thought of as the position vector for a point on the unit circle. As t ranges through the domain $(-\infty, \infty)$, the tip of the vector $f(t)$ traces the unit circle infinitely often, counterclockwise, once in each t-interval of length 2π.

The tail is pinned at the origin.

vector-valued = real # → vector
scalar-valued = 3-vector → real #

If, say, we wanted to trace only the right half of the unit circle, we restrict the parameter interval:

$$x = \cos t; \qquad y = \sin t; \qquad -\pi/2 \le t \le \pi/2.$$

In the language of vector-valued functions, this amounts to restricting the domain; we can write this economically as $\boldsymbol{f} : [-\pi/2, \pi/2] \to \mathbb{R}^2$. □

One idea, two views. As the preceding example shows, the difference between a pair of parametric equations and a vector-valued function is slight; we'll use both points of view and their corresponding notations more or less interchangeably.

When using vector-valued functions, it's sometimes useful to think of a curve as traced out by by a collection of position vectors $\boldsymbol{f}(t)$, one for each input t; all tails are pinned at the origin. The following picture shows several of these position vectors for the function $\boldsymbol{f}(t) = (t, 3 + \sin t)$:

Position vectors tracing out a curve

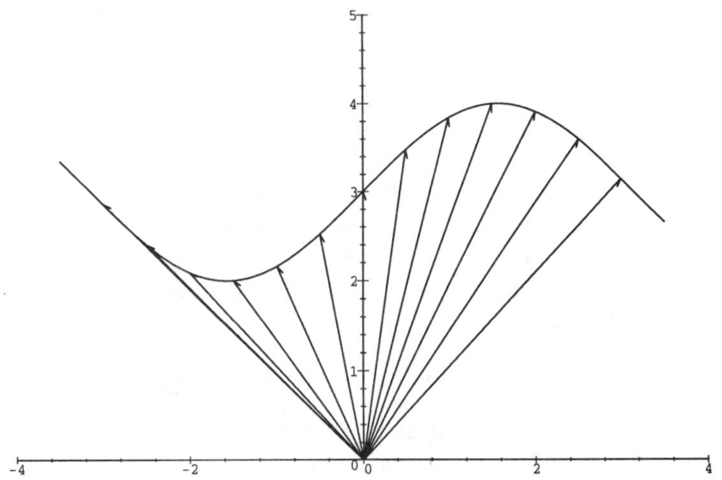

Lines, line segments, and vector-valued functions

Vectors and vector-valued functions make quick, efficient work of describing lines and line segments.

Lines in the plane. Any line ℓ in the xy-plane can be determined by two data: (i) a point $P_0 = (x_0, y_0)$ through which ℓ passes; and (ii) a **direction vector** $\boldsymbol{v} = (a, b)$ that points in the direction of ℓ. Geometrically, the situation looks like this:

Recall that $-\boldsymbol{v}$ points in the opposite direction to $\boldsymbol{v}$.

Determining a line

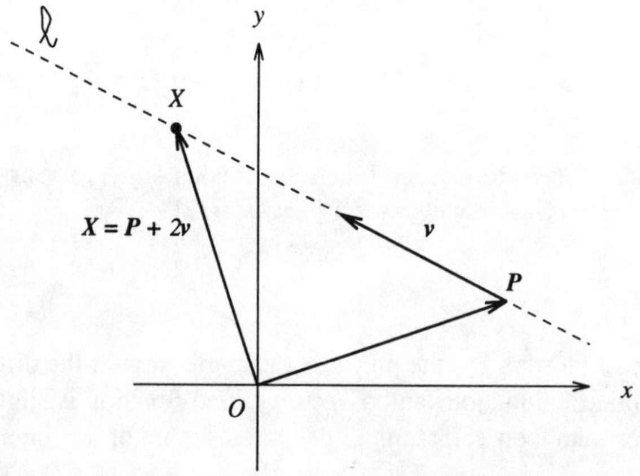

As the picture illustrates, a point $X = (x, y)$ lies on ℓ if and only if the vector $\boldsymbol{X}$ is of the form $\boldsymbol{P} + t\boldsymbol{v}$, for some (positive or negative) scalar t. (For the point X shown, $t = 2$.) Notice next that $\boldsymbol{X} = (x, y)$ and $\boldsymbol{P} = (x_0, y_0)$. Therefore, points on the line are those that satisfy the vector equation

$$\boldsymbol{X} = (x, y) = (x_0, y_0) + t(a, b)$$

for some real number t. This means that we can think of the line as the image of the vector-valued function $\boldsymbol{X} : \mathbb{R} \to \mathbb{R}^2$, defined for all real numbers t by $\boldsymbol{X}(t) = (x_0, y_0) + t(a, b)$.

In parametric equation language, ℓ can be parametrized by

$$x(t) = x_0 + at; \qquad y(t) = y_0 + bt; \qquad -\infty < t < \infty.$$

Line segments. Suppose we want only part of a line, say the segment from $P = (x_0, y_0)$ to $Q = (x_1, y_1)$. In this case we can use either P or Q as our fixed point, and either $\overrightarrow{PQ} = (x_1 - x_0, y_1 - y_0)$ or $\overrightarrow{QP} = (x_0 - x_1, y_0 - y_1)$ as our direction vector. Let's use the first option in each case. Then (reasoning as above), the line through P and Q has the form

$$\boldsymbol{X} = (x, y) = (x_0, y_0) + t(x_1 - x_0, y_1 - y_0),$$

where t can be any real number. Restricting t to the interval $0 \le t \le 1$ gives only the segment in question—$t = 0$ corresponds to P and $t = 1$ to Q.

Lines and segments in space. A line ℓ in xyz-space is determined by a point $P_0 = (x_0, y_0, z_0)$ and a direction vector $\boldsymbol{v} = (a, b, c)$. A general point (x, y, z) lies on ℓ if and only if the vector $\boldsymbol{X} = (x, y, z)$ is of the form

$$\boldsymbol{X} = \boldsymbol{P_0} + t\boldsymbol{v},$$

where t is a real number and $\boldsymbol{P_0}$ is the position vector from the origin to (x_0, y_0, z_0). If we want only part of the line, such as a segment, we can restrict t to a subset of the real numbers. If we take, say, $0 \le t \le 1$, then the vector equation above describes the segment that starts at P_0 and runs along ℓ, for one length of the vector $\boldsymbol{v}$.

If we prefer, we can think of the line above as the image of a vector-valued function $\boldsymbol{X} : \mathbb{R} \to \mathbb{R}^3$, defined by

$$\boldsymbol{X}(t) = \boldsymbol{P_0} + t\boldsymbol{v} = (x_0, y_0, z_0) + t(a, b, c).$$

Equivalently, we can consider the line to be determined by three (linear) parametric equations:

$$x(t) = x_0 + at; \quad y(t) = y_0 + bt; \quad z(t) = z_0 + ct.$$

Derivatives of vector-valued functions

In single-variable calculus derivatives describe rates of change. A vector-valued function $\boldsymbol{f}$ is, in one sense, just a list of separate, scalar-valued functions. It's reasonable, therefore, to describe the rate of change of $\boldsymbol{f}$ by differentiating the component functions separately. From this point of view, the following definition is natural:

> **Definition: (Derivative of a vector-valued function)** Let the vector-valued function $\boldsymbol{f} : \mathbb{R} \to \mathbb{R}^2$ be defined by $\boldsymbol{f}(t) = \big(f_1(t), f_2(t)\big)$. The derivative of **f** is the vector-valued function $\boldsymbol{f}' : \mathbb{R} \to \mathbb{R}^2$ defined by
>
> $$\boldsymbol{f}'(t) = \frac{d}{dt}\big(f_1(t), f_2(t)\big) = \big(f_1'(t), f_2'(t)\big).$$

The definition is similar in three variables: If $\boldsymbol{f} : \mathbb{R} \to \mathbb{R}^3$ is defined by $\boldsymbol{f}(t) = \big(f_1(t), f_2(t), f_3(t)\big)$, then $\boldsymbol{f}'(t) = \big(f_1'(t), f_2'(t), f_3'(t)\big)$.

Calculating this new derivative is no harder than finding a string of ordinary derivatives. For example,

$$\boldsymbol{f}(t) = (\cos t, \sin t) \implies \boldsymbol{f}'(t) = (-\sin t, \cos t);$$

differentiating each component function separately is all it takes.

The definition isn't surprising, but what does the derivative mean, geometrically? We'll discuss this question in detail below, but here's the short answer:

> **Fact:** Let I be an interval and $\boldsymbol{f} : I \to \mathbb{R}^2$ (or $\mathbb{R}^3$) a vector-valued function. Let $\boldsymbol{f}$ describe a curve C, and suppose that $\boldsymbol{f}'(t) \neq (0, 0)$ for all t in (a, b). For all t_0 in (a, b), the vector $\boldsymbol{f}'(t_0)$ is tangent to C at the point $\boldsymbol{f}(t_0)$, and points in the direction of increasing t. The scalar $|\boldsymbol{f}'(t_0)|$ (the magnitude of the derivative vector) tells the instantaneous speed (in units of distance per unit of time t) with which $\boldsymbol{f}(t)$ moves along C.

Putting its parts together, the Fact says that if $\boldsymbol{f}(t)$ describes the position of a moving particle (a vector quantity) at time t, then the derivative $\boldsymbol{f}'(t)$ describes the particle's **velocity**—another vector quantity—at the same time. That the derivative links position and velocity is not a new idea—we saw the same phenomenon in single-variable calculus. But that the same connection persists in the multivariable setting is a remarkable fact; it accounts for some of the most important applications of multivariable calculus.

We used the Fact to calculate the velocity vectors of the ant in Example 1, page 28. The next example illustrates the process for a less earth-bound insect.

■ **Example 2.** A fly buzzes through xyz-space along the spiral curve shown:

Velocity vectors on a space curve

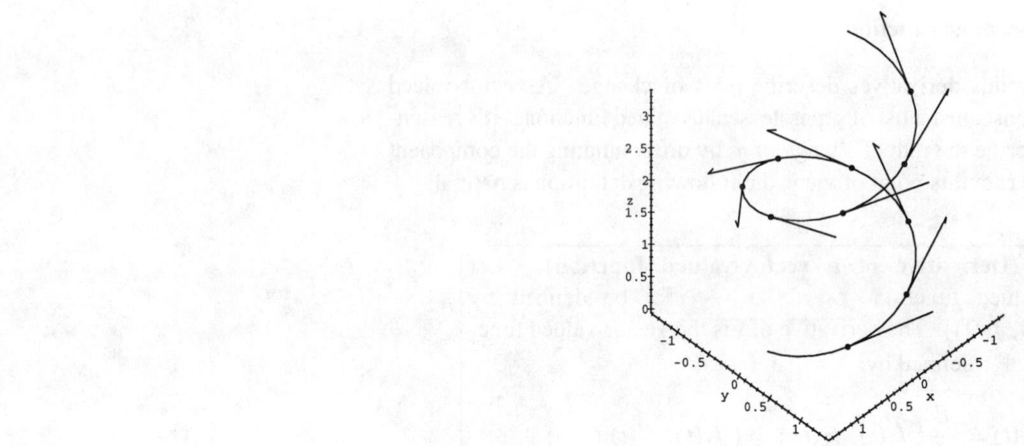

At time t seconds, for $0 \le t \le 10$, the fly's position $\boldsymbol{p}(t)$ is given by the vector-valued function $\boldsymbol{p}(t) = (\cos t, \sin t, t/3)$. How are the tangent vectors computed? What do they mean?

Solution: The fly's tangent vectors are found by differentiation. Since $\boldsymbol{p}(t) = (\cos t, \sin t, t/3)$, $\boldsymbol{p}'(t) = (-\sin t, \cos t, 1/3)$. For each t, the velocity vector $\boldsymbol{p}'(t)$ is tangent to the curve at $\boldsymbol{p}(t)$. At $t = 5$, for instance (the middle of the curve shown), we have $\boldsymbol{p}(5) = (\cos 5, \sin 5, 5/3) \approx (0.28, -0.96, 1.67)$, and $\boldsymbol{p}'(5) = (-\sin 5, \cos 5, 1/3) \approx (0.96, 0.28, 0.33)$; this vector, which tells velocity, has magnitude $|(0.96, 0.28, 0.33)| \approx 1.05$. Thus, at time 5 seconds, the fly was at position $(0.28, -0.96, 1.67)$, moving in the direction of the vector $(0.96, 0.28, 0.33)$, at about 1.05 units per second. (The picture was drawn by plotting several such tangent vectors along with the curve.) □

■ **Example 3.** What does the Fact above say about the vector-valued function $\boldsymbol{L}(t) = (x_0, y_0) + t(a, b)$? (As we saw above, this function describes the line ℓ through (x_0, y_0) with direction vector (a, b).)

Solution: Since $\boldsymbol{L}(t) = (x_0, y_0) + t(a, b) = (x_0 + at, y_0 + bt)$, it follows that $\boldsymbol{L}'(t) = (a, b)$. In other words, the derivative function has the *constant* vector value (a, b). This result is consistent with the Fact, in two ways:

(i) The direction vector (a, b) is tangent to ℓ at every point of ℓ, and points in the direction of increasing t.

(ii) In each unit of time, the position $\boldsymbol{L}(t)$ adds one more multiple of (a, b), i.e., $\boldsymbol{L}(t)$ moves through a distance of $|(a, b)| = \sqrt{a^2 + b^2}$. □

Understanding vector derivatives

Our definition makes differentiating a vector-valued function easy—we just differentiate each coordinate function separately. This makes calculations easy, but does it make good sense? Why does the vector function $\boldsymbol{f}'(t) = \left(f_1'(t), f_2'(t)\right)$ really deserve to be called the derivative of $\boldsymbol{f}$? The following Fact shows that a vector derivative, just like an ordinary one, is a limit of difference quotients. (Indeed, the vector derivative can be *defined* as such a limit.)

> **Fact: (The derivative as a limit.)** Let $\boldsymbol{f}$ and $\boldsymbol{f}'$ be as above. Then
> $$\boldsymbol{f}'(t) = \lim_{h\to 0} \frac{\boldsymbol{f}(t+h) - \boldsymbol{f}(t)}{h}.$$

In the next few paragraphs we'll "unpack" this Fact.

Interpreting the difference quotient. The ratio

$$\frac{\boldsymbol{f}(t+h) - \boldsymbol{f}(t)}{h}$$

resembles the difference quotient of one-variable calculus—except that here the numerator is a vector and the denominator a scalar. (The quotient therefore makes sense; it's a vector.) The numerator, being a difference of vectors, can be drawn as an arrow joining two nearby points on the curve:

The difference of nearby vectors

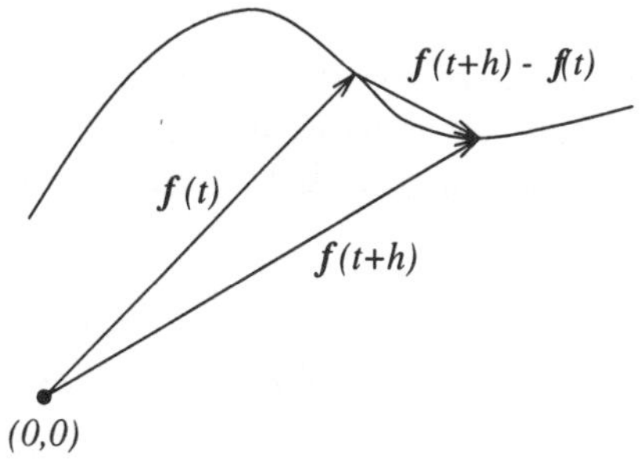

In particular, for small h, the vector $\boldsymbol{f}(t+h) - \boldsymbol{f}(t)$ points approximately in the direction of the curve. Because the quotient vector

$$\frac{\boldsymbol{f}(t+h) - \boldsymbol{f}(t)}{h}$$

is a scalar multiple of the numerator, the quotient vector *also* points approximately in the direction of the curve. (In other words, the quotient vector is approximately tangent to the curve at $\boldsymbol{f}(t)$.)

The multiplier is $1/h$.

We can apply similar reasoning to the magnitude of the difference quotient vector. The picture illustrates that $|\boldsymbol{f}(t+h) - \boldsymbol{f}(t)|$ (the length of the numerator) is approximately the length of the small curve segment from t to $t+h$, i.e., the distance traveled by a point moving along the curve from t to $t+h$. Therefore, dividing by h (the "time" elapsed) is approximately the average speed of $\boldsymbol{f}(t)$ from time t to time $t+h$.

In the next section we'll see how to use integrals to calculate "arclengths" i.e., distances along a curve.

In the limit. The preceding paragraphs imply that as $h \to 0$, the difference quotient vector tends to a vector that's (i) tangent to the curve at time t; and (ii) has length equal to the instantaneous speed at time t.

In components. Writing $\boldsymbol{f}$ as $\boldsymbol{f} = (f_1, f_2)$ shows why the Fact holds:

$$\begin{aligned}
\lim_{h\to 0} \frac{\boldsymbol{f}(t+h) - \boldsymbol{f}(t)}{h} &= \lim_{h\to 0} \frac{\big(f_1(t+h), f_2(t+h)\big) - \big(f_1(t), f_2(t)\big)}{h} \\
&= \lim_{h\to 0} \frac{\big(f_1(t+h) - f_1(t), f_2(t+h) - f_2(t)\big)}{h} \\
&= \lim_{h\to 0} \left(\frac{f_1(t+h) - f_1(t)}{h}, \frac{f_2(t+h) - f_2(t)}{h} \right) \\
&= \left(\lim_{h\to 0} \frac{f_1(t+h) - f_1(t)}{h}, \lim_{h\to 0} \frac{f_2(t+h) - f_2(t)}{h} \right) \\
&= \big(f_1'(t), f_2'(t)\big).
\end{aligned}$$

Derivatives and tangent lines. Elementary calculus teaches that, for scalar-valued functions $y = f(x)$, the derivative $f'(x_0)$ tells the *slope* of the line tangent to the graph of f at $x = x_0$. A similar relation holds between derivatives and tangent lines for vector-valued functions $\boldsymbol{f}(t)$. The derivative vector $\boldsymbol{f}'(t_0)$ is tangent at $t = t_0$ (i.e., at the point $\boldsymbol{f}(t_0)$) to the curve with position vector $\boldsymbol{f}(t)$.

■ **Example 4.** Use vector derivatives to find a vector equation for the line tangent to the unit circle at the point $(1/\sqrt{2}, 1/\sqrt{2})$.

Solution: We can use the position function $\boldsymbol{f}(t) = (\cos t, \sin t)$ to describe the unit circle; the point in question corresponds to $t = \pi/4$. Now $\boldsymbol{f}'(t) = (-\sin t, \cos t)$, so $\boldsymbol{f}'(\pi/4) = (-\sqrt{2}/2, \sqrt{2}/2)$. Thus the tangent line passes through $(1/\sqrt{2}, 1/\sqrt{2})$, with direction vector $\boldsymbol{v} = (-1/\sqrt{2}, 1/\sqrt{2})$. Knowing a point and a direction vector, we can describe the line in vector notation:

$$\boldsymbol{X}(t) = \left(1/\sqrt{2}, 1/\sqrt{2}\right) + t\left(-1/\sqrt{2}, 1/\sqrt{2}\right).$$

This one's easy to draw by hand.

A simple diagram shows that the line and the circle are indeed tangent at the target point. □

Algebraic rules for vector derivatives

Scalar- and vector-valued functions can be combined in various ways to form new functions. If $\boldsymbol{f}$ and $\boldsymbol{g}$ are vector-valued functions, and a is a scalar-valued function, then we can form such combinations as $\boldsymbol{f}(t) + \boldsymbol{g}(t)$ and $a(t)\boldsymbol{g}(t)$. Derivatives of such combinations can be found using rules that closely resemble the ordinary sum and product rules.

Theorem 2. (Sum and product rules) Let $\boldsymbol{f}$ and $\boldsymbol{g}$ be differentiable vector-valued functions, and let a be a differentiable scalar-valued function. Then the combined functions $\boldsymbol{f} + \boldsymbol{g}$ and $a\boldsymbol{g}$ are differentiable, with derivatives as follows:

- $\big(\boldsymbol{f}(t) + \boldsymbol{g}(t)\big)' = \boldsymbol{f}'(t) + \boldsymbol{g}'(t)$ **(sum rule)**
- $\big(a(t)\,\boldsymbol{f}(t)\big)' = a'(t)\,\boldsymbol{f}(t) + a(t)\,\boldsymbol{f}'(t)$ **(scalar multiple rule)**

Proofs. The theorem is proved by writing $\boldsymbol{f}$ and $\boldsymbol{g}$ in components and using the corresponding derivative rule for scalar-valued functions. To prove the sum rule, for example,➻ we first write $\boldsymbol{f} = (f_1, f_2)$ and $\boldsymbol{g} = (g_1, g_2)$. Then

We prove it for 2-dimensional vectors; the idea for the 3-dimensional case is exactly the same.

$$(\boldsymbol{f}+\boldsymbol{g})' = \big((f_1+g_1)', (f_2+g_2)'\big) = (f_1'+g_1', f_2'+g_2') = (f_1', f_2')+(g_1', g_2') = \boldsymbol{f}'+\boldsymbol{g}',$$

as desired. □

■ **Example 5.** Differentiate $\boldsymbol{g}(t) = t(\cos t, \sin t)$.

Solution: The scalar multiple rule applies, with $a(t) = t$ and $\boldsymbol{f}(t) = (\cos t, \sin t)$. It says:

$$\boldsymbol{g}'(t) = \big(t\,(\cos t, \sin t)\big)' = 1\,(\cos t, \sin t) + t\,(-\sin t, \cos t).$$

Not too exciting, perhaps—after all, we could have multiplied through by t before differentiating—but it works. □

Antiderivatives and integrals

Derivatives of vector-valued functions are found component-by-component. For the same reason, vector antiderivatives are found the same way. (In this section we'll just calculate some vector antiderivatives. In the next section we'll use them to model physical motion.)

Put tersely, in symbols: If $\boldsymbol{f}(t) = \big(f_1(t), f_2(t)\big)$, then

$$\int \boldsymbol{f}(t)\,dt = \int \big(f_1(t), f_2(t)\big)\,dt = \left(\int f_1(t)\,dt, \int f_2(t)\,dt,\right).$$

The next example shows the idea.

■ **Example 6.** Let $\boldsymbol{f}(t) = (1, 2) + t(3, 4)$. Show that

$$\int \boldsymbol{f}(t)\,dt = t(1, 2) + \frac{t^2}{2}(3, 4) + (C_1, C_2),$$

where C_1 and C_2 are arbitrary constants.

Solution: If we first write $\boldsymbol{f}(t) = (1 + 3t, 2 + 4t)$, then antidifferentiating separately in each component gives

$$\int \boldsymbol{f}(t)\,dt = \int (1 + 3t, 2 + 4t)\,dt = (t + 3t^2/2 + C_1, 2t + 2t^2 + C_2),$$

where C_1 and C_2 are arbitrary constants. Factoring like powers of t out of both components gives another way to write the answer:

$$(t + 3t^2/2 + C_1, 2t + 2t^2 + C_2) = t(1, 2) + \frac{t^2}{2}(3, 4) + (C_1, C_2).$$

(Both forms of the answer are correct, but, arguably, the second form shows more clearly where the answer came from.) □

Exercises

1. Let ℓ be the line through the point $P(1, 2)$ in the direction of the vector $\boldsymbol{v} = (2, 3)$.

   (a) The line ℓ is the image of the vector-valued function $\boldsymbol{L}(t) = (1, 2) + t(2, 3)$, for all real t. Draw a picture to explain why; label the points on ℓ that correspond to $t = 0$, $t = 1$, $t = 2$, and $t = -1$.

   (b) What is the image of the function $\boldsymbol{L}(t) = (1, 2) + t(2, 3)$, if the domain is restricted to $t \geq 0$?

   (c) What is the image of the function $\boldsymbol{L}(t) = (1, 2) + t(2, 3)$, if the domain is restricted to $-1 \leq t \leq 1$?

2. Repeat the previous exercise, but let ℓ be the line through (a, b) in the direction of (c, d). (Assume that c and d are not both zero.)

3. Explain why each of the following rules holds for vectors in space:

   (a) the sum rule for vector derivatives;

   (b) the scalar multiple rule for vector derivatives;

4. Suppose that the motion of a particle is described by the parametric equations $x(t) = 1 - 2t + t^3$, $y(t) = 2t^4$. What is the particle's speed at time $t = 1$?

1.5 Derivatives, antiderivatives, and motion

On March 13, 1986, the Giotto spacecraft approached within 600 kilometers of Halley's comet. This fly-by had been planned since Giotto's launch, on July 2, 1985. An unplanned event occurred, too—Giotto was severely damaged by space dust. (For a color photograph of the nucleus of Halley's comet, taken by Giotto, see `http://nssdc.gsfc.nasa.gov/planetary/giotto.html` on the World Wide Web.)

Halley's comet had last been seen near Earth in 1910. On its third previous pass, in 1682, the comet had been observed by the English astronomer Edmond Halley (1656–1742), a contemporary and supporter (intellectual and financial) of Isaac Newton. Based on Halley's theory of orbits, he correctly predicted that the comet of 1682 would return to Earth's vicinity on a 76-year cycle.

Predicting the orbit of a comet and arranging a rendezvous with a spacecraft would be impossible without calculus-based tools and techniques. So, for that matter, would much humbler tasks, such as estimating that the 206-mile drive from Chattanooga to Tuscaloosa takes 3 hours and 49 minutes—unless one takes the longer route through Talladega, which adds about 57 minutes to the trip.

Comets and spacecraft are hard or expensive to observe directly, so it's essential to *model* their motion mathematically. To model a physical phenomenon mathematically means first to describe it in mathematical language, using mathematical objects such as vectors, functions, derivatives, and integrals. Once this is done, mathematical consequences can be deduced, and their physical meaning interpreted.

Modeling physical motion was historically among the most important applications of calculus. To a large extent, in fact, the invention and development of calculus in the 17th and 18th centuries was spurred by the desire to understand and predict motion, whether the orbits of planets or the trajectories of mortar shells. Modeling motion remains just as important today, whether for predicting when we'll get to Grandma's or for programming a robot arm.

How many variables? One-variable calculus does fine modeling 1-dimensional motion, such as that of a car on a straight east–west road, like highway I-94 through North Dakota. But the curves and mountains of Montana aren't far away. When we get there, an extra variable or two will prove very handy.

The physical world is 3-dimensional, so it might seem that we'd always need three variables to describe motion realistically. Many types of motion, however, are essentially 1- or 2-dimensional. A westbound car on the North Dakota highway moves, for practical purposes, on a (1-dimensional) straight line. If the driver turns south at Mandan and takes the scenic loop through Carson, Elgin, Mott, and Richardton, rejoining I-94 at Dickinson, the car's motion is essentially 2-dimensional—nothing much happens in the third dimension, altitude. The third dimension will certainly be harder to ignore on the mountainous stretch from Billings to Butte. (But even there we might *choose* to ignore altitude for the sake of convenience or approximation.)

Vector derivatives and antiderivatives: the basics

Vector derivatives and antiderivatives, suitably employed, are basic tools for modeling phenomena of motion. We've seen that if C is a curve described by a position vector function $\boldsymbol{p}(t)$, and we think of t as time, then, for any $t = t_0$, the vector derivative

$$\boldsymbol{p}'(t_0) = \big(x'(t_0), y'(t_0)\big),$$

carries two useful types of information:

Speed. The scalar $|\boldsymbol{p}'(t_0)| = \sqrt{x'(t)^2 + y'(t)^2}$, i.e., the length of the derivative vector, tells the instantaneous speed (in units of distance per unit of time t) with which $\boldsymbol{p}(t)$ moves along C.

Direction. If $\boldsymbol{p}'(t_0) \neq (0, 0)$, then $\boldsymbol{p}'(t_0)$ is tangent to the curve C at $\big(x(t_0), y(t_0)\big)$ and points in the direction of increasing t.

As opposed to the opposite direction.

These two properties justify calling the derivative $\boldsymbol{p}'(t_0)$ the **velocity vector** at $t = t_0$. The name can also be justified on geometric grounds.

■ **Example 1.** Consider the curve C parametrized by

$$\boldsymbol{p}(t) = (\cos t, 2\sin t); \qquad 0 \leq t \leq 2\pi.$$

Find a *linear* curve $\boldsymbol{l}(t)$ that passes through $\boldsymbol{p}(\pi/3)$ with direction vector $\boldsymbol{p}'(\pi/3)$. Plot both curves together.

Solution: It's easy to see that

Check for yourself.

$$\boldsymbol{p}(\pi/3) = (1/2, \sqrt{3}) \quad \text{and} \quad \boldsymbol{p}'(\pi/3) = (-\sqrt{3}/2, 1).$$

Therefore, the linear curve

$$\boldsymbol{l}(t) = \boldsymbol{p}(\pi/3) + t\,\boldsymbol{p}'(\pi/3) = \big(1/2 - t\sqrt{3}/2, \sqrt{3} + t\big)$$

"matches" $\boldsymbol{p}(t)$ in the desired sense. Here are both curves:

A curve and a tangent line

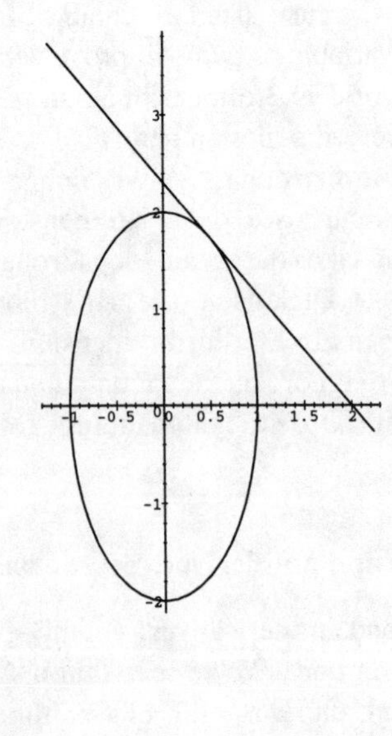

The line appears, as advertised, to be tangent to the curve (an ellipse) at the point $\boldsymbol{p}(\pi/3)$. □

Speed and arclength

How *long* is the curve C given by

$$\boldsymbol{p}(t) = \big(x(t), y(t)\big); \qquad a \le t \le b \quad ?$$

We said above that the speed at which C is traversed is given, at time t, by

$$|\boldsymbol{p}'(t)| = \sqrt{x'(t)^2 + y'(t)^2}.$$

This formula shows that, for most curves, the speed is a nonconstant function of t. In the (rare) exceptional case, we speak of a **constant-speed parametrization**.

Recall from single-variable calculus that to find the total distance a moving object travels over the t-interval $a \le t \le b$ we *integrate the speed function.* In the present setting, this means that we can compute **arclength** as follows:

Do you remember this fact? Does it sound reasonable?

> **Fact: (Arclength by integration.)** Let C be a curve as above, parametrized on the interval $a \le t \le b$. Then
>
> $$\text{arclength of } C = \int_a^b \sqrt{x'(t)^2 + y'(t)^2}\, dt.$$

The next two examples illustrate what can go right—and wrong.

■ **Example 2.** Find the length of the upper half of the unit circle; use the parametrization $\boldsymbol{p}(t) = (\cos t, \sin t)$, $0 \le t \le \pi$.

Solution: Since $x'(t) = -\sin t$ and $y'(t) = \cos t$, we see that $\sqrt{x'(t)^2 + y'(t)^2} = 1$—in this case, therefore, the speed is always 1.

$$\text{arclength} = \int_0^\pi 1\, dt = \pi.$$

What could be simpler? □

■ **Example 3.** Do the preceding example again, but this time with another parametrization: $\boldsymbol{p}(t) = \big(\cos(t^2), \sin(t^2)\big)$, $0 \le t \le \sqrt{\pi}$.

Solution: This time the speed is not constant: $x'(t) = -2t \sin(t^2)$ and $y'(t) = 2t \cos(t^2)$. But it doesn't make much difference in the end:

$$\sqrt{x'(t)^2 + y'(t)^2} = \sqrt{4t^2 \sin^2(t^2) + 4t^2 \cos^2(t^2)} = 2t,$$

and we see, to our relief, that

$$\text{arclength} = \int_0^{\sqrt{\pi}} 2t\, dt = \pi.$$

It *is* a relief, not just because the calculations worked out, but because we got the same answer both times. That we get the same answer (and will also do so for other sensible curves parametrized in different ways) means that the arclength of a curve is a property of the curve itself, not of a particular parametrization of the curve. □

Easy or hard? From the last two examples it might seem that finding arclength exactly is a piece of cake. Alas, it isn't. Very often, the pesky square root in the integrand foils us. In that case, we may need to resort to numerical integration methods.

■ **Example 4.** How long—exactly—is the curve $\mathbf{p}(t) = (t, \sin t)$ for $0 \le t \le \pi$? (The curve is one arch of a sine function.) How long is it approximately?

Check it!

Solution: The arclength integral is easy to write down:

$$\text{arclength} = \int_0^\pi \sqrt{1+\cos^2 t}\, dt.$$

The exact command your computer or calculator uses may be different; the result is what matters.

The integral, unfortunately, doesn't yield in symbolic form even to *Maple* or *Mathematica*, since there's no convenient antiderivative. But numerical methods, such as the midpoint rule with 20 subdivisions, are always available for definite integrals:

```
> midpointsum( sqrt(1+cos(x)^2), x=0..Pi,20);

                    3.820197791
```

The result means that our curve is approximately 3.82 units long. □

Position, velocity, and acceleration

"Particle" is the quaint-sounding term for any object that can be thought of as a point.

In one variable. Imagine a particle moving along the x-axis, with its position at time t seconds given by a **position function** $p(t)$. The particle has a corresponding **velocity function** $v(t)$ and an **acceleration function** $a(t)$. As we saw in elementary calculus, these functions are related by derivatives:

$$v(t) = p'(t); \quad a(t) = v'(t); \quad a(t) = p''(t).$$

We'll always assume this unless something is said to the contrary.

(We're assuming, of course, that the derivatives exist.) We used these facts to model various phenomena of motion. Here's a typical example; starting from an acceleration function, we *antidifferentiate* to work "back" to find the velocity and position functions.

■ **Example 5.** A particle has constant acceleration $a(t) = -3$ units per second per second. At time $t = 0$, the particle has velocity $v(0) = 12$ units per second and position $p(0) = 0$. Find formulas for $v(t)$ and $p(t)$. Describe the particle's journey over the next 8 seconds.

Solution: From the derivative relations above, we have

$$v'(t) = a(t) = -3 \implies v(t) = \int -3\, dt = -3t + C,$$

where C is some constant. Since $v(0) = 12$, we must have $C = 12$, so $v(t) = -3t + 12$. Arguing similarly for position gives

$$p'(t) = v(t) = -3t + 12 \implies p(t) = \int (-3t+12)\, dt = \frac{-3t^2}{2} + 12t + D,$$

where D is a constant. The condition $p(0) = 0$ implies that $D = 0$. Putting the pieces together gives

$$v(t) = -3t + 12 \quad \text{and} \quad p(t) = \frac{-3t^2}{2} + 12t.$$

The formulas➡ show that $t = 4$ is especially important: $v(4) = 0$, $p(4) = 24$, and $v(t)$ changes sign (from positive to negative) at $t = 4$. Now we can describe the particle's itinerary: It moves to the right until $t = 4$, reaching $x = 24$; then it moves to the left, returning to the origin, with velocity -12 units per second, at $t = 8$. □

Or their graphs.

In several variables. The position, velocity, and acceleration of a particle moving in the plane or in space (rather than along a line) are related in exactly the same way—by derivatives:

$$\boldsymbol{v}(t) = \boldsymbol{p}'(t); \quad \boldsymbol{a}(t) = \boldsymbol{v}'(t); \quad \boldsymbol{a}(t) = \boldsymbol{p}''(t).$$

There is a difference,➡ however, between these equations and their earlier counterparts: The position, velocity, and acceleration are now all *vector-valued* functions of t, and the derivatives must now be interpreted in the sense appropriate for such functions.

The boldface letters should be a clue.

Modeling motion

Using the vector derivative tools and techniques we've developed, we can model many types of 2- and 3-dimensional motion. (For simplicity we'll stick mainly to plane motion, but the ideas are essentially identical in 3 variables.) The key idea is that vector-valued acceleration, velocity, and position functions are all derivatives and antiderivatives of each other.

Starting with acceleration. In modeling physical motion, it's often most natural to begin with information about acceleration and to work from there to formulas for velocity and position. Acceleration crops up naturally in practice because of its close connection to force. **Newton's second law** puts it something like this:

> *A force acting on an object produces an acceleration that is directly proportional to the force and inversely proportional to the object's mass.*

Forces (gravitational force, air drag, sliding friction, a rocket's thrust, etc.) can often be measured directly and (thanks to Newton's law) converted into information about acceleration. That's why the "given" in physical modeling problems is often acceleration.

Gravity and acceleration. As everyone knows, earth's gravity pulls everything toward the center of the earth, producing a "downward" acceleration. It's also true (but not so obvious, until around 1600, when Galileo tossed stuff off the Leaning Tower of Pisa) that the acceleration due to gravity has the same magnitude for all objects, regardless of their mass. (Heavier objects fall *harder*, but not *faster*.)

As a result, gravity induces the same acceleration on all objects near the surface of the earth. Considered as a vector, the acceleration due to gravity is an arrow pointing toward the center of the earth; it has the same magnitude for all objects. (This magnitude is usually denoted by g; it is called the **acceleration due to gravity**.) In the metric system, $g \approx 9.8$ meters per second per second. (In the English system, $g \approx 32$ feet per second per second.)

Zero acceleration. In the simplest possible case, an object's acceleration is the zero vector. In this case it's easy to find velocity and position by antidifferentiation. For velocity,

$$\boldsymbol{a}(t) = (0, 0) \implies \boldsymbol{v}(t) = \int (0, 0)\, dt = \left(\int 0\, dt, \int 0\, dt\right) = (C_1, C_2),$$

Notice the antiderivative of a vector-valued function.

where C_1 and C_2 are constants. For position,

$$\boldsymbol{v}(t) = (C_1, C_2) \implies \boldsymbol{p}(t) = \int (C_1, C_2)\, dt = \left(\int C_1\, dt, \int C_2\, dt \right) = (C_1 t + C_3, C_2 t + C_4),$$

where C_3 and C_4 are arbitrary constants.

In specific cases, the four constants are evaluated using additional information, such as the velocity and position of the particle at time $t = 0$. Notice, however, that regardless of the numerical values of these constants, the calculation illustrates an interesting physical fact: In the absence of outside forces, an object has *zero* acceleration, *constant* velocity (and speed), and a *linear* position function.

■ **Example 6.** At time $t = 0$, a particle in the plane has velocity vector $(1, 2)$ and position vector $(3, 4)$. (These stipulations are called **initial conditions.**) No external force acts, so the acceleration is $\boldsymbol{a}(t) = (0, 0)$ at all times. Describe the particle's movement. Where is the particle at time $t = 100$?

Solution: As we just calculated, the particle has constant velocity function $\boldsymbol{v}(t) = (C_1, C_2)$ and linear position function $\boldsymbol{p}(t) = t(C_1, C_2) + (C_3, C_4)$. The initial conditions give $C_1 = 1$, $C_2 = 2$, $C_3 = 3$, and $C_4 = 4$, so

$$\boldsymbol{v}(t) = (1, 2) \quad \text{and} \quad \boldsymbol{p}(t) = (3, 4) + t(1, 2).$$

Thus the particle moves away from $(3, 4)$ with constant speed $\sqrt{5}$, in the direction of the vector $(1, 2)$. At $t = 100$ the particle's position is $\boldsymbol{p}(100) = (103, 204)$. □

Displacement and definite integrals. In the preceding example we antidifferentiated the velocity function $\boldsymbol{v}$ to get the position function $\boldsymbol{p}$; then we found $\boldsymbol{p}(100)$.

A slightly different approach to the same problem is to *integrate* the velocity function over the time interval in question. More precisely, in symbols:

If $\boldsymbol{v}(t)$ is a particle's velocity for $a \le t \le b$, and $\boldsymbol{v}(t)$ is the particle's position, then

$$\int_a^b \boldsymbol{v}(t)\, dt = \boldsymbol{p}(t)\Big]_a^b = \boldsymbol{p}(b) - \boldsymbol{p}(a).$$

The vector quantity on the right above is the **displacement** of the particle over the interval $a \le t \le b$; it tells how the particle's position *changes* over the time interval.

■ **Example 7.** In the situation of the preceding example, use a definite integral to find the particle's displacement over the interval $0 \le t \le 100$ and (again) its final position.

Solution: Note first that if $\boldsymbol{v}(t) = (1, 2)$, then $t\,(1, 2) + (C_1, C_2)$ is an antiderivative for *any* constants C_1 and C_2. Since we're working with a *definite* integral, we can safely set both constants to zero. Therefore,

$$\text{displacement} = \int_0^{100} (1, 2)\, dt = t\,(1, 2)\Big]_0^{100} = (100, 200).$$

Since the particle starts at $(3, 4)$, adding the displacement gives the particle's final position. That is, $\boldsymbol{p}(100) = (3, 4) + (100, 200) = (103, 204)$. □

The same answer we found earlier.

Constant acceleration. In the next simplest case, an object's acceleration is a nonzero constant vector. This case is especially important; it arises, for instance, when a constant force (such as gravity, under appropriate conditions) acts on an object, producing a constant acceleration. Suppose, then, that an object has constant acceleration vector $\boldsymbol{a}(t) = (a_1, a_2)$. Then

$$\boldsymbol{v}(t) = \int (a_1, a_2)\, dt = t(a_1, a_2) + (C_1, C_2),$$

where C_1 and C_2 are arbitrary constants. Since $\boldsymbol{v}(0) = (C_1, C_2)$, the constants represent the object's initial velocity.

To find position we antidifferentiate again:

Convince yourself that these antiderivatives are correct.

$$\boldsymbol{p}(t) = \int \big(t(a_1, a_2) + (C_1, C_2) \big)\, dt = \frac{t^2}{2}(a_1, a_2) + t(C_1, C_2) + (C_3, C_4),$$

where C_3 and C_4 are again arbitrary constants. Since $\boldsymbol{p}(0) = (C_3, C_4)$, these constants represent the object's initial position.

We see, in particular, that *constant* acceleration begets *linear* velocity and *quadratic* position.

Unless, perchance, some of the constants are zero.

■ **Example 8.** An object starts from rest at the origin and has constant acceleration $\boldsymbol{a} = (1, 2)$. What happens? What are the object's position and velocity at $t = 100$ seconds?

Solution: The calculations above, together with the initial conditions, mean that the object's velocity and position functions are

$$\boldsymbol{v}(t) = t(1, 2) \quad \text{and} \quad \boldsymbol{p}(t) = \frac{t^2}{2}(1, 2).$$

Thus the particle moves along the curve defined by $\boldsymbol{p}(t)$. At $t = 100$ the particle's velocity is $100(1, 2) = (100, 200)$; it's speed is $100\sqrt{5} \approx 223.6$ units per second. The particle's position is $\boldsymbol{p}(100) = 5000(1, 2) = (5000, 10000)$. □

It's gone a long way!

Free fall

A moving object that is affected only by gravity is said to be in **free fall**. In the real world, falling objects are always affected to some degree by other forces—air drag, passing breezes, stray pigeons, etc. Sometimes these forces are negligible, sometimes not. In any event, modeling free fall is useful as a first step toward understanding more complex combinations of forces.

Free fall is conveniently modeled in the xy-plane; we'll let the negative y-direction represent "downward" (i.e., toward the center of the earth); the x- and y-units will be meters. Then the acceleration due to gravity is of the form $\boldsymbol{a}(t) = (0, -g)$, where $g \approx 9.8$ meters per second per second is the (constant) acceleration due to gravity. Using the calculations above we can model free-fall phenomena.

A natural choice . . .

Trajectories. What shape of path does a shotput or baseball follow (ignoring wind resistance and other stray forces) after being thrown or hit? Galileo knew the answer: a parabola. He deduced this from the fact that gravity's acceleration is constant. So can we.

■ **Example 9.** A free-falling projectile leaves the origin at time $t = 0$, with initial speed 100 meters per second, at an angle α to the horizontal. What path does the projectile follow? Where does it land? (Let the x-axis represent ground level.)

Do you see why $\boldsymbol{v}(0)$ is the initial velocity vector?

Solution: The given conditions say:

$$\boldsymbol{a}(t) = (0, -g), \quad \boldsymbol{v}(0) = 100(\cos\alpha, \sin\alpha), \quad \text{and} \quad \boldsymbol{p}(0) = (0, 0).$$

Therefore, using the calculations above,

$$\boldsymbol{v}(t) = t(0, -g) + 100(\cos\alpha, \sin\alpha); \quad \boldsymbol{p}(t) = \frac{t^2}{2}(0, -g) + 100t(\cos\alpha, \sin\alpha).$$

The curve $\boldsymbol{p}(t)$ has parametric equations

$$x(t) = 100t\cos\alpha; \quad y(t) = \frac{-gt^2}{2} + 100t\sin\alpha.$$

Here's a sample of such curves for various values of α:

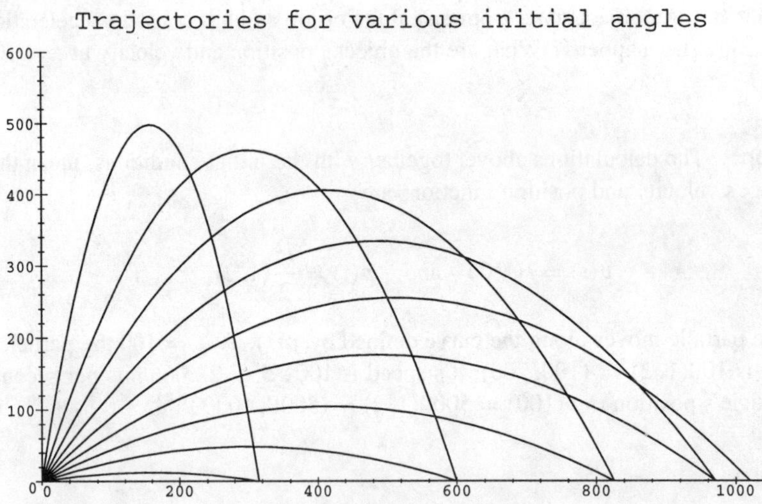

The picture suggests Galileo's answer: The trajectories are parabolas. (Exercise 7 explores the reason for this.)

The projectile lands when $y(t) = 0$. Solving for t (we want the positive value) gives

$$\frac{-gt^2}{2} + 100t\sin\alpha = 0 \implies t = \frac{200\sin\alpha}{g}.$$

For this t,

$$x(t) = 100t\cos\alpha = 100\frac{200\sin\alpha}{g}\cos\alpha;$$

this is the projectile's range. If, say, $\alpha = \pi/4$, then the range is $10000/g \approx 1020$ meters. □

Exercises

1. In Example 5, page 50, we claimed: ... *the particle moves to the right until* $t = 4$, *reaching* $x = 24$; *then it moves to the left, returning to the origin, with velocity* -12, *at* $t = 8$. Plot the function $p(t)$ for $0 \le t \le 8$. Then explain how the graph "agrees with" all parts of the italicized statement.

2. This exercise uses the same notation as Example 5, page 50. In each part, find equations for $p(t)$ and $v(t)$. What happens over the interval $t = 0$ to $t = 10$? Find and interpret $v(10)$ and $p(10)$.

   (a) Assume that $a(t) = 0$, $v(0) = 0$, and $p(0) = 1$.

   (b) Assume that $a(t) = 0$, $v(0) = 1$, and $p(0) = 0$.

   (c) Assume that $a(t) = 1$, $v(0) = 0$, and $p(0) = 0$.

   (d) Assume that $a(t) = t$, $v(0) = 0$, and $p(0) = 0$.

3. This exercise is about Example 8, page 53.

   (a) By hand, sketch the "curve" defined by $\mathbf{p}(t)$ for $0 \le t \le 100$. What very simple shape does the "curve" seem to have? (Hint: The curve is parametrized by $x(t) = p_1(t)$; $y(t) = p_2(t)$; $0 \le t \le 100$.)

   (b) Eliminate the variable t in the parametric equations for the curve $\mathbf{p}(t)$; what's the resulting equation? Does it agree with what you found in the previous part?

   (c) Find the arclength of the curve $\mathbf{p}(t)$ from $t = 0$ to $t = 100$.

4. Redo the previous exercise, but assume that the acceleration vector is $(1, -1)$.

5. At time $t = 0$ seconds a particle is at the origin, and has velocity vector $(4, 4)$. It undergoes constant acceleration $\boldsymbol{a}(t) = (0, -1)$.

   (a) Find formulas for the velocity function $\boldsymbol{v}(t)$ and the position function $\mathbf{p}(t)$.

   (b) Plot the path taken by the particle $\mathbf{p}(t)$ for $0 \le t \le 10$. What familiar shape does the path seem to have?

   (c) Eliminate the variable t in the parametric equations for the curve $\mathbf{p}(t)$; what's the resulting equation? Does it agree with what you found in the previous part?

   (d) Find the arclength of the curve $\mathbf{p}(t)$ from $t = 0$ to $t = 10$. (Hint: Set up the integral by hand; either solve it exactly using a table of integrals, or use technology to approximate the integral numerically.)

6. Example 9, page 54, shows that a projectile with initial speed 100 meters per second and initial angle α follows path given parametrically by

$$x(t) = 100t\cos\alpha; \quad y(t) = \frac{-gt^2}{2} + 100t\sin\alpha.$$

   (a) By eliminating t, find an xy-equation for the parabola. (Hint: Since $x = 100t\cos\alpha$, $t = x/(100\cos\alpha)$. Substitute this into the equation for y.)

(b) Show that the maximum range is obtained if $\alpha = \pi/4$. What *is* the maximum range?

7. A projectile is at $(0, 0)$ at time 0 seconds. It has initial speed 100 meters per second, initial angle $\pi/3$, and travels under free fall conditions.

   (a) Find equations for the velocity and position functions.

   (b) When does the projectile touch down?

   (c) Plot the projectile's trajectory from takeoff to landing.

   (d) At what time is the projectile at maximum height? How high is this? (Hint: Find the time at which the velocity vector is horizontal.)

   (e) Find the speed and the velocity at the moment the projectile is highest.

8. A projectile is at $(0, 0)$ at time 0 seconds. It has initial speed s_0 meters per second, initial angle α, and travels under free fall conditions.

   (a) Find equations for the velocity and position functions.

   (b) At what time is the projectile highest? (Hint: The answer depends on both s_0 and α.)

   (c) Find the speed and the velocity at the moment when the projectile is highest.

   (d) Find the speed and the velocity at the moment when the projectile lands.

9. Suppose that the velocity of an object at time t seconds is $(5t^2 + 3t - 4, 1 - t)$ meters/second. At time $t = 2$ the object is at the point with coordinates $(1, 0)$.

   (a) Find the object's acceleration vector at time $t = 1$.

   (b) Find the position of the object at time $t = 0$.

   (c) Express the distance traveled by the object between time $t = 0$ and time $t = 2$ as an integral.

10. Suppose that the motion of a particle is described by the parametric equations $x = t^3 - 3t$, $y = t^2 - 2t$.

    (a) Does the particle ever come to a stop? If so, when and where?

    (b) Is the particle ever moving straight up or down? If so, when and where?

    (c) Is the particle ever moving horizontally left or right? If so, when and where?

11. A particle moves in the xy-plane with constant acceleration vector $\boldsymbol{a}(t) = (0, -1)$. At time $t = 0$ the particle is at the point $\boldsymbol{p}(0) = (0, 0)$ and has velocity $\boldsymbol{v}(0) = (1, 0)$.

    (a) Give a formula for the particle's velocity function $\boldsymbol{v}(t)$.

    (b) Give a formula for the particle's position function $\boldsymbol{p}(t)$.

    (c) Write down an integral whose value is the distance traveled by the particle between $t = 0$ and $t = 5$. (Don't evaluate this integral.)

12. An astronaut is flying along the path described by $\boldsymbol{r}(t) = (t^2 - t, 2 + t, -3/t)$, where t is given in hours. The engines are shut off when the spacecraft reaches the point $(6, 5, -1)$. Where is the astronaut two hours later?

13. An object moving with constant velocity passes through the point $(1, 1, 1)$ and then passes through the point $(2, -1, 3)$ five seconds later.

    (a) Find the object's velocity vector.

    (b) Find the object's acceleration vector.

14. Suppose that the velocity of an object at time t seconds is $\left(4e^{-t}, 3t^2, 5\cos(\pi t)\right)$ meters/second. At time $t = 1$ the object is at the point with coordinates $(1, 2, 3)$.

    (a) Find the object's speed at time $t = 0$.

    (b) Find the object's acceleration vector at time $t = 1$.

    (c) Find the position of the object at time $t = 2$.

15. In each part below, plot the given curve and estimate its arclength by eye. Set up the appropriate integral. If possible, evaluate it exactly by antidifferentiation; otherwise, estimate the answer using a midpoint approximating sum with 20 subdivisions.

    (a) $\boldsymbol{r}(t) = (3 + t, 2 + 3t), 0 \le t \le 1$. (Do this first without any calculus.)

    (b) $\boldsymbol{r}(t) = (\cos(2t), \sin(2t)); 0 \le t \le \pi$

    (c) $\boldsymbol{r}(t) = (\sin(3t), \cos(3t)); 0 \le t \le 2\pi/3$

    (d) $\boldsymbol{r}(t) = (3\sin t, \cos t); 0 \le t \le 2\pi$

    (e) $\boldsymbol{r}(t) = (t\cos t, t\sin t); 0 \le t \le 4\pi$

16. Do this exercise in the spirit of Example 1, page 48. In each part, find a linear curve $\boldsymbol{l}(t)$ that passes through $\boldsymbol{r}(t_0)$, with direction vector $\boldsymbol{r}'(t_0)$. Check your work by plotting.

    (Notes: (i) Use *Maple* or other technology to plot the curves and lines. (ii) Choose any convenient parameter interval; a good choice will show enough of both curve and line to reveal what's going on. (iii) When plotting, be sure units are the same size in x- and y-directions.)

    (a) $\boldsymbol{r}(t) = (t^2, t^3); t_0 = 1$

    (b) $\boldsymbol{r}(t) = (\cos t, \sin t); t_0 = \pi/4$

    (c) $\boldsymbol{r}(t) = (t, \sin t); t_0 = \pi/4$

    (d) $\boldsymbol{r}(t) = (t\sin t, t\cos t); t_0 = \pi$

17. Consider the vector-valued function $\boldsymbol{p}(t) = (\cos t, \sin t, t)$.

    (a) Draw (by hand) a rough plot of the curve defined by $\boldsymbol{p}(t)$; let $0 \le t \le 4\pi$.

    (b) Find the position, velocity, and acceleration vectors at $t = 0$.

    (c) Find a vector equation for the tangent line ℓ at $t = \pi$ to the curve defined by $\boldsymbol{p}(t)$.

    (d) On one set of axes, plot both the curve and the tangent line from the preceding part. (Use technology!)

    (e) Show that $\boldsymbol{p}(t) = (\cos t, \sin t, t)$ has constant speed. Find the arclength from $t = 0$ to $t = 4\pi$.

18. Repeat all parts of the preceding exercise, but use the vector-valued function $\boldsymbol{p}(t) = \big(\cos(t/2), \sin(t/2), t\big)$.

19. Suppose that the position of a particle at time t is $\boldsymbol{p}(t) = (3\cos t, 3\sin t, t)$. If the particle flies off on a tangent at $t = \pi/2$, where is the particle at $t = \pi$?

20. Find the length of the curve $\boldsymbol{x}(t) = \big(t, \sin(4t), \cos(4t)\big)$ between $t = 0$ and $t = \pi$.

21. Find the length of the logarithmic sprial $r = e^{\theta}$ between $\theta = 0$ and $\theta = \pi$.

22. Which spring requires more material, one of radius 5 cm and height 4 cm that makes three complete turns or one of radius 3 cm and height 4 cm that makes five complete turns? Justify your answer.

Project: Beyond Free Fall

In the real world, most falling objects do *not* undergo what physicists call free fall (i.e., falling under the *sole* influence of gravity). And a good thing, too—otherwise we'd all be dead, brained by falling raindrops. Luckily, most falling objects are influenced by some "resistive" force, such as air drag, that acts in the direction *opposite* to the direction of falling. (For objects falling in other media than air, such as a ball bearing in oil, the viscosity of the medium has similar resistive effects.) This project explores the life-saving effects of air drag.

To model the situation, we'll let the positive y-direction represent "up," with the units on both axes in meters. We'll use the usual notations $\boldsymbol{a}(t)$, $\boldsymbol{v}(t)$, and $\boldsymbol{p}(t)$ to denote the acceleration, velocity, and position vectors for an object falling through the xy-plane. These are vector-valued functions; we'll sometimes write them using components, in the form

$$\boldsymbol{a}(t) = (a_1(t), a_2(t)), \quad \boldsymbol{v}(t) = (v_1(t), v_2(t)), \quad \text{and} \quad \boldsymbol{p}(t) = (p_1(t), p_2(t)).$$

For a small, light, object falling at moderate speed—such as a raindrop in air—the resistive force of air drag is approximately *proportional to the velocity*. This means that the acceleration function $\boldsymbol{a}(t)$ has the form

$$\boldsymbol{a}(t) = (a_1(t), a_2(t)) = (0, -g) - k(v_1(t), v_2(t)),$$

where $k > 0$ is some positive constant, and $g \approx 9.8$ is (as always) the magnitude of the acceleration due to gravity. The vector equation above is equivalent to the two scalar equations

$$a_1(t) = -kv_1(t) \quad \text{and} \quad a_2(t) = -g - kv_2(t).$$

Throughout this project we'll assume for simplicity that the falling object starts at the origin (so $p_1(0) = 0$ and $p_2(0) = 0$) and that it's initial velocity is *horizontal*, so $v_2(0) = 0$. (This might occur, e.g., if the raindrop were blown horizontally by a sudden gust of wind.)

Problems

1. The acceleration is the derivative of the velocity. This means that the two scalar equations above are equivalent to the two differential equations (DEs)

$$v_1'(t) = -kv_1(t) \quad \text{and} \quad v_2'(t) = -g - kv_2(t).$$

Show that for any constants C_1 and C_2, the functions

$$v_1(t) = C_1 e^{-kt} \quad \text{and} \quad v_2(t) = \frac{C_2 e^{-kt} - g}{k}$$

are solutions of the DEs above. (Note: You can just *check* these claims by differentiation. But students who've studied separable differential equations should also try to *derive* the solutions by separation of variables.)

2. Explain why the constant C_1 represents the initial horizontal velocity. Also, we're assuming that $v_2(0) = 0$; use this to show that

$$v_2(t) = -\frac{g}{k}(1 - e^{-kt}).$$

Are v_1 and v_2 positive or negative for $t > 0$?

3. Integrate the formulas for v_1 and v_2 to show that

$$p_1(t) = -\frac{C_1}{k}e^{-kt} + D_1 \quad \text{and} \quad p_2(t) = -\frac{g}{k}(t + e^{-kt}/k) + D_2,$$

where D_1 and D_2 are constants of integration. Then use the fact that $\boldsymbol{p}(0) = (0, 0)$ to rewrite the position functions as

$$p_1(t) = \frac{C_1}{k}(1 - e^{-kt}) \quad \text{and} \quad p_2(t) = \frac{g}{k^2}(1 - e^{-kt} - kt).$$

4. Find the limits

$$\lim_{t\to\infty} v_1(t); \quad \lim_{t\to\infty} v_2(t); \quad \lim_{t\to\infty} p_1(t); \quad \lim_{t\to\infty} p_2(t).$$

(The answers involve the constants.) What do these numbers mean about the physical situation?

5. For an average-sized raindrop, the value $k = 1$ is reasonable. Let $s(t)$ be the object's speed at time t. The limit $\lim_{t\to\infty} s(t)$ is called the **terminal speed**. Find the raindrop's terminal speed in meters per second. (The result explains among other things, why we don't die in a rainstorm.)

6. Plot the position function for each initial velocity condition below. (Assume always that $k = 1$ and use $g = 9.8$.) Always use the plotting window $[0, 100] \times [-100, 0]$. If possible, plot all the position functions on the same axes.

   (a) $\boldsymbol{v}(0) = (0, 0)$
   (b) $\boldsymbol{v}(0) = (10, 0)$
   (c) $\boldsymbol{v}(0) = (40, 0)$
   (d) $\boldsymbol{v}(0) = (100, 0)$

1.6 The dot product

We've already met two algebraic operations on vectors, both of which produce new *vectors* from old ingredients. If $\boldsymbol{v}$ and $\boldsymbol{w}$ are vectors and r is a scalar, then both the sum $\boldsymbol{v} + \boldsymbol{w}$ and the scalar multiple $r\boldsymbol{v}$ are vectors. In this section we meet a new operation, the **dot product**, also known, sometimes, as the **scalar product**. The latter name stems from an important property: The dot product $\boldsymbol{v} \cdot \boldsymbol{w}$ of two vectors is a *scalar*. And a very interesting scalar at that .... Indeed, few other mathematical tools reveal as much information, at such modest cost in calculation, as the dot product. Any effort spent in understanding it will be richly rewarded.

Here endeth today's sermon.

Defining the dot product

The algebraic definition is essentially the same in 2 and 3 variables:

> **Definition: (The dot product.)** Let $\boldsymbol{v} = (v_1, v_2)$ and $\boldsymbol{w} = (w_1, w_2)$ be vectors. The dot product of $\boldsymbol{v}$ and $\boldsymbol{w}$ is the scalar defined by
>
> $$\boldsymbol{v} \cdot \boldsymbol{w} = v_1 w_1 + v_2 w_2.$$
>
> If $\boldsymbol{v} = (v_1, v_2, v_3)$ and $\boldsymbol{w} = (w_1, w_2, w_3)$, then
>
> $$\boldsymbol{v} \cdot \boldsymbol{w} = v_1 w_1 + v_2 w_2 + v_3 w_3.$$

The definition is very simple. What the dot product means, algebraically and geometrically, is our main question. We'll study it first in two variables, then in three.

■ **Example 1.** Consider the vectors $\boldsymbol{i} = (1, 0)$, $\boldsymbol{j} = (0, 1)$, $\boldsymbol{v} = (2, 3)$, and $\boldsymbol{w} = (-3, 2)$. Find some dot products.

Solution: The calculations could hardly be simpler. Here are a few:

$$\begin{aligned}
\boldsymbol{i} \cdot \boldsymbol{j} &= (1, 0) \cdot (0, 1) = 0 + 0 = 0; \\
\boldsymbol{i} \cdot \boldsymbol{i} &= (1, 0) \cdot (1, 0) = 1; \\
\boldsymbol{i} \cdot \boldsymbol{v} &= (1, 0) \cdot (2, 3) = 2; \\
\boldsymbol{j} \cdot \boldsymbol{v} &= (0, 1) \cdot (2, 3) = 3; \\
\boldsymbol{v} \cdot \boldsymbol{v} &= (2, 3) \cdot (2, 3) = 2^2 + 3^2 = 13; \\
\boldsymbol{v} \cdot \boldsymbol{w} &= (2, 3) \cdot (-3, 2) = -6 + 6 = 0.
\end{aligned}$$

□

Do you see any patterns emerging? When, in particular, is the answer zero? Let's look at some more general examples.

■ **Example 2.** Let $\boldsymbol{v} = (a, b)$ and $\boldsymbol{w} = (-b, a)$. Find all possible dot products. What do the answers mean geometrically?

Solution: Finding the dot products is easy:

$$\begin{aligned}
\boldsymbol{v} \cdot \boldsymbol{v} &= (a, b) \cdot (a, b) = a^2 + b^2 = |\boldsymbol{v}|^2; \\
\boldsymbol{v} \cdot \boldsymbol{w} &= (a, b) \cdot (-b, a) = -ab + ba = 0; \\
\boldsymbol{w} \cdot \boldsymbol{w} &= (-b, a) \cdot (-b, a) = b^2 + a^2 = |\boldsymbol{w}|^2.
\end{aligned}$$

Here's a picture:

Perpendicular vectors

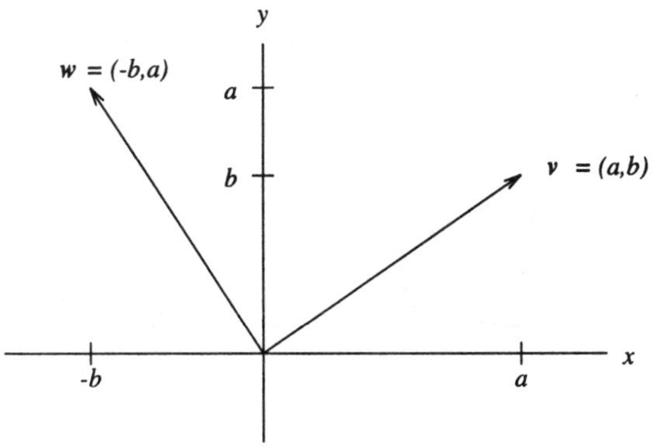

The picture suggests an important fact about perpendicular vectors and the dot product, which we'll state formally in a moment. ☐

But can you guess it now?

The following theorem summarizes what we've seen, and adds a bit more. All parts are readily checked using the definition.

> **Theorem 3 (Algebraic properties of the dot product).** Let $\boldsymbol{u}$, $\boldsymbol{v}$, and $\boldsymbol{w}$ be vectors, and let r be any scalar. Then
>
> (a) $\boldsymbol{u} \cdot \boldsymbol{v} = \boldsymbol{v} \cdot \boldsymbol{u}$;
>
> (b) $(r\boldsymbol{u}) \cdot \boldsymbol{v} = \boldsymbol{u} \cdot (r\boldsymbol{v}) = r(\boldsymbol{u} \cdot \boldsymbol{v})$;
>
> (c) $\boldsymbol{u} \cdot (\boldsymbol{v} + \boldsymbol{w}) = \boldsymbol{u} \cdot \boldsymbol{v} + \boldsymbol{u} \cdot \boldsymbol{w}$; → Distributes
>
> (d) $\boldsymbol{u} \cdot \boldsymbol{u} = |\boldsymbol{u}|^2$.

Geometry of the dot product

In studying geometric properties of the dot product, **unit vectors** turn out to be especially useful.

Unit vectors in all directions. The following picture shows unit vectors in 12 directions, corresponding to angles θ that are multiples of $\pi/6$:

Unit vectors in various directions

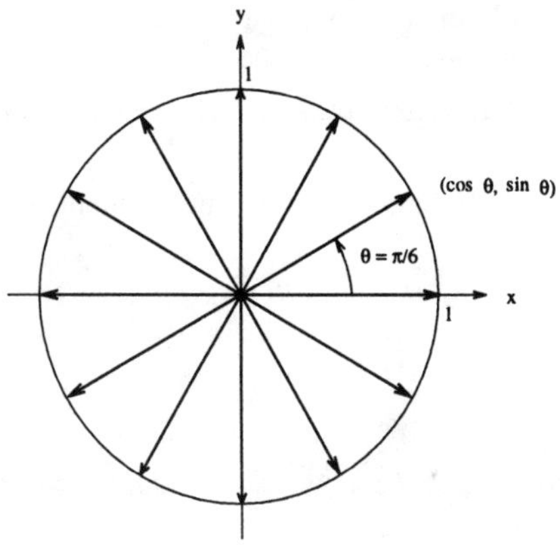

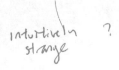

For any angle θ, the vector $\boldsymbol{u}_\theta$ that makes angle θ with the x-axis has components $(\cos\theta, \sin\theta)$. Every vector of this form, moreover, has length 1. The dot product shows why:

$$|\boldsymbol{u}_\theta|^2 = (\cos\theta, \sin\theta) \cdot (\cos\theta, \sin\theta) = \cos^2\theta + \sin^2\theta = 1.$$

We'll call the vector $\boldsymbol{u}_\theta$ the **unit vector in the θ-direction**. For example, if $\theta = \pi/6$, then

$$\boldsymbol{u}_\theta = (\cos(\pi/6), \sin(\pi/6)) = (\sqrt{3}/2, 1/2).$$

In fact, *any* non-zero vector $\boldsymbol{v} = (a, b)$ can be written as a scalar multiple of some $\boldsymbol{u}_\theta$. The idea is the same as that for polar coordinates. If the Cartesian point (a, b) has polar coordinates (r, θ), then

$$(a, b) = (r\cos\theta, r\sin\theta) = r(\cos\theta, \sin\theta) = r\boldsymbol{u}_\theta,$$

where $r = \sqrt{a^2 + b^2}$ and $\tan\theta = b/a$.

■ **Example 3.** Write the vector $\boldsymbol{v} = (4, 3)$ in the form $r\boldsymbol{u}_\theta$. What are r and θ?

Solution: The vector $(4, 3)$ has length 5, so the unit vector in the same direction is $\boldsymbol{u} = (4/5, 3/5)$. The point $(x, y) = (4, 3)$ has polar coordinates $r = 5$ and $\theta = \arctan(3/4) \approx 0.644$; this means that $\boldsymbol{v}$ makes angle $\theta \approx 0.644$ with the x-axis. □

Recall: $\theta = \arctan(y/x)$.

Another example, a little more subtle than the last two, will complete our exploratory tour of the dot product.

■ **Example 4.** Let α and β be angles; consider the unit vectors

$$\boldsymbol{u}_\alpha = (\cos\alpha, \sin\alpha) \quad \text{and} \quad \boldsymbol{u}_\beta = (\cos\beta, \sin\beta)$$

in the directions of α and β, respectively. Find $\boldsymbol{u}_\alpha \cdot \boldsymbol{u}_\beta$. What does *this* answer mean geometrically?

Solution: The calculation is easy enough:

$$\boldsymbol{u}_\alpha \cdot \boldsymbol{u}_\beta = (\cos\alpha, \sin\alpha) \cdot (\cos\beta, \sin\beta) = \cos\alpha\,\cos\beta + \sin\alpha\,\sin\beta.$$

But what does it mean? A standard trigonometric identity—

$$\cos(\alpha - \beta) = \cos\alpha\,\cos\beta + \sin\alpha\,\sin\beta$$

leads to the key insight:

For unit vectors $\boldsymbol{u}_\alpha$ and $\boldsymbol{u}_\beta$, the dot product is the cosine of the angle between $\boldsymbol{u}_\alpha$ and $\boldsymbol{u}_\beta$.

The picture looks something like this:

The angle between two vectors

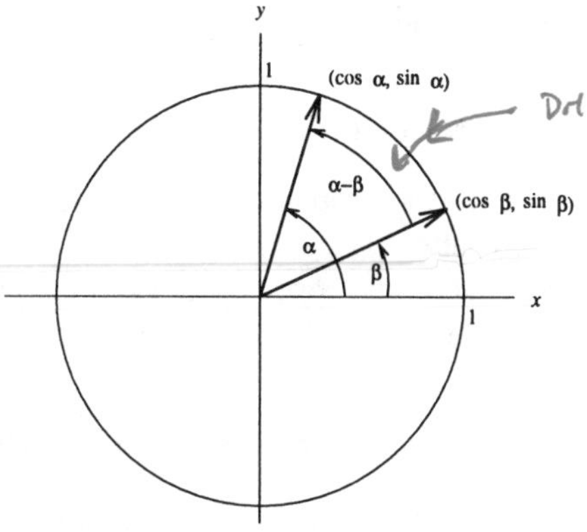

□

Now we can state—and even prove—the main geometric property of the dot product.

We don't prove everything, but this proof is instructive.

> **Theorem 4. (The dot product geometrically.)** Let $\boldsymbol{v}$ and $\boldsymbol{w}$ be any two vectors, and let θ be the (smaller) angle between $\boldsymbol{v}$ and $\boldsymbol{w}$. Then
> $$\boldsymbol{v} \cdot \boldsymbol{w} = |\boldsymbol{v}|\,|\boldsymbol{w}|\,\cos\theta.$$

Proof. Suppose that $\boldsymbol{v}$ and $\boldsymbol{w}$ makes angles α and β, respectively, with the x-axis, and that $\theta = \alpha - \beta$. In polar form, we have

$$\boldsymbol{v} = |\boldsymbol{v}|\boldsymbol{u}_\alpha \quad \text{and} \quad \boldsymbol{w} = |\boldsymbol{w}|\boldsymbol{u}_\beta,$$

where $\boldsymbol{u}_\alpha$ and $\boldsymbol{u}_\beta$ are the unit vectors in the α- and β-directions, respectively. By Theorem 3 (b) and results of the preceding example,

$$\boldsymbol{v} \cdot \boldsymbol{w} = (|\boldsymbol{v}|\boldsymbol{u}_\alpha) \cdot (|\boldsymbol{w}|\boldsymbol{u}_\beta) = |\boldsymbol{v}|\,|\boldsymbol{w}|\,(\boldsymbol{u}_\alpha \cdot \boldsymbol{u}_\beta) = |\boldsymbol{v}|\,|\boldsymbol{w}|\,\cos\theta,$$

as desired. □

Special angles. Theorem 4 makes it easy to detect whether two non-zero vectors meet at *right* angles. (If either vector is the zero vector, they don't meet at any sensible angle.) In this case $\theta = \pi/2$, so $\cos\theta = 0$. In general:

> **Fact:** Two nonzero vectors $\boldsymbol{v}$ and $\boldsymbol{w}$ are perpendicular if and only if $\boldsymbol{v} \cdot \boldsymbol{w} = 0$.

In math-speak, any vectors $\boldsymbol{v}$ and $\boldsymbol{w}$ (even if one or both are zero) that satisfy the condition $\boldsymbol{v} \cdot \boldsymbol{w} = 0$ are called **orthogonal**. This term is slightly more general than "perpendicular," and is sometimes useful when vectors don't have a convenient geometric meaning.

At the opposite extreme from being perpendicular, two vectors might be identical. In this case, the angle between them is 0, and Theorem 4 says (as did Theorem 3) that

$$\boldsymbol{v} \cdot \boldsymbol{v} = |\boldsymbol{v}||\boldsymbol{v}|\cos 0 = |\boldsymbol{v}|^2.$$

The sign of the dot product. Theorem 4 says that the *sign* of the dot product $\boldsymbol{v} \cdot \boldsymbol{w}$ depends entirely on $\cos\theta$. More precisely:

The angle between $\boldsymbol{v}$ and $\boldsymbol{w}$ is acute if $\boldsymbol{v} \cdot \boldsymbol{w} > 0$; it's obtuse if $\boldsymbol{v} \cdot \boldsymbol{w} < 0$.

Projecting one vector onto another

The dot product is useful for calculating various quantities that depend on the lengths of vectors and the angles between them. **Projections** offer one type of example. They arise in physics, for instance, when one wishes to express a vector quantity, such as a force, as a sum of forces in perpendicular directions.

■ **Example 5.** Consider the vectors $\boldsymbol{v}$, $\boldsymbol{w}$, $\boldsymbol{a}$, and $\boldsymbol{b}$ shown in the following figure.

Projecting one vector onto another

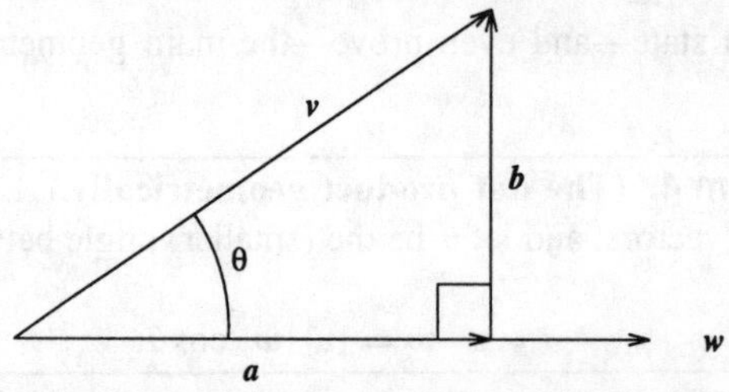

Use the dot product to give expressions for $\boldsymbol{a}$, $\boldsymbol{b}$, and $|\boldsymbol{a}|$ in terms of $\boldsymbol{v}$ and $\boldsymbol{w}$. The scalar $|\boldsymbol{a}|$ is the **scalar projection of $\boldsymbol{v}$ onto $\boldsymbol{w}$**. The vector $\boldsymbol{a}$ is the **component of $\boldsymbol{v}$ in the $\boldsymbol{w}$-direction**; the vector $\boldsymbol{b}$ is the **component of $\boldsymbol{v}$ perpendicular to the $\boldsymbol{w}$-direction.**

Solution: A look at the picture reveals the promising equation $|\boldsymbol{a}| = |\boldsymbol{v}|\cos\theta$. On the other hand, we know that $\boldsymbol{v} \cdot \boldsymbol{w} = |\boldsymbol{v}||\boldsymbol{w}|\cos\theta$. Combining the two equations shows that

$$|\boldsymbol{a}| = |\boldsymbol{v}|\cos\theta = \frac{\boldsymbol{v} \cdot \boldsymbol{w}}{|\boldsymbol{w}|}.$$

To find $\boldsymbol{a}$ itself, we can multiply $|\boldsymbol{a}|$ by the unit vector in the appropriate direction, namely $\boldsymbol{w}/|\boldsymbol{w}|$, to get

$$\boldsymbol{a} = \frac{\boldsymbol{v} \cdot \boldsymbol{w}}{|\boldsymbol{w}|}\frac{\boldsymbol{w}}{|\boldsymbol{w}|} = \frac{\boldsymbol{v} \cdot \boldsymbol{w}}{|\boldsymbol{w}|^2}\boldsymbol{w}.$$

The picture shows, too, that $\boldsymbol{a} + \boldsymbol{b} = \boldsymbol{v}$, so

$$\boldsymbol{b} = \boldsymbol{v} - \boldsymbol{a} = \boldsymbol{v} - \frac{\boldsymbol{v} \cdot \boldsymbol{w}}{|\boldsymbol{w}|^2}\boldsymbol{w}. \qquad \square$$

These formulas may look scary, but in concrete cases everything reduces to numbers.

■ **Example 6.** Let $\boldsymbol{v} = (2, 3)$ and $\boldsymbol{w} = (1, 1)$. Find perpendicular vectors $\boldsymbol{a}$ and $\boldsymbol{b}$ as above.

Solution: Working through the formulas above with these particular vectors gives

$$\boldsymbol{a} = \frac{\boldsymbol{v}\cdot\boldsymbol{w}}{|\boldsymbol{w}|^2}\boldsymbol{w} = \frac{(2,3)\cdot(1,1)}{2}(1,1) = (5/2, 5/2).$$

It follows that $\boldsymbol{b} = (-1/2, 1/2)$; notice that, just as advertised, $\boldsymbol{a}$ and $\boldsymbol{b}$ are perpendicular. □

The distance from a point to a line. The idea of using the dot product to project one vector onto another offers an easy way of calculating the shortest distance (i.e., the perpendicular distance) from any point to any line. (We illustrate the process in two variables, but the same formulas apply in three variables.)

Suppose that the situation is as shown:

Distance from a point to a line

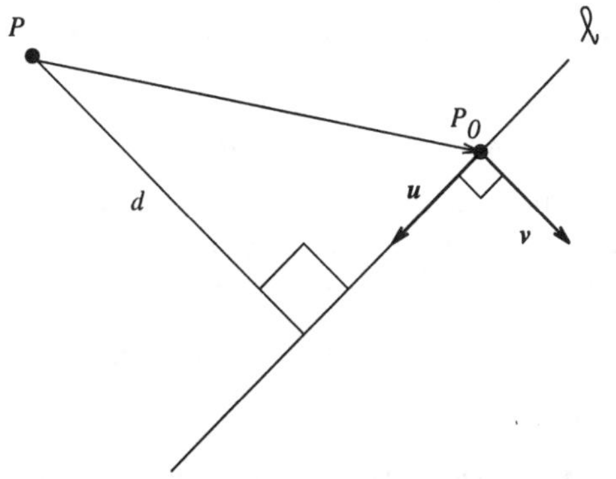

The line ℓ is determined, as shown, by a point P_0 and a direction vector $\boldsymbol{u}$; P is a point not on ℓ; we want the perpendicular distance d. A look at the picture shows that d is is the scalar projection of the vector $\overrightarrow{PP_0}$ in the direction of $\boldsymbol{v}$, where $\boldsymbol{v}$ is any vector perpendicular to $\boldsymbol{u}$. We'll call $\boldsymbol{v}$ a **normal vector** to the line ℓ. (Finding a vector $\boldsymbol{v}$ perpendicular to $\boldsymbol{u}$ is no problem: if $\boldsymbol{u} = (a, b)$, then $\boldsymbol{v} = (-b, a)$ will do.) In symbols,

Or the opposite thereof, if the answer is negative.

$$d = \left|\frac{\overrightarrow{PP_0}\cdot\boldsymbol{v}}{|\boldsymbol{v}|}\right|.$$

■ **Example 7.** Find the distance from the point $P(2, 3)$ to the line through $(1, 1)$ with direction vector $(-2, 3)$.

Solution: Given these data, we have $\boldsymbol{u} = (-2, 3)$ and $\overrightarrow{PP_0} = (1, 1) - (2, 3) = (-1, -2)$. As mentioned above, we can use $\boldsymbol{v} = (3, 2)$ as a vector perpendicular to $\boldsymbol{u}$, and therefore to ℓ. The rest is a straightforward calculation:

Draw a picture to convince yourself the answer is reasonable.

$$d = \left|\frac{(-1,-2)\cdot(-3,-2)}{\sqrt{13}}\right| = \frac{7}{\sqrt{13}} \approx 1.94.$$ □

The dot product in three variables

We defined the dot product of 3-vectors $\boldsymbol{v} = (v_1, v_2, v_3)$ and $\boldsymbol{w} = (w_1, w_2, w_3)$ as

$$\boldsymbol{v}\cdot\boldsymbol{w} = v_1w_1 + v_2w_2 + v_3w_3.$$

Given the similarity to the two-dimensional formula, it's not surprising that the same *algebraic* properties hold for dot products in three variables:

$$\boldsymbol{v} \cdot \boldsymbol{v} = |\boldsymbol{v}|^2; \qquad \boldsymbol{u} \cdot (\boldsymbol{v} + \boldsymbol{w}) = \boldsymbol{u} \cdot \boldsymbol{v} + \boldsymbol{u} \cdot \boldsymbol{w}; \qquad a\boldsymbol{v} \cdot \boldsymbol{w} = \boldsymbol{v} \cdot a\boldsymbol{w} = a(\boldsymbol{v} \cdot \boldsymbol{w}).$$

Like the corresponding formulas in two variables, these are easily proved by rewriting everything in terms of coordinates.

It's more remarkable (and a little less obvious) that the same *geometric* properties of the dot product continue to hold—word for word—in three dimensions:

> **Theorem 5. (The dot product geometrically)** Let $\boldsymbol{v}$ and $\boldsymbol{w}$ be any vectors, and let θ be the angle between $\boldsymbol{v}$ and $\boldsymbol{w}$. Then
>
> $$\boldsymbol{v} \cdot \boldsymbol{w} = |\boldsymbol{v}|\,|\boldsymbol{w}| \cos\theta.$$
>
> In particular, $\boldsymbol{v}$ and $\boldsymbol{w}$ are perpendicular if and only if $\boldsymbol{v} \cdot \boldsymbol{w} = 0$.

The law of cosines generalizes the Pythagorean rule; see the exercises at the end of this section.

Proof. The theorem follows from the **law of cosines:**

If a, b, and c are the sides of any triangle, and θ is the angle between sides a and b, then $c^2 = a^2 + b^2 - 2ab\cos\theta$.

Draw your own picture of this triangle.

To prove the theorem, assuming the law of cosines, we consider the triangle whose sides are the vectors $\boldsymbol{v}$, $\boldsymbol{w}$, and $\boldsymbol{w} - \boldsymbol{v}$; the sides have lengths $a = |\boldsymbol{v}|$, $b = |\boldsymbol{w}|$, and $c = |\boldsymbol{w} - \boldsymbol{v}|$. Applying the law of cosines gives

$$|\boldsymbol{w} - \boldsymbol{v}|^2 = |\boldsymbol{v}|^2 + |\boldsymbol{w}|^2 - 2|\boldsymbol{v}||\boldsymbol{w}|\cos\theta.$$

If we now substitute

$$|\boldsymbol{w} - \boldsymbol{v}|^2 = (\boldsymbol{w} - \boldsymbol{v}) \cdot (\boldsymbol{w} - \boldsymbol{v}) = \boldsymbol{w} \cdot \boldsymbol{w} - 2\boldsymbol{w} \cdot \boldsymbol{v} + \boldsymbol{v} \cdot \boldsymbol{v},$$

and let the symbol-dust settle, the theorem follows immediately. □

Unit vectors, the dot product, and direction cosines. A space vector of length one is called a **unit vector**. If $\boldsymbol{v} = (a, b, c)$ is *any* nonzero vector, then

$$\frac{\boldsymbol{v}}{|\boldsymbol{v}|} = \frac{\boldsymbol{v}}{\sqrt{a^2 + b^2 + c^2}}$$

is a unit vector, parallel to $\boldsymbol{v}$.

The dot product can help us interpret the coordinates of a unit vector geometrically. Let $\boldsymbol{u} = (u_1, u_2, u_3)$ be any unit vector. Then $\boldsymbol{u} \cdot \boldsymbol{i} = u_1 = \cos(\theta_1)$, where θ_1 is the angle between $\boldsymbol{u}$ and $\boldsymbol{i}$. Similarly, u_2 and u_3 are the cosines of the angles between $\boldsymbol{i}$ and $\boldsymbol{j}$, respectively. (These are also the angles between $\boldsymbol{u}$ and the x-, y-, and z-axes, respectively.) For this reason the coordinates of a unit vector $\boldsymbol{u}$ are sometimes called the **direction cosines** of $\boldsymbol{u}$.

■ **Example 8.** What angles does the vector $\boldsymbol{v} = (1, 1, 1)$ make with the coordinate axes?

Solution: Dividing $\boldsymbol{v}$ by $|\boldsymbol{v}| = \sqrt{3}$ produces the unit vector

$$\boldsymbol{u} = \left(\frac{1}{\sqrt{3}}, \frac{1}{\sqrt{3}}, \frac{1}{\sqrt{3}}\right),$$

which has the same direction as $\boldsymbol{v}$. If we interpret the coordinates of $\boldsymbol{u}$ as cosines, and observe that $\arccos(1/\sqrt{3}) \approx 0.96$, then we see that $\boldsymbol{v}$ makes an angle of (approximately) 0.96 radians with each of three coordinate axes. □

Try to visualize these angles.

Projections and components. The same formula—involving the dot product—works regardless of dimension for finding the projection of one vector onto another. If $\boldsymbol{v}$ and $\boldsymbol{w}$ are any vectors, then

$$\frac{\boldsymbol{v}\cdot\boldsymbol{w}}{|\boldsymbol{w}|} \quad \text{and} \quad \frac{\boldsymbol{v}\cdot\boldsymbol{w}}{|\boldsymbol{w}|^2}\boldsymbol{w}$$

are, respectively, the **scalar projection** of $\boldsymbol{v}$ onto $\boldsymbol{w}$ and the **component** of $\boldsymbol{v}$ in the direction of $\boldsymbol{w}$.

■ **Example 9.** Find the scalar projection and the component of $\boldsymbol{v} = (a, b, c)$ in each of the directions $\boldsymbol{i}$, $\boldsymbol{j}$, and $\boldsymbol{k}$.

Solution: The formulas above are especially simple if $\boldsymbol{w}$ is any of the standard basis vectors $\boldsymbol{i}$, $\boldsymbol{j}$, and $\boldsymbol{k}$, each of which has unit length. If $\boldsymbol{w} = \boldsymbol{i}$, then

$$\frac{\boldsymbol{v}\cdot\boldsymbol{i}}{|\boldsymbol{i}|} = (a, b, c)\cdot(1, 0, 0) = a, \quad \text{and} \quad \frac{\boldsymbol{v}\cdot\boldsymbol{i}}{|\boldsymbol{i}|^2}\boldsymbol{i} = (a, 0, 0) = a\boldsymbol{i}.$$

Similarly, the scalar projections of $\boldsymbol{v}$ on $\boldsymbol{j}$ and $\boldsymbol{k}$ are b and c, respectively, and the respective components are $b\boldsymbol{j}$ and $c\boldsymbol{k}$. The following picture shows how the parts fit together:

Projections and components

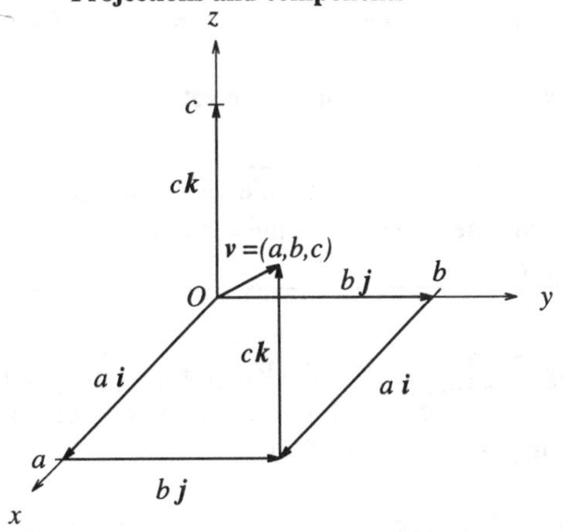

Observe:

Components. The picture shows that $\boldsymbol{v} = (a, b, c)$ is a very simple linear combination of the standard basis $\boldsymbol{i}$, $\boldsymbol{j}$, and $\boldsymbol{k}$, with coefficients a, b, and c. To wit:

$$\boldsymbol{v} = a\boldsymbol{i} + b\boldsymbol{j} + c\boldsymbol{k}.$$

We see, too, that the summands on the right are, respectively, the *components* of $\boldsymbol{v}$ in the $\boldsymbol{i}$-, $\boldsymbol{j}$-, and $\boldsymbol{k}$-directions.

Projections. As the picture shows, the scalar projections of $\boldsymbol{v} = (a, b, c)$ on $\boldsymbol{i}$, $\boldsymbol{j}$, and $\boldsymbol{k}$, are simply a, b, and c—the same as the *coordinates* of $\boldsymbol{v}$.

Derivatives and the dot product

If $\boldsymbol{f}$ and $\boldsymbol{g}$ are vector-valued functions, then their dot product, $\boldsymbol{f}\cdot\boldsymbol{g}$, is a new scalar-valued function. It's a satisfying fact that the derivative of the product function can be found by a "dot product rule" that closely resembles the familiar product rule of elementary calculus:

> **Theorem 6. (Dot product rule for derivatives)** Let $\boldsymbol{f}$ and $\boldsymbol{g}$ be differentiable vector-valued functions. Then the dot product function $\boldsymbol{f} \cdot \boldsymbol{g}$ is is differentiable, and
>
> $$(\boldsymbol{f}(t) \cdot \boldsymbol{g}(t))' = \boldsymbol{f}'(t) \cdot \boldsymbol{g}(t) + \boldsymbol{f}(t) \cdot \boldsymbol{g}'(t).$$

Check each step.

Proof. The theorem is proved by writing $\boldsymbol{f}$ and $\boldsymbol{g}$ in components and using the derivative rules of elementary calculus:

$$\begin{aligned}(\boldsymbol{f} \cdot \boldsymbol{g})' &= (f_1 g_1 + f_2 g_2)' = f_1' g_1 + f_1 g_1' + f_2' g_2 + f_2 g_2' \\ &= f_1' g_1 + f_2' g_2 + f_1 g_1' + f_2 g_2' = \boldsymbol{f}' \cdot \boldsymbol{g} + \boldsymbol{f} \cdot \boldsymbol{g}'.\end{aligned}$$

Notice that we used the ordinary product rule—twice—in the calculation. □

Vector-valued functions of constant radius. As a striking application of the dot product rule, consider a vector-valued function $\boldsymbol{f}$ for which the magnitude $|\boldsymbol{f}(t)|$ is a constant, say k; then $\boldsymbol{f}(t) \cdot \boldsymbol{f}(t) = k^2$. The right side is constant, so its derivative is zero. The left side can be differentiated using the dot product rule. The result is

$$\boldsymbol{f}'(t) \cdot \boldsymbol{f}(t) + \boldsymbol{f}(t) \cdot \boldsymbol{f}'(t) = 2\boldsymbol{f}(t) \cdot \boldsymbol{f}'(t) = 0.$$

The last equality says something quite interesting:

> **Fact:** Let $\boldsymbol{f}$ be a differentiable vector-valued function with constant magnitude $|\boldsymbol{f}(t)|$. Then the vectors $\boldsymbol{f}(t)$ and $\boldsymbol{f}'(t)$ are perpendicular for all t.

The following picture illustrates the Fact for the function $\boldsymbol{f}(t) = (\cos t, \sin t)$. As predicted, the velocity vector, $\boldsymbol{f}'(t)$, at each point on the circle is perpendicular to the corresponding position vector, $\boldsymbol{f}(t)$.

Position and velocity vectors on the unit circle

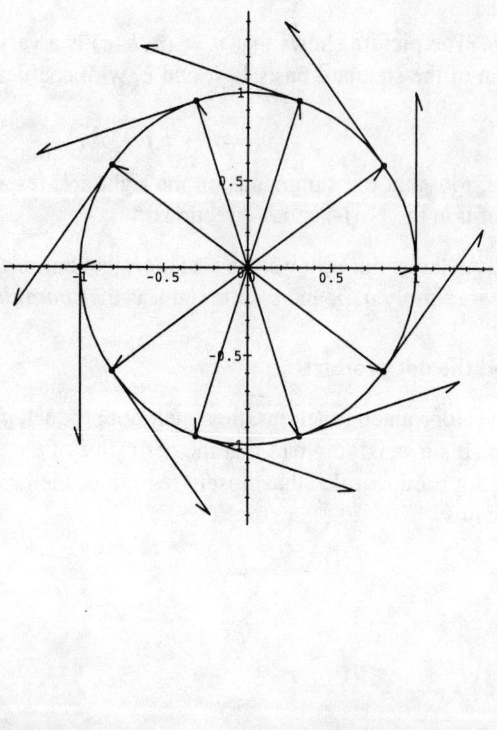

Work and the dot product

Work (in the physicist's sense) occurs when a force, say $\boldsymbol{F}$, acts along a displacement, say $\boldsymbol{d}$. (Both $\boldsymbol{F}$ and $\boldsymbol{d}$ are vector quantities.) In the simplest possible case, such as lifting a weight vertically, a constant force is exerted in the same direction as the displacement; then the work done is the ordinary product of the force and the distance through which it acts.

Often, however, the force $\boldsymbol{F}$ acts at some angle θ to the displacement $\boldsymbol{d}$. This occurs, for instance, when the force of gravity acts to slide an object down an inclined plane—gravity acts downward but the displacement is oblique. Sailors have relied on the same phenomenon for thousands of years. Thanks to its keel and rudder, a boat under sail normally travels at some angle to the force of the wind on its sail. (Otherwise, boats could move only straight downwind.)

The dot product $\boldsymbol{F} \cdot \boldsymbol{d}$ is exactly the quantity we want when calculating the work done by an oblique force $\boldsymbol{F}$ along a displacement $\boldsymbol{d}$. In this case, work is done only by the component of the force that's parallel to the displacement. This component of force, as we've seen, has magnitude $|\boldsymbol{F}| \cos\theta$, and it acts through distance $|\boldsymbol{d}|$. Thus the work done is precisely

$$\boldsymbol{F} \cdot \boldsymbol{d} = |\boldsymbol{d}||\boldsymbol{F}| \cos\theta.$$

■ **Example 10.** An inclined plane makes an angle of $\pi/3$ radians with the downward vertical. How much work, in foot-pounds, is done by gravity as a 10-pound object slides 10 feet down the plane?

Solution: The force $\boldsymbol{F}$ acts in the straight downward direction, with magnitude 10 pounds. The displacement $\boldsymbol{d}$ has magnitude 10 feet, and makes an angle of $\pi/3$ with $\boldsymbol{F}$. The work done, therefore, is

$$\boldsymbol{F} \cdot \boldsymbol{d} = 10 \times 10 \times \cos(\pi/3) = 50 \text{ foot-pounds.}$$ □

Exercises

1. (For students who have taken linear algebra.) In this section we referred to the vectors $\boldsymbol{i} = (1, 0)$ and $\boldsymbol{j} = (0, 1)$ as the **standard basis vectors for** $\mathbb{R}^2$. This problem is about why that terminology is justified.

   Recall that, by definition, a basis for $\mathbb{R}^2$ is any set of vectors that is (i) linearly independent; and (ii) "spans" $\mathbb{R}^2$. We saw in this section why (ii) holds.

   (a) What does it mean for a set of two vectors to be linearly independent?

   (b) Explain carefully why the set $\{\boldsymbol{i}, \boldsymbol{j}\}$ is linearly independent.

   (c) Is the set $\{(1, 1), (0, 1)\}$ linearly independent? Why or why not?

   (d) Does the set $\{(1, 1), (0, 1)\}$ span $\mathbb{R}^2$? Why or why not?

   (e) Is the set $\{(1, 1), (0, 1), (2, 3)\}$ linearly independent? Does this set span $\mathbb{R}^2$? Explain your answers.

2. For each θ below, find the components of the unit vector in the θ-direction. (Do these by hand—only famous values of θ are involved.)

(a) $\theta = 0$

(b) $\theta = \pi/4$

(c) $\theta = -\pi/4$

(d) $\theta = 2\pi/3$

(e) $\theta = 1234\pi/3$

3. In each part below, find a number $r > 0$ and an angle θ such that $\boldsymbol{v} = r\boldsymbol{u}_\theta$. (Do these by hand—only famous values of θ are involved.)

   (a) $\boldsymbol{v} = (2, 0)$

   (b) $\boldsymbol{v} = (2, 2)$

   (c) $\boldsymbol{v} = (\sqrt{3}, 1)$

   (d) $\boldsymbol{v} = (1, -\sqrt{3})$

4. This exercise is about Theorem 3, page 61. Using the notation $\boldsymbol{u} = (u_1, u_2)$, $\boldsymbol{v} = (v_1, v_2)$, and $\boldsymbol{w} = (w_1, w_2)$, part (a) is easy to show:

$$\boldsymbol{u} \cdot \boldsymbol{v} = u_1 v_1 + u_2 v_2 = v_1 u_1 + v_2 u_2 = \boldsymbol{v} \cdot \boldsymbol{u}.$$

   (The middle step works because ordinary multiplication is commutative.) Do the parts below in the same spirit.

   (a) Show part (b) of the theorem.

   (b) Show part (c) of the theorem.

5. Use the formulas and notation of Example 5, page 64 in this exercise.

   (a) Let $\boldsymbol{v} = (3, 2)$ and $\boldsymbol{w} = (1, 0)$. Find $\boldsymbol{a}$ and $\boldsymbol{b}$. Note that $\boldsymbol{a}$ and $\boldsymbol{b}$ should be perpendicular, and $\boldsymbol{a} + \boldsymbol{b}$ should add up to $\boldsymbol{v}$.

   (b) Let $\boldsymbol{v} = (3, 2)$ and $\boldsymbol{w} = (0, 1)$. Find $\boldsymbol{a}$ and $\boldsymbol{b}$. Note that $\boldsymbol{a}$ and $\boldsymbol{b}$ should be perpendicular, and $\boldsymbol{a} + \boldsymbol{b}$ should add up to $\boldsymbol{v}$.

   (c) Let $\boldsymbol{v} = (3, 2)$ and $\boldsymbol{w} = (1, 1)$. Find $\boldsymbol{a}$ and $\boldsymbol{b}$. Note that $\boldsymbol{a}$ and $\boldsymbol{b}$ should be perpendicular, and $\boldsymbol{a} + \boldsymbol{b}$ should add up to $\boldsymbol{v}$. Sketch all the vectors in question—are the appropriate vectors really perpendicular?

6. In each case below, a 10-pound object is moved by gravity down a 10-foot inclined plane, tilted at angle $\pi/4$ to the vertical.

   (a) How much work is done by gravity?

   (b) How far, vertically, does the object descend?

   (c) How much work would have been done if the ramp were twice as long, but had the same vertical drop?

7. In each part below, a curve is defined by a vector-valued position function $\boldsymbol{f}(t)$, as described in the preceding section. Show that for each value of t, the position vector and the velocity vector are perpendicular.

   (a) $\boldsymbol{f}(t) = (\cos t, \sin t)$; $\quad 0 \le t \le 2\pi$.

   (b) $\boldsymbol{f}(t) = (t, \sqrt{4 - t^2})$; $\quad -2 < t < 2$.

   (c) $\boldsymbol{f}(t) = (\sin(2t), \cos(2t))$; $\quad 0 \le t \le \pi$.

(d) What do all the curves in this exercise have in common geometrically? (Plot them to see the relationships.)

8. Let $\mathbf{u} = (1, -2)$ and $\mathbf{v} = (-3, 5)$. Find the angle between the vectors $\mathbf{u}$ and $\mathbf{v}$. [HINT: Write your answer as the value of an inverse trigonometric function.]

9. (a) Find a vector of length 3 that is perpendicular to $\boldsymbol{u} = (1, 2)$. How many such vectors are there?

   (b) Find a vector of length 4 that is perpendicular to $\boldsymbol{u} = (1, 2, 3)$. How many such vectors are there?

10. (a) Suppose that the vectors $\boldsymbol{u}$ and $\boldsymbol{v}$ have the same length. Show that the vectors $\boldsymbol{u} + \boldsymbol{v}$ and $\boldsymbol{u} - \boldsymbol{v}$ are orthogonal.

    (b) Prove that an angle inscribed in a semicircle is a right angle. (Such an angle has its vertex on the circle and its sides pass through the ends of a diameter of the circle.) [HINT: Let $\boldsymbol{x}$ be the vector from the center of the circle to the vertex of the angle and let $\boldsymbol{y}$ be the vector from the center of the circle to the point on the circle intersected by one of the sides of the angle.]

11. Show that the line segment connecting the midpoints of two sides of a triangle is parallel to, and half as long as, the third side.

12. Suppose that $\boldsymbol{x}(t)$ and $\boldsymbol{x}'(t)$ are always perpendicular. Show that $|\boldsymbol{x}(t)|$ is constant. [HINT: Show that $|\boldsymbol{x}(t)|^2$ is constant.]

13. A rectangle has dimensions 2 by 3. Find the angles between its sides and a diagonal.

14. Show that the diagonals of a parallelogram are perpendicular if and only if all the sides of the parallelogram are equal.

15. Show that an equilateral triangle also has all angles equal.

16. Suppose that $\boldsymbol{r}(t)$ is a differentiable vector-valued function and that $\boldsymbol{r}(t) \neq \boldsymbol{0}$ for all t. Show that

$$\frac{d}{dt}\left(\frac{\boldsymbol{r}}{|\boldsymbol{r}|}\right) = \frac{\boldsymbol{r}'}{|\boldsymbol{r}|} - \frac{\boldsymbol{r} \cdot \boldsymbol{r}'}{|\boldsymbol{r}|^3}\boldsymbol{r}.$$

17. Suppose that the path of a particle in space is described by a differentiable curve and that the particle's speed is always nonzero. Show that the particle's velocity and acceleration vectors are perpendicular whenever the particle's speed is a local maximum or a local minimum.

18. Are the vectors $\mathbf{u} = (1, -2, 3)$ and $\mathbf{v} = (-3, 0, 1)$ perpendicular? Justify your answer.

19. (a) Find a vector of length 3 that is perpendicular to $\boldsymbol{u} = (1, 2)$. How many such vectors are there?

    (b) Find a vector of length 4 that is perpendicular to $\boldsymbol{u} = (1, 2, 3)$. How many such vectors are there?

20. Find the component of $\boldsymbol{u} = (2, -3, 4)$ in the direction $\boldsymbol{v} = (1, 1, \sqrt{2})$.

21. Find the angle between the diagonal of a cube and any one of its edges.

22. The curves $\boldsymbol{x}(t) = (1 + t, t^2, t^3)$ and $\boldsymbol{y}(s) = (\sin s, \cos s, s - \pi/2)$ intersect at the point $(1, 0, 0)$. Find the angle between the curves at this point (i.e., find the angle between their tangent vectors).

23. Suppose that P, Q, and R are points in space. Explain how the dot product can be used to determine whether these points are collinear. [HINT: Consider the vectors $\overrightarrow{PQ}$ and $\overrightarrow{PR}$.]

24. Is the angle between the vectors $(1, -2, 3)$ and $(-3, 2, 1)$ less than $\pi/2$? Justify your answer.

25. Find the distance from the point P to the line ℓ in each case following.

    (a) from $P(2, 3)$ to the line ℓ through $(0, 0)$ with direction vector $\boldsymbol{i}$

    (b) from $P(2, 3)$ to the line ℓ through $(1, 1)$ with direction vector $(-1, 1)$

    (c) from $P(2, 3)$ to the line ℓ through $(1, 1)$ and $(0, 2)$

    (d) from the origin to the line with Cartesian equation $y = mx + b$; assume that $b \neq 0$. (Could you do this without vectors?)

26. This exercise is about applying the Fact near the end of this section to the vector-valued function $\boldsymbol{f}(t) = (\cos t, \sin t)$.

    (a) Find $\boldsymbol{f}(t)$ and $\boldsymbol{f}'(t)$ for $t = 0$, $t = \pi$, $t = \pm\pi/2$, and $t = \pi/4$.

    (b) Plot all the vectors you found in the previous part, along with the unit circle. (Plot the vectors $\boldsymbol{f}(t)$ with tails at the origin; plot each vector $\boldsymbol{f}'(t)$ with its tail at the appropriate point on the unit circle.) Is your picture consistent with the Fact?

27. Redo the previous exercise, but use the vector-valued function $\boldsymbol{f}(t) = (\sin t, \cos t)$.

28. Redo the previous exercise, but use the vector-valued function $\boldsymbol{f}(t) = (t, \sin(t))$.

29. Let $\boldsymbol{v} = (1, 2, 3)$.

    (a) Find the length $|\boldsymbol{v}|$.

    (b) Find the cosine of the angle that $\boldsymbol{v}$ makes with each of the standard basis vectors $\boldsymbol{i}$, $\boldsymbol{j}$, and $\boldsymbol{k}$.

30. Repeat the preceding exercise using $\boldsymbol{v} = (1, -2, 3)$.

31. Let $\boldsymbol{v} = (1, 2, 3)$ and $\boldsymbol{w} = (2, 3, -4)$.

    (a) Find the cosine of the angle θ between $\boldsymbol{v}$ and $\boldsymbol{w}$. Is θ obtuse or acute (i.e., more or less than a right angle)? How do you know?

    (b) Find the component of $\boldsymbol{v}$ in the direction of $\boldsymbol{w}$.

    (c) Find a nonzero vector $\boldsymbol{x} = (x_1, x_2, x_3)$ that's perpendicular to *both* $\boldsymbol{v}$ and $\boldsymbol{w}$. (Hints: To be perpendicular to $\boldsymbol{v}$, $\boldsymbol{x}$ must satisfy $x_1 + 2x_2 + 3x_3 = 0$. Use this idea to set up two equations in the three unknowns x_1, x_2, and x_3. Then solve the two equations simultaneously. There are infinitely many solutions.)

32. Redo the preceding exercise, but use $\boldsymbol{v} = (1, 2, 0)$ and $\boldsymbol{w} = (2, 3, 0)$.

33. Let $\boldsymbol{v} = (v_1, v_2, v_3)$ and $\boldsymbol{w} = (w_1, w_2, w_3)$ be two vectors; define a third vector $\boldsymbol{u}$, concocted from $\boldsymbol{v}$ and $\boldsymbol{w}$, by the formula

$$\boldsymbol{u} = (v_2 w_3 - v_3 w_2, v_3 w_1 - v_1 w_3, v_1 w_2 - v_2 w_1)$$

   (a) Show that $\boldsymbol{u}$ is perpendicular to $\boldsymbol{v}$ and to $\boldsymbol{w}$.

   (b) Show that

$$|\boldsymbol{u}|^2 = |\boldsymbol{v}|^2 |\boldsymbol{w}|^2 - |v \cdot w|^2 = |\boldsymbol{v}|^2 |\boldsymbol{w}|^2 |\sin\theta|^2,$$

   where θ is the (positive) angle between $\boldsymbol{v}$ and $\boldsymbol{w}$. (Hint: It's a slightly long-winded calculation, but it works out in the end.)

Note: The vector $\boldsymbol{u}$ in this exercise is called the **cross product** of $\boldsymbol{v}$ and $\boldsymbol{w}$. We'll study the cross product in detail later in this chapter.

34. Let $\boldsymbol{v} = (v_1, v_2, v_3)$ and $\boldsymbol{w} = (w_1, w_2, w_3)$ be vectors, and a any scalar.

   (a) Show that $|a\boldsymbol{v}| = |a||\boldsymbol{v}|$.

   (b) Show that $\boldsymbol{v} \cdot \boldsymbol{v} = |\boldsymbol{v}|^2$.

   (c) Show that $|\boldsymbol{v} + \boldsymbol{w}| \leq |\boldsymbol{v}| + |\boldsymbol{w}|$. (Hint: Show first that $|\boldsymbol{v} + \boldsymbol{w}|^2 = (\boldsymbol{v} + \boldsymbol{w}) \cdot (\boldsymbol{v} + \boldsymbol{w}) \leq (|\boldsymbol{v}| + |\boldsymbol{w}|)^2$.)

   (d) Show that $\boldsymbol{u} \cdot (\boldsymbol{v} + \boldsymbol{w}) = \boldsymbol{u} \cdot \boldsymbol{v} + \boldsymbol{u} \cdot \boldsymbol{w}$.

   (e) Show that $a(\boldsymbol{v} \cdot \boldsymbol{w}) = (a\boldsymbol{v}) \cdot \boldsymbol{w}$.

35. Prove the **law of cosines**: If a triangle has sides a, b, and c, with angle θ between sides a and b, then $c^2 = a^2 + b^2 - 2ab\cos\theta$. [HINT: The sides of the triangle can be represented as vectors $\boldsymbol{a}, \boldsymbol{b}, \boldsymbol{c}$ such that $\boldsymbol{a} = \boldsymbol{b} + \boldsymbol{c}$, $|\boldsymbol{a}| = a$, $|\boldsymbol{b}| = b$, and $|\boldsymbol{c}| = c$.]

36. Let $\boldsymbol{v}$ and $\boldsymbol{w}$ be any nonzero vectors. Let $\boldsymbol{a}$ be the component of $\boldsymbol{v}$ in the direction of $\boldsymbol{w}$, and let $\boldsymbol{b} = \boldsymbol{v} - \boldsymbol{a}$. Draw a picture to show the situation. Then show using the dot product that $\boldsymbol{b}$ is perpendicular to $\boldsymbol{w}$.

37. Suppose that **u** and **v** are the vectors in the xz-plane pictured below, $|\mathbf{u}| = 3$, $|\mathbf{v}| = 4$, and $\phi = \pi/3$.

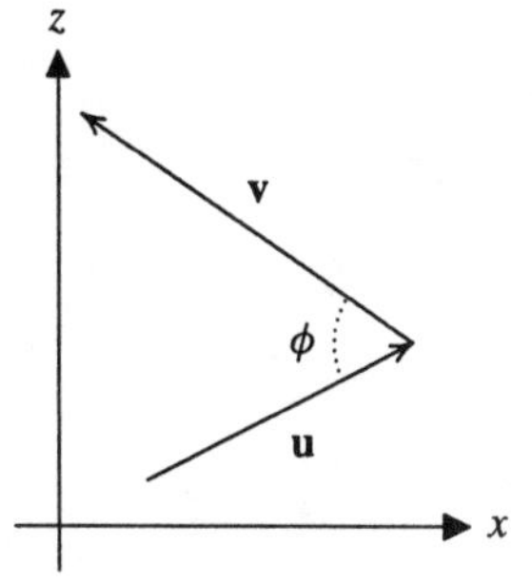

   (a) Evaluate $\mathbf{u} \bullet \mathbf{v}$.

   (b) Evaluate $|\mathbf{u} + \mathbf{v}|$.

38. Suppose that **u** and **v** are the vectors in the yz-plane pictured below, $|\mathbf{u}| = 3$, $|\mathbf{v}| = 4$, and $\phi = \pi/3$.

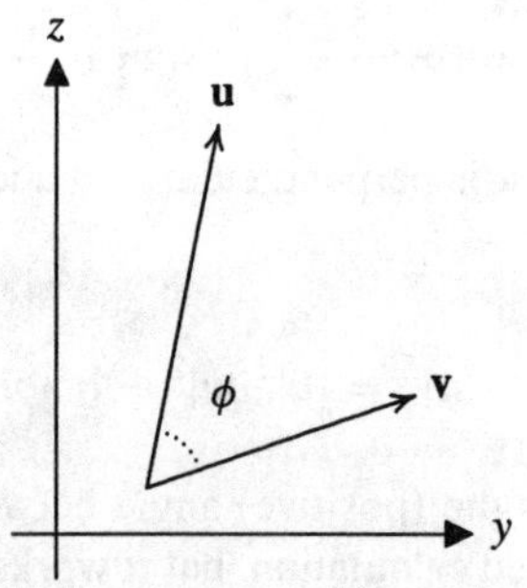

(a) Evaluate $\mathbf{u} \cdot \mathbf{v}$.
(b) Evaluate $|\mathbf{u} - \mathbf{v}|$.

39. Suppose that $\boldsymbol{x}(t) = (\cos t, \sin t, t)$ and $\mathbf{y}(t) = (2, t^2, -4t)$. Evaluate $\dfrac{d}{dt}\boldsymbol{x}(t) \cdot \mathbf{y}(t)$.

1.7 Lines and planes in three dimensions

Lines and planes have already appeared informally, in several settings. We've used derivatives to calculate tangent *lines* to curves, and we've seen that the graph of a linear equation $ax + by + cz = d$ is a *plane* in xyz-space. In this section we look more systematically at lines and planes in space. Our newly-acquired vector tools will simplify and streamline the work.

In Section 1.5, for example.

Lines

A line in *any* dimension is completely determined by (i) a point P_0 through which it passes; and (ii) a direction vector $\boldsymbol{v}$. In practice, one or both of these data may need to be ferreted out from other sorts of information. The direction vector, for instance, is sometimes deduced from knowledge of *two* points on the line, rather than being given directly.

Lines and segments in $\mathbb{R}^3$ can be described in various ways. We collect several below, with examples.

Vector equations for lines. Let ℓ be the line through $P_0 = (x_0, y_0, z_0)$ in the direction of the vector $\boldsymbol{v} = (a, b, c)$. A point $X = (x, y, z)$ lies on ℓ if and only if the vector joining X to P_0 is a scalar multiple, say $t\boldsymbol{v}$, of the direction vector. This is another way of saying that the line is the image of the vector-valued function $\boldsymbol{X}(t)$, defined for all real numbers t by

$$\boldsymbol{X}(t) = \boldsymbol{P_0} + t\boldsymbol{v},$$

where $\boldsymbol{P_0} = (x_0, y_0, z_0)$ is the position vector of the point P_0. This equation is called a **vector equation**, or a **vector parametrization**, for ℓ.

The difference between the point P_0 and the position vector $\boldsymbol{P_0}$—note the boldface type for the latter—is mainly in our point of view.

We said "a" rather than "the" in the preceding sentence because more than one vector equation can determine the same line. This should not be too surprising, in hindsight; as we've seen, *every* curve can be parametrized in many different ways.

■ **Example 1.** Explain why the two vector equations

$$\boldsymbol{X}(t) = (0, 0, 0) + t(1, 2, 3) \quad \text{and} \quad \boldsymbol{Y}(t) = (1, 2, 3) + t(-2, -4, -6)$$

describe the same line. Is there *any* difference between the two equations?

Solution: Let ℓ_1 and ℓ_2 be the lines determined by the equations above. To show that ℓ_1 and ℓ_2 are the same line, it's enough to find any two points in common. In fact, it's easy to see that $(0, 0, 0)$ lies on both ℓ_1 and ℓ_2 (set $t = 0$ in the first equation and $t = 1/2$ in the second). Similarly, $(1, 2, 3)$ lies on both lines (set $t = 1$ in the first equation and $t = 0$ in the second).

Only one line passes through any two points.

Though both vector-valued functions define the same line, they differ in other ways. One difference is in the respective velocities—

$$\boldsymbol{X}'(t) = (1, 2, 3) \quad \text{and} \quad \boldsymbol{Y}'(t) = (-2, -4, -6),$$

respectively. In effect, the second equation parametrizes the line at twice the speed of the first, and in the opposite direction. □

Parametric equations for lines. Let ℓ (again) be the line through $P_0 = (x_0, y_0, z_0)$ in the direction of the vector $\boldsymbol{v} = (a, b, c)$. Writing out the vector equation $\boldsymbol{X}(t) = \boldsymbol{P_0} + t\boldsymbol{v}$ in its three coordinates gives **parametric equations** for ℓ:

The parameter is t.

$$x = x_0 + at; \quad y = y_0 + bt; \quad z = z_0 + ct.$$

■ **Example 2.** At what point does the line ℓ through $(1, 2, 3)$ and $(3, 5, 7)$ intersect the xy-plane? Where does ℓ intersect the plane $z = x + y - 4$?

Solution: To describe ℓ we need a point and a direction vector. Given the information at hand, we may as well use $(1, 2, 3)$ as our point, and the difference vector $(3, 5, 7) - (1, 2, 3) = (2, 3, 4)$ as our direction vector. Using these data we can give the following parametric equations for ℓ:

$$x = 1 + 2t; \quad y = 2 + 3t; \quad z = 3 + 4t.$$

The intersection of ℓ with the xy-plane occurs where $z = 3 + 4t = 0$, i.e., at $t = -3/4$. At this value of t, the parametric equations give

$$x = 1 - 2 \cdot \frac{3}{4} = -\frac{1}{2}, \quad \text{and} \quad y = 2 - 3 \cdot \frac{3}{4} = -\frac{1}{4}.$$

Thus ℓ intersects the xy-plane at $(-1/2, -1/4, 0)$.

The intersection of ℓ with the plane $z = x + y - 4$ is found similarly. Substituting $x = 1 + 2t$, $y = 2 + 3t$, and $z = 3 + 4t$ in the plane equation gives a new equation in just one variable, t; its solution is $t = 4$. Thus the intersection occurs at $(9, 14, 19)$. □

Check these details for yourself.

Symmetric scalar equations. Given the parametric form

$$x = x_0 + at; \quad y = y_0 + bt; \quad z = z_0 + ct$$

of a line ℓ through (x_0, y_0, z_0) with direction vector (a, b, c), we can solve all three equations for t, to get

$$t = \frac{x - x_0}{a} = \frac{y - y_0}{b} = \frac{z - z_0}{c}.$$

The value of t is now immaterial, so we drop t. (This is called "eliminating the variable" t.) The remaining equations—

$$\frac{x - x_0}{a} = \frac{y - y_0}{b} = \frac{z - z_0}{c}.$$

are called **symmetric scalar equations** for ℓ.

The symmetric scalar form exhibits some minor behavioral quirks:

No t. There's no parameter t in the form above. In other words, ℓ is defined as the solution set of two *equations* in x, y, and z, rather than as the image of a function of t.

How many equations? The symmetric scalar form consists of *two* (linear) equations in *three* unknowns. To students with some linear algebra background, this should make some sense: *One* linear equation determines a plane in $\mathbb{R}^3$, so *two* linear equations determine the intersection of two planes—i.e., a line in $\mathbb{R}^3$.

Vanishing denominators. If any of a, b, and c happens to be zero, then the symmetric form needs a little help, but no real harm is done. If, say, $a = 0$, then the parametric equations become

$$x = x_0; \quad y = y_0 + bt; \quad z = z_0 + ct,$$

eliminating t from the last two equations gives

$$x = x_0; \quad \frac{y - y_0}{b} = \frac{z - z_0}{c}.$$

As before, the result is two linear equations in x, y, and z.

Not unique. As with the vector and parametric forms of a line, and for exactly the same reasons, different sets of symmetric scalar equations can be given for the same line. In math-speak, one says that the symmetric scalar equations are "not unique."

■ **Example 3.** Find and compare symmetric scalar equations for the following lines:

ℓ_1: through $(0, 0, 0)$ in the direction of $(1, 2, 3)$

ℓ_2: with vector equation $\boldsymbol{X}(t) = (1, 2, 3) + t(-2, -4, -6)$

Solution: Line ℓ_1 perfectly fits the pattern above; its symmetric scalar equations are

$$\frac{x - 0}{1} = \frac{y - 0}{2} = \frac{z - 0}{3}.$$

Line ℓ_2 passes through $(1, 2, 3)$, with direction vector $(-2, -4, -6)$, so its symmetric scalar equations are

$$\frac{x - 1}{-2} = \frac{y - 2}{-4} = \frac{z - 3}{-6}.$$

Thus ℓ_1 and ℓ_2 have different-looking equations. Nevertheless (as we saw in Example 1, page 75), ℓ_1 and ℓ_2 are actually the same line. □

Do two lines meet? Not likely. Almost every pair of lines in the xy-plane intersects somewhere. Unless they happen to be parallel or equal, two lines in the plane have exactly *one* point of intersection. The situation for lines in space is quite different; there's plenty of "room" in xyz-space for lines to miss each other. Lines that miss each other are called **skew lines**. Basic equation-counting says the same thing: One line in xyz-space corresponds to *two* equations in three variables, so two lines determine *four* equations in three variables. When equations outnumber variables, solutions are unlikely. But not impossible—the next example illustrates both alternatives.

■ **Example 4.** Three lines are given in vector form as follows:➡

Why use three variables r, s, and t? Read on.

ℓ_1: $\boldsymbol{X}(r) = (1, 1, 1) + r(1, 2, 3)$;

ℓ_2: $\boldsymbol{Y}(s) = (-3, 2, 4) + s(2, 1, 1)$;

ℓ_3: $\boldsymbol{Z}(t) = (0, 0, 0) + t(1, 3, 1)$.

Do any of the lines intersect? If so, where?

Solution: Notice first the *three* variable names; we used r, s, and t to avoid assuming, unnecessarily, that the lines meet at equal values of the parameter.

The lines ℓ_1 and ℓ_2 meet if $\boldsymbol{X}(r) = \boldsymbol{X}(s)$ for some values r and s. Writing this condition in vector form and simplifying slightly gives

$$(1,1,1)+r(1,2,3) = (-3,2,4)+s(2,1,1) \iff (4,-1,-3) = -r(1,2,3)+s(2,1,1).$$

The last vector equation comes apart to give three scalar equations in two unknowns:

$$4 = -r + 2s; \quad -1 = -2r + s; \quad -3 = -3r + s.$$

Just do it.

Routine algebra shows that $r = 2$ and $s = 3$ is a solution. This means that ℓ_1 and ℓ_2 meet at the common point $(1,1,1)+2(1,2,3) = (-3,2,4)+3(2,1,1) = (3,5,7)$. (But they meet at different parameter values!)

The same method applied to ℓ_1 and ℓ_3 leads to the vector equations

$$(1,1,1)+r(1,2,3) = (0,0,0)+t(1,3,1) \iff (1,1,1) = -r(1,2,3)+t(1,3,1),$$

and thus to another set of three scalar equations in two unknowns:

$$1 = -r + t; \quad 1 = -2r + 3t; \quad 1 = -3r + t.$$

Just do this, too.

This time, more routine algebra reveals that *no* solutions exist. □

Planes

Straight lines in the xy-plane are the simplest "curves." They're useful, therefore, for describing and approximating more complicated curves. We take this point of view often in elementary calculus, when we approximate a smooth curve by its tangent line at a point. Indeed, it's an important fact of single-variable calculus that *every* smooth curve can be closely approximated near every point by a straight line.

This sentence can be made more precise, e.g., by explaining exactly what's meant by "smooth," "closely," and "near." Making such figurative language concrete and precise is an important goal of more advanced courses.

Planes in xyz-space—tangent planes, in particular—play similar roles in multivariable calculus. Planes, being "flat," are the simplest surfaces in $\mathbb{R}^3$. They're useful, therefore, for modeling and approximating more complicated surfaces and functions. First, however, we need simple, convenient ways to represent planes, using such familiar ingredients as algebraic formulas and parametric equations. Planes can be described in various ways; the remainder of this section describes some of them.

Planes described by points and normal vectors. One way to determine a plane Π in $\mathbb{R}^3$ is to specify (i) a point P_0 through which Π passes; and (ii) a **normal vector** $\boldsymbol{v}$, which is perpendicular to Π. The following picture shows how the ingredients

"Normal" is a rough synonym for "perpendicular."

fit together:

Determining a plane by a point and a normal vector

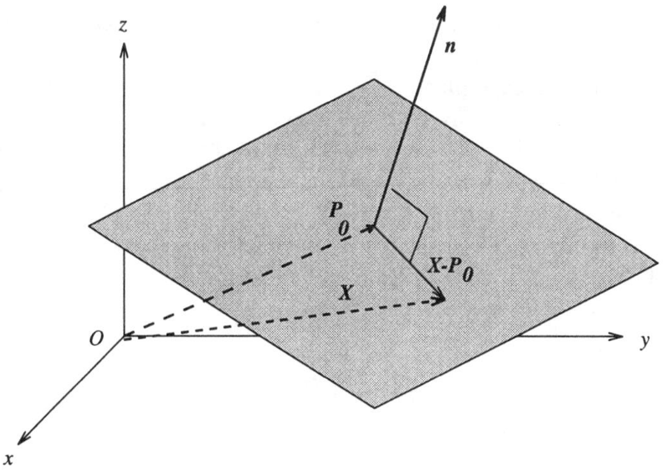

As the picture suggests, the plane consists of all points for which the vector $\boldsymbol{X} = (x, y, z)$ has the property that $\boldsymbol{X} - \boldsymbol{P_0}$ is perpendicular to $\boldsymbol{n}$. The plane's defining equation, therefore, involves the dot product, which detects perpendicularity. Here are the formalities:

> **Definition:** Suppose that the plane Π passes through $P_0 = (x_0, y_0, z_0)$, and has normal vector $\boldsymbol{n} = (a, b, c)$. Then
>
> $$(\boldsymbol{X} - \boldsymbol{P_0}) \cdot \boldsymbol{n} = 0.$$
>
> is a **vector equation** for Π. Rewriting the equation in scalar form gives
>
> $$a(x - x_0) + b(y - y_0) + c(z - z_0) = 0,$$
>
> a **scalar equation** for Π.

Observe:

One linear equation. The scalar equation for Π can be rewritten in the form $ax + by + cz = d$ (where d is the constant $ax_0 + by_0 + cz_0$). In Section 1.1 we called such equations **linear equations**, and observed that every linear equation in x, y, and z determines a plane in xyz-space.

Not unique. These equations are *not* uniquely determined by a plane Π. Any point P_0 on Π does as well as any other, and we can multiply the normal vector $\boldsymbol{n}$ by any nonzero scalar without changing its direction. It's easy to see, for instance, that the equations $x + y + z = 1$ and $2x + 2y + 2z = 2$ define the same plane.➡

It's also easy to see how to change one equation into the other.

Reading the equation. The discussion above includes a very useful property of a linear equation $ax + by + cz = d$:

The coefficients of x, y, and z are the components of the normal vector (a, b, c).

We'll use this remark in the next example.

■ **Example 5.** Planes Π_1 and Π_2 are defined by the linear equations $x+2y+3z=6$ and $y=3$. For each plane, find a normal vector and a point on the plane; use them to write a vector equation for each plane. Describe Π_1 and Π_2 geometrically.

Solution: The italicized remark above lets us simply read off suitable normal vectors: $\boldsymbol{n} = (1,2,3)$ is normal to Π_1 and $\boldsymbol{n} = (0,1,0) = \boldsymbol{j}$ is normal to Π_2. Finding a point on each plane is also quite easy—one linear equation in three unknowns has infinitely many solutions, and any one will do as well as another. For $x+2y+3z=6$, $(1,1,1)$ is one convenient solution; for the equation $y=3$, the solution $(0,3,0)$ works fine. Thus Π_1 is the plane through $(1,1,1)$, perpendicular to $(1,2,3)$; a vector equation for Π_1 is $\big(\boldsymbol{X} - (1,1,1)\big) \cdot (1,2,3) = 0$. Similarly, Π_2 passes through $(0,3,0)$ and is perpendicular to $\boldsymbol{j}$, so a vector equation for Π_2 is $\big(\boldsymbol{X} - (0,3,0)\big) \cdot (0,0,1) = 0$. Here's a first-quadrant picture of Π_1:

There are many others.

The plane x+2y+3z=6

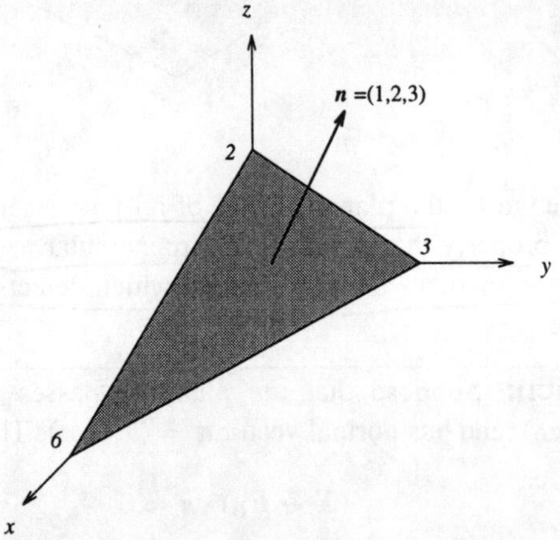

The plane Π_2 should be easy to visualize, so we leave it to the reader's imagination.□

Parametrizing planes and parts of planes. We've seen how to describe *lines* in space both by equations in x, y, and z and in parametric form. One big advantage of the parametric form is in describing only *part* of a line, such as a ray or a line segment. (We describe such subsets by restricting the parameter interval.)

Planes (and parts of planes) can also be described parametrically—but we'll need *two* parameters, not one. The following picture illustrates the idea:

Parametrizing a plane patch

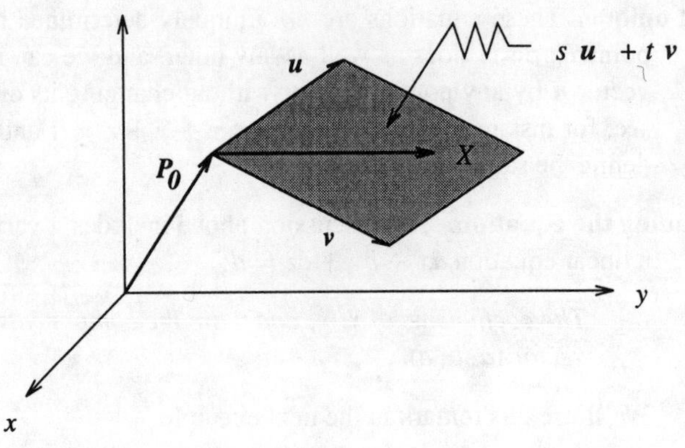

In the picture, a piece (or "patch") of a plane (the shaded region) is determined by a point P_0 (indicated by the position vector $\boldsymbol{P_0}$) and two **spanning vectors** $\boldsymbol{u}$ and $\boldsymbol{v}$. A point on the plane has a position vector of the form

$$\boldsymbol{X}(s,t) = \boldsymbol{P_0} + s\boldsymbol{u} + t\boldsymbol{v},$$

where s and t are real numbers. (One such point X is shown in the picture.) In other words, points in the plane differ from $\boldsymbol{P_0}$ by a **linear combination** of $\boldsymbol{u}$ and $\boldsymbol{v}$. As s and t range through all real numbers, the vector $\boldsymbol{P_0} + s\boldsymbol{u} + t\boldsymbol{v}$ traces out the entire plane spanned by $\boldsymbol{u}$ and $\boldsymbol{v}$. Restricting the values of s and t, on the other hand, gives various *subsets* of the plane. The shaded region shown, for instance, corresponds to restricting s and t to the intervals $0 \le s \le 1$ and $0 \le t \le 1$.

What difference would it make if we took $0 \le s \le 2$ and $0 \le t \le 2$?

The equation

$$\boldsymbol{X}(s,t) = \boldsymbol{P_0} + s\boldsymbol{u} + t\boldsymbol{v},$$

is called a **vector parametric equation** for a plane; looking at the components separately gives the corresponding **scalar parametric equations**:

$$x = x_0 + su_1 + tv_1; \quad y = y_0 + su_2 + tv_2; \quad z = z_0 + su_3 + tv_3.$$

■ **Example 6.** Write parametric equations for the plane Π_1 shown in Example 5, page 80.

Solution: We'll choose a point and two spanning vectors on Π_1. (There are many ways to do this—our choices aren't sacred.) As the picture shows, the plane passes through the points $(6,0,0)$, $(0,3,0)$, and $(0,0,2)$. We'll use $(6,0,0)$ as our fixed point, and the vectors joining $(6,0,0)$ to the other two known points as our spanning vectors. These spanning vectors are, respectively,

$$\boldsymbol{u} = (-6,3,0) \quad \text{and} \quad \boldsymbol{v} = (-6,0,2).$$

Thus our plane can be written in vector parametric form as

$$\boldsymbol{X}(s,t) = (6,0,0) + s(-6,3,0) + t(-6,0,2).$$

In scalar form, we get

$$x = 6 - 6s - 6t; \quad y = 3s; \quad z = 2t.$$

By restricting the values of s and t we can parametrize as much or as little of the plane as we wish. □

Planes vs. lines. Notice that the function $\boldsymbol{X}(s,t)$ above is a vector-valued function of *two* variables s and t. Notice the close resemblance between the vector parametrizations of lines and planes:

$$\text{line: } \boldsymbol{X}(t) = \boldsymbol{P_0} + t\boldsymbol{v}; \quad \text{plane: } \boldsymbol{X}(s,t) = \boldsymbol{P_0} + s\boldsymbol{u} + t\boldsymbol{v}.$$

The main difference is in the number of parameters. To put it another way, it takes *one* vector to span a line, but *two* vectors to span a plane.

Exercises

1. Find a vector parametric equation for each line below. If possible, plot your line over a convenient parameter interval.

   (a) The y-axis.

   (b) The line through $(1, 2, 3)$ and $(2, 3, 4)$.

   (c) The line with symmetric scalar equations

   $$\frac{x-1}{2} = \frac{y-2}{3} = \frac{z-3}{4}.$$

   (d) The tangent line to the curve $x = \cos t$, $y = \sin t$, $z = t$ at the point $(1, 0, 0)$ (i.e., where $t = 0$).

2. In each part below, decide whether the given lines intersect. If so, find the point of intersection; if not, explain why not.

   (a) The lines with parametric equations $\boldsymbol{X}(t) = (1, 2, 3) + t(1, 1, 1)$ and $\boldsymbol{Y}(s) = (5, 3, 5) + s(-3, 6, 3)$.

   (b) The x-axis and the line with symmetric scalar equations

   $$\frac{x-1}{2} = \frac{y-2}{3} = \frac{z-3}{4}.$$

3. *(For students with linear algebra.)* We said in this section:

   > *Almost every pair of lines in the xy-plane intersects somewhere. Unless the lines happen to be parallel or equal, they have exactly one point of intersection.*

   Let two lines be given by linear equations $ax + by = c$ and $dx + ey = f$; all of the letters a through f are constants. Recall that the **coefficient matrix** has rows (a, b) and (d, e), and the **augmented matrix** has rows (a, b, c) and (d, e, f).

   (a) Under what conditions on a through f (or on an appropriate matrix) does the pair of equations have *one* solution? Give a specific example.

   (b) Under what conditions does the pair of equations have *no* solutions? Give a specific example.

   (c) Under what conditions does the pair of equations have *infinitely many* solutions? Give a specific example.

4. Do the lines ℓ_2 and ℓ_3 of Example 4 (page 77) have an intersection point? If so, where? If not, why not?

5. In each part below, find vector and scalar equations for the given plane.

   (a) Through $(1, 2, 3)$, perpendicular to $\boldsymbol{n} = (3, 4, 5)$.

   (b) The xz-plane. (Hint: Find a suitable normal vector first.)

   (c) Through $(1, 2, 3)$, with normal vector parallel to the line through $(0, 1, 2)$ and $(3, 3, 3)$.

(d) Through $(0, 0, 0)$, perpendicular to the line with symmetric scalar equations

$$\frac{x-1}{2} = \frac{y-2}{3} = \frac{z-3}{4}.$$

(e) The graph of the function $L(x, y) = 2x + 3y + 5$.

6. Does the line with vector equation $\boldsymbol{X} = (1, 1, 2) + t(2, 3, 4)$ intersect the plane $x + 2y - 2z = 0$? If so, where? If not, why not?

7. Find the line of intersection of the two planes $x+2y+3z = 6$ and $x+y+z = 3$. Write the answer in parametric form. (Hint: First find two points that are on both planes.)

8. In Example 6, (page 81), we found the scalar parametric equations

$$x = 6 - 6s - 6t; \quad y = 3s; \quad z = 2t.$$

for a certain plane. Eliminate the variable s and t in the three equations above to get one equation in x, y, and z. Does the equation look familiar?

9. The plane $x + 2y + 3z = 6$ (shown in Example 5) passes through $(1, 1, 1)$. Using this point as $\boldsymbol{P}_0$, write vector and scalar parametric equations for the plane. (Use the method of Example 6.)

10. Write parametric equations for each plane or plane piece described below.

(a) The xy-plane.

(b) Through $(1, 2, 3)$, spanned by $\boldsymbol{i}$ and $\boldsymbol{j}$.

(c) Through $(1, 2, 3)$, spanned by $\boldsymbol{i}$ and $\boldsymbol{i} + \boldsymbol{j}$.

(d) The first quadrant of the xy-plane.

(e) The rectangle $0 \leq x \leq 1$, $0 \leq y \leq 2$ in the xy-plane.

11. Let $\boldsymbol{u}$ be a vector and $\boldsymbol{v}$ be a unit vector.

(a) Show that the vector $\boldsymbol{w} = \boldsymbol{u} - (\boldsymbol{u} \cdot \boldsymbol{v})\boldsymbol{v}$ is perpendicular to $\boldsymbol{v}$.

(b) Let $\boldsymbol{x} = (1, 2, 3)$. Use part (a) to write the vector $(4, 5, 6)$ as the sum of a vector parallel to $\boldsymbol{x}$ and a vector perpendicular to $\boldsymbol{x}$.

12. Does the line $(2, -3, 1) + t(1, 2, -3)$ intersect the line $x = 1 - 2s$, $y = 2 + 3s$, $z = 4 + 6s$? Justify your answer.

13. Find an equation for the line through $(1, 1, 3)$ that is parallel to the line $x = 2 - t$, $y = -t$, $z = 3 + 3t$.

14. Are the lines $(0, -2, 0) + t(1, -2, -3)$ and $(0, 1, 0) + s(-1, -2, 1)$ perpendicular? Justify your answer.

15. Find the scalar equation of the plane through $(1, -2, 5)$ perpendicular to $\boldsymbol{n} = (3, -4, 1)$.

16. The line through the origin perpendicular to a plane intersects the plane at the point $(2, -1, 1)$. Find an equation of the plane.

17. A plane through $(2, -2, 5)$ is perpendicular to the line through $(1, 1, 1)$ and $(-2, 3, 1)$. Find a scalar equation of the plane.

18. Find an equation of the plane through $(1, 2, 3)$ that is parallel to the plane described by the equation $-x + 3y - 4z = 2$.

19. Let ℓ be the line $(2, 1, 1) + t(-1, 3, 2)$ and Π be the plane $x - 3y - 2z = 11$.

 (a) Find the point where ℓ and Π intersect.

 (b) Is ℓ perpendicular to Π? Justify your answer.

20. Does the line $(3, 1, -2) + t(1, -1, 3)$ intersect the plane $2x - y - z = 5$? Justify your answer.

21. Prove that the line $\boldsymbol{x}_0 + t\boldsymbol{v}$ is parallel to the plane $\boldsymbol{n} \cdot (\boldsymbol{x} - \boldsymbol{x}_1) = 0$ if and only if $\boldsymbol{n} \cdot \boldsymbol{v} = 0$.

22. Let $\boldsymbol{P}_0$ and $\boldsymbol{P}_1$ be points in space, $\boldsymbol{n}$ a unit vector, and Π be the plane $\boldsymbol{n} \cdot (\boldsymbol{X} - \boldsymbol{P}_0)$. Show that the (perpendicular) distance between $\boldsymbol{P}_1$ and Π is $d = |\boldsymbol{n} \cdot \boldsymbol{P}_0 - \boldsymbol{n} \cdot \boldsymbol{P}_1|$.

23. Find an equation for the plane containing the lines $(2, 0, -1) + s(3, -3, -1)$ and $(3, -1, 1) + t(1, -1, 2)$.

24. Find an equation of the plane that contains the line $(2, 3, 0) + t(-1, 2, 4)$ and is perpendicular to the plane $x + 2y - z = 3$.

25. Show that the distance D from the point $P_0 = (x_0, y_0, z_0)$ to the plane $ax + by + cz = d$ is

$$D = \frac{|ax_0 + by_0 + cz_0 - d|}{\sqrt{a^2 + b^2 + c^2}}.$$

26. Let Π_1 denote the plane $ax + by + cz = d_1$ and let Π_2 denote the plane $ax + by + cz = d_2$.

 (a) Explain why the planes Π_1 and Π_2 are parallel.

 (b) Find the distance between the planes Π_1 and Π_2.

27. Show that any two lines in space that do not intersect must lie in parallel planes.

28. Find an equation of the line through the point $(3, 2, 1)$ that is perpendicular to the plane $x + y = 0$.

29. Let ℓ_1 be the line $\mathbf{x} = (1, 1, 2) + t(3, -1, 4)$ and ℓ_2 be the line $\dfrac{x-1}{6} = \dfrac{y}{-2} = \dfrac{z-3}{8}$.

 (a) Are the lines ℓ_1 and ℓ_2 parallel? Explain.

 (b) The point $P(1, 1, 2)$ is on ℓ_1 and the point $Q(1, 0, 3)$ is on ℓ_2. Find $|\overrightarrow{PQ}|$.

 (c) Find the scalar projection $\overrightarrow{PQ}$ in the direction of $\mathbf{v} = (3, -1, 4)$.

 (d) Find the distance between ℓ_1 and ℓ_2.

30. Find the distance between the point $(8, 7, 9)$ and the line $\dfrac{x-1}{6} = \dfrac{y-2}{5} = \dfrac{z+3}{4}$.

31. Let ℓ be the line $(2, 1, 0) + t(1, 1, 1)$ and Π be the plane $x - 3y + 2z = 4$. Do ℓ and Π intersect? Justify your answer.

32. Find an equation of the plane that contains the points $(1, 2, 3)$, $(1, 1, 1)$, and $(1, 0, 2)$.

33. Let Π_1 and Π_2 be planes in $\mathbb{R}^3$ that both contain the origin. Furthermore, suppose that the vector $\boldsymbol{n}_1 = (1, 2, 1)$ is orthogonal to the plane Π_1 and that the vector $\boldsymbol{n}_2 = (-3, -5, 0)$ is orthogonal to the plane Π_2. Find an equation for the line of intersection of Π_1 and Π_2.

34. The position of a particle in $\mathbb{R}^3$ at time t is $\mathbf{x}(t) = (t^2, 3t, t^3 - 12t + 5)$.

   (a) Find the velocity of the particle at time t.

   (b) Find all times at which the particle is traveling parallel to the xy-plane.

   (c) Find all times at which the particle's velocity vector is perpendicular to the plane $2x + 6y + 5z = 7$. If there are no such times, carefully explain how you know this.

1.8 The cross product

This section introduces a new operation, the **cross product**, on vectors in $\mathbb{R}^3$. If $\boldsymbol{v}$ and $\boldsymbol{w}$ are any vectors in space, then their cross product, denoted by $\boldsymbol{v} \times \boldsymbol{w}$, is another vector, whose definition and properties we're about to discuss.

$\boldsymbol{v} \times \boldsymbol{w}$ is a vector; $\boldsymbol{v} \cdot \boldsymbol{w}$ is a scalar.

Before getting into details, let's acknowledge some peculiarities of the cross product:

- $\boldsymbol{v} \times \boldsymbol{w}$ is defined *only* for vectors in $\mathbb{R}^3$ (by contrast, $\boldsymbol{v} \cdot \boldsymbol{w}$ is defined for vectors in any dimension);
- the cross product is *not commutative*—on the contrary, $\boldsymbol{v} \times \boldsymbol{w} = -(\boldsymbol{w} \times \boldsymbol{v})$;
- the recipe for finding $\boldsymbol{v} \times \boldsymbol{w}$ from the components of $\boldsymbol{v}$ and $\boldsymbol{w}$ is slightly complicated.

This is not a fatal objection—many good recipes are complicated.

Despite these apparent drawbacks, the cross product is a surprisingly useful basic tool—useful enough to be "known" to mathematical software ranging from *Maple* and *Mathematica* to modest graphing calculators.

The idea and the definition

Let $\boldsymbol{v}$ and $\boldsymbol{w}$ be any two vectors. Is there a third vector that's perpendicular to *both* $\boldsymbol{v}$ and $\boldsymbol{w}$? If so, how can we find one?

For vectors in the plane, there *is* no answer: it's impossible for a nonzero *plane* vector to be perpendicular to two different directions. In $\mathbb{R}^3$, by contrast, there's more "room." Any two nonparallel vectors $\boldsymbol{v}$ and $\boldsymbol{w}$ determine a plane Π in $\mathbb{R}^3$; any vector that's perpendicular to Π is perpendicular to both $\boldsymbol{v}$ and $\boldsymbol{w}$. In fact, there are infinitely many such space vectors, because if $\boldsymbol{n}$ is perpendicular to both $\boldsymbol{v}$ and $\boldsymbol{w}$, then so is $k\boldsymbol{n}$, for any scalar k.

Unless $\boldsymbol{v}$ and $\boldsymbol{w}$ happen to be parallel.

In linear algebra terminology, the set of vectors perpendicular to both $\boldsymbol{v}$ and $\boldsymbol{w}$ is a subspace.

The cross product chooses—from among the infinitely many possibilities—one particular vector $\boldsymbol{v} \times \boldsymbol{w}$ that's perpendicular to both $\boldsymbol{v}$ and $\boldsymbol{w}$. As we'll see, the choice is made in such a way that both the length and the direction of $\boldsymbol{v} \times \boldsymbol{w}$ give valuable information.

> **Definition: (Cross product)** Let $\boldsymbol{v} = (v_1, v_2, v_3)$ and $\boldsymbol{w} = (w_1, w_2, w_3)$ be vectors. Their **cross product** is the vector
>
> $$\boldsymbol{v} \times \boldsymbol{w} = (v_2w_3 - v_3w_2,\ v_3w_1 - v_1w_3,\ v_1w_2 - v_2w_1).$$

This somewhat unlikely-looking contraption has many interesting algebraic and geometric properties. Here are several:

Perpendicularity It's easy to check, by direct calculation, that $\boldsymbol{v} \times \boldsymbol{w}$ is indeed perpendicular to both $\boldsymbol{v}$ and $\boldsymbol{w}$. For instance:

$$\boldsymbol{v} \cdot (\boldsymbol{v} \times \boldsymbol{w}) = v_1(v_2w_3 - v_3w_2) + v_2(v_3w_1 - v_1w_3) + v_3(v_1w_2 - v_2w_1) = 0,$$

because all the terms cancel in pairs.

Check this for yourself.

Anti-commutativity Unlike the dot product, which is commutative, the cross product is *anti-commutative*:

$\boldsymbol{v} \cdot \boldsymbol{w} = \boldsymbol{w} \cdot \boldsymbol{v}$

$$\boldsymbol{v} \times \boldsymbol{w} = -(\boldsymbol{w} \times \boldsymbol{v}).$$

A close look at the definition explains why. Thanks to the minus signs in the definition, reversing the roles of $\boldsymbol{v}$ and $\boldsymbol{w}$ changes the sign of each component, and therefore reverses the product vector. Notice, in particular, what this means about the cross product of a vector with *itself*:

$$\boldsymbol{v} \times \boldsymbol{v} = -(\boldsymbol{v} \times \boldsymbol{v}) = (0, 0, 0).$$

(The last equation holds because only the zero vector is its own opposite.)

Standard basis vectors. Taking cross products of the standard basis vectors gives especially simple results:

$$\boldsymbol{i} \times \boldsymbol{j} = \boldsymbol{k}; \quad \boldsymbol{j} \times \boldsymbol{k} = \boldsymbol{i}; \quad \boldsymbol{k} \times \boldsymbol{i} = \boldsymbol{j}.$$

As observed in the previous paragraph, reversing the order of factors introduces a minus sign:

$$\boldsymbol{j} \times \boldsymbol{i} = -\boldsymbol{k}; \quad \boldsymbol{k} \times \boldsymbol{j} = -\boldsymbol{i}; \quad \boldsymbol{i} \times \boldsymbol{k} = -\boldsymbol{j}.$$

One way to remember these rules is to think of the basis vectors as a repeating cycle: $\boldsymbol{i}\,\boldsymbol{j}\,\boldsymbol{k}\,\boldsymbol{i}\,\boldsymbol{j}\,\boldsymbol{k}\ldots.$ Then, reading from left to right, the cross product of any two adjacent vectors is the next one.

Remembering the formula. There are various useful devices for remembering the cross product formula. One of them is to think of the cross product as the 3 × 3 **determinant**➸

Matrices and determinants are reviewed in an Appendix.

$$\begin{vmatrix} \boldsymbol{i} & \boldsymbol{j} & \boldsymbol{k} \\ v_1 & v_2 & v_3 \\ w_1 & w_2 & w_3 \end{vmatrix},$$

expanded along the first row. Another memory device begins with this "double" array:

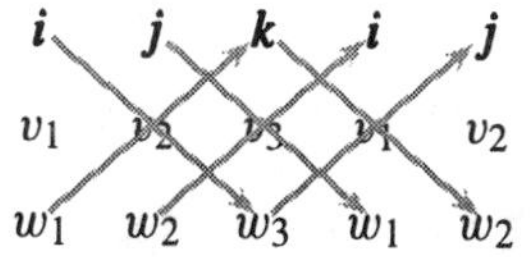

Then the cross product is the sum of six *threefold diagonal products*, with southeast diagonals counted positive and northeast diagonals negative.➸

See for yourself that this device gives the cross product.

For still another method of remembering the cross product formula, see the exercises.

Algebra with the cross product

The cross product enjoys many of the algebraic properties one hopes for from any well-mannered mathematical operation.➸ The next theorem collects several such properties.

But not all—the cross product isn't commutative.

> **Theorem 7. (Algebra with the cross product.)** Let $\boldsymbol{u}$, $\boldsymbol{v}$, and $\boldsymbol{w}$ be vectors in $\mathbb{R}^3$, and a a scalar. Then
>
> - (Scalars factor out) $a\boldsymbol{v} \times \boldsymbol{w} = \boldsymbol{v} \times a\boldsymbol{w} = a(\boldsymbol{v} \times \boldsymbol{w})$
> - (Anti-commutativity) $\boldsymbol{v} \times \boldsymbol{w} = -(\boldsymbol{w} \times \boldsymbol{v})$
> - (Distributivity) $(\boldsymbol{u} + \boldsymbol{v}) \times \boldsymbol{w} = \boldsymbol{u} \times \boldsymbol{w} + \boldsymbol{v} \times \boldsymbol{w}$

All parts of the theorem can be proved by straightforward (but slightly tedious) algebraic calculations with the vectors' components, using the cross product definition. We'll give one example and leave others to the exercises.

■ **Example 1.** Prove the first part of Theorem 7.

Solution: Let $\boldsymbol{v} = (v_1, v_2, v_3)$ and $\boldsymbol{w} = (w_1, w_2, w_3)$. Then

$$\begin{aligned} a\boldsymbol{v} \times \boldsymbol{w} &= (av_1, av_2, av_3) \times (w_1, w_2, w_3) \\ &= (av_2w_3 - av_3w_2, av_3w_1 - av_1w_3, av_1w_2 - av_2w_1) \\ &= a(v_2w_3 - v_3w_2, v_3w_1 - v_1w_3, v_1w_2 - v_2w_1) = a(\boldsymbol{v} \times \boldsymbol{w}). \end{aligned}$$

(The proof that $\boldsymbol{v} \times a\boldsymbol{w} = a(\boldsymbol{v} \times \boldsymbol{w})$ is almost identical.) □

Parallel vectors. The dot product detects perpendicularity: $\boldsymbol{v}$ and $\boldsymbol{w}$ are perpendicular if $\boldsymbol{v} \cdot \boldsymbol{w} = 0$. In a similar way, the cross product detects *parallelism*:

Vectors $\boldsymbol{v}$ and $\boldsymbol{w}$ are scalar multiples if and only if $\boldsymbol{v} \times \boldsymbol{w} = (0, 0, 0)$.

To see why, suppose that $\boldsymbol{w} = a\boldsymbol{v}$, for some scalar a. Then

$$\boldsymbol{v} \times \boldsymbol{w} = \boldsymbol{v} \times a\boldsymbol{v} = a(\boldsymbol{v} \times \boldsymbol{v}) = (0, 0, 0).$$

The proof that $\boldsymbol{v} \times \boldsymbol{w} = (0, 0, 0)$ implies that $\boldsymbol{v}$ and $\boldsymbol{w}$ are parallel is left as an exercise.

The cross product is often just what's needed to find a normal vector to a plane.

■ **Example 2.** Find scalar and vector equations for the plane Π through the three points $A = (1, 1, 1)$, $B = (1, 2, 3)$, and $C = (4, 5, 6)$.

There's nothing sacred about this choice.

Solution: We'll use A as our fixed point. The vectors

$$\overrightarrow{AB} = (1, 2, 3) - (1, 1, 1) = (0, 1, 2) \quad \text{and} \quad \overrightarrow{AC} = (4, 5, 6) - (1, 1, 1) = (3, 4, 5)$$

lie in Π; computing their cross product gives $\overrightarrow{AB} \times \overrightarrow{AC} = (-3, 6, -3)$—a suitable normal vector. Thus Π has the vector equation

$$\big(\boldsymbol{X} - (1, 1, 1)\big) \cdot (-3, 6, -3) = 0.$$

The scalar version of this equation is $-3x+6y-3z = 0$, or, equivalently, $x-2y+z = 0$. □

Geometry of the cross product

Compare with the analogues for the dot product.

Like the dot product, the cross product has important geometric interpretations:

> **Fact:** Let $\boldsymbol{v}$ and $\boldsymbol{w}$ be vectors in $\mathbb{R}^3$, and let θ be the angle between them. Then $\boldsymbol{v} \times \boldsymbol{w}$ has these geometric properties:
>
> **Length:** $|\,\boldsymbol{v} \times \boldsymbol{w}\,| = |\boldsymbol{v}||\boldsymbol{w}| \sin\theta$
>
> **Direction:** If $\boldsymbol{v}$ and $\boldsymbol{w}$ are not parallel, then $\boldsymbol{v} \times \boldsymbol{w}$ is perpendicular to both $\boldsymbol{v}$ and $\boldsymbol{w}$, and points in the direction determined by the **right-hand rule**: If the fingers of the right hand sweep from $\boldsymbol{v}$ to $\boldsymbol{w}$, then the thumb points in the direction of $\boldsymbol{v} \times \boldsymbol{w}$.

A proof, based on straightforward calculation, is outlined in the exercises—it's better worked through than read. The rest of this section illustrates the Fact's uses.

Both parts of the Fact need some comment:

About θ. Since the angle between $\boldsymbol{v}$ and $\boldsymbol{w}$ never exceeds π radians, $\sin\theta \geq 0$. In particular, $\sin\theta = 0$ if θ is either 0 or π—in either case, $\boldsymbol{v}$ and $\boldsymbol{w}$ must be parallel.➻

We saw this earlier by algebraic means.

Interpreting length. As the following picture shows—

The area spanned by v and w

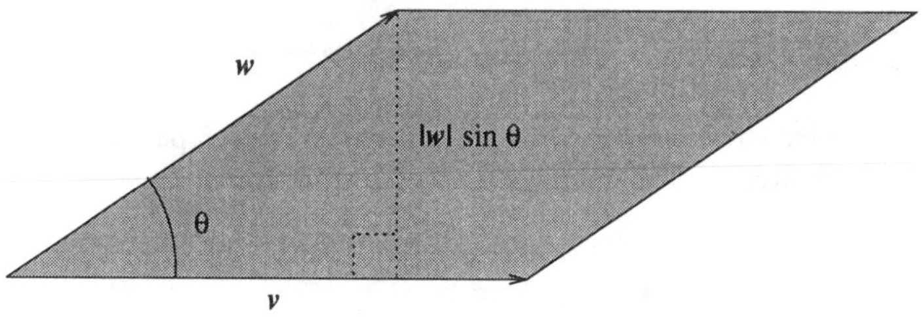

the quantity $|\boldsymbol{v}||\boldsymbol{w}|\sin\theta$ (i.e., the magnitude of $\boldsymbol{v} \times \boldsymbol{w}$) is the area of the parallelogram spanned by $\boldsymbol{v}$ and $\boldsymbol{w}$.

More on the right-hand rule. The right-hand rule, applied to the picture above, shows that $\boldsymbol{v} \times \boldsymbol{w}$ points straight up.➻ If $\boldsymbol{v}$ and $\boldsymbol{w}$ were switched, then the area spanned wouldn't change, but the direction of $\boldsymbol{v} \times \boldsymbol{w}$ would reverse.

Toward you, the reader . . . try it!

Here's another way to think of the right-hand rule. Imagine driving an ordinary wood screw into the plane of $\boldsymbol{v}$ and $\boldsymbol{w}$, as shown:

Turning a screw

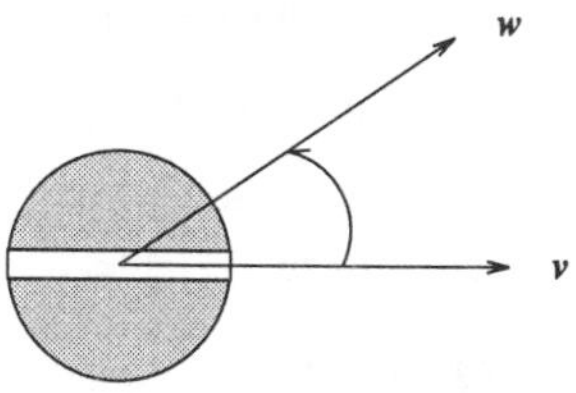

Turning the screw head from $\boldsymbol{v}$ to $\boldsymbol{w}$ causes the screw to move either into or out of the plane—the direction of motion is that of $\boldsymbol{v} \times \boldsymbol{w}$.➻

Ordinary bolts and screws are sometimes called "right-threaded." Left-threaded bolts and screws are occasionally used for special purposes.

■ **Example 3.** Find the area of the triangle with corners at $A = (1, 1, 1)$, $B = (1, 2, 3)$, and $C = (4, 5, 6)$.

Solution: The triangle is *half* the parallelogram spanned by the vectors $\overrightarrow{AB} = (0, 1, 2)$ and $\overrightarrow{AC} = (3, 4, 5)$. The parallelogram has area

$$\left|\overrightarrow{AB} \times \overrightarrow{AC}\right| = |(-3, 6, -3)| = \sqrt{54},$$

so the triangle has area $\sqrt{54}/2 \approx 3.67$. □

The cross product: a physical interpretation

The cross product $\boldsymbol{v} \times \boldsymbol{w}$ is often interpreted by physicists as **torque**, i.e., a vector that describes the tendency of a force to twist a body about an axis. From this point of view, we can think of the vector $\boldsymbol{v}$ as a rigid bar, pinned at its tail, and the vector $\boldsymbol{w}$ as a force exerted at the *head* of $\boldsymbol{v}$. For physical reasons, the twisting tendency at the tail of $\boldsymbol{v}$ is the product lf, where l is the length of $\boldsymbol{v}$ and f is the scalar component of the force $\boldsymbol{w}$ perpendicular to $\boldsymbol{v}$. (This is essentially the principle of the lever, which has been known since Archimedes.) In other words, the torque at the tail of $\boldsymbol{v}$ has magnitude $|\boldsymbol{v}||\boldsymbol{w}|\sin\theta$. The direction of the torque vector is given by the right-hand rule.

Exercises

1. Find (by hand) the cross product $\boldsymbol{v} \times \boldsymbol{w}$ of each pair of vectors below. (Check answers with technology if you like.)

   (a) $\boldsymbol{v} = (1, 2, 3)$; $\boldsymbol{w} = (4, 5, 6)$
   (b) $\boldsymbol{v} = (4, 5, 6)$; $\boldsymbol{w} = (2, 3, 4)$
   (c) $\boldsymbol{v} = (1, 2, 3)$; $\boldsymbol{w} = (40, 50, 60)$
   (d) $\boldsymbol{v} = \boldsymbol{i} + \boldsymbol{j}$; $\boldsymbol{w} = \boldsymbol{j} + \boldsymbol{k}$
   (e) $\boldsymbol{v} = (v_1, v_2, v_3)$; $\boldsymbol{w} = i$
   (f) $\boldsymbol{v} = (v_1, v_2, v_3)$; $\boldsymbol{w} = j$
   (g) $\boldsymbol{v} = (v_1, v_2, v_3)$; $\boldsymbol{w} = k$

2. Find a scalar equation for each plane described below. (Use the cross product as needed to find a normal vector.)

   (a) Through $(0, 0, 0)$, $(1, 2, 3)$, and $(-1, 1, 2)$.
   (b) Through $(1, 2, 3)$, with spanning vectors $\boldsymbol{i}$ and $\boldsymbol{i} + \boldsymbol{j}$.
   (c) With parametric equations $x = 1+2s+3t$, $y = -1+3s-2t$, $z = s+t$.

3. Let $\boldsymbol{v} = (v_1, v_2, v_3)$ and $\boldsymbol{w} = (w_1, w_2, w_3)$. Show from the definition that $\boldsymbol{v} \times \boldsymbol{w}$ is perpendicular to $\boldsymbol{w}$.

4. Let $\boldsymbol{v} = (v_1, v_2, v_3)$ and $\boldsymbol{w} = (w_1, w_2, w_3)$. Show from the definition that $\boldsymbol{v} \times \boldsymbol{w} = -(\boldsymbol{w} \times \boldsymbol{v})$.

5. Is the cross product associative? In other words, is $\boldsymbol{u} \times (\boldsymbol{v} \times \boldsymbol{w}) = (\boldsymbol{u} \times \boldsymbol{v}) \times \boldsymbol{w}$ for all vectors $\boldsymbol{u}$, $\boldsymbol{v}$, and $\boldsymbol{w}$? Either show that the cross product is associative or give an example to show that it is not.

6. Let $\boldsymbol{v} = (v_1, v_2, v_3)$ and $\boldsymbol{w} = (w_1, w_2, w_3)$ be nonzero vectors. Show that if $\boldsymbol{v} \times \boldsymbol{w} = (0, 0, 0)$, then $\boldsymbol{v}$ and $\boldsymbol{w}$ are parallel, i.e., each is a scalar multiple of the other. (Hints: Since $\boldsymbol{v} \neq (0, 0, 0)$, it's OK to assume that some component, say v_1, is not zero. Let $t = w_1/v_1$. Now use the assumption that $\boldsymbol{v} \times \boldsymbol{w} = (0, 0, 0)$ to show that $\boldsymbol{w} = t\boldsymbol{v}$.)

7. *(For students with linear algebra.)* We said in this section that the cross product $\boldsymbol{v} \times \boldsymbol{w}$ can be thought of as the determinant

$$\begin{vmatrix} \boldsymbol{i} & \boldsymbol{j} & \boldsymbol{k} \\ v_1 & v_2 & v_3 \\ w_1 & w_2 & w_3 \end{vmatrix},$$

   expanded along the first row. This exercise explores some other links between determinants and the cross product.

(a) Expand the determinant along the first column; what do you get?

(b) What is the effect on a determinant of exchanging any two rows? How is this related to a property of the cross product?

(c) What can be said about the determinant of a matrix with two equal rows? How is this related to a property of the cross product?

(d) What can be said about the determinant of a matrix if one row is a multiple of another? How is this related to a property of the cross product?

8. Here's another way to make sense of the algebraic formula for the cross product. We will assume three basic cross products:

$$\boldsymbol{i} \times \boldsymbol{j} = \boldsymbol{k}; \quad \boldsymbol{j} \times \boldsymbol{k} = \boldsymbol{i}; \quad \boldsymbol{k} \times \boldsymbol{i} = \boldsymbol{j}.$$

We'll also assume the algebraic properties of Theorem 7, page 87. From these assumptions we'll deduce the general formula for $\boldsymbol{v} \times \boldsymbol{w}$.

Let $\boldsymbol{v} = (v_1, v_2, v_3)$ and $\boldsymbol{w} = (w_1, w_2, w_3)$. Cite a brief reason for each identity below:

$$\begin{aligned}
\boldsymbol{v} \times \boldsymbol{w} &= (v_1\boldsymbol{i} + v_2\boldsymbol{j} + v_3\boldsymbol{k}) \times (w_1\boldsymbol{i} + w_2\boldsymbol{j} + w_3\boldsymbol{k}) \\
&= v_1w_1\boldsymbol{i} \times \boldsymbol{i} + v_2w_1\boldsymbol{j} \times \boldsymbol{i} + v_3w_1\boldsymbol{k} \times \boldsymbol{i} + v_1w_2\boldsymbol{i} \times \boldsymbol{j} + v_2w_2\boldsymbol{j} \times \boldsymbol{j} + \\
&\quad v_3w_2\boldsymbol{k} \times \boldsymbol{j} + v_1w_3\boldsymbol{i} \times \boldsymbol{k} + v_2w_3\boldsymbol{j} \times \boldsymbol{k} + v_3w_3\boldsymbol{k} \times \boldsymbol{k} \\
&= v_2w_1\boldsymbol{j} \times \boldsymbol{i} + v_3w_1\boldsymbol{k} \times \boldsymbol{i} + v_1w_2\boldsymbol{i} \times \boldsymbol{j} + v_3w_2\boldsymbol{k} \times \boldsymbol{j} + \\
&\quad v_1w_3\boldsymbol{i} \times \boldsymbol{k} + v_2w_3\boldsymbol{j} \times \boldsymbol{k} \\
&= (v_2w_3 - v_3w_2)\boldsymbol{i} + (v_3w_1 - v_1w_3)\boldsymbol{j} + (v_1w_2 - v_2w_2)\boldsymbol{k}.
\end{aligned}$$

9. The Fact on page 88 says that $|\boldsymbol{v} \times \boldsymbol{w}| = |\boldsymbol{v}|\,|\boldsymbol{w}| \sin\theta$. where θ is the angle between $\boldsymbol{v}$ and $\boldsymbol{w}$. This exercise outlines a proof; the underlying idea is to relate the cross product to the dot product.

(a) Show (by direct calculation) that $|\boldsymbol{v} \times \boldsymbol{w}|^2 = |\boldsymbol{v}|^2\,|\boldsymbol{w}|^2 - (\boldsymbol{v} \cdot \boldsymbol{w})^2$.

(b) Use the result of (a) to show that $|\boldsymbol{v} \times \boldsymbol{w}| = |\boldsymbol{v}|\,|\boldsymbol{w}| \sin\theta$, as the Fact claims.

(Hints: In (b), write $(\boldsymbol{v} \cdot \boldsymbol{w})^2 = |\boldsymbol{v}|^2|\boldsymbol{w}|^2 \cos^2\theta$. Then take the square root of both sides of (a).)

10. Let $\boldsymbol{u}$, $\boldsymbol{v}$, and $\boldsymbol{w}$ be vectors in $\mathbb{R}^3$. The quantity $\boldsymbol{u} \cdot \boldsymbol{v} \times \boldsymbol{w}$ is called the **triple scalar product** of $\boldsymbol{u}$, $\boldsymbol{v}$, and $\boldsymbol{w}$, in that order. We'll investigate the following useful property of the triple product:

The absolute value $|\boldsymbol{u} \cdot \boldsymbol{v} \times \boldsymbol{w}|$ is the volume of the parallelepiped in $\mathbb{R}^3$ (i.e., the three-dimensional solid) spanned by $\boldsymbol{u}$, $\boldsymbol{v}$, and $\boldsymbol{w}$.

Here's a possible picture:

The solid spanned by three vectors

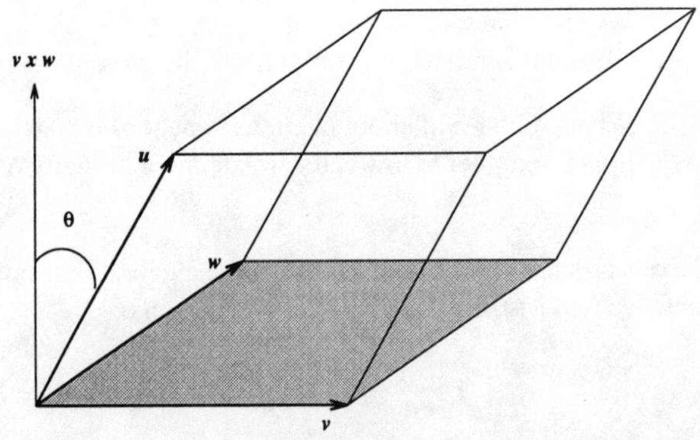

(a) Calculate the triple scalar product of $\boldsymbol{i}$, $\boldsymbol{j}$, and $\boldsymbol{k}$. Is the result consistent with the italicized fact above?

(b) Calculate the triple scalar product of $a\boldsymbol{i}$, $b\boldsymbol{j}$, and $c\boldsymbol{k}$, where a, b, and c are positive constants. Is the result consistent with the italicized fact above?

(c) Suppose that $\boldsymbol{u}$ is in the plane spanned by $\boldsymbol{v}$ and $\boldsymbol{w}$. What's the value of $\boldsymbol{u} \cdot \boldsymbol{v} \times \boldsymbol{w}$? What does this mean about the parallelepiped?

(d) Find the area of the base of the solid, shown shaded.

(e) Explain why $|\boldsymbol{u}||\cos\theta|$ is the height (perpendicular to the shaded plane) of the solid.

(f) Explain why $|\boldsymbol{u} \cdot \boldsymbol{v} \times \boldsymbol{w}|$ is the claimed volume. (Hint: Volume = base times height.)

11. Let $\boldsymbol{u} = (1, 2, 3)$ and $\boldsymbol{v} = (4, 5, 6)$. Evaluate $\boldsymbol{u} \times \boldsymbol{v}$.

12. Let $\boldsymbol{u}$ and $\boldsymbol{v}$ be vectors in $\mathbb{R}^3$. Give a geometric explanation of the identity $(\boldsymbol{u} \times \boldsymbol{v}) \bullet \boldsymbol{v} = 0$.

13. Let L be the line through the point $(1, 1, 2)$ that is perpendicular to the plane Π described by the equation $3x + 5y + 8z = 13$. Find the point where L and Π intersect.

14. Find an equation for the plane that contains the points $(1, 2, 3)$, $(1, 1, 1)$, and $(1, 0, 2)$.

15. Find an equation for the line that passes through the point $(2, 3, 1)$ and is parallel to the line of intersection of the planes with equations $2x - 2y - 3z = 2$ and $3x - 2y + z = 1$. [HINT: The line of intersection is perpendicular to normal lines for either plane.]

16. Let ℓ be the line through the points $(1, 2, 3)$ and $(4, 5, 6)$. Write an equation for the plane that is perpendicular to ℓ and passes through the point $(7, 8, 9)$.

17. Let $\boldsymbol{u} = (1, 2, 3)$ and $\boldsymbol{v} = (4, 5, 6)$. Find a unit vector that is perpendicular to both $\boldsymbol{u}$ and $\boldsymbol{v}$.

18. Find the area of the parallelogram in space formed by the vectors $\boldsymbol{u} = (1, 5, 1)$ and $\boldsymbol{v} = (-2, 1, 3)$.

19. Prove that $\boldsymbol{u} \bullet \boldsymbol{v} \times \boldsymbol{w} = \boldsymbol{u} \times \boldsymbol{v} \bullet \boldsymbol{w}$.

20. Suppose that $\boldsymbol{X}_0 + s\boldsymbol{u}$ and $\boldsymbol{X}_1 + t\boldsymbol{v}$ are two skew (i.e., nonintersecting, nonparallel) lines in space. Find an expression for the distance between the lines. [HINT: The vector $\boldsymbol{u} \times \boldsymbol{v}$ is perpendicular to both lines.]

21. Show that $\frac{d}{dt}\big(\boldsymbol{x}(t) \times \boldsymbol{y}(t)\big) = \boldsymbol{x}'(t) \times \boldsymbol{y}(t) + \boldsymbol{x}(t) \times \boldsymbol{y}'(t)$.

22. Suppose that $\boldsymbol{x}(t) = (\cos t, \sin t, t)$ and $\boldsymbol{y}(t) = (2, t^2, -4t)$. Evaluate $\frac{d}{dt}\boldsymbol{x}(t) \times \boldsymbol{y}(t)$.

23. Show that if $\boldsymbol{u}$, $\boldsymbol{v}$, and $\boldsymbol{w}$ are distinct nonzero vectors, then $\boldsymbol{u} \times \boldsymbol{v} = \boldsymbol{u} \times \boldsymbol{w}$ if and only if $\boldsymbol{u}$ is parallel to $\boldsymbol{v} - \boldsymbol{w}$.

24. Show that $(\boldsymbol{v} + \boldsymbol{w}) \times (\boldsymbol{v} - \boldsymbol{w}) = 2\boldsymbol{w} \times \boldsymbol{v}$.

25. Prove that if three vectors form a triangle, then the lengths of their pairwise cross products are all equal.

26. Explain why the vector $\boldsymbol{u} \times (\boldsymbol{v} \times \boldsymbol{w})$ must lie in the plane spanned by $\boldsymbol{v}$ and $\boldsymbol{w}$.

27. Let L be a line with direction vector $\boldsymbol{v}$ and P a point not on L. Let $\boldsymbol{w}$ be a vector from some point on L to P. Show that the distance from P to L is given by $|\boldsymbol{w} \times \boldsymbol{v}| / |\boldsymbol{v}|$.

28. Suppose that the plane spanned by the nonzero vectors $\boldsymbol{a}$ and $\boldsymbol{b}$ and the plane spanned by the nonzero vectors $\boldsymbol{c}$ and $\boldsymbol{d}$ are not parallel. Explain why the vector $(\boldsymbol{a} \times \boldsymbol{b}) \times (\boldsymbol{c} \times \boldsymbol{d})$ is parallel to the line of intersection of the planes.

29. Find a unit vector orthogonal to *both* $\boldsymbol{i} + \boldsymbol{k}$ and $2\boldsymbol{j} - 3\boldsymbol{k}$.

30. Suppose that $\mathbf{u}$ and $\mathbf{v}$ are the vectors in the xz-plane pictured below, $|\mathbf{u}| = 3$, $|\mathbf{v}| = 4$, and $\phi = \pi/3$.

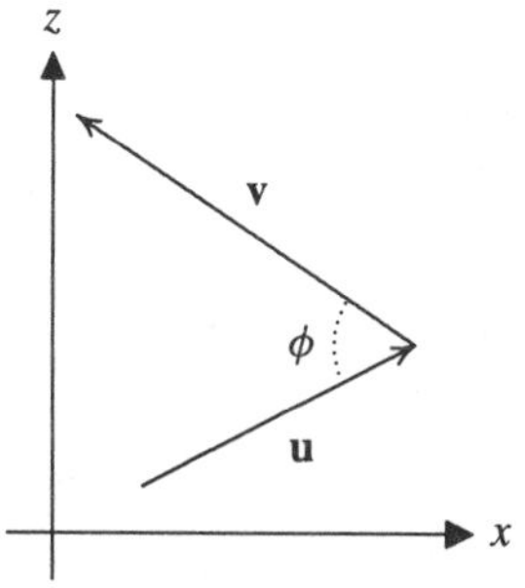

Evaluate $\mathbf{u} \times \mathbf{v}$.

31. Let $\mathbf{u} = (1, 2, 3)$ and $\mathbf{v} = (1, 3, 5)$. Find a unit vector that is perpendicular to both $\mathbf{u}$ and $\mathbf{v}$.

32. Suppose that **u** and **v** are the vectors in the yz-plane pictured below, $|\mathbf{u}| = 3$, $|\mathbf{v}| = 4$, and $\phi = \pi/3$.

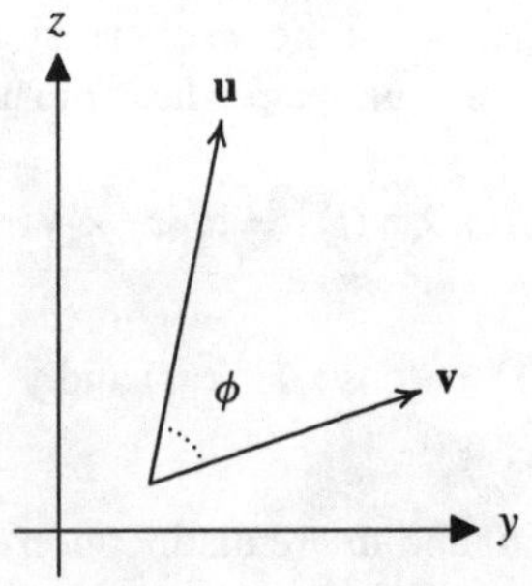

Evaluate $\mathbf{u} \times \mathbf{v}$.

33. Let $\boldsymbol{u} = (1, 2, 3)$ and $\boldsymbol{v} = (2, 3, 4)$. Find a unit vector that is orthogonal to both $\boldsymbol{u}$ and $\boldsymbol{v}$.

34. (a) Explain how the cross product can be used to determine whether three points in space are collinear.

 (b) Describe a method for determining whether four points lie in the same plane.

Chapter 2

Derivatives

2.1 Functions of several variables

The derivative and the integral are the most important concepts of calculus—in one *and* several variables. This chapter is about derivatives of functions of several variables. We use the plural "derivatives" intentionally, because there is more than one way to generalize the basic idea of derivative from functions of one variable setting to the multivariable setting. Indeed, we have already seen derivatives in one of their multivariable forms. In Chapter 1, we found derivatives of vector-valued functions; we interpreted them either geometrically, as tangent vectors to curves, or physically, in the language of velocity and speed.

Most of the functions of Chapter 1 produced vectors as *outputs*, but they were all functions of only one *input* variable.➽ Thus, in a sense, the functions of Chapter 1 are merely *lists* of ordinary, single-variable calculus functions. In this chapter, by contrast, we study derivatives of functions of *more than one* input variable.➽

We often used t as our (single) input variable.

We'll stress functions of two variables.

Functions of one variable are the basic objects of single-variable calculus. Functions of two (or more) variables play a similar role in multivariable calculus. In this section we meet such functions in their own right, and consider some of their rudimentary properties.

Functions of one or more variables

The squaring function, defined for all real numbers x by $f(x) = x^2$, is typical of the functions of beginning calculus: f accepts *one* number, x, as input, and assigns another number, x^2, as output. If, say, $x = 2$, then $f(2) = 4$.

Consider, by contrast, the function g defined by $g(x, y) = x^2 + y^2$. Unlike f, g accepts a *pair* (x, y) of real numbers as inputs. The output is a third real number, $x^2 + y^2$. If, say, $x = 2$ and $y = 1$, then $g(2, 1) = 4 + 1 = 5$.

Naturally enough, f is called a function of one variable, and g a function of two variables.➽ The difference has to do with domains: The domain of f is the one-dimensional real number line; the domain of g is the two-dimensional xy-plane.

"One" and "two" count the input variables to f and g.

The function f corresponds to the equation $y = x^2$; x is the **independent variable** and y is the **dependent variable**. The function g corresponds to the **equation** $z = x^2 + y^2$; now both x and y are **independent variables** and z is the **dependent variable**.

We'll use f and g below to illustrate various similarities and differences between functions of one and two variables. Notice, however, that there's nothing sacred about *two* variables—we could (and will) discuss such functions as

$$h(x, y, z) = x^2 + y^2 + z^2 \quad \text{and} \quad k(x, y, z, w) = x^2 + y^2 + z^2 + w^2,$$

which accept three or more variables. For now, however, we'll keep things simple by sticking mainly to functions of two variables.

Multivariable functions: why bother?

Single-variable calculus is challenging enough. Why complicate things by adding more variables?

There are purely mathematical answers, too.

It's a fair question. One good practical answer is that functions of several variables are essential for describing and predicting phenomena we care about, both natural and human-made. In economics, for instance, a manufacturer's profit depends on many "input" variables: labor costs, distance to markets, tax rates, etc. In physics, a satellite's motion through space depends on a variety of forces. In biology, populations rise and fall with variations in climate, food supply, predation, and other factors. The weather varies with both longitude and latitude. Our world, in short, is multidimensional: modeling it successfully requires multivariable tools.

By Upper Midwest standards . . .

The National Weather Service plots multivariable functions every day—they may appear on the back page of your newspaper. Here, for instance, are noon surface temperatures on a relatively warm winter day:

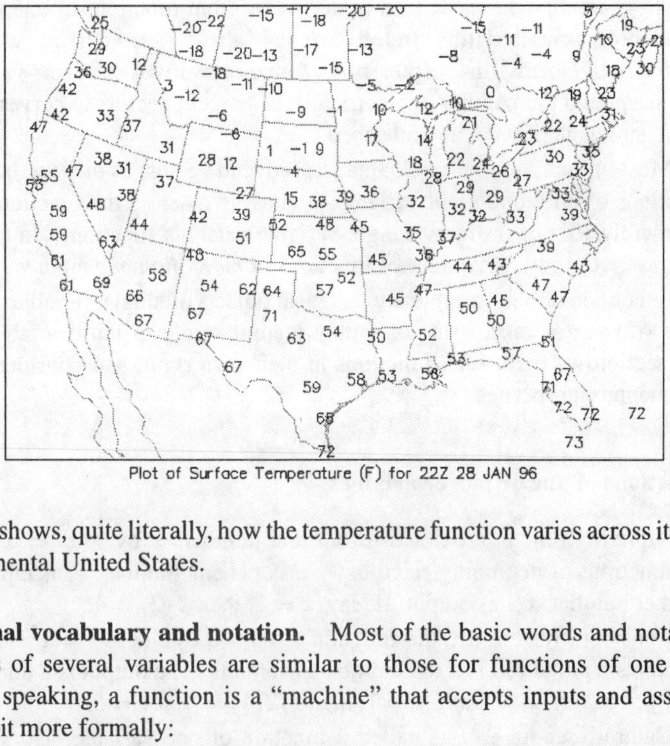

Plot of Surface Temperature (F) for 22Z 28 JAN 96

The map shows, quite literally, how the temperature function varies across its domain—the continental United States.

Functional vocabulary and notation. Most of the basic words and notations for functions of several variables are similar to those for functions of one variable. Roughly speaking, a function is a "machine" that accepts inputs and assigns outputs. A bit more formally:

Domain. The **domain** of a function is the set of permissible inputs. The function $g(x, y) = x^2 + y^2$, for instance, accepts any 2-tuple (x, y) as input, so its domain is the set $\mathbb{R}^2$.

Range. The **range** of a function is the set of possible outputs. For g above, the range is the set $[0, \infty)$ of nonnegative real numbers. (The sum of squares can't be negative.)

Rule. The **rule** of a function is the method for assigning inputs to outputs. For g, the rule is given by the algebraic formula $g(x, y) = x^2 + y^2$. Not every function has a simple symbolic rule. We'll often meet functions given by tables, by graphs, or in other ways.

"Arrow" notations. Above we used the notation $g(x, y) = x^2 + y^2$ to describe a certain function of two variables. Other notations, involving arrows, are sometimes convenient. The notation

$$g : \mathbb{R}^2 \to \mathbb{R}$$

says that g is a function that accepts *two* real numbers as input and produces *one* real number as output. If we want to specify the rule by which g sends inputs to outputs, we can write

$$g : (x, y) \to x^2 + y^2.$$

These notations remind us that a function *begins* with an input (a 2-tuple, in this case) and *ends* with an output (a single number, in this case).

Vector variables. It's sometimes convenient to use vector notation in describing functions of several variables. For instance, if we write $\boldsymbol{X} = (x, y)$, then $\boldsymbol{X} \cdot \boldsymbol{X} = (x, y) \cdot (x, y) = x^2 + y^2$, so the function $g(x, y) = x^2 + y^2$ could also be written, using vectors, in the form $g(\boldsymbol{X}) = \boldsymbol{X} \cdot \boldsymbol{X}$. In three variables, if we write $\boldsymbol{X} = (x, y, z)$, and $\boldsymbol{A} = (1, 2, 3)$, then the notations

$$f(x, y, z) = x + 2y + 3z + 4 \quad \text{and} \quad f(\boldsymbol{X}) = \boldsymbol{A} \cdot \boldsymbol{X} + 4$$

convey the same information.

Graphs in one and several variables

The graph of f is the set of all points (x, y) for which $y = f(x) = x^2$. For example, $f(2) = 4$, so the point $(2, 4)$ lies on the graph. Geometrically, this graph is a curve—a parabola, in this case—in the the xy-plane. Like a straight line, the curve $y = x^2$ is a *one-dimensional* object➼ that lives in a *two-dimensional* space—the xy-plane. Here's part of the familiar graph:

To an ant walking along it, the parabola "looks" like a (one-dimensional) straight line.

Graph of $f(x) = x^2$: a parabola

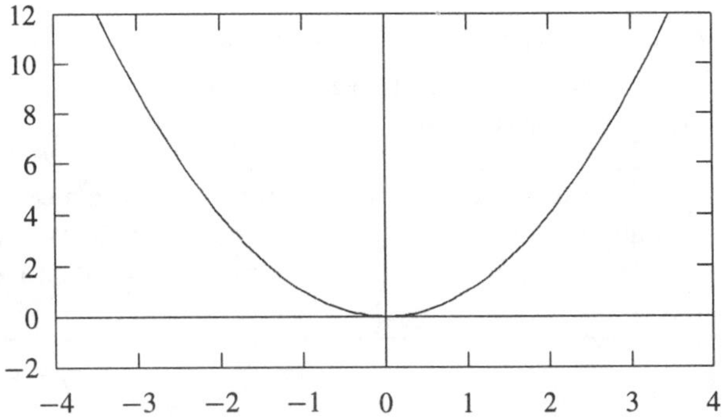

The graph of g is the set of all points (x, y, z) for which $z = g(x, y) = x^2 + y^2$. For example, since $g(2, 1) = 5$, the point $(2, 1, 5)$ lies on the g-graph. Here's a portion of the g-graph:

Graph of $g(x, y) = x^2 + y^2$: a paraboloid

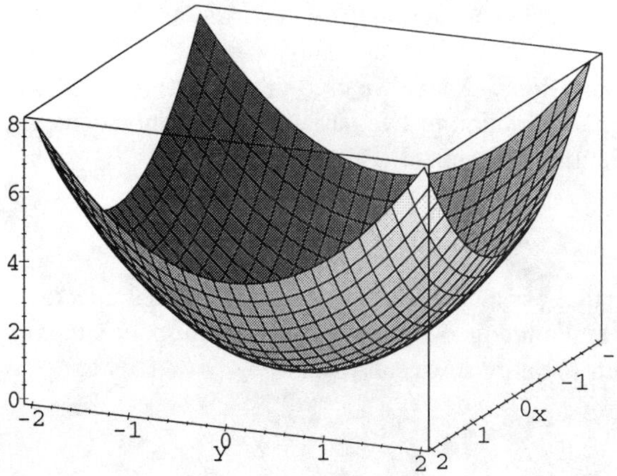

We discussed this paraboloid in Section 1.1.

To an ant walking along it, the g-graph "looks" like a (two-dimensional) flat plane—just as Earth's surface looks flat from a human vantage point.

The graph, a paraboloid, is quite different from the parabola shown earlier. Notice especially the dimensions involved: The graph of g is a *two-dimensional* surface hovering in *three-dimensional* xyz-space. As a rule, the "dimension" of any graph is the number of *input* variables the function accepts. (This rule of thumb applies to almost every function met in calculus courses. For the record, though, some very ill-behaved functions do violate this rule.)

In other respects, the graphs of f and g are quite similar to each other. For both functions the *height* of the graph above a given domain point (a typical input is x_0 for f and (x_0, y_0) for g) tells the corresponding output value (typical outputs are $y_0 = f(x_0)$ and $z_0 = g(x_0, y_0)$, respectively).

But it also adds to the fun . . . .

Multivariable graphs: beware . . . Graphs are at least as important in multivariable calculus as in elementary calculus, but multivariable graphs are usually more complicated, and so need extra care in handling. Choosing a "good" viewing window, for example, takes some care even for functions of one variable, and the problem can be stickier still for functions of two variables. The fact that multivariable graphs "live" naturally in three-dimensional (or even higher-dimensional) space—not on a flat page or computer screen—only adds to the problem. For this and other reasons, we'll look at functions from as many points of view as possible.

Level curves and contour maps

Occasionally this curve is just a point.

Let $f(x, y)$ be a function of two variables, and let c be a number in the range of f. Then $f(x, y) = c$ is an equation in x and y; its graph is (usually) a curve in the xy-plane. Such curves have a special name:

> **Definition:** Let $f(x, y)$ be a function and c a constant. The set of all (x, y) for which $f(x, y) = c$ is called a **level curve** of f. A collection of level curves drawn together is called a **contour map** of f.

Observe:

Why "level"? The word "level" makes good sense here, because $f(x, y)$ has the same value, namely c, at each point along the level curve. In other words, the graph of f is "level" above the level curve $f = c$.

Which level curves to draw? Each number c in the range of f has its own level curve. Since the range of a function is usually infinite, we can't possibly draw all the level curves. In practice, we draw some convenient selection of curves, corresponding to *evenly-spaced* values of c. The spacing between curves reflects how fast the function increases or decreases.

Labels. We'll label level curves with their corresponding output values. Typical labels, therefore, might be of the form $z = c$, $f = c$, or even simply c.

■ **Example 1.** Here's a sample of level curves for the function $g(x, y) = x^2 + y^2$. Label each level curve with the appropriate output value. The graph of g is a paraboloid. How is this shape reflected in the level curves?

Level curves of g(x,y)=x^2+y^2

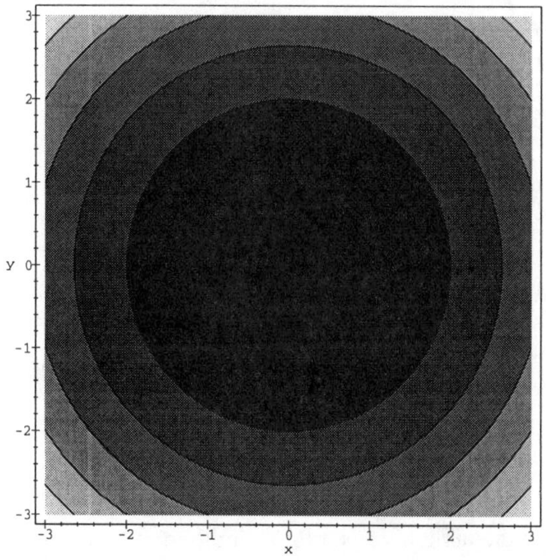

Solution: All the level curves are circles about the origin;➻ each one is the graph of an equation $x^2 + y^2 = c$, for some c. The circles, from smallest to largest, correspond to the levels $z = 1, 4, 7, 10, 13, 16, 19$. (The shading corresponds to levels—darker shades are "lower.")

Some of the circles aren't fully visible.

Notice that level curves of g get closer and closer together as we move outward from the origin. This reflects the fact that the g-graph is a paraboloid—it gets steeper and steeper as we move away from the origin. □

■ **Example 2.** Drawing level curves on a temperature map makes the map easier to read and interpret. (Newspapers usually do this.) Here's another version of the map above; the boundaries of shaded regions correspond to the level curves of the temperature function; they're called **isotherms**.➻

Compare this temperature map to the other one.

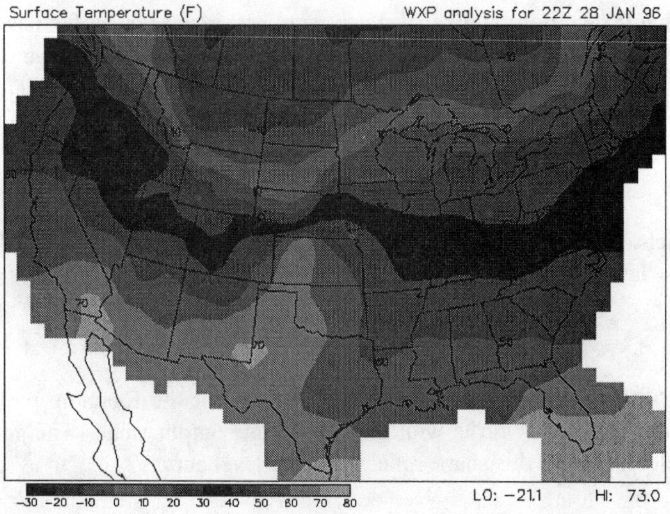

Functions as tables

In other cases we may have only a table of values.

Sometimes information about a function is given numerically, in tabular form, rather than graphically or symbolically. Here, for instance, are some output values of the function $g(x, y) = x^2 + y^2$ for inputs near (2, 1):

Values of $g(x, y) = x^2 + y^2$ near (1, 2)

$y \backslash x$	0.8	0.9	1.0	1.1	1.2
2.2	5.48	5.65	5.84	6.05	6.28
2.1	5.05	5.22	5.41	5.62	5.85
2.0	4.64	4.81	5.00	5.21	5.44
1.9	4.25	4.42	4.61	4.82	5.05
1.8	3.88	4.05	4.24	4.45	4.68

The table alone reveals nothing about how g behaves anywhere else—for that we need a graph or formula.

A close look at the table reveals a lot about the function g for inputs near (2, 1). For instance, the table's entries *increase* as we move either to the right or upward. This means that, for inputs (x, y) near (2, 1), the function's output values increase as either x or y increases. We see, too, that the table entries increase *faster and faster* as we move to the right or upward. Notice that the graph and the contour map convey the same information, in their own (geometric) ways.

Linear functions

A linear function of one variable is one that can be written in the form $L(x) = a + bx$, where a and b are constants. A linear function in several variables has a similar algebraic form:

> **Definition:** A **linear function** in two variables is one that can be written in the form
>
> $$L(x, y) = a + bx + cy,$$
>
> where a, b, and c are any constants.

Linear functions in three (or more) variables are similar. A linear function of three variables has the form $L(x, y, z) = a + bx + cy + dz$, where a, b, c, and d are constants. In vector notation, with $\mathbf{X} = (x, y, z)$ and $\mathbf{A} = (a, b, c)$, a linear function has the simple formula $L(\mathbf{X}) = \mathbf{A} \cdot \mathbf{X} + d$, where d is a constant.

"Linear": What's in the name? Why do linear functions deserve that name? In the xy-plane, there's no mystery: The graph of a linear function $y = a + bx$ is a straight line. In xyz-space, however, the graph of a linear function $z = a + bx + cy$ is not a line, but a plane. (In $xyzw$-space, the graph of a linear function $w = a+bx+cy+dz$ is even less like a line. It's a three-dimensional solid, called a hypersurface.) Nevertheless, we'll call any function linear that involves only constants and the first power of each variable, regardless of the number of variables. In the same spirit, we'll call a function "quadratic" (no matter how many variables are involved) if the variables appear in nonnegative integer powers, none higher than two.➡

The function $f(x, y) = x^2 + xy + y^2$ is quadratic in this sense.

Readers who have studied linear algebra➡ will recall that the word "linear" (as in "linear transformation") is used in still another way in that subject. (The functions we call "linear" in this book are sometimes referred to, in linear algebra jargon, as "affine" functions.) Like other very useful English words, "linear" seems to have spawned a whole family of related but not identical meanings.

There's another use of "linear"!

Planes and linear functions. We saw in Chapter 1 that every plane Π in $\mathbb{R}^3$ has a scalar equation of the form $ax + by + cz = d$, with a, b, c, and d all constants. We saw, too, that the vector (a, b, c) is normal to Π at any point of Π. These facts can help us understand and visualize planes that arise as the graphs of linear functions, rather than being specified from the outset in terms of points and normals.

Let $L(x, y) = ax + by + c$ be a linear function. The graph of L is the set of points (x, y, z) that satisfy the equation

$$z = ax + by + c, \quad \text{or, equivalently,} \quad ax + by - z = -c.$$

The form of the last equation lets us simply read off the normal vector. Here's the conclusion:

Fact: Let $L(x, y) = ax+by+c$ define a linear function $L : \mathbb{R}^2 \to \mathbb{R}$. The graph of L is the plane with scalar equation $ax+by-z = -c$. This plane passes through the point $(0, 0, c)$, and has normal vector $(a, b, -1)$.

■ **Example 3.** Describe the graphs of the linear functions $L(x, y) = 3$ and $M(x, y) = -x/3 - 2y/3 + 2$.

Solution: By the Fact, the graph of L, a plane, has normal vector $(a, b, -1) = (0, 0, -1) = -\mathbf{k}$. This means that the graph is perpendicular to the z-axis. This is easy to visualize, since the plane has equation $z = 3$, so is parallel to the xy-plane.

The graph of M is the plane with equation $z = -x/3-2y/3+2$, or, equivalently, $x+2y+3z = 6$.➡ By the Fact, the vector $(-1/3, -2/3, -1)$ is normal to the plane; therefore, $-3(-1/3, -2/3, -1) = (1, 2, 3)$ is also normal to the plane.➡ This plane and the normal vector $(1, 2, 3)$ are shown in Example 5, page 80. □

Multiply both sides by 3.

Multiplying a normal by a nonzero constant doesn't affect perpendicularity.

Why linear functions matter. Linear functions are simple, useful, and easy to work with. Most important for us, linear functions are prototypes or models for all differentiable functions. Indeed, any differentiable function, in any number of variables, might be called "almost linear," or "locally linear," in much the same sense that an ordinary calculus function $y = f(x)$ looks like a straight line if we zoom in repeatedly on a typical point on its graph. We'll return often to this theme.

Vector-valued functions of several variables; matrix notations

So far in this section we've considered only *scalar-valued* functions of several variables, i.e., functions of the form $f : \mathbb{R}^n \to \mathbb{R}$ (with $n = 2$ or $n = 3$), which accept several variables as input, and produce scalars as output.

But *vector-valued* functions of several variables are also perfectly possible. For instance, consider the function $\boldsymbol{f} : \mathbb{R}^2 \to \mathbb{R}^2$ defined by $\boldsymbol{f}(x, y) = (x + 2y + 3, 4x + 5y + 6)$, which accepts two variables as input and produces a 2-vector as output. Using vectors and matrices, we could, instead, write $\boldsymbol{f}(x, y)$ in the form

$$\boldsymbol{f}(\boldsymbol{X}) = \begin{bmatrix} 1 & 2 \\ 4 & 5 \end{bmatrix} \begin{bmatrix} x \\ y \end{bmatrix} + \begin{bmatrix} 3 \\ 6 \end{bmatrix}.$$

For a brief review, see the appendices.

Matrix multiplication can simplify and unclutter treatment of some multivariable ideas; we'll use it later in this book. For the moment the main point is simply to acknowledge the variety of functions that multivariable calculus can treat.

Vector-valued functions of several variables have important practical applications, especially in modeling physical phenomena, such as forces that vary over space. We'll return to vector-valued functions and their uses in later chapters. For the moment, our main purpose will be to study the components from which functions like $\boldsymbol{f}$ are built—scalar-valued functions of several variables.

Limits and continuity in several variables (optional)

The theory of calculus—in one *and* several variables—relies on certain tacit assumptions about the functions under study. For example, the mean value theorem of one-variable calculus requires the function under study to be *continuous* on a closed interval $[a, b]$ and *differentiable* on the open interval (a, b). Not every function has these properties. For example, the absolute value function $f(x) = |x|$ is continuous on every interval, but *not* differentiable at $x = 0$. On the other hand, the standard elementary functions, such as polynomial, trigonometric, and exponential functions, are continuous and differentiable wherever they're defined, so we can often use them without undue fuss.

Similar technical requirements apply in multivariable calculus as well. We'll often assume, sometimes without explicit mention, that the functions we use are "well-behaved" in the sense of being continuous or differentiable, having limits, etc. As it turns out, these technical concepts are substantially harder to define and study rigorously in several variables than in one variable; doing so is a matter for more advanced courses. As a rule, we'll treat such matters informally, with occasional glances at the theoretical side of the subject.

We'll see multivariable limits, briefly, later in this chapter.

Consider, for example, the question of **continuity** for a function $f(x, y)$. A rigorous definition involves a limit—which itself needs rigorous definition for a function of several variables. Intuitively, though, a continuous function of two variables is one whose graph is a surface free of holes or tears; a differentiable function is one whose graph is "smooth," without kinks or corners. (We'll define differentiability in more detail later in this Chapter.) As one might expect, multivariable functions such as $g(x, y) = x^2 + y^2$, that are built from standard elementary function ingredients, are differentiable and continuous wherever they're defined.

Exercises

1. Find the range and domain of each function.

   (a) $g(x, y) = x^2 + y^2$

   (b) $h(x, y) = x^2 + y^2 + 3$

   (c) $j(x, y) = 1/(x^2 + y^2)$

   (d) $k(x, y) = x^2 - y^2$

   (e) $m(x, y) = \sqrt{1 - x^2 - y^2}$

2. Let $f(x, y) = y - x^2$ and let $g(x, y) = x - y^2$.

   (a) In the rectangle $[-3, 3] \times [-3, 3]$, draw and label the level curves of f that correspond to $z = -3, z = -2, \ldots, z = 2$, and $z = 3$. What is the shape of each level curve?

   (b) In the rectangle $[-3, 3] \times [-3, 3]$, draw and label the level curves of g that correspond to $z = -3, z = -2, \ldots, z = 2$, and $z = 3$. What is the shape of each level curve?

   (c) How are the results of parts (a) and (b) similar? How are they different?

   (d) Use technology to plot the graphs $z = f(x, y)$ and $z = g(x, y)$, for (x, y) in $[-3, 3] \times [-3, 3]$. Describe briefly, in words, how the two graphs are related to each other.

3. Let $f(x, y) = x^2 + y^2$ and let $g(x, y) = x^2 + y^2 + 1$.

   (a) In the rectangle $[-3, 3] \times [-3, 3]$, draw and label the level curves of f that correspond to $z = 0, z = 2, z = 4, z = 6$, and $z = 8$. What is the shape of each level curve?

   (b) In the rectangle $[-3, 3] \times [-3, 3]$, draw and label the level curves of g that correspond to $z = 1, z = 3, z = 5, z = 7$, and $z = 9$. What is the shape of each level curve?

   (c) How are the results of parts (a) and (b) similar? How are they different?

   (d) Use technology to plot the graphs $z = f(x, y)$ and $z = g(x, y)$, for (x, y) in $[-3, 3] \times [-3, 3]$. Describe briefly, in words, how the two graphs are related to each other.

4. Let f and g be the linear functions $f(x, y) = 2x - 3y$ and $g(x, y) = -2x + 3y$.

   (a) In the rectangle $[-3, 3] \times [-3, 3]$, draw and label the level curves of f that correspond to $z = -5, z = -3, z = -1, z = 1, z = 3$, and $z = 5$. What is the shape of each level curve?

   (b) In the rectangle $[-3, 3] \times [-3, 3]$, draw and label the level curves of g that correspond to $z = -5, z = -3, z = -1, z = 1, z = 3$,and $z = 5$. What is the shape of each level curve?

(c) How are the results of parts (a) and (b) similar? How are they different?

(d) What special properties do the level curves of a *linear* function have?

(e) Use technology to plot the graphs $z = f(x, y)$ and $z = g(x, y)$, for (x, y) in $[-3, 3] \times [-3, 3]$. Describe briefly, in words, how the two graphs are related to each other.

5. Let f and g be the functions $f(x, y) = 2 + x^2$ and $g(x, y) = 2 + y^2$.

(a) In the rectangle $[-3, 3] \times [-3, 3]$, draw and label the level curves of f that correspond to $z = 0$, $z = 2$, $z = 4$, $z = 6$, and $z = 8$. What is the shape of each level curve?

(b) In the rectangle $[-3, 3] \times [-3, 3]$, draw and label the level curves of g that correspond to $z = 0$, $z = 2$, $z = 4$, $z = 6$, and $z = 8$. What is the shape of each level curve?

(c) How are the results of parts (a) and (b) similar? How are they different?

(d) Use technology to plot the graphs $z = f(x, y)$ and $z = g(x, y)$, for (x, y) in $[-3, 3] \times [-3, 3]$. Describe briefly, in words, how the two graphs are related to each other.

(e) The graphs of f and g are both "cylinders" (in the sense we defined in Section 14.1). How do the contour maps of f and g reflect this fact?

6. Imagine a map of the United States in the usual position. The positive x-direction is east, and the positive y-direction is north. Suppose that the units of x and y are miles and that Los Angeles, California, has coordinates $(0, 0)$. (Several approximations are involved here. The earth's surface is not flat, and Los Angeles occupies more than a single point.) Let $T(x, y)$ be the temperature, in degrees Celsius, at the location (x, y) at noon, Central Standard Time, on January 1, 1996.

(a) What does it mean in weather language to say that $T(0, 0) = 15$?

(b) What do the level curves of T mean in weather language? As a rule, would you expect, level curves of T to run north and south or east and west? Why?

(c) International Falls, Minnesota, is about 1400 miles east and 1100 miles north of Los Angeles. The noon temperature in International Falls on January 1, 1996, was -15 degrees Celsius. What does this mean about $T(x, y)$?

(d) Suppose that International Falls was the coldest spot in the country at the time in question. How would you expect the level curves to look near International Falls?

7. For any point (x, y) in the xy-plane, let $f(x, y)$ be the distance from (x, y) to the origin. Then f has the formula $f(x, y) = \sqrt{x^2 + y^2}$.

(a) Find the range and domain of f.

(b) Plot f. Describe the graph in words.

(c) Draw the level curve of f that passes through $(3, 4)$.

(d) All level curves of f have the same shape. What is it?

8. For any point (x, y) in the xy-plane, let $f(x, y)$ be the distance from (x, y) to the line $x = 1$.

   (a) Find a formula for $f(x, y)$.

   (b) Plot f. Describe the graph in words.

   (c) Draw the level curve that passes through $(3, 4)$.

   (d) All level curves of f have the same shape. What is it?

9. Here are some values of a linear function $L(x, y)$. (No explicit symbolic formula for L is given.) Use the table to answer the following questions.

Values of $L(x, y)$

$y \backslash x$	−3	−2	−1	0	1	2	3
3	−15	−12	−9	−6	−3	0	3
2	−13	−10	−7	−4	−1	2	5
1	−11	−8	−5	−2	1	4	7
0	−9	−6	−3	0	3	6	9
−1	−7	−4	−1	2	5	8	11
−2	−5	−2	1	4	7	10	13
−3	−3	0	3	6	9	12	15

   (a) All the level curves of L are straight lines. Using this fact, draw (all in the rectangle $[-3, 3] \times [-3, 3]$) and label the level curves $z = -12$, $z = -8$, $z = -4$, $z = 0$, $z = 4$, $z = 8$, $z = 12$.

   (b) Find an equation in x and y for the level line $z = 0$.

   (c) Because L is a linear function, its formula has the form $L(x, y) = a + bx + cy$, for some constants a, b, and c. Find numerical values for a, b, and c. [HINT: The table says that $L(0, 0) = 0$. Therefore, $L(0, 0) = a + b \cdot 0 + c \cdot 0 = 0$, so $a = 0$. Use similar reasoning to find values for b and c.]

   (d) Use technology to plot $L(x, y)$ over the rectangle $[-3, 3] \times [-3, 3]$. Is the shape of the graph consistent with the level curves you plotted in part (a)?

10. Let $f(x, y) = \ln(x^2 + y^2)$.

   (a) What is the domain of f?

   (b) What is the range of f?

   (c) Explain why the level curves of f are circles centered at the origin of the xy-plane.

11. Carefully, but briefly, describe the idea of a ***level curve*** of a function $f(x, y)$.

12. Let $f(x, y) = x - y^2 + 4$. Draw and label the level curves of f corresponding to $z = -4$, $z = 0$, and $z = 4$ on one set of axes.

13. Suppose that the level curves of a function $g(x, y)$ are horizontal lines. What does this imply about g?

14. Indicate whether each of the following statements ***must*** be true, ***might*** be true, or ***cannot*** be true. Justify your answers.

    (a) The level curve of the function $f(x, y)$ corresponding to $z = 7$ consists of the lines $y = -2x$ and $y = 5x$.

    (b) The level curve of the function $g(x, y)$ that corresponds to $z = -3$ is the ellipse $x^2/25 + y^2/9 = 1$ and the level curve that corresponds to $z = 2$ is the circle $x^2 + y^2 = 16$.

    (c) The level curve of the linear function $h(x, y)$ that corresponds to $z = 0$ is the parabola $y = 1 - x^2$.

15. Suppose that $f(x, y) = x^2 + y$. Describe the level curves of f.

16. Throughout this problem, let $\boldsymbol{X} = (x, y)$. For each of the following functions, describe the level sets $f(\boldsymbol{X}) = 0$ and $f(\boldsymbol{X}) = 3$. (The level sets are subsets of $\mathbb{R}^2$.)

    (a) $f(\boldsymbol{X}) = \boldsymbol{X} \cdot \boldsymbol{i}$

    (b) $f(\boldsymbol{X}) = \boldsymbol{X} \cdot \boldsymbol{X}$

17. Throughout this problem, let $\boldsymbol{X} = (x, y)$. For each of the following functions, describe the level sets $f(\boldsymbol{X}) = 0$ and $f(\boldsymbol{X}) = 3$. (The level sets are subsets of $\mathbb{R}^2$.)

    (a) $f(\boldsymbol{X}) = \boldsymbol{X} \cdot (\boldsymbol{i} + \boldsymbol{j})$

    (b) $f(\boldsymbol{X}) = \boldsymbol{X} \cdot (2, 3)$

18. Throughout this problem, let $\boldsymbol{X} = (x, y, z)$. For each of the following functions, describe the level sets $f(\boldsymbol{X}) = 0$ and $f(\boldsymbol{X}) = 3$. (The level sets are subsets of $\mathbb{R}^3$.)

    (a) $f(\boldsymbol{X}) = \boldsymbol{X} \cdot \boldsymbol{i}$

    (b) $f(\boldsymbol{X}) = \boldsymbol{X} \cdot \boldsymbol{X}$

19. Throughout this problem, let $\boldsymbol{X} = (x, y, z)$. For each of the following functions, describe the level sets $f(\boldsymbol{X}) = 0$ and $f(\boldsymbol{X}) = 3$. (The level sets are subsets of $\mathbb{R}^3$.)

    (a) $f(\boldsymbol{X}) = \boldsymbol{X} \cdot (\boldsymbol{i} + \boldsymbol{j} + \boldsymbol{k})$

    (b) $f(\boldsymbol{X}) = \boldsymbol{X} \cdot (1, 2, 3) + 4$

2.2 Partial derivatives

Derivatives in one variable—interpretations

Let f be a function of one variable. Recall some familiar properties of the derivative function f':

We assume, to avoid distractions, that f and f' are continuous functions.

Slope. For any fixed input $x = x_0$, the derivative $f'(x_0)$ tells the *slope* of the f-graph at the point $(x_0, f(x_0))$. The sign of $f'(x_0)$, in particular, tells whether f is increasing or decreasing at $x = x_0$.

Rising or falling, in everyday speech.

Rate of change. The derivative f' can also be interpreted as the *rate function* associated to f, as follows: For any input x_0, $f'(x_0)$ tells the instantaneous rate of change of $f(x)$ with respect to x. If, say, $f(x)$ gives the *position* of a moving object at time x, then $f'(x)$ gives the corresponding *velocity* at time x.

In a specific example we'd need to specify appropriate units for everything.

Limit. The derivative $f'(x_0)$ is defined as a *limit* of difference quotients:

$$f'(x_0) = \lim_{h \to 0} \frac{f(x_0 + h) - f(x_0)}{h}.$$

For any $h > 0$, the difference quotient can be thought of either as the *average rate of change* of f over the interval $[x_0, x_0 + h]$ or as the *slope of a secant line* on the f-graph, over the same interval. Taking the limit as $h \to 0$ corresponds to finding the *instantaneous* rate of change of f, or, equivalently, the slope of the *tangent* line to the f-graph at $x = x_0$.

Similar interpretations hold if $h < 0$, but $h = 0$ is taboo.

See the exercises for a brief refresher on these ideas.

Linear approximation. At a point $(x_0, f(x_0))$ on the curve $y = f(x)$, the tangent line has slope $f'(x_0)$. This tangent line is the graph of the linear function L, with equation

$$y = L(x) = f(x_0) + f'(x_0)(x - x_0).$$

The function L is called the **linear approximation to f at x_0.** The name makes sense because the graphs of f and L are close together near x_0. In symbols:

$$L(x) \approx f(x) \quad \text{when} \quad x \approx x_0.$$

See the exercises for more on linear approximation.

Derivatives in several variables

Derivatives are just as important in multivariable calculus as in one-variable calculus, but the idea is—not surprisingly—more complicated for functions of several variables.

Take slope, for instance. The graph of a one-variable function $y = f(x)$ is a curve in the xy-plane. The slope at (x_0, y_0)—a single number—completely describes the graph's "direction" at (x_0, y_0). The graph of a two-variable function $z = f(x, y)$, by contrast, is a *surface*, which has no single "slope" at a point. A surface's steepness at a point depends on the direction (uphill, downhill, along the "contour," etc.) that one follows from the point. To put it another way, the graph of a one-variable function can be approximated near a given point by a one-dimensional tangent *line*. The graph of a two-variable function, as we'll see, can be approximated near a fixed point by a two-dimensional tangent *plane*.

Every hiker knows this. How steep a mountain "feels" depends on the direction of the trail.

We'll say more soon (and repeatedly throughout this book) about tangent planes.

Here's the moral: To suit *multivariable* calculus, our notion of derivative must go beyond the simple idea of slope. As we'll see later (and often), the idea of linear approximation—not slope—turns out to be the key to extending the idea of derivative to functions of more than one variable.

In this section we start to extend the derivative idea to functions of several variables. **Partial derivatives** are the simplest multivariable analogues of ordinary derivatives.

We'll continue the project over several sections.

"Partial" suggests—correctly—that there's more to the derivative story.

Partial derivatives: the idea. The basic idea of a partial derivative is to differentiate with respect to *one* variable, holding all the others constant. An easy example will illustrate the idea and introduce some useful terminology and notation. The fine print will come later.

■ **Example 1.** Let $f(x, y) = x^2 - 3xy + 6$. Find $f_x(x, y)$ and $f_y(x, y)$, the partial derivatives of f with respect to x and y, respectively. Find the numerical values $f_x(2, 1)$ and $f_y(2, 1)$.

Solution: To find f_x, we differentiate $f(x, y) = x^2 - 3xy + 6$ with respect to x, *treating y as a constant*:

Check this calculation and the next carefully. Do you agree?

$$f_x(x, y) = 2x - 3y.$$

To find f_y, we *treat x as a constant*:

$$f_y(x, y) = -3x.$$

Setting $x = 2$ and $y = 1$ in these formulas gives

$$f_x(2, 1) = 2 \cdot 2 - 3 \cdot 1 = 1; \quad f_y(2, 1) = -3 \cdot 2 = -6.$$ □

The calculations were easy, but what do the results mean? What do they say about how f behaves near $(x, y) = (2, 1)$? Can we interpret the results graphically and numerically? Understanding multivariable derivatives fully is a long-term proposition, but here are some starters.

Partial derivatives are usually easy to calculate. What they mean is more important.

Holding variables constant. To find the partial derivative f_x of a function f with respect to x, we treat all the *other* variables as constants. This produces a function of just one variable, x, which we differentiate in the "usual" way.

For example, suppose we fix $y = 3$ in $f(x, y) = x^2 - 3xy + 6$. Then $f(x, y) = f(x, 3) = x^2 - 9x + 6$, and

$$\frac{d}{dx}(f(x, 3)) = f_x(x, 3) = 2x - 9.$$

This agrees, as it should, with the general formula $f_x(x, y) = 2x - 3y$, found above.

Directional rates of change. Partial derivatives (like ordinary derivatives) can be interpreted as rates or slopes—but with an important proviso about directions. For a function $f(x, y)$ and a point (x_0, y_0) in its domain, the partial derivative $f_x(x_0, y_0)$ tells the rate of change of $f(x, y)$ with respect to x, i.e., how fast $f(x, y)$ increases as the input (x, y) moves away from (x_0, y_0) *in the positive x-direction*. The other partial derivative, $f_y(x_0, y_0)$, tells how fast f increases near (x_0, y_0) as y increases. The next example illustrates what this means numerically.

x is the only variable that's free to move.

■ **Example 2.** The function $f(x, y) = x^2 - 3xy + 6$ has partial derivatives $f_x(x, y) = 2x - 3y$ and $f_y(x, y) = -3x$. What does this say about rates of change of f at $(x, y) = (2, 1)$? At $(x, y) = (1, 2)$?

We saw this above.

Solution: We'll start at (2, 1). The formulas give $f_x(2,1) = 1$ and $f_y(2,1) = -6$. These numbers represent rates of change of f with respect to x and y, respectively, at (2, 1). A table of f-values "centered" at (2, 1) shows what this means:

The boxed row and column meet at (2, 1), our target point.

Values of $f(x, y) = x^2 - 3xy + 6$ near (2, 1)

$y \backslash x$	1.97	1.98	1.99	2.00	2.01	2.02	2.03
1.03	3.7936	3.8022	3.8110	3.8200	3.8292	3.8386	3.8482
1.02	3.8527	3.8616	3.8707	3.8800	3.8895	3.8992	3.9091
1.01	3.9118	3.9210	3.9304	3.9400	3.9498	3.9598	3.9700
1.00	3.9709	3.9804	3.9901	4.0000	4.0101	4.0204	4.0309
0.99	4.0300	4.0398	4.0498	4.0600	4.0704	4.0810	4.0918
0.98	4.0891	4.0992	4.1095	4.1200	4.1307	4.1416	4.1527
0.97	4.1482	4.1586	4.1692	4.1800	4.1910	4.2022	4.2136

Reading *up* the boxed column (at each step, y *increases* by 0.01) shows successive corresponding values of f *decreasing* by about 0.06. Thus −6 is the rate of change of f with respect to y at (2, 1)—equivalently, $f_y(2,1) = -6$. Similarly, the fact that $f_x(2,1) = 1$ suggests that values of $f(x, 1)$ should increase at about the same rate as x, if $x \approx 2$. Reading across the boxed *row* confirms this expectation.

Convince yourself of this.

Similar reasoning applies at (1, 2). The values $f_x(1,2) = -4$ and $f_y(1,2) = -3$ mean that f *decreases* (at different rates) with respect to both x and y near (1, 2). Numerical f-values bear this out:

Use the formulas to check these values.

Values of $f(x, y) = x^2 - 3xy + 6$ near (1, 2)

$y \backslash x$	0.98	0.99	1.00	1.01	1.02
2.02	1.0216	0.9807	0.9400	0.8995	0.8592
2.01	1.0510	1.0104	0.9700	0.9298	0.8898
2.00	1.0804	1.0401	1.0000	0.9601	0.9204
1.99	1.1098	1.0698	1.0300	0.9904	0.9510
1.98	1.1392	1.0995	1.0600	1.0207	0.9816

Reading either across or up shows f-values decreasing, at rates of about −4 and −3, respectively. □

Formal definitions

Partial derivatives are defined formally as limits, much like ordinary derivatives:

Definition: Let $f(x, y)$ be a function of two variables. The partial derivative with respect to x of f at (x_0, y_0), denoted $f_x(x_0, y_0)$, is defined by

$$f_x(x_0, y_0) = \lim_{h \to 0} \frac{f(x_0 + h, y_0) - f(x_0, y_0)}{h},$$

if the limit exists. The partial derivative with respect to y at (x_0, y_0), denoted $f_y(x_0, y_0)$, is defined by

$$f_y(x_0, y_0) = \lim_{h \to 0} \frac{f(x_0, y_0 + h) - f(x_0, y_0)}{h},$$

if the limit exists.

The definition says, in effect, that $f_x(x_0, y_0)$ is the ordinary derivative at $x = x_0$ of an ordinary function of one variable—namely, the function given by the rule $x \to f(x, y_0)$. Here are some further comments and observations:

Do they exist? If either limit does not exist, then neither does the corresponding partial derivative. It's possible, for instance, that $f_x(0, 0)$ exists but $f_y(0, 0)$ doesn't. For most functions we'll see in this course, however, both partial derivatives *do* exist.

The exercises explore this a little further.

About domains. In order for the limits in the definition above to exist, $f(x, y)$ must be defined at (x_0, y_0) and at nearby points (x, y). In practice, this condition seldom causes trouble. In particular, there's no problem if (x_0, y_0) lies in the interior of the domain of f, i.e., if f is defined both at and near (x_0, y_0). (For the record, trouble is likeliest if (x_0, y_0) lies on the edge of the domain of f.)

Other notations. As with ordinary derivatives, various notations are used to denote partial derivatives. If, say, $z = f(x, y) = x \sin y$, then all of the expressions

$$f_x, \quad \frac{\partial f}{\partial x}, \quad \frac{\partial z}{\partial x}, \quad \text{and} \quad \frac{\partial}{\partial x}(x \sin y)$$

mean the same thing, as do

$$f_x(x_0, y_0), \quad \frac{\partial f}{\partial x}(x_0, y_0), \quad \left.\frac{\partial z}{\partial x}\right|_{(x_0, y_0)}, \quad \text{and} \quad \left.\frac{\partial}{\partial x}(x \sin y)\right|_{(x_0, y_0)}.$$

Partial derivatives with respect to y use similar notations. Notice especially the "curly-d" symbol ∂—it's read aloud as "partial."

Partial derivatives and contour maps

We saw in Example 2, page 108, how to estimate partial derivatives from a table of function values. Contour maps can be used for the same purpose. Thinking of $f_x(x_0, y_0)$ and $f_y(x_0, y_0)$ as rates suggests how to estimate partial derivatives: Use the contour map to measure how fast $z = f(x, y)$ rises or falls near (x_0, y_0) as either x or y increases.

On a topographic map, oriented the "usual" way, with north pointing "up," the partial derivatives f_x and f_y describe the steepness of the terrain in, respectively, the eastward and northward directions.

■ **Example 3.** Below is a contour map of the function $f(x, y) = x^2 - 3xy + 6$, centered at $(2, 1)$. Use the contour map to estimate the partial derivatives $f_x(2, 1)$ and $f_y(2, 1)$.

Level curves of $f(x, y) = x^2 - 3xy + 6$

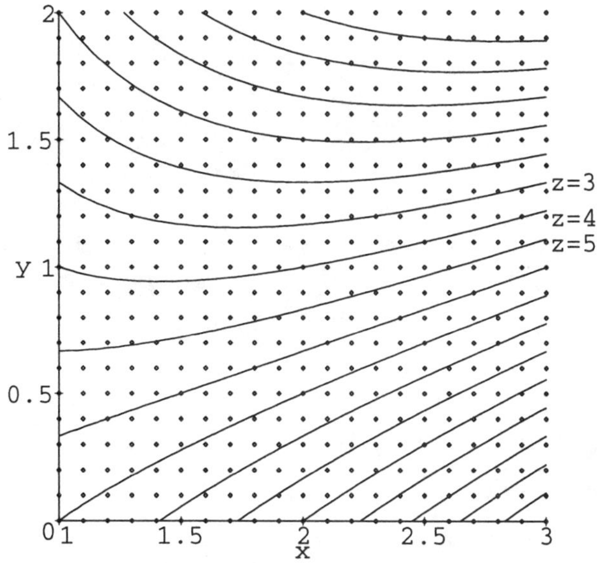

Solution: The level curves represent successive integer values of $z = f(x, y)$. In general, values of $f(x, y)$ increase as (x, y) moves toward the lower right.

Take a close look at the contour map to convince yourself of these claims.

First we'll estimate $f_x(2, 1)$. A close look at the picture suggests that $f(2.1, 1) \approx 4.1$. Increasing x by 0.1 increases f by 0.1; this suggests that $f_x(2, 1) \approx 1$. Similarly $f(2, 1.1) \approx 3.4$, so increasing y by 0.1 increases f by -0.6, and so $f_y(2, 1) \approx -6$. □

Partial derivatives, slicing, and traces

Fixing $y = y_0$, as we do when finding a partial derivative $f_x(x_0, y_0)$, can be thought of geometrically as slicing the surface $z = f(x, y)$ with the plane $y = y_0$—the curve of intersection between the plane and the surface is called a **trace.** Then the partial derivative $f_x(x_0, y_0)$ is the *slope* of the trace at $x = x_0$. We illustrate the idea with an example.

We discussed traces in Section 1.1.

■ **Example 4.** Consider the function $f(x, y) = 3 + \cos(x)\sin(2y)$, shown below.

The surface z=3+cos(x)*sin(2y)

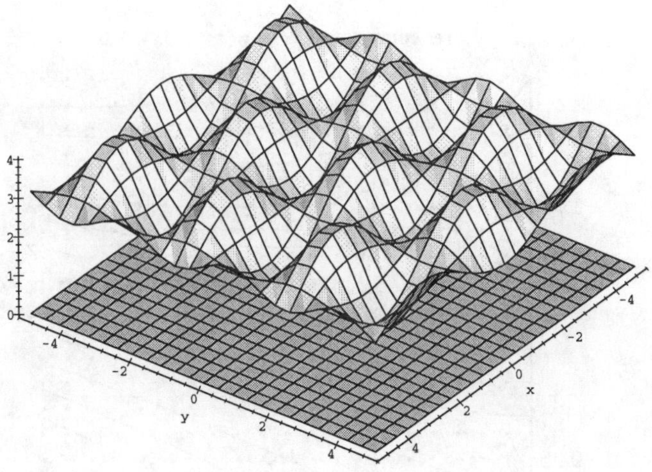

Calculate the partial derivatives. What are their numerical values at the origin? What do the answers mean about traces?

Solution: The symbolic calculations are easy:

$$\frac{\partial f}{\partial x}(x, y) = -\sin(x)\sin(2y) \quad \text{and} \quad \frac{\partial f}{\partial y}(x, y) = 2\cos(x)\cos(2y).$$

At the origin, therefore,

$$\frac{\partial f}{\partial x}(0, 0) = 0 \quad \text{and} \quad \frac{\partial f}{\partial y}(0, 0) = 2.$$

Both answers can be interpreted as slopes. Have a close look: Slicing the surface with the plane $y = 0$ produces the curve $z = 3$, which is "flat" at $x = 0$—as the partial derivative $f_x(0, 0) = 0$ suggests. Similarly, slicing the surface with the plane $x = 0$ produces the curve $z = 3 + \sin(2y)$, which has slope 2 at $y = 0$; correspondingly, $f_y(0, 0) = 2$. □

Partial derivatives and stationary points

In several variables as in one, points where derivatives vanish are of special interest.

In one variable. For a one-variable function $f(x)$, a domain point x_0 at which $f'(x_0) = 0$ is called a **stationary point**. Stationary points are natural places to look for maximum and minimum values of f—but it's possible that f may have neither a maximum nor a minimum point at x_0. (For example, $f(x) = x^3$ is stationary at $x = 0$, but $f(x)$ takes neither a maximum nor a minimum value at $x = 0$.) On the other hand, if a function f does assume a local maximum or local minimum value at x_0, and $f'(x_0)$ exists, then, necessarily, $f'(x_0) = 0$.

In several variables. The same terminology (and some of the same reasoning) applies to functions of of several variables. A domain point (x_0, y_0) is called a **stationary point** for a function $f(x, y)$ if *both* partial derivatives vanish there:

$$f_x(x_0, y_0) = 0; \quad f_y(x_0, y_0) = 0.$$

(For a function of three variables, all three partial derivatives must vanish.)

As in the one-variable case, a stationary point in several variables need not correspond either to a local maximum or to a local minimum of the function at hand. (The phrases "local maximum" and "local minimum" mean that $f(x_0, y_0)$ is either larger or smaller than $f(x, y)$ for all nearby values of (x, y)—exactly as for functions of one variable.) Indeed, the extra "room" in several variables permits $f(x, y)$ to assume many different types of behavior near a stationary point. On the other hand:

> **Theorem 1. (Extreme points and partial derivatives)** Suppose that $f(x, y)$ has a local maximum or a local minimum at (x_0, y_0). If both partial derivatives exist, then
>
> $$f_x(x_0, y_0) = 0 \quad \text{and} \quad f_y(x_0, y_0) = 0.$$

The result probably sounds plausible, but why is it true? Consider the case of a local maximum. Geometrically, the surface $z = f(x, y)$ has a "peak" above the domain point (x_0, y_0). If we slice the surface with any vertical plane, say the plane $y = y_0$, then the resulting curve—i.e., the trace of the surface $z = f(x, y)$ in the plane $y = y_0$—has the equation $z = f(x, y_0)$, and this curve must have a peak at $x = x_0$. From one-variable calculus we know, therefore, that if the function $x \to f(x, y_0)$ is differentiable, then its derivative must be zero at x_0. In other words,

$$\frac{dz}{dx}(x_0) = f_x(x_0, y_0) = 0.$$

For similar reasons, $f_y(x_0, y_0) = 0$.

The following example begins to suggest the variety of behaviors that are possible for a multivariable function near a stationary point.

■ **Example 5.** Let $f(x, y) = x^2 + y^2$ and $g(x, y) = xy$. Find all the stationary points of f and g. What happens at each one?

Solution: Finding the partial derivatives is easy:

$$f_x(x, y) = 2x; \quad f_y(x, y) = 2y; \quad g_x(x, y) = y; \quad g_y(x, y) = x.$$

Both f and g, therefore, are stationary only at the origin $(0, 0)$. For f, the origin is a minimum point, since $f(x, y) = x^2 + y^2 \geq 0$ for all (x, y). For g, however, the origin is *neither* a maximum nor a minimum point, since $g(x, y)$ assumes both positive and negative values near $(0, 0)$. (For example, $g(0.1, -0.1) < 0$, but $g(0.1, 0.1) > 0$.) Contour maps of f and g illustrate their very different behavior near the stationary point. (In the following pictures the level curves are the edges of the shaded regions. Note the "key" to the right of each contour map.) Here's f:

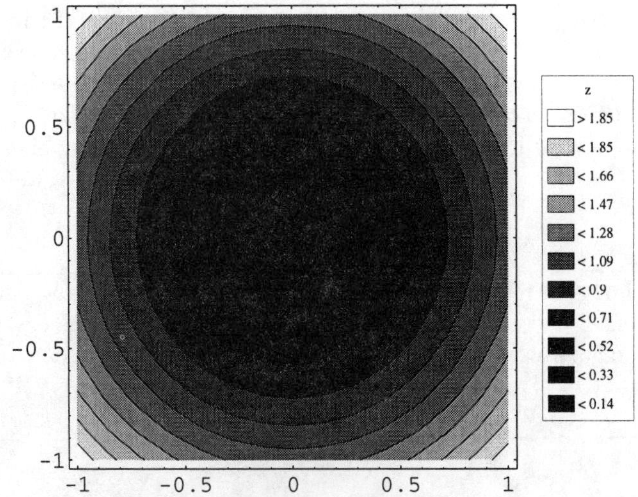

But take a careful look. Notice, in particular, that darker regions are "lower."

Level curves of f are circles, centered on the "basin" at $(0, 0)$. Thus the picture suggests that f has a local *minimum* at the stationary point $(x, y) = (0, 0)$. The suggestion is correct: $f(0, 0) = 0$, and $f(x, y) \geq 0$ for all (x, y) so f assumes a local (and even global) minimum value at $(0, 0)$.

Now look at g near its stationary point:

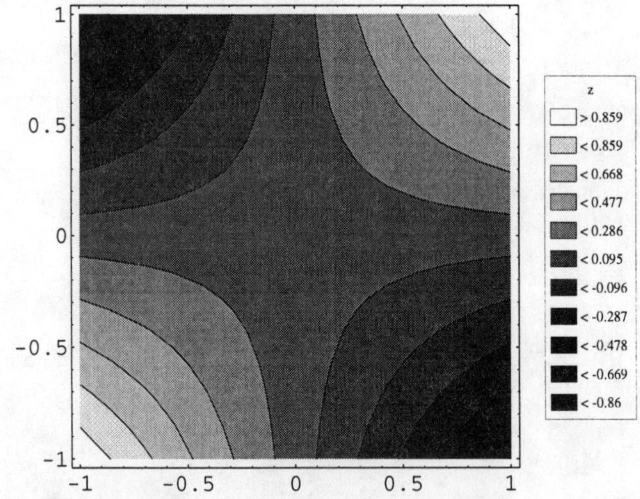

Think about this carefully. Can you "see" the saddle in the contour map?

Level curves of g show a **saddle point** at $(0, 0)$. (If the surface were literally a saddle, the horse would be walking either "northeast" or "southwest".) The surface rises above the first and third quadrants, and falls below the second and fourth quadrants.□

Maxima and minima—more to the story. The full story of finding maximum and minimum values of functions of several variables is more complicated than these simple examples convey. We'll return to it later in this chapter, after developing some more sophisticated calculus tools.

Exercises

1. For each of the following functions, find the partial derivative with respect to each variable.

   (a) $f(x, y) = x^2 - y^2$

   (b) $f(x, y) = x^2y^2$

   (c) $f(x, y) = \dfrac{x^2}{y^2}$

   (d) $f(x, y) = \cos(xy)$

   (e) $f(x, y) = \cos(x)\cos(y)$

   (f) $f(x, y) = \dfrac{\cos(x)}{\cos(y)}$

   (g) $f(x, y, z) = xy^2z^3$

   (h) $f(x, y, z) = \cos(xyz)$

2. Let $f(x) = x^2$.

   (a) Since $f'(x) = 2x$, $f'(3) = 6$. What does this mean about the graph of f? (Use a graph of f for $2.5 \le x \le 3.5$ to illustrate your answer.)

   (b) Plot f over the interval $2.5 \le x \le 3.5$. (It's OK to use a calculator or computer, but copy the graph onto paper.) On your graph, draw the secant lines from $x = 3$ to $x = 3.5$ and from $x = 3$ to $x = 3.1$. Find their slopes.

   (c) Find the average rate of change $\Delta y/\Delta x$ of f over the intervals $[3, 3.5]$ and $[3, 3.1]$. [HINT: The answers should be familiar from part (b).]

   (d) Find the limit $\lim_{h\to 0} \dfrac{f(3+h) - f(3)}{h}$. Interpret the answer as a derivative. Does the answer agree with other information already given?

3. Redo Exercise 2, but use $f(x) = x^2 - x$.

4. Some values of a function $g(x, y)$. (No explicit symbolic formula for g is given.) Use the table to answer the following questions.

Values of $g(x, y)$

$y\backslash x$	−0.0100	0.0000	0.0100	0.0200	...	0.9900	1.0000	1.0100	1.0200
1.02	2.0603	2.0604	2.0603	2.0600	...	1.0803	1.0604	1.0403	1.0200
1.01	2.0300	2.0301	2.0300	2.0297	...	1.0500	1.0301	1.0100	0.9897
1.00	1.9999	2.0000	1.9999	1.9996	...	1.0199	1.0000	0.9799	0.9596
0.99	1.9700	1.9701	1.9700	1.9697	...	0.9900	0.9701	0.9500	0.9297
⋮	⋮	⋮	⋮	⋮	⋮	⋮	⋮	⋮	⋮
0.02	0.0203	0.0204	0.0203	0.0200	...	−0.9597	−0.9796	−0.9997	−1.0200
0.01	0.0100	0.0101	0.0100	0.0097	...	−0.9700	−0.9899	−1.0100	−1.0303
0.00	−0.0001	0.0000	−0.0001	−0.0004	...	−0.9801	−1.0000	−1.0201	−1.0404
−0.01	−0.0100	−0.0099	−0.0100	−0.0103	...	−0.9900	−1.0099	−1.0300	−1.0503

$f'(1) = \frac{f(1.01) - f(1)}{.01}$

   (a) Use the table to estimate the partial derivatives $g_x(1, 1)$ and $g_y(1, 1)$.

   (b) It's true that $g_x(0, 0) = 0$ and $g_y(0, 0) = 1$. How do the table entries reflect these facts?

(c) Consider the linear function $L(x, y) = 0 + 0x + 1y = y$. Show that $L_x(0, 0) = g_x(0, 0) = 0$, $L_y(0, 0) = g_y(0, 0) = 1$, and $L(0, 0) = g(0, 0) = 0$.

(d) Fill in the following table of values for the function L from part (c).

Values of $L(x, y)$

$y \setminus x$	−0.02	−0.01	0.00	0.01	0.02
0.02					
0.01					
0.00					
−0.01					
−0.02					

Compare your results with the tabulated values of g. [The results show how L linearly approximates g near $(0, 0)$.]

(e) Find the linear function $M(x, y)$ such that (i) $M(1, 1) = g(1, 1)$, (ii) $M_x(1, 1) = g_x(1, 1)$, and (iii) $M_y(1, 1) = g_y(1, 1)$. [HINT: One approach is to write $M(x, y) = a + b(x - 1) + c(y - 1)$ and then use the conditions to find values for a, b, and c.]

5. Let $f(x, y) = \sin x$.

(a) Draw a contour map of f in the rectangle $-\pi \leq x \leq \pi$, $-2 \leq y \leq 2$. Show the level curves that correspond to $z = \pm 1$, $z = \pm 0.75$, $z = \pm 0.5$, $z = \pm 0.25$, and $z = 0$.

(b) Use the level curve diagram to estimate $f_x(0, 0)$ and $f_y(0, 0)$.

(c) Use the level curve diagram to estimate $f_x(\pi/2, 0)$ and $f_y(\pi/2, 0)$.

(d) The formula shows that $f_y(x, y) = 0$ for all (x, y). How does the contour map reflect this fact?

(e) The formula shows that $f_x(x, y)$ is independent of y. How does the contour map of f reflect this fact?

6. Let $f(x, y) = \cos(y)$.

(a) Draw a contour map of f in the rectangle $-\pi \leq x \leq \pi$, $-2 \leq y \leq 2$. Show the level curves that correspond to $z = \pm 1$, $z = \pm 0.75$, $z = \pm 0.5$, $z = \pm 0.25$, and $z = 0$.

(b) Use the level curve diagram to estimate $f_x(0, 0)$ and $f_y(0, 0)$.

(c) Use the level curve diagram to estimate $f_x(\pi/2, 0)$ and $f_y(\pi/2, 0)$.

(d) The formula for f shows that $f_x(x, y) = 0$ for all (x, y). How does the contour map of f reflect this fact?

(e) The formula for f shows that $f_y(x, y)$ is independent of x. How does the contour map of f reflect this fact?

7. Let $f(x, y) = 2x - 3y$.

(a) Draw a contour map of f in the rectangle $[-3, 3] \times [-3, 3]$. Show the level curves that correspond to $z = -5$, $z = -4$, $z = -3$, $\ldots$, $z = 4$, and $z = 5$.

(b) Use your contour map (not the formula) to find $f_x(0,0)$ and $f_y(0,0)$.

(c) The formula for f implies that both f_x and f_y are constant functions. How does the contour map of f reflect this fact?

(d) The formula for f implies that for any (x, y), $f_x(x, y) = 2$ and $f_y(x, y) = -3$. How does the contour map reflect the fact that $f_x(x, y)$ is positive but $f_y(x, y)$ is negative?

8. Let $f(x, y) = 2y - x$.

   (a) Draw a contour map of f in the rectangle $[-3, 3] \times [-3, 3]$. Show the level curves that correspond to $z = -5$, $z = -4$, $z = -3, \ldots, z = 4$, and $z = 5$.

   (b) Use your contour map (not the formula) to find $f_x(0,0)$ and $f_y(0,0)$.

   (c) The formula for f implies that $f_x(x, y) = -1$ and $f_y(x, y) = 2$ for all (x, y). How does the contour map reflect these facts? In particular, how does the contour map show that $f_x(x, y)$ is negative but $f_y(x, y)$ is positive?

9. Let $f(x, y) = |y| \cos x$. This exercise explores the fact that the partial derivatives of a function may or may not exist at a given point.

   (a) Use technology to plot $z = f(x, y)$ over the rectangle $[-5, 5] \times [-5, 5]$. The graph suggests that there may be trouble with partial derivatives where $y = 0$, i.e., along the x-axis. How does the graph suggest this? Which partial derivative (f_x or f_y) seems to be in trouble?

   (b) Use the definition to show that $f_y(0,0)$ does not exist. In other words, explain why the limit

$$\lim_{h \to 0} \frac{f(0, h) - f(0, 0)}{h}$$

   does not exist.

   (c) Show that $f_x(0,0)$ *does* exist; find its value. How does the result appear on the graph?

   (d) Use the limit definition to show that $f_y(\pi/2, 0)$ *does* exist; find its value.

   (e) How does the graph reflect the result of part (d)? (You may need to do some experimenting with the graph to answer this.)

   (f) Find a function $g(x, y)$ for which $g_y(0,0)$ exists but $g_x(0,0)$ does not. Use technology to plot its graph.

10. Let $f(x, y) = |x| y$.

   (a) By experimenting with graphs (use technology!), try to guess where $f_x(x, y)$ and $f_y(x, y)$ do exist and where they don't. (No proofs are needed.)

   (b) Use the limit definition to find $f_x(0,0)$.

   (c) Using the limit definition, explain why $f_x(0,1)$ does not exist.

11. Suppose that the intersection of the surface $z = f(x, y)$ and the plane $x = 3$ is the curve $z = 3 - y^3$. Suppose also that the intersection of the surface $z = f(x, y)$ and the plane $y = -2$ is the curve $z = 2 + x^2$. Evaluate $f_x(3, -2)$.

12. On a certain mountain, the elevation z above a point (x, y) in a horizontal xy-plane at sea level is $z = 2500 - 3x^2 - 4y^2$ ft. The positive x-axis points east; the positive y-axis points north. Suppose that a climber is at the point $(15, -10, 1425)$. If the climber walks due west, will the climber begin to ascend or descend? Justify your answer.
13. Find all stationary points of the function $f(x, y) = -x^3 + 4xy - 2y^2 + 1$.
14. Does the function $f(x, y) = 2y - \sin x$ have any stationary points? If so, find one. If not, explain why not.
15. Suppose that the level curves of the function $h(x, y)$ are described by equations of the form $y = x^2 + k$, where k is a real number. Furthermore, suppose that $h(2, 1) = 10$ and $h(1, 2) = 7$.

    (a) Evaluate $h(1, -2)$.

    (b) Estimate $h_y(0, -3)$.
16. Find each of the indicated partial derivatives.

    (a) $\dfrac{\partial}{\partial T}\left(\dfrac{2\pi r}{T}\right)$ (c) $\dfrac{\partial}{\partial y}\sin(3x^4y + 2x^3y^5)$

    (b) $\dfrac{\partial F}{\partial m_1}$ if $F = \dfrac{Gm_1m_2}{r^2}$ (d) $\dfrac{\partial}{\partial x}\left(xe^{\sqrt{xy}}\right)$
17. The equation $x^3y + y^3z + xz^3 = 1$ implicitly defines a function $z = f(x, y)$ near the point $(1, -2, 3)$. Compute $f_x(1, -2)$ and $f_y(1, -2)$. [HINT: Use implicit differentiation techniques from single-variable calculus.]
18. The equation $PV = nRT$ describes the relationship between the pressure P, volume V, and temperature T of n moles of an ideal gas; R is a number called the ideal gas constant. (This equation allows each variable to be expressed as a function of the other two.) Show that
$$\frac{\partial V}{\partial T}\cdot\frac{\partial T}{\partial P}\cdot\frac{\partial P}{\partial V} = -1.$$
19. The partial differential equation
$$\frac{\partial u}{\partial t} + \frac{\partial u}{\partial x} = ku$$
is used in population modeling. Here $u = u(x, t)$ is the number of individuals of age x at time t, and k is the mortality rate. Show that the function $u(x, t) = e^{\alpha x + \beta t}$ is a solution to this partial differential equation if the constants α and β satisfy the equation $\alpha + \beta = k$.
20. Suppose that the function $f : \mathbb{R}^2 \to \mathbb{R}$ has the values shown in the following table.

$y \backslash x$	1.5	2.0	2.5	3.0
3.0	4	6	9	6
2.5	6	9	7	5
2.0	4	8	6	4
1.5	3	5	5	7

Use the table to estimate $f_y(2.5, 2.5)$.

2.3 Partial derivatives and linear approximation

For a differentiable function $y = f(x)$ of one variable, the ordinary derivative can be interpreted in terms of linear approximation. For any point (x_0, y_0) on the f-graph, there's a certain line through this point—the tangent line—that best "fits" the graph near $x = x_0$. The derivative $f'(x_0)$ gives the slope of this tangent line. Knowing this, it's easy to find an equation for the tangent line in point-slope form:

$$y - y_0 = f'(x_0)(x - x_0).$$

Equivalently, we can think of the tangent line as the graph of the linear function L defined by➸

Recall: $f(x_0) = y_0$.

$$y = L(x) = y_0 + f'(x_0)(x - x_0).$$

The function L is called the **linear approximation to f at x_0**. The name is appropriate for three good reasons:

$$\text{(i) } L \text{ is linear;} \quad \text{(ii) } L(x_0) = f(x_0); \quad \text{(iii) } L'(x_0) = f'(x_0).$$

In short, L "agrees" with f at x_0 as closely as any linear function can—both functions have the same value and the same (first) derivative.➸

The idea of quadratic approximation is similar, except that we'd require agreement in the second derivative, too.

Linear approximation in several variables. The idea of linear approximation is essentially the same for functions of two (or more) variables: Given a function $f(x, y)$ and a point (x_0, y_0) in the domain of f, we look for a *linear* function $L(x, y)$ that has the same value and partial derivatives as those of f at (x_0, y_0). In other words, we want a linear function L such that

$$\text{(i) } L(x_0, y_0) = f(x_0, y_0); \quad \text{(ii) } L_x(x_0, y_0) = f_x(x_0, y_0); \quad \text{(iii) } L_y(x_0, y_0) = f_y(x_0, y_0).$$

In the one-variable case the graph of a linear approximation function is called a **tangent line**; for a function of two variables, the graph of the linear approximation function is called a **tangent plane**.

To give a geometric sense of what tangent planes mean, here is the tangent plane to the surface $z = f(x, y) = 3 + \cos(x)\sin(2y)$ at the point $(0, 0, 3)$ (a small part of the curved surface is shown, too):

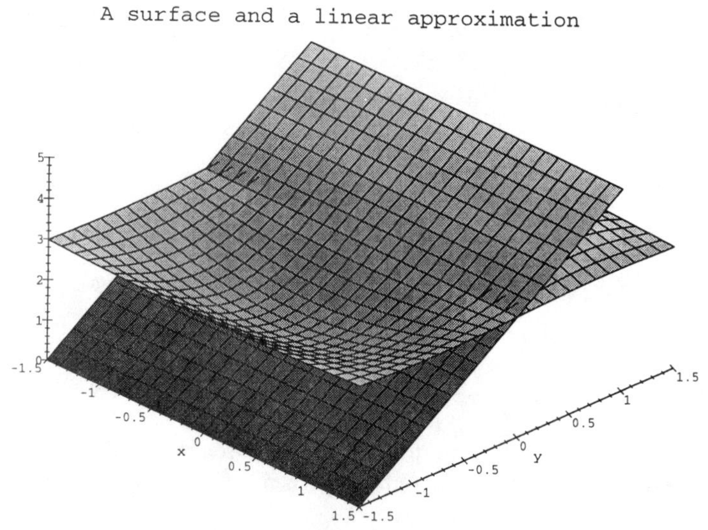

Observe how closely the flat plane and the curved surface fit together near the target point $(0, 0, 3)$. The next example shows how to find a tangent plane's formula.

Elsewhere, they don't fit so well.

Further calculations for the plane above are left to the exercises.

■ **Example 1.** Find the linear approximation function L to $f(x, y) = x^2 + y^2$ at $(x_0, y_0) = (2, 1)$. How are the graphs of f and L related? How are their contour plots related?

Solution: The partial derivatives of f are $f_x(x, y) = 2x$ and $f_y(x, y) = 2y$. Thus, at our base point $(x_0, y_0) = (2, 1)$:

$$f(2, 1) = 5; \quad f_x(2, 1) = 4; \quad f_y(2, 1) = 2.$$

Let's find a linear function L to "match" these values.

It's easiest to write L in the convenient form

We'll see why it's convenient in a moment.

$$L(x, y) = a(x - x_0) + b(y - y_0) + c = a(x - 2) + b(y - 1) + c,$$

and then to choose appropriate values for a, b, and c. Since $L(x, y) = a(x - 2) + b(y - 1) + c$, it's easy to see that

But check for yourself, especially for the derivatives.

(ie, point-slope form)

$$L(2, 1) = c; \quad L_x(2, 1) = a; \quad L_y(2, 1) = b.$$

To "match" the value and partial derivatives of f we must have $c = 5$, $a = 4$, and $b = 2$, so

$$L(x, y) = a(x - 2) + b(y - 1) + c = 4(x - 2) + 2(y - 1) + 5.$$

Here are views of f and L together:

Graphs of f and L together

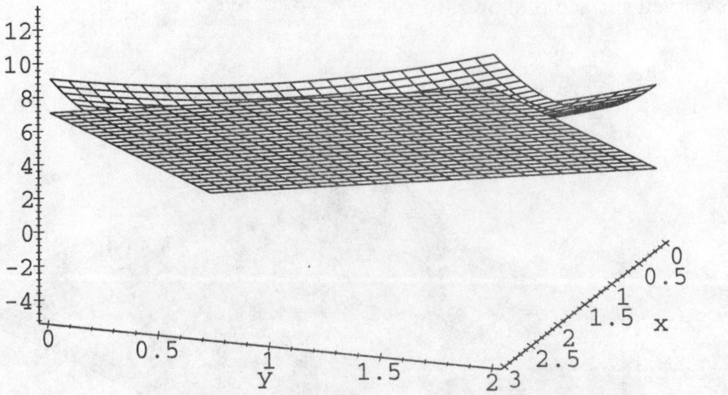

This picture, too, illustrates the phrases "tangent plane" and "linear approximation": The plane $z = L(x, y)$ touches the surface $z = f(x, y)$ at $(2, 1, 5)$; at this point, moreover, the flat plane "fits" the curved surface as well as possible. A closer look at both functions near $(2, 1)$, this time using contour maps, shows how good the fit really is:

The dotted lines are contours of L.

Level curves of f and L together

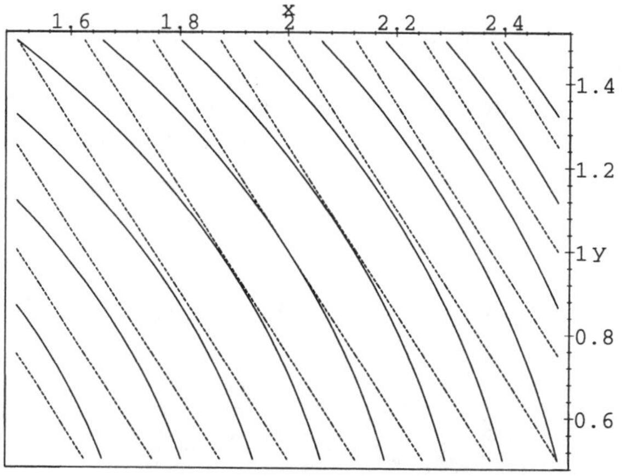

The two contour maps are almost identical near (2, 1). □

Linear approximation: the general formula. The procedure in the previous example works the same way for any function of several variables, as long as the necessary partial derivatives exist. Here are definitions for two and three variables:

The same idea works for any number of variables.

> **Definition: (Linear approximation)** Let $f(x, y)$ and $g(x, y, z)$ be functions, and suppose that all the partial derivatives mentioned below exist. The linear approximation to f at (x_0, y_0) is the function
>
> $$L(x, y) = f(x_0, y_0) + f_x(x_0, y_0)(x - x_0) + f_y(x_0, y_0)(y - y_0).$$
>
> The linear approximation to g at (x_0, y_0, z_0) is the function
>
> $$\begin{aligned} L(x, y, z) &= g(x_0, y_0, z_0) + g_x(x_0, y_0, z_0)(x - x_0) \\ &+ g_y(x_0, y_0, z_0)(y - y_0) + g_z(x_0, y_0, z_0)(z - z_0). \end{aligned}$$

■ **Example 2.** Find the linear approximation to $g(x, y, z) = x + yz^2$ at (1, 2, 3). Does the answer make numerical or graphical sense?

Solution: Easy calculations give

$$g(1, 2, 3) = 19; \quad g_x(1, 2, 3) = 1; \quad g_y(1, 2, 3) = 9; \quad g_z(1, 2, 3) = 12.$$

The linear approximation function, therefore, has the form

$$\begin{aligned} L(x, y, z) &= g_x(1, 2, 3)(x - 1) + g_y(1, 2, 3)(y - 2) + g_z(1, 2, 3)(z - 3) + g(1, 2, 3) \\ &= 1(x - 1) + 9(y - 2) + 12(z - 3) + 19. \end{aligned}$$

To see the situation numerically, we'll tabulate some values of each function:

Values of L and g near (1, 2, 3)

(x, y, z)	(1, 2, 3)	(1.1, 2, 3)	(1, 2.1, 3)	(1, 2, 3.1)	(1.1, 2.1, 3.1)	(3, 4, 5)
$g(x, y, z)$	19	19.1	19.9	20.22	21.281	103
$L(x, y, z)$	19	19.1	19.9	20.2	21.2	63

As the numbers illustrate, $L(x, y, z)$ and $g(x, y, z)$ are close together if, but only if, (x, y, z) is near $(1, 2, 3)$.

The last column illustrates what happens "far" from $(1, 2, 3)$.

To plot ordinary graphs of g and L would require four dimensions. Instead we'll plot, for comparison, the level surfaces $L(x, y, z) = 19$ and $g(x, y, z) = 19$, both of which pass through the base point $(1, 2, 3)$. (A **level surface** is like a level curve—it's a set of inputs along which a function has constant output value.) The pictures suggest, again, how similarly $g(x, y, z)$ and $L(x, y, z)$ behave when $(x, y, z) \approx (1, 2, 3)$.

One for each input variable and one for the output.

Level surfaces of g and L

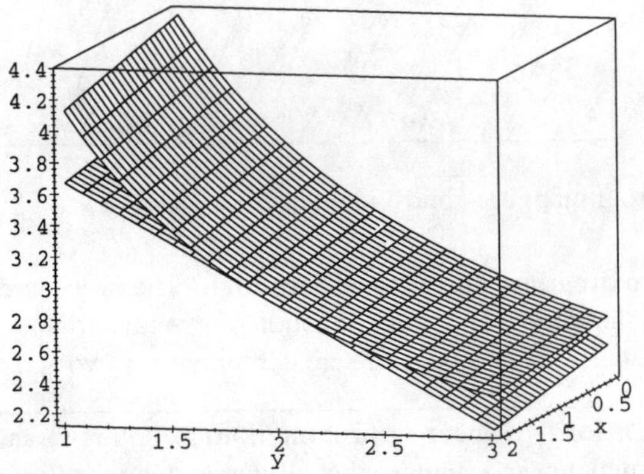

□

Partial derivatives, the cross product, and the tangent plane

Here's another way to think of the tangent plane to a surface $z = f(x, y)$ at a point (x_0, y_0, z_0). Consider again the curves formed by intersecting the surface with planes either of the form $y = y_0$ or of the form $x = x_0$. Look carefully again at the graph of $f(x, y) = 2 + \cos(x)\sin(2y)$. It shows many such curves; together, these curves give the "wireframe" appearance of the surface $z = f(x, y)$.

Let's parametrize the curve of intersection between a surface $z = f(x, y)$ (with $a \le x \le b$ and $c \le y \le d$) and a plane $y = y_0$. All points on this curve are of the form $(x, y_0, f(x, y_0))$. As a vector-valued function of t, therefore, the curve can be parametrized by

$$x = t; \quad y = y_0; \quad z = f(t, y_0); \quad a \le t \le b.$$

At $t = x_0$, the curve has position (x_0, y_0, z_0)—i.e., the curve passes through the point of interest, where $t = x_0$. The velocity vector at this point, therefore, is

$$\boldsymbol{v}(x_0) = (1, 0, f_x(x_0, y_0)).$$

(The last coordinate is found by differentiating with respect to t—but t plays exactly the same role in $f(t, y_0)$ as x does in $f(x, y_0)$.) We know that a velocity vector is tangent to its curve at the point in question. Here, the curve lies "in" the surface $z = f(x, y)$, so the velocity vector $(1, 0, f_x(x_0, y_0))$ is tangent to the surface at (x_0, y_0, z_0).

The name of the variable—t, x, s, etc.—is immaterial.

Now we replay the same game, but this time we use the curve of intersection between the surface $z = f(x, y)$ and the plane $x = x_0$. Reasoning just as before, we find *another* vector, $(0, 1, f_y(x_0, y_0))$, that is also tangent to the surface at (x_0, y_0, z_0). The results are worth putting together:

> **Fact:** Let $z = f(x, y)$ define a surface, with $z_0 = f(x_0, y_0)$. Then the two vectors
>
> $$(1, 0, f_x(x_0, y_0)) \quad \text{and} \quad (0, 1, f_y(x_0, y_0))$$
>
> are tangent to the surface at (x_0, y_0, z_0).

Having found *two* non-collinear vectors, each tangent to our surface at the "target" point, we can now write equations—in any form we like—for the tangent plane. In vector parametric form, we get

$$\boldsymbol{X}(s,t) = (x_0, y_0, z_0) + s(1, 0, f_x(x_0, y_0)) + t(0, 1, f_y(x_0, y_0));$$

the scalar parametric equations are

$$x = x_0 + s; \quad y = y_0 + t; \quad z = z_0 + s f_x(x_0, y_0) + t f_y(x_0, y_0).$$

We can also, if we prefer, write the plane in ordinary scalar form. To find a normal vector, we can take the cross product➸ of the two tangent vectors just found:

In either order—the answers are opposites, but both work equally well.

$$\boldsymbol{n} = (1, 0, f_x(x_0, y_0)) \times (0, 1, f_y(x_0, y_0)) = (-f_x(x_0, y_0), -f_y(x_0, y_0), 1).$$

Thus the tangent plane has vector equation

$$(\boldsymbol{X} - (x_0, y_0, z_0)) \cdot (-f_x(x_0, y_0), -f_y(x_0, y_0), 1) = 0,$$

or, in scalar form,➸

We did a little simplification.

$$(x - x_0) f_x(x_0, y_0) + (y - y_0) f_y(x_0, y_0) = z - z_0.$$

The last form should look familiar; we've derived it before.

■ **Example 3.** Use the preceding Fact to describe the tangent plane to the surface $f(x, y) = x^2 - 2y$ at the point $(3, 4, 1)$.

Solution: It's easy to see that $f_x(3, 4) = 6$ and $f_y(3, 4) = -2$. Therefore, by Fact, the vectors $(1, 0, 6)$ and $(0, 1, -2)$ are tangent to the surface at $(3, 4, 1)$. Their cross product, which is normal to the tangent plane, is $\boldsymbol{n} = (-6, 2, 1)$. It follows that the plane has scalar equation $-6x + 2y + z = -9$. In vector parametric form, the plane is given by

$-6(x-3) +2(y-4)+ z-$

$$\boldsymbol{X}(s,t) = (3, 4, 1) + s(1, 0, 6) + t(0, 1, -2).$$

Here's a rough picture of the tangent plane "patch" corresponding to the parameter values $0 \le s \le 2$ and $0 \le t \le 2$.

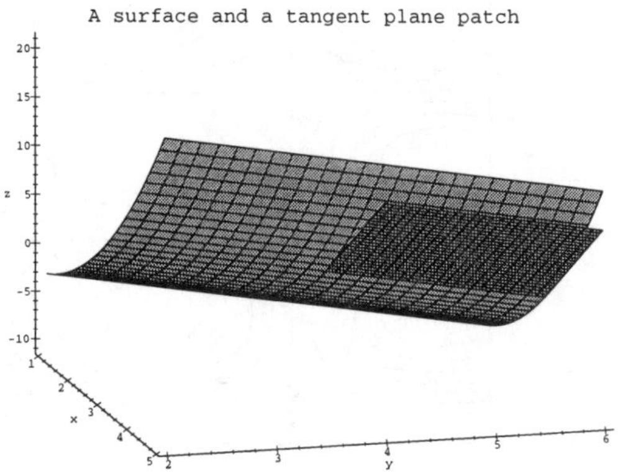

□

Exercises

1. Let $f(x, y) = 3+\cos(x)\sin(2y)$. Some values of f appear in the table below:

Values of $f(x, y) = 3 + \cos(x)\sin(2y)$					
$y \setminus x$	−0.2	−0.1	0.0	0.1	0.2
0.2	3.382	3.388	3.389	3.388	3.382
0.1	3.195	3.198	3.199	3.198	3.195
0.0	3.000	3.000	3.000	3.000	3.000
−0.1	2.805	2.802	2.801	2.802	2.805
−0.2	2.618	2.613	2.611	2.613	2.618

(a) Use the table to estimate the partial derivatives $f_x(0.1, 0.1)$ and $f_y(0.1, 0.1)$.

(b) Differentiate symbolically to find $f_x(0.1, 0.1)$ and $f_y(0.1, 0.1)$ exactly. Compare with results from (a).

2. The tangent plane at $(0, 0, 3)$ to the surface $z = f(x, y) = 3 + \cos(x)\sin(2y)$ is shown in this section.

(a) Find the linear approximation function $L(x, y)$ to f at $(0, 0)$. (Its graph is the tangent plane.)

(b) Write a vector equation for the tangent plane. (First find a normal vector.)

3. Let $f(x, y) = 3 + \cos(x)\sin(2y)$. A contour plot of f appears below:

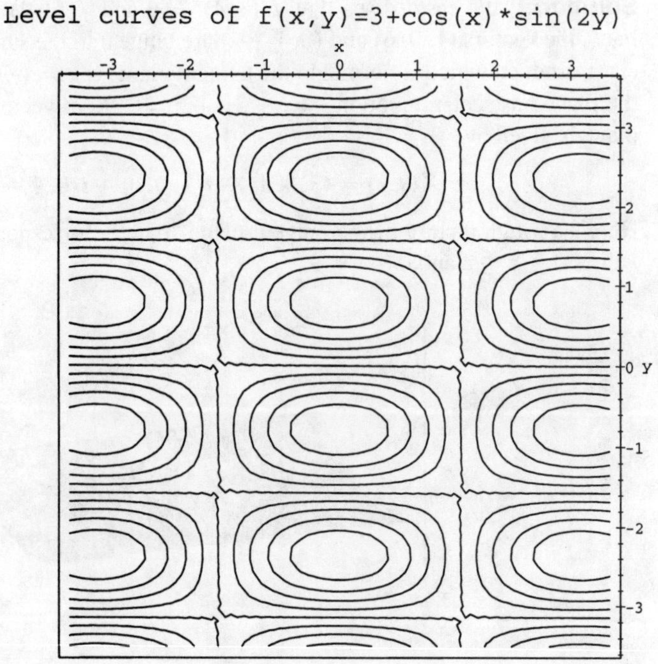

(a) Explain why f has a stationary point at every point of the form $(\pm\pi/2, k\pi/2)$, where k is any integer.

(b) How do these points appear on the contour plot?

(c) What kind of stationary point (a local maximum, a local minimum, or a saddle point) is $(\pi/2, 0)$? Indicate "uphill" and "downhill" directions as one moves through this point from southwest to northeast. What happens when moving from southeast to northwest?

4. In each part below, find the tangent plane to the surface $z = f(x, y)$ at the given point. ("Find" means give (i) a scalar equation, involving a normal vector; and (ii) parametric equations, involving two spanning vectors. For both parts, use the Fact on page 123).

If possible, use technology to plot the surface and the plane together. (Finding the "right" window may take some experimenting.)

(a) $z = x^2 + y^2$, at the point $(2, 1, 5)$

(b) $z = x^2 - y^2$, at the point $(2, 1, 3)$

(c) $z = \cos(xy)$ at the point $(0, 0, 1)$

(d) $z = 1 + \sin(xy)$ at the point $(0, 0, 0)$

5. Are the two vectors mentioned in the Fact on page 123 ever collinear? Why or why not?

6. Find an equation for a line that's perpendicular to each of the following surfaces at the given point. If possible, use technology to plot both the surface and the line.

(a) $z = x^2 + y^2$, at $(2, 1, 5)$

(b) $z = x - y^2$, at $(2, 1, 1)$

7. Let $f(x) = x^2$ and let $x_0 = 3$.

(a) Let L be the linear approximation to f at x_0. Show that $L(x) = 9 + 6(x - 3)$.

(b) Plot L and f on the same axes. (Choose your own plotting window.) Supposedly, "L linearly approximates f near x_0." How do the graphs illustrate this?

(c) Find an interval $a \le x \le b$ on which $|f(x) - L(x)| < 0.01$. (On this interval, $L(x)$ approximates $f(x)$ to within 0.01.) [HINT: This can be done either graphically, by zooming, or symbolically, by solving inequalities.]

8. Redo Exercise 7, but work with $f(x) = \sqrt{x}$ and $x_0 = 9$.

9. Let $f(x, y) = 2 + \sin y$. (The formula is independent of x.)

(a) Plot f; use the domain $-5 \le x \le 5$, $-5 \le y \le 5$. How does the shape of the graph reflect the fact that f is independent of x? (In Section 1.1 we called such graphs **cylinders**.)

(b) Find $f_x(x, y)$ and $f_y(x, y)$. How do the answers reflect the fact that f is independent of x?

(c) Find the linear approximation function L for f at the point $(0, 0)$. How does its form reflect the fact that f is independent of x?

10. This exercise is about the situation described in Example 3 (page 110). Use the function f and the contour map given there.

    (a) Use the contour map to estimate the partial derivatives $f_x(1.5, 1.5)$ and $f_y(1.5, 1.5)$.
    (b) Use the formula $f(x, y) = x^2 - 3xy + 6$ to find $f_x(1.5, 1.5)$ and $f_y(1.5, 1.5)$ exactly.
    (c) Use results of part (b) to find the linear approximation $L(x, y)$ to $f(x, y)$ at $(1.5, 1.5)$.

11. Let $f(x, y) = xy$. This exercise explores ideas like those of Example 1 (page 120).

    (a) Find the linear approximation function L to f at $(x_0, y_0) = (2, 1)$.
    (b) (Do this part by hand.) On one set of xy-axes, draw the level curves $L(x, y) = k$ for $k = 1, 2, 3, 4, 5$. On another set of axes, draw the level curves $f(x, y) = k$ for $k = 1, 2, 3, 4, 5$. (In each case, draw the curves into the square $[0, 3] \times [0, 3]$.)
    (c) How do the contour maps in part (b) reflect the fact that L is the linear approximation to f at the point $(2, 1)$? Explain briefly in words.
    (d) Use technology to plot contour maps of f and L in the window $1.8 \leq x \leq 2.2$, $0.8 \leq y \leq 1.2$. (This small window is centered at $(2, 1)$.) Explain what you see.

12. Repeat Exercise 11 using the function $f(x, y) = x^2 - y^2$.

13. Let $f(x, y) = x^2 + y^2$.

    (a) Use the contour map of f on page 99 to estimate the partial derivatives $f_x(1, 2)$ and $f_y(1, 2)$.
    (b) Check your answers to part (a) by symbolic differentiation.
    (c) Use your answers from part (a) to find the linear approximation $L(x, y)$ to $f(x, y)$ at $(1, 2)$.
    (d) On one set of axes, plot the level curves $L(x, y) = k$ and $f(x, y) = k$ for $k = 3, 4, 5, 6, 7$. (Use the window $[0, 3] \times [0, 3]$.) What's special about the point $(1, 2)$?

14. Let $f(x, y) = \sin(x) + 2y + xy$.

    (a) Find the partial derivatives $f_x(x, y)$ and $f_y(x, y)$; then evaluate $f_x(0, 0)$ and $f_y(0, 0)$.
    (b) Find a linear function $L(x, y) = a + bx + cy$ such that $L_x(0, 0) = f_x(0, 0)$, $L_y(0, 0) = f_y(0, 0)$, and $L(0, 0) = f(0, 0)$.
    (c) Complete the following table (report answers to 4 decimals).

(x, y)	$(0, 0)$	$(0.01, 0.01)$	$(0.1, 0.1)$	$(1, 1)$
$f(x, y)$				
$L(x, y)$				

How do the answers reflect the fact that L approximates f closely near $(0, 0)$?

(d) Use technology to draw contour plots of both f and L on the rectangle $-1 \leq x \leq 1$, $-1 \leq y \leq 1$. Label several contours on each. How do the pictures reflect the fact that L approximates f closely near $(0, 0)$?

15. For each function f, find the linear function L that linearly approximates f at the given point (x_0, y_0). (If possible, check your answers graphically by plotting both f and L near (x_0, y_0).)

    (a) $f(x, y) = x^2 + y^2$; $(x_0, y_0) = (2, 1)$.

    (b) $f(x, y) = x^2 + y^2$; $(x_0, y_0) = (0, 0)$.

    (c) $f(x, y) = \sin(x) + \sin(y)$; $(x_0, y_0) = (0, 0)$.

    (d) $f(x, y) = \sin(x)\sin(y)$; $(x_0, y_0) = (0, 0)$.

16. Let $f(x, y)$ be a differentiable function of two variables, let (x_0, y_0) any point in its domain, and let $L(x, y)$ be the linear approximation to f at (x_0, y_0). Show that if f is independent of one of the variables—say, x—then so is L.

17. Suppose we know that for a certain function f, $f(3, 4) = 25$, $f_x(3, 4) = 6$, $f_y(3, 4) = 8$, and $f(4, 5) = 41$.

    (a) Find a linear function $L(x, y)$ that approximates f as well as possible near $(3, 4)$.

    (b) Use L to estimate $f(2.9, 3.9)$, $f(3.1, 4.1)$, and $f(4, 5)$.

    (c) Could f itself be a linear function? Why or why not?

18. Suppose we know that for a certain function g, $g(3, 4) = 5$, $g_x(3, 4) = 3/5$, $g_y(3, 4) = 4/5$, and $g(4, 5) = \sqrt{41}$.

    (a) Find a linear function $L(x, y)$ that approximates g as well as possible near $(3, 4)$.

    (b) Use L to estimate $g(2.9, 4.1)$ and $g(4, 5)$.

    (c) Could g be a linear function? Why or why not?

19. Let $f(x, y) = 3x^2 - 4y^5$. Find the function $L(x, y)$ that is the linear approximation to f at $(2, 1)$.

20. Let $g(u, v) = uv\cos(u + v)$. Find the function $L(u, v)$ that is the linear approximation to g at (π, π).

21. When two electrical resistors R_1 and R_2 are connected in parallel, the equivalent resistance R is

$$R = \left(\frac{1}{R_1} + \frac{1}{R_2}\right)^{-1}.$$

Suppose that $R_1 = 300\,\Omega$ within 6% and it is desired that $R = 75\,\Omega$ within 3%. Use a linear approximation of the function $R(R_1, R_2)$ to estimate the interval of acceptable values for R_2.

22. Find a point on the surface $x^2 + y^2 + 3z^2 = 8$ where the tangent plane is parallel to the plane $2x + y + 3z = 0$.

2.4 The gradient and directional derivatives

A function $f(x, y)$ has two partial derivatives at a point (x_0, y_0) of its domain, one for each variable. Given their common origin, it's natural to "store" both derivatives as components of a single vector

Assuming (as we do) that both partial derivatives exist.

$$\nabla f(x_0, y_0) = \left(f_x(x_0, y_0), f_y(x_0, y_0) \right),$$

called the **gradient** of f at (x_0, y_0).

The preceding sections were about partial derivatives—the separate *components* of gradient vectors. We interpreted them in several ways: as directional rates of change, as slopes of certain curves, and as ingredients in finding linear approximation functions and tangent planes.

But what does the gradient mean as a *vector*? What do its length and direction tell us? How is the gradient related to other geometric objects we've already seen, such as level curves, tangent planes, and the surface $z = f(x, y)$? We study such questions in this section.

A vector of partial derivatives

Here is the formal definition:

> **Definition: (Gradient of a function at a point)** Let $f(x, y)$ be a function of two variables and (x_0, y_0) a point of its domain; assume that both partial derivatives exist at (x_0, y_0). The gradient of f at (x_0, y_0) is the plane vector
>
> $$\nabla f(x_0, y_0) = \left(f_x(x_0, y_0), f_y(x_0, y_0) \right).$$
>
> For a function $g(x, y, z)$ of three variables, the gradient is the space vector
>
> $$\nabla g(x_0, y_0, z_0) = \left(g_x(x_0, y_0, z_0), g_y(x_0, y_0, z_0), g_z(x_0, y_0, z_0) \right).$$

Notice:

Counting dimensions. Keeping track of dimensions can be tricky in multivariable calculus. For example, the graph of a function $f : \mathbb{R}^2 \to \mathbb{R}$ is a 2-dimensional object (a surface) hovering in 3-dimensional space. To keep things in their proper spaces, it's important to remember that for any function f and any input $\boldsymbol{X_0}$:

We're thinking of $\boldsymbol{X_0}$ as a point in either $\mathbb{R}^2$ or $\mathbb{R}^3$.

The gradient $\nabla f(\boldsymbol{X_0})$ is a vector with the same dimension as the domain of f.

Indeed, gradient vectors naturally "live" in the domain of f. We'll often draw them there, with the tail of $\nabla f(\boldsymbol{X_0})$ pinned at the point $\boldsymbol{X_0}$.

The gradient as a function. The ordinary derivative $f'(x)$ of a function of one variable is a new function of one variable. For a function $f(x, y)$, the gradient $\nabla f(x, y)$ can also be thought of as a new function—a vector-valued function—of x and y: For each domain point (x, y), the gradient recipe produces an associated vector. Notice one new feature of the multivariable setting: Although f is scalar-valued, the gradient function ∇f is a *vector*-valued function, sometimes known as a **vector field**.

The same principle applies for functions of three variables.

Calculating gradients is very easy. Developing intuition for what they mean is a little more challenging; we'll approach the matter through examples and pictures.

■ **Example 1.** Let $f(x, y) = x^2 - y^2$. Calculate some gradient vectors in the vicinity of the origin. What do their lengths and directions say?

Solution: For any input (x, y), $\nabla f(x, y) = (2x, -2y)$. For instance,

$$\nabla f(0, 0) = (0, 0); \quad \nabla f(2, 2) = (4, -4); \quad \nabla f(-2, 4) = (-4, -8).$$

We could keep this up forever, but it's much more enlightening to plot the results, pinning the tail of each vector arrow at the corresponding domain point. Here is a sample; level curves of f are added for reference:

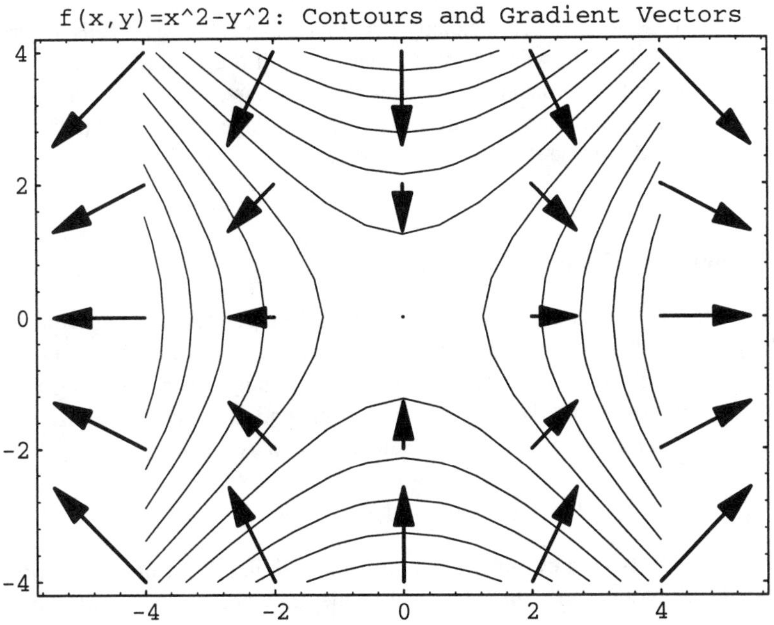

The picture deserves a very close look:

Just a sample. The gradient function assigns a vector to *every* point in the domain. For obvious reasons, the picture can show only a tiny sample. But even the 25 arrows shown here give some sense of how the gradient function varies over the domain.

Lengths not to scale. The gradient arrows point in the right directions, but lengths are not drawn to the same scale as shown on the axes. This is necessary to avoid arrows overlapping each other in the picture. Exactly *how* the lengths are scaled depends on the software.➻ In any event, arrows that should be longer *are* longer.

The details won't matter to us.

Gradient vectors point "uphill." Notice carefully the vectors along the coordinate axes. On the x-axis,➻ $f(x, y) = f(x, 0) = x^2$, so $f(x, y)$ *increases*, first slowly and then faster and faster, as (x, y) moves either east or west. For similar reasons, $f(x, y)$ *decreases* as (x, y) moves either north or south along the y-axis.

Where $y = 0$

All of this information (and much more) appears in the picture. Along the x-axis, all gradient vectors point away from the origin, and get longer as their base points get farther from the origin. On the y-axis, exactly the opposite happens. In short, gradient vectors point "uphill," and their lengths reflect how "steeply" the function increases.

A stationary point. At a stationary point for f, the gradient is the zero vector; the dot at the origin represents this situation. The picture also shows the sense in which the origin is a *saddle point* of f: Above the origin, the surface $z = f(x, y)$ resembles a saddle, with the horse heading east or west.

Not a maximum point or a minimum point.

Gradient vectors and contour lines. We've saved the most striking observation for last:

> *At each point* (x_0, y_0) *of the domain, the gradient vector is perpendicular to the level curve through* (x_0, y_0).

We'll soon discuss exactly *why* this occurs, though a fully rigorous proof will have to wait a few sections. But all the pictures in this section offer strong circumstantial evidence *that* it occurs. (Intuitively, the phenomenon seems reasonable: It says, in hiking language, that a level path runs perpendicular to the steepest uphill direction.)

□

The gradient of a linear function

A linear function, of the form $L(x, y) = ax + by + c$, has constant partial derivatives. Its gradient, therefore, is a *constant* vector: $\nabla L(x, y) = (a, b)$ for all (x, y).

This result looks especially simple and pleasing when everthing is written in vector notation, using the dot product. To this end, let's write $\mathbf{A} = (a, b)$ and $\mathbf{X} = (x, y)$. Then the observation above can be written as follows:

$$\text{If } L(\mathbf{X}) = \mathbf{A} \cdot \mathbf{X} + c, \quad \text{then} \quad \nabla L(\mathbf{X}) = \mathbf{A}.$$

The same fact—and even the same notation!—holds for a linear function of three variables. Notice, too, the similarity to the analogous *one*-variable fact: $L(x) = ax + b \implies L'(x) = a$.

Plotting gradient vectors and level curves together shows what it means, geometrically, for a function to be linear. Here's a sample of gradient vectors for the linear function $f(x, y) = 3x + 2y$:

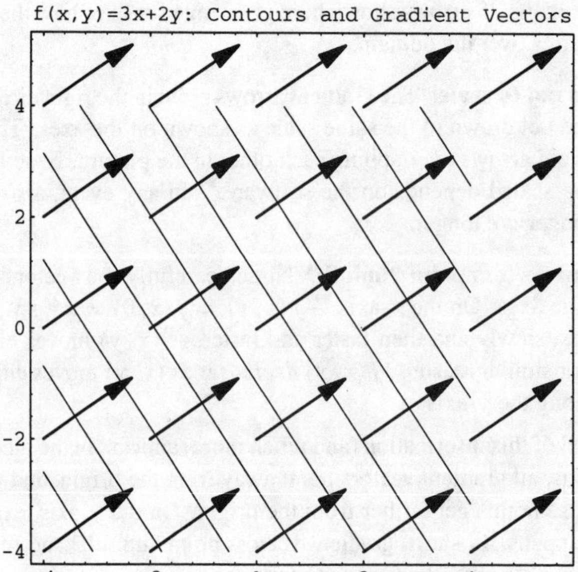

Again, the gradient vectors $(3, 2)$ look perpendicular to the level curves (lines, in this case). Indeed they are: For this function, every level line has the form $3x + 2y = k$, for some constant k. Each such line has slope $-3/2$, so $(2, -3)$ is a tangent vector, and the line is perpendicular to $(3, 2)$, as claimed.

Linear functions in three variables: gradients and level sets. Let $L(x, y, z) = ax+by+cz+d$ be a linear function of three variables. The gradient is $\nabla L(x, y, z) = (a, b, c)$—a constant 3-vector. Let (x_0, y_0, z_0) be any point of the domain, and suppose that $L(x_0, y_0, z_0) = w_0$. Consider the level set of L passing through any point (x_0, y_0, z_0), i.e., the set of points (x, y, z) such that $L(x, y, z) = ax+by+cz+d = w_0$. Since d and w_0 are constants, this set is a *plane* Π, with equation $ax+by+cz = w_0 - d$. As we've seen, Π has normal vector (a, b, c). This shows that for linear functions of three variables (as for two):

> *The gradient vector at any point is perpendicular to the level set through that point.*

Gradient vectors and linear approximations

Let $f(x, y)$ be any function and (x_0, y_0) a point of its domain. We've defined the linear approximation to f at (x_0, y_0) to be the linear function $L(x, y)$ that has the same value and the same partial derivatives as f at (x_0, y_0). That is,

$$L(x, y) = f(x_0, y_0) + f_x(x_0, y_0)(x - x_0) + f_y(x_0, y_0)(y - y_0).$$

We can write this in vector form, using the gradient vector and the notations $\boldsymbol{X_0} = (x_0, y_0)$ and $\boldsymbol{X} = (x, y)$:

$$L(x, y) = f(x_0, y_0)+\nabla f(x_0, y_0)\cdot(x-x_0, y-y_0) = f(\boldsymbol{X_0})+\nabla f(\boldsymbol{X_0})\cdot(\boldsymbol{X} - \boldsymbol{X_0}).$$

(For functions of three variables, the vector formula looks identical.)

As we've seen (and will return to more formally in the next section), every differentiable function of several variables—linear or not—can be closely approximated near a domain point $\boldsymbol{X_0}$ by its linear approximation function there. That's why the property of gradient vectors being perpendicular to level sets holds not only for linear functions, but for *all* differentiable functions of several variables. The following picture shows this principle in action again, this time for the *nonlinear* function

$f(x, y) = x^2 + y^2$:

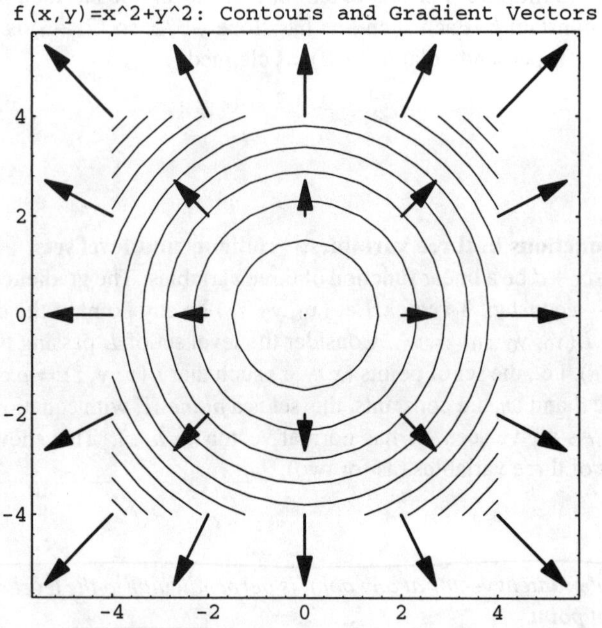

As in earlier pictures, all the gradient vectors point "uphill," perpendicular to the level curves.

Finding tangent planes. The perpendicularity property of gradient vectors makes it easy to find the tangent plane to a surface in $\mathbb{R}^3$.

Try to picture the situation.

■ **Example 2.** Find an equation for the tangent plane Π to the sphere $x^2+y^2+z^2 = 6$ at the point (2, 1, 1).

Solution: We can think of the sphere as the level surface $f(x, y, z) = 6$, where $f(x, y, z) = x^2+y^2+z^2$. The gradient is $\nabla f(x, y, z) = (2x, 2y, 2z)$; in particular, $\nabla f(2, 1, 1) = (4, 2, 2)$. This is a suitable normal vector for Π; an xyz-equation, therefore, is $4(x-2)+2(y-1)+2(z-1) = 0$, or $4x+2y+2z = 12$. □

Directional derivatives

Partial derivatives tell the rates of change of a function as inputs vary in the coordinate directions. But there's nothing sacred about the coordinate directions, and it's reasonable to ask about rates of change of f as inputs vary in *any* direction. **Directional derivatives** get at just this question. The following definition works in any dimension:

Definition: (Directional derivative) Let f be a function, $\boldsymbol{X_0}$ a point of its domain, and $\boldsymbol{u}$ a unit vector. The derivative of f at $\boldsymbol{X_0}$ in the $\boldsymbol{u}$-direction, denoted by $D_{\boldsymbol{u}} f(\boldsymbol{X_0})$, is defined by

$$D_{\boldsymbol{u}} f(\boldsymbol{X_0}) = \lim_{h\to 0} \frac{f(\boldsymbol{X_0}+h\boldsymbol{u}) - f(\boldsymbol{X_0})}{h},$$

if the limit exists.

Notice that with $\boldsymbol{u} = \boldsymbol{i}$, the limit above is the same one that defines $f_x(\boldsymbol{X_0})$. In fact,

$$D_{\boldsymbol{i}} f(\boldsymbol{X_0}) = f_x(\boldsymbol{X_0}), \quad D_{\boldsymbol{j}} f(\boldsymbol{X_0}) = f_y(\boldsymbol{X_0}), \quad \text{and} \quad D_{\boldsymbol{k}} f(\boldsymbol{X_0}) = f_z(\boldsymbol{X_0}).$$

Calculating directional derivatives with the gradient. As with other derivatives, the limit definition, useful as it is for explaining what the directional derivative is, is almost useless in calculations. Fortunately, the gradient comes to our rescue. For technical reasons,➥ we assume that f has continuous partial derivatives at $\boldsymbol{X_0}$. (This is so for virtually all the functions we'll encounter.)

More details in the next section.

> **Fact:** Let f, $\boldsymbol{X_0}$, and $\boldsymbol{u}$ be as above. Then
>
> $$D_{\boldsymbol{u}} f(\boldsymbol{X_0}) = \nabla f(\boldsymbol{X_0}) \cdot \boldsymbol{u}.$$

To see why the Fact is true, consider first what happens if f is a *linear* function of two variables, i.e., if $f(x, y) = ax + by + c = \nabla f \cdot (x, y) + c$. In this case,➥

Watch carefully as things drop away by subtraction.

$$\begin{aligned}\frac{f(\boldsymbol{X_0} + h\boldsymbol{u}) - f(\boldsymbol{X_0})}{h} &= \frac{\nabla f \cdot (\boldsymbol{X_0} + h\boldsymbol{u}) - \nabla f \cdot \boldsymbol{X_0}}{h} \\ &= \frac{\nabla f \cdot h\boldsymbol{u}}{h} = \frac{h \nabla f \cdot \boldsymbol{u}}{h} \\ &= \nabla f \cdot \boldsymbol{u}.\end{aligned}$$

This shows that the Fact holds for linear functions. Nonlinear functions are closely approximated by their linear approximations (the technical assumption stated just before the Fact guarantees this); it follows that the Fact holds for nonlinear functions as well.

Interpreting the gradient vector. Let $\boldsymbol{u}$ be any unit vector. By a property of the dot product,

because u has length 1.

$$D_{\boldsymbol{u}} f(\boldsymbol{X_0}) = \nabla f(\boldsymbol{X_0}) \cdot \boldsymbol{u} = |\nabla f(\boldsymbol{X_0})| \cos\theta,$$

where θ is the angle between $\nabla f(\boldsymbol{X_0})$ and $\boldsymbol{u}$. In particular,

$$D_{\boldsymbol{u}} f(\boldsymbol{X_0}) \le |\nabla f(\boldsymbol{X_0})|;$$

equality holds if and only if $\boldsymbol{u}$ points in the same direction as ∇f. Thus two important properties of the gradient follow from the Fact above:

Steepest ascent. The gradient ∇f points in the direction of fastest increase of f.

As fast as possible. The magnitude $|\nabla f(\boldsymbol{X_0})|$ is the largest possible rate of change of f.

■ **Example 3.** Find the directional derivatives of $f(x, y) = x^2 + y^2$ in various directions at $(2, 1)$. In which direction does f increase fastest? Decrease fastest?

Solution: The gradient is $\nabla f(2, 1) = (4, 2)$. In this direction, therefore, f increases at the rate of $|(4, 2)| = \sqrt{20}$ units of output per unit of input. In the opposite direction (i.e., the direction of $(-4, -2)$), $\cos\theta = -1$, so the directional derivative is $-\sqrt{20}$. In other directions, such as $\boldsymbol{u} = (1/\sqrt{2}, 1/\sqrt{2})$, the directional derivative is found from the gradient:

$$D_{\boldsymbol{u}} f(2, 1) = (4, 2) \cdot \left(\frac{1}{\sqrt{2}}, \frac{1}{\sqrt{2}}\right) = \frac{6}{\sqrt{2}}.$$

□

Exercises

1. In each part below, draw (by hand) a gradient map on the rectangle $[0, 2] \times [0, 2]$. At each point with integer coordinates, calculate and draw the gradient vector. Also draw the level curve through each such point.

   (a) $f(x, y) = (x + y)/2$
   (b) $f(x, y) = (x^2 - y)/2$
   (c) $f(x, y) = (y - x^2)/2$

2. In each part below, find the gradient at the given point. Then find the level curve through the given point and show that the gradient vector is perpendicular to the level curve at that point. (Hint: You'll need to find a tangent vector to the level curve at the given point. One way to do so is to parametrize the curve, and then use the parametrization to find a velocity vector to the curve at the given point. Another way is to use ordinary derivatives to find the slope of the curve at the given point, and then use the slope to find a suitable tangent vector. Once a tangent vector is found, check that it's perpendicular to the gradient vector.)

   (a) $f(x, y) = x$; $(x_0, y_0) = (2, 3)$.
   (b) $f(x, y) = x^2 - y$; $(x_0, y_0) = (1, 1)$.
   (c) $f(x, y) = x^2 - y$; $(x_0, y_0) = (2, 1)$.
   (d) $f(x, y) = x^2 + y^2$; $(x_0, y_0) = (2, 1)$.

3. Let $f(x, y) = ax + by + c$ be a linear function and (x_0, y_0) a point of the domain.

   (a) Write an equation for the level line through (x_0, y_0).
   (b) Show that the level line is perpendicular to the gradient.

4. In each part following, use the method of Example 2 (page 132) to find the tangent plane to the given surfaces at the given point. If possible, use technology to check your answer, i.e., to plot both the surface and the plane in an appropriate window.

   (a) $x^2 + y^2 + z^2 = 1$ at $(1, 0, 0)$ (Note: You should be able to check your answer using geometric intuition.)
   (b) $x^2 + y^2 + z^2 = 1$ at $(1/\sqrt{3}, 1/\sqrt{3}, 1/\sqrt{3})$
   (c) $x^2 - 2y = z$ at $(3, 4, 1)$ (Note: Compare Example 2 in the preceding section.)
   (d) $z = x^2 + y^2$ at $(2, 1, 5)$

5. Let $f(x, y) = x^2 + y^2$.

   (a) Find the directional derivatives of f at $(2, 0)$ in each of the 8 directions $\theta = 0, \theta = \pi/4, \theta = \pi/2, \ldots, \theta = 7\pi/4$. Express answers as decimal numbers.
   (b) Plot the data found in (a) as a function of θ. What general shape does the graph have?
   (c) Find a direction (is there more than one?) in which, at $(2, 0)$, f increases at the rate of 3 output units per input unit .

6. Consider the function $f(x, y) = x + y + \sin y$ and the gradient diagram below:

f(x,y)=x+y+sin(y): Contours, Gradient Vectors

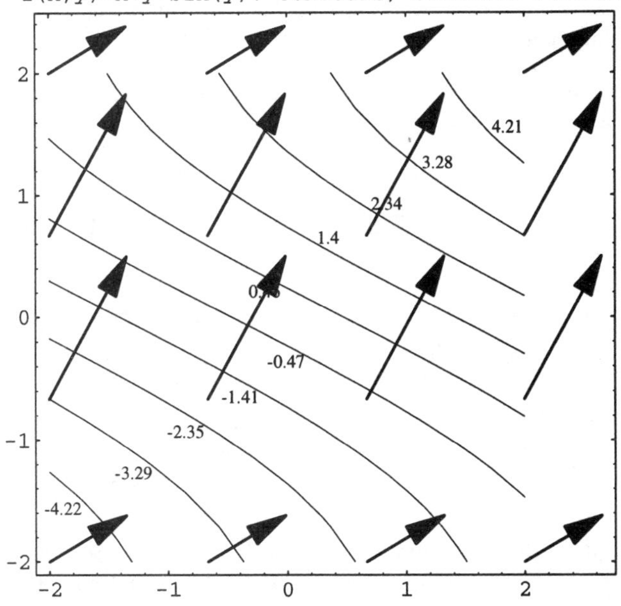

(a) Using the diagram alone (pretend you don't know the formula), estimate the partial derivatives $f_x(0, 0)$ and $f_y(0, 0)$. (Hint: Estimate the rates of change of f in the x- and y-directions.)

(b) Use the formula for f to check your guess in (a).

(c) Does f have any stationary points? If so, what type are they? If not, why not?

7. Let $f(x, y) = \cos(xy) + xe^y$.

(a) Find $\nabla f(x, y)$.

(b) Find an equation of the plane tangent to the surface $z = f(x, y)$ at the point $(1, 0, 2)$.

8. On a certain mountain, the elevation z above a point (x, y) in a horizontal xy-plane at sea level is $z = 2500 - 3x^2 - 4y^2$ ft. The positive x-axis points east; the positive y-axis points north. Suppose that a climber is at the point $(15, -10, 1425)$.

(a) If the climber walks due west, will the climber begin to ascend or descend? At what rate?

(b) If the climber walks southeast, will the climber begin to ascend or descend? At what rate?

(c) In what direction should the climber walk to travel a level path?

9. Suppose that at the point $(4, 5, 6)$ the function $f(x, y, z)$ increases most rapidly in the direction $(2, -1, -3)$ and that the rate of increase of f in this direction is 7. What is the rate of change of f at the point $(4, 5, 6)$ in the direction $(-1, 1, -2)$?

10. Find the minimum value of the directional derivative of $f(x, y) = 2x + \cos(xy)$ at the point $(\pi/4, 1)$. What direction corresponds to this value?

11. Suppose that the depth of the ocean at (x, y) is $h(x, y) = 2x^2+3y^2$ kilometers. If a ship sails from the point $(-1, 2)$ in the direction $(4, 1)$, is the water getting shallower or deeper? At what rate? (Assume that x and y are measured in kilometers.)

12. At the point $(1, 3)$, a function $f(x, y)$ has directional derivatives of $+2$ in the direction from $(1, 3)$ to $(2, 3)$ and -2 in the direction from $(1, 3)$ to $(1, 4)$.

    (a) Determine the gradient vector at $(1, 3)$.

    (b) Compute the directional derivative of f in the direction from $(1, 3)$ to $(3, 6)$.

13. Captain Kork is in trouble. His ship, the Exitprize, is too near the sun. The temperature of the hull is $T(x, y, z) = e^{-x^2-2y^2-3z^2}$ when the Exitprize is located at (x, y, z). (x, y, and z are measured in meters.) The Exitprize is currently at $(1, 1, 1)$.

    (a) In what direction should the Exitprize travel to decrease the temperature of its hull most rapidly?

    (b) If the ship travels at e^8 meters per second, how fast will the temperature decrease if the ship moves in the direction found in part (a)?

    (c) Unfortunately, the hull will crack if it cools at a rate greater than $\sqrt{14}e^2$ degrees per second. Describe the set of directions in which the ship can travel safely to cool its hull.

14. Let $f(x, y) = e^{xy} + y^2$.

    (a) Find $\nabla f(x, y)$.

    (b) Find a scalar equation for the plane tangent to the surface $z = f(x, y)$ at the point $(0, 1, 2)$.

15. Suppose that at the point $(-4, 2, 4)$ the function $f(x, y, z)$ decreases most rapidly in the direction $(3, 6, 2)$ and that the rate of change of f in this direction is -5. What is the rate of change of f at the point $(-4, 2, 4)$ in the direction $(4, 0, -3)$?

16. Suppose that $\nabla f(x, y) = (y^2 + x, 2xy)$ for all $(x, y) \in \mathbb{R}^2$ and that $f(1, 0) = 5$. Find $f(2, 1)$.

17. (a) Describe the direction in which the directional derivative of a function is a maximum.

    (b) Describe the direction in which the directional derivative of a function is a minimum.

    (c) Describe the direction in which the directional derivative of a function is zero.

2.5 Local linearity: theory of the derivative

Our discussion of multivariable derivatives has been, so far, mainly "practical." We've seen how to calculate partial derivatives and gradients, what they mean geometrically and numerically, and how to use them for such purposes as finding linear approximations to nonlinear functions.

This section is a brief excursion into the *theory* of multivariable derivatives. We'll consider more carefully than before just what it means for a function to "be differentiable" (in the multivariable sense) or to "have a linear approximation"—phrases we've used a little cavalierly up to now. We won't aim to dot every i and cross every t; rather, we'll point out some important theoretical questions (and a pitfall or two) and suggest approaches to handling them.

For simplicity, we'll discuss only functions of *two* variables. This is only a convenience—the same questions and answers arise in any dimension.

Linear approximation and differentiable functions

Let $f(x, y)$ be a function and (x_0, y_0) a point of its domain. We've defined the **linear approximation** to f at (x_0, y_0) to be the linear function

$$\begin{aligned} L(x, y) &= f(x_0, y_0) + f_x(x_0, y_0)(x - x_0) + f_y(x_0, y_0)(y - y_0) \\ &= f(x_0, y_0) + \nabla f(x_0, y_0) \cdot (\mathbf{X} - \mathbf{X_0}). \end{aligned}$$

(The third expression is the vector form of the second.)

To write L down, all we need from f is to have a value and both partial derivatives at (x_0, y_0). If this happens, then L and f have the same value and the same partial derivatives at (x_0, y_0). For these reasons it's natural to expect—and usually true in practice, as we've seen in examples—that $L(x, y)$ approximates $f(x, y)$ closely, not only *at* (x_0, y_0) but nearby as well.

The next example shows, however, that things don't always work quite so well.

■ **Example 1.** Consider the function f defined by

$$f(x, y) = \begin{cases} \dfrac{xy}{x^2 + y^2} & \text{if } (x, y) \neq (0, 0); \\ 0 & \text{if } (x, y) = (0, 0). \end{cases}$$

Find the linear approximation L to f at $(0, 0)$. Does L approximate f closely near $(0, 0)$?

Solution: In a word, no. Notice first that $f(x, y) = 0$ when (x, y) is on *either* coordinate axis; it follows that $f_x(0, 0) = 0$ and $f_y(0, 0) = 0$. Therefore the linear approximation function at $(0, 0)$ is the constant, zero:

Not surprising, probably . . . Otherwise we wouldn't be making such a fuss over it.

$$L(x, y) = f(0) + f_x(0, 0)(x - 0) + f_y(0, 0)(y - 0) = 0.$$

However, $f(x, y)$ is *not* near zero for (x, y) near the origin. For instance, if (x, y) lies on the line $y = x$, and $(x, y) \neq (0, 0)$, then

$$f(x, y) = \frac{xy}{x^2 + y^2} = \frac{x^2}{x^2 + x^2} = \frac{1}{2}.$$

If (x, y) is on the line $y = -x$, then

$$f(x, y) = \frac{xy}{x^2 + y^2} = \frac{-x^2}{x^2 + x^2} = -\frac{1}{2}.$$

In short, f behaves quite bizarrely near the origin, assuming different constant values on various lines through the same point. In fact, f is **discontinuous** at $(0,0)$, so there's no hope of approximating f closely with any linear function. □

Defining differentiability in one and several variables. The preceding example shows that for functions of several variables, having both partial derivatives at (x_0, y_0) is not enough to guarantee that f is continuous—let alone differentiable—at (x_0, y_0). That's why the definition we'll give below, which involves a limit, must require something more than mere existence of partial derivatives.

Recall that for a function $f(x)$ of *one* variable, the derivative $f'(x_0)$ is usually defined as the limit

$$\lim_{x\to x_0} \frac{f(x)-f(x_0)}{x-x_0} = f'(x_0),$$

if the limit exists. For a function $f(x, y)$ of two variables, which accepts vectors as inputs, we might first guess at something like this limit—

$$\lim_{\boldsymbol{X}\to\boldsymbol{X_0}} \frac{f(\boldsymbol{X})-f(\boldsymbol{X_0})}{\boldsymbol{X}-\boldsymbol{X_0}},$$

obtained by replacing the old scalar inputs x and x_0 with vectors $\boldsymbol{X}$ and $\boldsymbol{X_0}$. Alas, this strategy fails; we have no way of dividing a scalar (the numerator inside the limit) by a vector (the denominator inside the limit).

We can skirt this problem in quite a clever way. Notice first that for a function of *one* variable, the difference quotient definition of derivative is equivalent to another limit statement:

$$\lim_{x\to x_0} \frac{f(x)-f(x_0)}{x-x_0} = f'(x_0) \iff \lim_{x\to x_0} \frac{f(x)-f(x_0)-f'(x_0)(x-x_0)}{x-x_0} = 0.$$

The last condition, in turn, is equivalent to requiring that for some number $f'(x_0)$,

$$\lim_{x\to x_0} \frac{f(x)-\left(f(x_0)+f'(x_0)(x-x_0)\right)}{|x-x_0|} = 0.$$

(The absolute value in the denominator is harmless here—but it will be essential in the multivariable definition to come.)

Notice the numerator in the preceding limit: It's the difference $f(x) - L(x)$, where $L(x)$ is a *linear* function. That the limit above is zero means that as x tends to x_0, the difference $f(x) - L(x)$ tends to zero *faster* than the denominator $|x - x_0|$—in other words, the linear approximation $L(x)$ "fits" $f(x)$ even better than x "fits" x_0.

This last condition is key—the multivariable definition makes it formal:

Definition: (The total derivative) Let $f(x, y)$ be a function and $\boldsymbol{X_0} = (x_0, y_0)$ a point of its domain. Let

$$\begin{aligned} L(x,y) &= f(x_0,y_0) + f_x(x_0,y_0)(x-x_0) + f_y(x_0,y_0)(y-y_0) \\ &= f(\boldsymbol{X_0}) + \nabla f(\boldsymbol{X_0}) \cdot (\boldsymbol{X}-\boldsymbol{X_0}) \end{aligned}$$

be the linear approximation to f at (x_0, y_0). If

$$\lim_{\boldsymbol{X}\to\boldsymbol{X_0}} \frac{f(\boldsymbol{X})-L(\boldsymbol{X})}{|\boldsymbol{X}-\boldsymbol{X_0}|} = 0,$$

then f is **differentiable** at $\boldsymbol{X_0}$, and the vector $\nabla f(\boldsymbol{X_0})$ is the **total derivative** of f at $\boldsymbol{X_0}$.

Observe:

Local linearity. Here, as in the one-variable situation, the limit in the definition guarantees that the linear approximation $L(\boldsymbol{X})$ "fits" $f(\boldsymbol{X})$ even more closely near $\boldsymbol{X}_0$ than $\boldsymbol{X}$ approximates $\boldsymbol{X}_0$. This condition is described by the phrase: f is **locally linear** at $\boldsymbol{X}_0$.

Limit subtleties. For the limit condition in the definition to hold, the quantity inside the limit must tend to zero *regardless* of the direction through which $\boldsymbol{X}$ tends to $\boldsymbol{X}_0$. This is what's wrong with the function in Example 1: The quantity inside the limit tends to different "destinations" along different lines through the origin.

A fully rigorous treatment of the theory of limits in several variables would have to take careful account of the infinitely many directions along which one vector quantity can tend to another.

Which functions have total derivatives? Fortunately, functions like the one in Example 1, which have partial derivatives but no total derivative, appear only rarely (and then, mainly as counterexamples) in courses like this one. The following handy Fact (we state it without proof) guarantees this for most of the functions one is likely to meet in calculus courses:

> **Fact: (A condition for local linearity)** If the partial derivatives f_x and f_y are continuous at (x_0, y_0), then f is differentiable at (x_0, y_0), and has total derivative $\nabla f(x_0, y_0)$.

Exercises

Notes. *Maple* or another technology may be useful in some exercises.

1. This exercise concerns the "pathological" function of Example 1 (page 137), defined by

$$f(x, y) = \begin{cases} \dfrac{xy}{x^2 + y^2} & \text{if } (x, y) \neq (0, 0); \\ 0 & \text{if } (x, y) = (0, 0); \end{cases}$$

   (a) Explain why $f(x, y)$ is constant along every line $y = mx$ through the origin (except *at* the origin). What is the value of f along the line $y = mx$?

   (b) Draw (by hand) the contour lines $f(x, y) = \pm 1/2$, $f(x, y) = \pm 2/5$, and $f(x, y) = \pm 3/10$. Compare, if possible, with what *Maple* or another technology shows. Why do you think *Maple* has trouble?

   (c) Let $\boldsymbol{u} = (1/\sqrt{2}, 1/\sqrt{2})$. Does the directional derivative $D_{\boldsymbol{u}} f(0, 0)$ exist? Why or why not?

   (d) In what directions $\boldsymbol{u}$, if any, does the directional derivative $D_{\boldsymbol{u}} f(0, 0)$ exist?

   (e) If technology permits, plot the surface $z = \dfrac{xy}{x^2 + y^2}$ near the origin. Does the surface's shape seem right? Why do you think *Maple* has trouble?

2. Throughout this exercise, use the function defined by

$$g(x, y) = \begin{cases} \dfrac{x^2}{x^2 + y^2} & \text{if } (x, y) \neq (0, 0); \\ 0 & \text{if } (x, y) = (0, 0); \end{cases}$$

(a) Does one or both of the partial derivatives $g_x(0, 0)$ and $g_y(0, 0)$ exist? Why or why not?

(b) Explain why $g(x, y)$ is constant along every line $y = mx$ through the origin (except *at* the origin). What is the value of g along the line $y = mx$?

(c) Draw (by hand) the contour lines $g(x, y) = \pm 1/2$, $g(x, y) = \pm 1/5$, and $g(x, y) = \pm 1/10$. Compare, if possible, with what *Maple* or another technology shows. Why do you think *Maple* has trouble?

(d) Let $\boldsymbol{u} = (1/\sqrt{2}, 1/\sqrt{2})$. Does the directional derivative $D_{\boldsymbol{u}}g(0, 0)$ exist? Why or why not?

(e) In what directions $\boldsymbol{u}$, if any, does the directional derivative $D_{\boldsymbol{u}}g(0, 0)$ exist?

(f) If technology permits, plot the surface $z = \dfrac{x^2}{x^2 + y^2}$ near the origin. Does the surface's shape seem right? Why do you think *Maple* has trouble?

3. In each part below, find the linear approximation to the given function f at $(0, 0)$. Then use *Maple* or other technology to plot the quantity

$$\frac{f(x, y) - L(x, y)}{|\sqrt{x^2 + y^2}|}$$

near the origin. (This is the quantity in the limit definition of the total derivative—for a differentiable function, it should tend to zero near $(0, 0)$.)

(a) $f(x, y) = \sin(x + y)$

(b) $f(x, y) = \sin(xy)$

(c) $f(x, y) = x^2 + y$

2.6 Higher order derivatives and quadratic approximation

For a well-behaved function f of one variable, higher-order derivatives f'', f''', $\ldots$, $f^{(17)}$, $\ldots$ are readily calculated. Higher derivatives have various uses and interpretations. The second derivative, f'', for instance, has an important geometric meaning: It tells how fast (and in what direction) the slope function f' varies, and therefore describes the concavity of the f-graph. One use of this information is in distinguishing among various kinds of stationary points of f. Suppose, for instance, that $f'(x_0) = 0$, and that $f''(x_0) < 0$. Then the f-graph is concave down at x_0, and so f must have achieve a local *maximum* at x_0. This sort of reasoning is collected in the **second derivative test** of single-variable calculus. (We'll follow this line of thought in the next section.) Another one-variable use of higher derivatives is in writing Taylor and Maclaurin polynomials to approximate a given nonpolynomial function f.

In this section and the next we'll briefly review these ideas in the one-variable setting, and then see how they extend to functions of several variables.

Second and higher derivatives

Functions of several variables have repeated (partial) derivatives, just like their one-variable relatives. The following calculations won't be surprising, but notice the various notations and terminology.

■ **Example 1.** Let $f(x, y) = x^2 + xy^2$. Find all possible second partial derivatives.

Solution: The first partial derivatives are

$$f_x = \frac{\partial f}{\partial x} = 2x + y^2; \quad f_y = \frac{\partial f}{\partial y} = 2xy.$$

Differentiating *again* gives four results:

$$f_{xx} = \frac{\partial^2 f}{\partial x^2} = 2; \quad f_{xy} = \frac{\partial^2 f}{\partial y \partial x} = 2y;$$

and

$$f_{yx} = \frac{\partial^2 f}{\partial x \partial y} = 2y; \quad f_{yy} = \frac{\partial^2 f}{\partial y^2} = 2x.$$

(Notice that the symbols f_{xy} and $\dfrac{\partial^2 f}{\partial y \partial x}$ mean the same thing—even though the order of symbols may seem reversed.) □

A matrix of second derivatives. All the second partial derivatives of a function $f(x, y)$ can be collected in a 2×2 matrix, called the **second derivative** (or **Hessian**) matrix of f at (x, y):

$$f''(x, y) = \begin{pmatrix} f_{xx} & f_{xy} \\ f_{yx} & f_{yy} \end{pmatrix} = \begin{pmatrix} 2 & 2y \\ 2y & 2x \end{pmatrix}.$$

(The Hessian matrix of f at $\boldsymbol{X_0}$ is also sometimes denoted by $H_f(\boldsymbol{X_0})$; it's named for the German mathematician Ludwig Otto Hesse (1811–1874).)

Observe some properties of the second derivative matrix:

Dimension. For a function $f(x_1, x_2, \ldots, x_n)$ of n variables, the Hessian is an $n \times n$ matrix. The j-th entry in the i-th row is $f_{x_i x_j}$, the result of differentiating first with respect to x_i and then x_j. If, say, $f : \mathbb{R}^3 \to \mathbb{R}$ is defined by $f(x, y, z) = xz^2 + yz$, then the Hessian of f is the 3×3 matrix

Do you agree? Check details.

$$f''(x, y, z) = \begin{pmatrix} f_{xx} & f_{xy} & f_{xz} \\ f_{yx} & f_{yy} & f_{yz} \\ f_{zx} & f_{zy} & f_{zz} \end{pmatrix} = \begin{pmatrix} 0 & 0 & 2z \\ 0 & 0 & 1 \\ 2z & 1 & 2x \end{pmatrix}.$$

Rows are gradients. Successive rows of the Hessian are gradients of the *first* partial derivatives of f. The middle row above, for instance, is the gradient $\nabla f_y(x, y, z)$.

Order of differentiation (usually) doesn't matter. In both examples above, the second derivative matrix is symmetric with respect to the main diagonal. In other words,

$$f_{xy} = f_{yx}; \quad f_{xz} = f_{zx}; \quad f_{yz} = f_{zy};$$

the order of differentiation of "mixed partials" doesn't seem to matter (at least in the cases seen so far). It's an interesting fact that this phenomenon, holds for *all* well-behaved functions of several variables. (It's a useful fact, too—it cuts the work of evaluating a Hessian matrix more or less in half.) We'll state the result for a function $f(x, y)$ (though the same result holds for functions of any number of variables):

Theorem 2. (Equality of cross partials) Let $f(x, y)$ be a function; assume that the second partial derivatives f_{xy} and f_{yx} are defined and continuous on the domain of f. Then for all (x, y),

$$f_{xy}(x, y) = f_{yx}(x, y).$$

There are several ways to prove this theorem. One of the simplest involves a "double integral." Since we haven't met this idea yet we'll defer the proof until Section 3.2.

It might be surprising that an integral is involved.

Taylor polynomials and quadratic approximation

Let $f(x)$ be a function of one variable. Recall the idea of **Taylor polynomials** for $f(x)$ at a point x_0, from single-variable calculus. The polynomials

$$p_1(x) = f(x_0) + f'(x_0)(x - x_0) \text{ and } p_2(x) = f(x_0) + f'(x_0)(x - x_0) + \frac{f''(x_0)}{2}(x - x_0)^2$$

are called, respectively, the first and second (or first-order and second-order) Taylor polynomials for f, based at x_0. We've seen p_1 recently—it's simply the linear approximation to f at x_0. In the same spirit, p_2 is the **quadratic approximation** to f at x_0. (We'll generalize to multivariable functions in a moment.) In general, the nth Taylor polynomial has the form

$$p_n(x) = f(x_0) + f'(x_0)(x - x_0) + \frac{f''(x_0)}{2}(x - x_0)^2 + \frac{f'''(x_0)}{3!}(x - x_0)^3 + \cdots + \frac{f^{(n)}(x_0)}{n!}(x - x_0)^n, \tag{2.6.1}$$

where $f^{(n)}$ denotes the nth derivative and $n!$ is the factorial of n. The number $f^{(n)}(x_0)/n!$ is called the nth **Taylor coefficient**.

■ **Example 2.** Let $f(x) = e^x$. Find several Taylor polynomials for f, based at $x_0 = 0$. Do the same for $g(x) = \sin x$. How are the functions related to their Taylor polynomials?

Solution: The function $f(x) = e^x$ is the simplest possible case. Since $f(x) = f'(x)$, it follows that $f^{(n)}(0) = 1$ for all n. Thus, for all n, f has Taylor polynomials of the form

$$p_n(x) = 1 + 1x + \frac{x^2}{2!} + \frac{x^3}{3!} + \cdots + \frac{x^n}{n!}.$$

A similar calculation shows that for $g(x) = \sin x$, the first few Taylor polynomials based at $x_0 = 0$ have the form➻

Convince yourself!

$$P_1(x) = x; \quad P_3(x) = x - \frac{x^3}{6}; \quad \text{and} \quad P_5(x) = x - \frac{x^3}{6} + \frac{x^5}{120}.$$

The following picture shows the sine function and several of its Taylor polynomials:

Several Taylor polynomial approximations to $f(x) = \sin x$

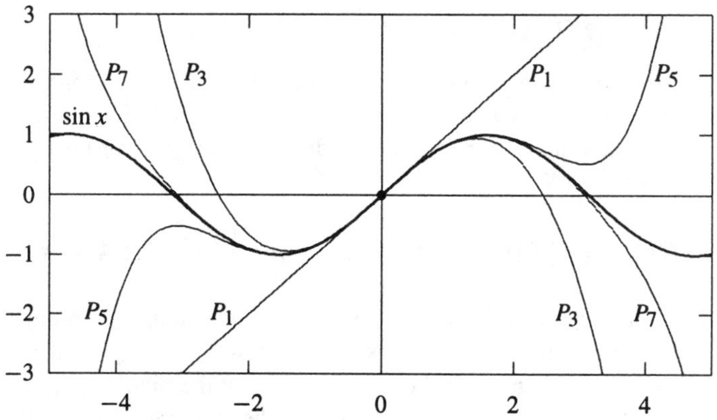

The key point, of course, is how closely the Taylor polynomials approximate the sine function at and near $x_0 = 0$. □

Matching derivatives. The key property of Taylor polynomials is that they match derivatives—from order zero through order n—with their "parent" functions at the base point.➻ If $n = 3$, for instance,

The value of a function is its "zeroth derivative."

$$p_3(x_0) = f(x_0); \quad p_3'(x_0) = f'(x_0); \quad p_3''(x_0) = f''(x_0); \quad p_3'''(x_0) = f'''(x_0).$$

This is easy to check; just differentiate the right side of Equation 2.6.1.➻ Because the function and the Taylor polynomial agree to this extent *at* $x = x_0$, they must behave very similarly nearby, as well.

Do this now; you'll see where the factorial comes from!

Taylor polynomials in several variables

The idea of Taylor polynomials makes excellent sense for functions of several variables, too. Indeed, we've already seen that for a function $f(x, y)$, the linear approximation at $\mathbf{X_0} = (x_0, y_0)$, which we've defined by

$$\begin{aligned} L(\mathbf{X}) &= f(\mathbf{X_0}) + \nabla f(\mathbf{X_0}) \cdot (\mathbf{X} - \mathbf{X_0}) \\ &= f(x_0, y_0) + f_x(x_0, y_0)(x - x_0) + f_y(x_0, y_0)(y - y_0), \end{aligned}$$

has the same value and the same (first) partial derivatives as f at (x_0, y_0).➻ Thus it's

Check this if you've forgotten why.

reasonable to call $L(x, y)$ the first-order Taylor polynomial for f at (x_0, y_0).

The next step—on to second partial derivatives—is now natural: The **quadratic approximation** to f at (x_0, y_0) is the function Q defined by

$$Q(x, y) = f(x_0, y_0) + f_x(x_0, y_0)(x - x_0) + f_y(x_0, y_0)(y - y_0) + \frac{f_{xx}(x_0, y_0)}{2}(x - x_0)^2 + f_{xy}(x_0, y_0)(x - x_0)(y - y_0) + \frac{f_{yy}(x_0, y_0)}{2}(y - y_0)^2.$$

Observe:

Partial derivatives match. The definition of $Q(x, y)$ guarantees that Q and f have the same value, the same *first* partial derivatives, and the same *second* partial derivatives at (x_0, y_0). To see, for instance, that $f_{xy}(x_0, y_0) = Q_{xy}(x_0, y_0)$, we differentiate twice, once with respect to each variable. We get

Notice that the order doesn't matter!

$$Q_x(x, y) = f_x(x_0, y_0) + f_{xx}(x_0, y_0)(x - x_0) + f_{xy}(x_0, y_0)(y - y_0);$$

therefore $Q_{xy}(x, y) = f_{xy}(x_0, y_0)$, as claimed.

In vector form. The second-order coefficients in the definition of Q come from the Hessian matrix of second partial derivatives f. It isn't surprising, therefore, that the vector form of the definition of $Q(x, y)$ involves the Hessian matrix $f''(\mathbf{X_0})$:

$$Q(\mathbf{X}) = f(\mathbf{X_0}) + \nabla f(\mathbf{X_0}) \cdot (\mathbf{X} - \mathbf{X_0}) + \frac{1}{2}\left[f''(\mathbf{X_0})(\mathbf{X} - \mathbf{X_0})\right] \cdot (\mathbf{X} - \mathbf{X_0}).$$

Both dots indicate dot products. There's also a matrix multiplication in the last term. Technically speaking, the first $(\mathbf{X} - \mathbf{X_0})$ in the last summand should be taken as a column vector, so that the matrix multiplication makes sense. The main point, however, is the typographical similarity with the one-variable form of the second Taylor polynomial.

More variables. A similar definition holds for functions f and Q of three or more variables. We'll stick mainly to two variables.

Appeal to higher powers. There's no need to stop with second-order Taylor polynomials. As in the one-variable case, the idea makes perfectly good sense for higher-order derivatives. The **cubic approximation** $C(x, y)$ to f at (x_0, y_0), for instance, includes all the terms in $Q(x, y)$, plus additional terms of the form

$$\frac{f_{xxx}(x_0, y_0)}{3!}(x - x_0)^3, \quad \frac{f_{xxy}(x_0, y_0)}{3!}(x - x_0)^2(y - y_0),$$

$$\frac{f_{xyy}(x_0, y_0)}{3!}(x - x_0)(y - y_0)^2, \text{ etc.}$$

At some point the bookkeeping becomes excessive for humans (though *Maple*, *Mathemtica*, and their relatives have no trouble); we'll content ourselves with second-order approximations.

■ **Example 3.** Find the quadratic approximation $Q(x, y)$ to $f(x, y) = xe^y$ at $(0, 0)$. How is it related to the Hessian matrix? How closely does Q appear to approximate f near $(0, 0)$?

Solution: Calculating partial derivatives of f gives

$$f(0,0) = 0; \quad \nabla f(0,0) = (1,0); \quad f''(0,0) = \begin{pmatrix} 0 & 1 \\ 1 & 0 \end{pmatrix}.$$

Therefore,

$$Q(x,y) = 0 + x + 0y + \frac{1}{2}(0x^2 + xy + yx + 0y^2) = x + xy.$$

The coefficients of the quadratic terms in Q, therefore, come directly from the Hessian matrix.

Like the linear approximation L, Q approximates f closely near the $(0,0)$. The following pictures show graphs of f, L, and Q for comparison. In each case, the darker surface is f:

Graphs of f and L

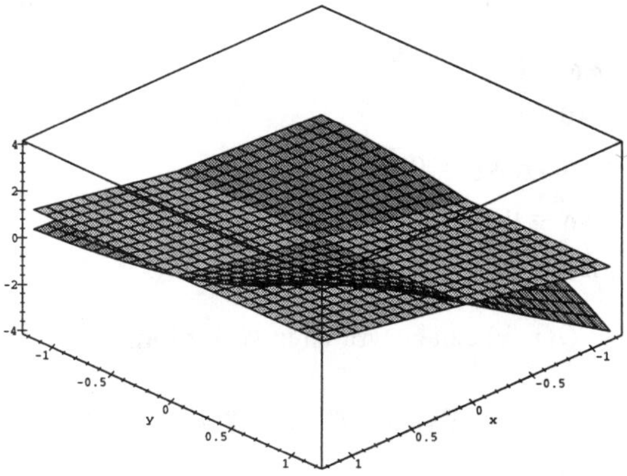

Graphs of f and Q

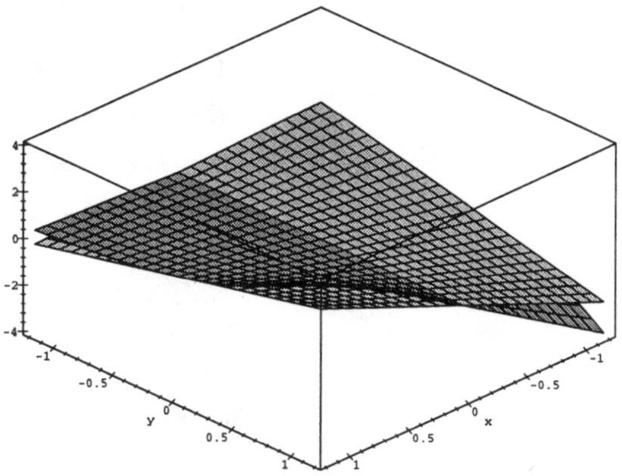

A close look shows that both L and Q approximate f, but that Q, being slightly curved, does a better job than L—as it should, given the extra matching derivatives.□

Exercises

1. Let $f(x) = \sin x$. Find $p_5(x)$, the 5th-order Taylor polynomial for f, based at $x_0 = 0$. Plot both f and p_5 on the same axes.

2. Find (by hand) the first, second, and third degree Taylor polynomials p_1, p_2, and p_3 for each of the following functions f, at the indicated base point x_0. indicated. If possible, use technology to plot f, p_1, p_2, and p_3 on the same axes, on an interval centered at x_0, to see their relation.

   (a) $f(x) = \cos x$; $x_0 = 0$

   (b) $f(x) = 1 + x + x^2 + x^3 + x^4$; $x_0 = 0$

   (c) $f(x) = \ln x$; $x_0 = 1$

3. In each part below, write (by hand) the Hessian matrix $f''(\boldsymbol{X_0})$ at the given value of $\boldsymbol{X_0}$; it's OK to check work using *Maple* or other technology.

(a) $f(x, y) = \sin(xy)$; $\boldsymbol{X_0} = (0, 0)$
(b) $f(x, y) = xy$; $\boldsymbol{X_0} = (0, 0)$
(c) $f(x, y) = \sin x + \cos(2y)$; $\boldsymbol{X_0} = (0, 0)$
(d) $f(x, y) = x^2 + y^2$; $\boldsymbol{X_0} = (0, 0)$
(e) $f(x, y) = x^2 - y^2$; $\boldsymbol{X_0} = (0, 0)$
(f) $f(x, y, z) = Ax + By + Cz + D$; $\boldsymbol{X_0} = (x_0, y_0, z_0)$
(g) $f(x, y) = Ax^2 + By^2 + Cxy + Dx + Ey + F$; $\boldsymbol{X_0} = (x_0, y_0)$
(h) $f(x, y) = Ax^3 + Bx^2y + Cxy^2 + Dy^3$; $\boldsymbol{X_0} = (0, 0)$
(i) $f(x, y) = (x + y)^3$; $\boldsymbol{X_0} = (0, 0)$ (Hint: First expand the formula for f in powers of x and y.)

4. In each part below, write out (by hand) the quadratic approximation function $Q(x, y)$ at the given value of $\boldsymbol{X_0}$. (Use results from the previous exercise.) It's OK to check work using *Maple* or other technology. If possible, plot both f and Q as surfaces and/or as contourplots, to see the connection between them.

(a) $f(x, y) = \sin(xy)$; $\boldsymbol{X_0} = (0, 0)$
(b) $f(x, y) = xy$; $\boldsymbol{X_0} = (0, 0)$
(c) $f(x, y) = \sin x + \cos(2y)$; $\boldsymbol{X_0} = (0, 0)$
(d) $f(x, y) = x^2 + y^2$; $\boldsymbol{X_0} = (0, 0)$
(e) $f(x, y) = x^2 - y^2$; $\boldsymbol{X_0} = (0, 0)$

5. We said in this section that $Q(x, y)$ can be written in vector form as

$$Q(\boldsymbol{X}) = f(\boldsymbol{X_0}) + \nabla f(\boldsymbol{X_0}) \cdot (\boldsymbol{X} - \boldsymbol{X_0}) + \frac{1}{2} f''(\boldsymbol{X_0})(\boldsymbol{X} - \boldsymbol{X_0}) \cdot (\boldsymbol{X} - \boldsymbol{X_0}).$$

where the dots indicate dot products, and the first $(\boldsymbol{X} - \boldsymbol{X_0})$ in the last summand should be written as a column vector. Work out the details to convince yourself of this.

6. Let $f(x, y, z)$ be a function of three variables, and let $\boldsymbol{X_0} = (0, 0, 0)$. Write out the formula for the quadratic approximation $Q(x, y, z)$ at $\boldsymbol{X_0}$. How many terms are there?

7. Let $f(x, y) = x^2y^2 + xy^3$. Find $f''(x, y)$.

2.7 Maxima, minima, and quadratic approximation

Optimization—finding maximum and minimum values of a function—is as important for multivariable functions as for one-variable functions. Functions of several variables are more complicated, of course, but derivatives remain the crucial tools.

A one-variable review

For a one-variable differentiable function $y = f(x)$ on an interval I, finding maximum and minimum values is relatively straightforward. Maxima and minima are found only at **stationary points**—where $f'(x) = 0$—or at the endpoints (if any) of the interval I. Usually, only a few such "candidate" points exist, and we can check directly which produces, say, the largest value of f.

A simple idea lies behind all talk of derivatives and extrema: At a local maximum or minimum point x_0, the graph of a differentiable function f must be "flat," so $f'(x_0) = 0$. However, a stationary point x_0 might be (i) a local maximum point; (ii) a local minimum point; or (iii) neither. (The function $f(x) = x^3$ at $x = 0$ illustrates the "neither" case.)

"Extremum" (singular) means "either maximum or minimum"; "extrema" is the plural.

For simple, one-variable functions, deciding which of (i)—(iii) actually holds is easy. One strategy is to check the sign of $f''(x_0)$. If, say, $f''(x_0) < 0$, then f is concave down at x_0, so f has a local maximum there. Alternatively, we might just plot f and see directly how it behaves near x_0.

*We'll review the **second derivative test** in more detail later.*

Without technology at hand, plotting f may be hard.

Local talk. When is a maximum or minimum "local"? When is it "global"? This graph shows the difference:

Graph of f: local vs. global extrema

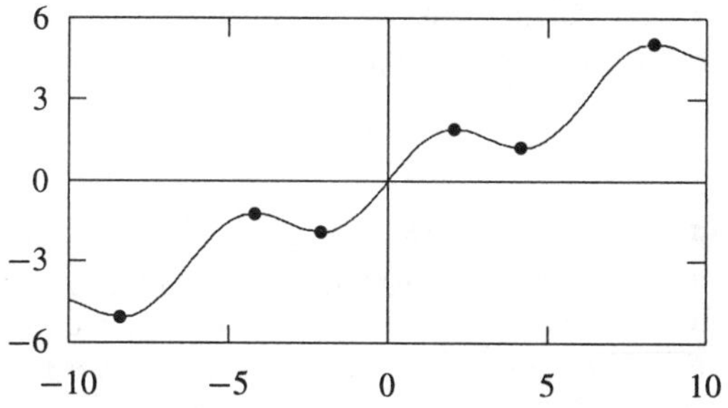

All six bulleted points (three maximum points and three minimum points), correspond to *local* extrema. At each bulleted point $(x_0, f(x_0))$, $f(x_0)$ is either highest or lowest among *nearby* points $(x, f(x))$. (For present purposes, exactly *how* nearby doesn't matter.) Only the first and last points, however, are *global* extrema, because they (and only they) represent the largest and smallest values of $f(x)$ among *all* possibilities shown in the picture. (On a larger domain interval, say $-15 \le x \le 15$, these extrema might *not* be global.)

Local maxima and minima are conceptually simpler than the global variety, readily recognizable on graphs, and often convenient to locate symbolically, using derivatives. With all these advantages, local extrema are usually the main tools in solving optimization problems.

Functions of several variables

In optimizing functions of several variables, it's natural (just as for functions of one variable) to look especially at stationary points of the domain, where all partial derivatives are zero. Indeed, as we observed in Section 2.2, local maximum and minimum values can occur *only* at stationary points:

See page 113 for the idea of a proof.

> **Fact: (Extreme points and partial derivatives)** Suppose that $f(x, y)$ has a local maximum or a local minimum at (x_0, y_0). If both partial derivatives exist, then
>
> $$f_x(x_0, y_0) = 0 \quad \text{and} \quad f_y(x_0, y_0) = 0.$$

But the Fact alone doesn't tell the whole story—different types of behavior are possible for a function at and near a stationary point.

Different types of stationary points. While the basic strategy for optimizing a function—find the stationary points and analyze them—is exactly the same for functions of one and of several variables, the situation is usually more complicated for functions of several variables. For one thing, finding stationary points may be harder; for another, functions of several variables can behave in more complicated ways near a stationary point. This makes multivariable optimization harder, but also more interesting.

As we did for functions of one variable, we'll identify three main types of stationary point for a function $f(x, y)$. (The definitions for a function $g(x, y, z)$ are almost identical.)

Local minimum point. A stationary point (x_0, y_0) is a **local minimum point** for f if $f(x, y) \geq f(x_0, y_0)$ for all (x, y) near (x_0, y_0). (A little more formally: $f(x, y) \geq f(x_0, y_0)$ for all (x, y) in some *rectangle* surrounding (x_0, y_0).) In this case we say that f **assumes a local minimum value** at (x_0, y_0). For example, $(0, 0)$ is a local minimum point for $f(x, y) = x^2 + y^2$.

Picture the graph—it's a paraboloid, opening upward.

Local maximum point. A stationary point (x_0, y_0) is a **local maximum point** for f if $f(x, y) \leq f(x_0, y_0)$ for all (x, y) near (x_0, y_0). In this case we say that f **assumes a local maximum value** at (x_0, y_0). For example, $(0, 0)$ is a local maximum point for $f(x, y) = 1 - x^2 - y^2$.

The graph is another paraboloid, opening downward.

Saddle point. A stationary point (x_0, y_0) is a **saddle point** for f if f assumes neither a local maximum nor a local minimum at (x_0, y_0).

The following contour plot shows that all three possibilities for stationary points can coexist in close proximity; the function in question is $f(x, y) = \cos(x)\sin(y)$:

Look in the picture for a stationary point of each type listed above.

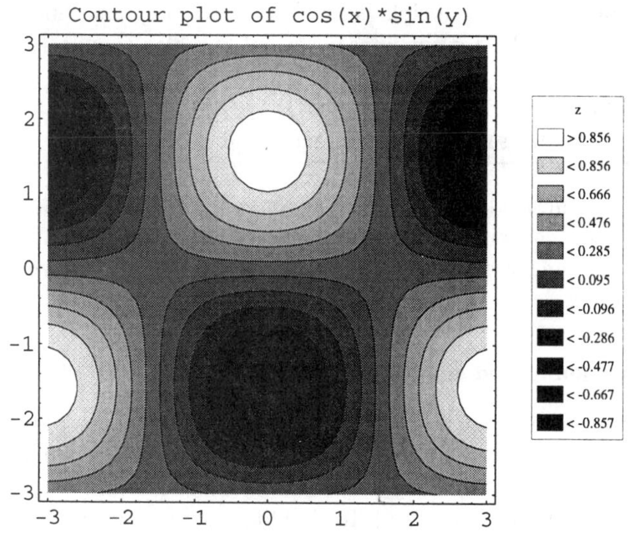

Second derivatives

For functions of one variable, we sometimes use the second derivative to distinguish among stationary points of different types. The situation is similar—but a little more complicated—for functions of two variables. (Things become more complicated still for functions of three or more variables—but the important ideas remain the same. We'll stick mainly to the two-variable case.)

Think about it "geographically." The surface $z = f(x, y)$ resembles a mountainous landscape; at a stationary point, the terrain is "level" in both the east–west and the north–south directions. As hikers know, this can happen in various ways. The point in question might be a mountain peak, the base of a bowl-shaped valley, a mountain pass between two peaks, or a point on a "terrace."

If technology is available, it's often possible to distinguish among different types of stationary points just by looking at a graph or contour map. The preceding contour plot, for instance, shows quite clearly the nature of the various stationary points.

But sometimes graphs aren't available, and in any case it's satisfying to relate stationary points to a function's symbolic formula. Second derivatives—in this case, the Hessian matrix of second partials—turn out to be helpful, and quadratic approximation is the key tool. At a stationary point x_0, the linear approximation is a constant function. This means that the quadratic approximation function has the form

$$Q(x) = f(x_0) + \frac{f''(x_0)}{2}(x - x_0)^2.$$

Therefore:

- If $f''(x_0) > 0$, then the Q-graph is a parabola opening upward, so Q (and therefore f) has a local minimum at $x = x_0$.

- If $f''(x_0) < 0$, then the Q-graph is a parabola opening downward, so Q (and therefore f) has a local maximum at $x = x_0$.

- If $f''(x_0) = 0$, then Q is constant, and so the quadratic approximation gives no information about the stationary point at x_0. (If the cubic term of the Taylor approximation is non-zero, then *it* may help determine the nature of the stationary point. But that's another story.)

The following key principle is implicit in the paragraphs above and in most of what follows; it holds for functions of several variables, too:

> **Fact:** Let f and Q be as above. If Q is not constant, then f and Q have the same type of stationary point at x_0.

This Fact means that quadratic functions are, for many other functions, simpler prototypes for checking behavior near a stationary point. It's especially important, therefore, to understand clearly how quadratic functions themselves behave near stationary points.

Stationary points and quadratic approximation

Let f be a function f of two variables; to avoid technical complications we assume that all the partial derivatives in question exist and are continuous. (This assures that for any input point (x_0, y_0), f *has* a quadratic approximation; without that, nothing following makes sense.) To simplify notation and save a little space, we'll use $(x_0, y_0) = (0, 0)$. Then the quadratic approximation has the form

This is only a convenience—we won't use any special properties of the origin.

$$Q(x, y) = f(0,0)+f_x(0,0)x+f_y(0,0)y+\frac{f_{xx}(0,0)}{2}x^2+f_{xy}(0,0)xy+\frac{f_{yy}(0,0)}{2}y^2.$$

If $(0, 0)$ is a stationary point, then the first-order terms disappear. Here's what's left:

$$Q(x, y) = f(0,0) + \frac{f_{xx}(0,0)}{2}x^2 + f_{xy}(0,0)xy + \frac{f_{yy}(0,0)}{2}y^2.$$

Our question, then, is how the values of f_{xx}, f_{xy}, and f_{yy} determine the type of stationary point.

The first term above is a constant, so what matters here are the last three terms, which have the form $Ax^2 + Bxy + Cy^2$. Let's decide how such a function behaves near $(0, 0)$; then we'll return and relate the answers to second derivatives. (For later reference, $A = f_{xx}(0,0)/2$, $B = f_{xy}(0,0)$, and $C = f_{yy}(0,0)/2$.)

Analyzing $Ax^2 + Bxy + Cy^2$

Let $f(x, y) = Ax^2 + Bxy + Cy^2$; then $(0, 0)$ is a stationary point of f regardless of the values of A, B, and C. To see how the *type* of stationary point depends on these values, we'll study some simple but important examples. In each case, we'll calculate the Hessian matrix at $(0, 0)$ for later reference.

Plot as many surfaces as possible as aids to intuition.

■ **Example 1.** Let $f(x, y) = x^2 + y^2$. How does f behave near the stationary point $(0, 0)$? Describe the surface $z = f(x, y)$. What difference would it make if, instead, $f(x, y) = -(x^2 + y^2)$?

Solution: The Hessian matrix is simple:

$$f''(0,0) = \begin{pmatrix} 2 & 0 \\ 0 & 2 \end{pmatrix}.$$

Clearly, f has a local minimum at $(0, 0)$, because for all (x, y),

$$f(x, y) = x^2 + y^2 \geq 0 = f(0, 0).$$

The surface $z = x^2 + y^2$ is a **circular paraboloid** with vertex at $(0, 0)$. The level curves of f are circles centered at $(0, 0)$.

Trading $f(x, y) = x^2 + y^2$ for $f(x, y) = -x^2 - y^2$ turns everything upside down. The local minimum becomes a local maximum at $(0, 0)$, the surface becomes a downward-opening paraboloid, and the Hessian matrix acquires negative signs. □

■ **Example 2.** Let $g(x, y) = 3x^2 + 2y^2$. How does g behave near the stationary point $(0, 0)$? Describe the surface $z = g(x, y)$.

Solution: The difference from the preceding example is only in the positive constants 2 and 3. So again, $(0, 0)$ is a local minimum point: For all (x, y),

$$g(x, y) = 3x^2 + 2y^2 \geq 0 = g(0, 0).$$

This time, however, the different coefficients of x^2 and y^2 mean that the level curves (which correspond to equations $g(x, y) = 3x^2 + 2y^2 = c$) are ellipses. Here is a contour map:

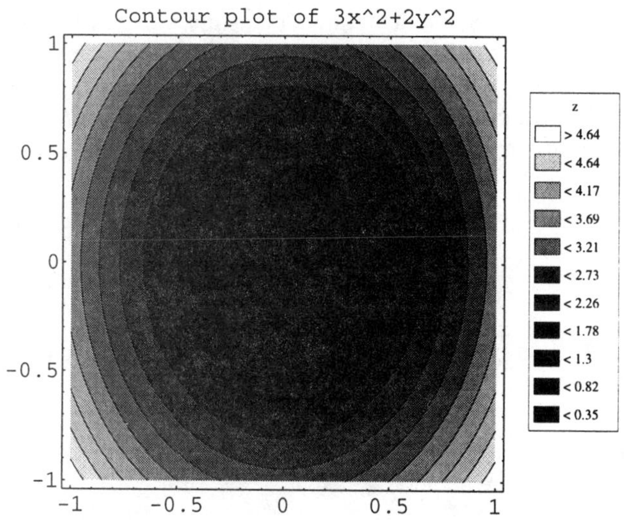

The surface $z = g(x, y)$ is called an **elliptic paraboloid**. (See the exercises at the end of this section for more on this terminology.) The Hessian matrix is

Note the factors of two.

$$g''(0, 0) = \begin{pmatrix} 6 & 0 \\ 0 & 4 \end{pmatrix}.$$ □

■ **Example 3.** Let $h(x, y) = 3x^2 - 2y^2$. How does h behave near the stationary point $(0, 0)$? Describe the surface $z = h(x, y)$.

Solution: Because the coefficients of x^2 and y^2 have different signs, the level curves (which correspond to equations of the form $h(x, y) = 3x^2 - 2y^2 = c$) are now hyperbolas, and the surface is called a **hyperbolic paraboloid**. Here is a contour map:

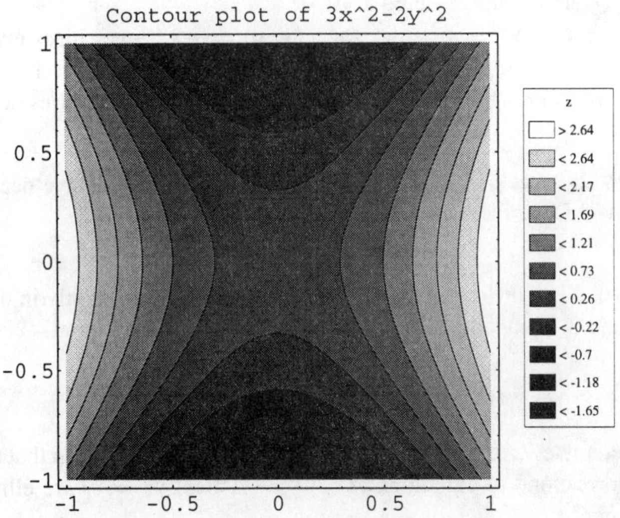

Or see the surface itself.

The contour map shows that in this case, $(0,0)$ is a **saddle point**, i.e., a point that represents *both* maximum and minimum values of h, depending on direction. If we fix $x = 0$, then $h(0,y) = -2y^2$; thus, this slice of the surface is a parabola opening downward, with a *maximum* at the origin. But if we fix $y = 0$, then $h(x,0) = 3x^2$—a parabola opening upward, with a *minimum* at the origin. The Hessian matrix is now

An ant walking south to north along the surface experiences the origin as a peak.

Note the minus sign.

$$h''(0,0) = \begin{pmatrix} 6 & 0 \\ 0 & -4 \end{pmatrix}. \qquad \square$$

■ **Example 4.** Let $j(x,y) = xy$. How does j behave near the stationary point $(0,0)$? Describe the surface $z = j(x,y)$.

Solution: The function j behaves similarly to h of the preceding example. Again, level curves of j are hyperbolas, this time of the form $xy = c$.

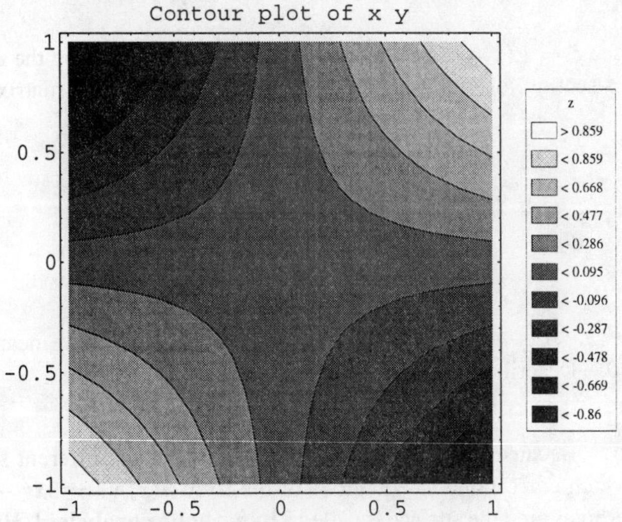

Slicing the surface with the plane $y = x$ gives $j(x, y) = x^2$, while if $y = -x$, then $j(x, y) = -x^2$. These opposite tendencies mean that, again, (0, 0) is a saddle point, and that the surface is another hyperbolic paraboloid.

The Hessian matrix, this time, is

$$j''(0,0) = \begin{pmatrix} 0 & 1 \\ 1 & 0 \end{pmatrix}. \qquad \square$$

A last example illustrates an important technique we'll use more generally in a moment.

■ **Example 5.** Let $k(x, y) = x^2+xy+y^2$. Discuss the stationary point at the origin.

Solution: Let's complete the square in x and y:

$$\begin{aligned} k(x,y) &= x^2 + xy + y^2 = (x + y/2)^2 - y^2/4 + y^2 \\ &= (x + y/2)^2 + \frac{3}{4}y^2. \end{aligned}$$

The last version shows that (0, 0) is a minimum point, since for all (x, y),

$$k(x, y) = (x + y/2)^2 + \frac{3}{4}y^2 \geq 0 = k(0, 0).$$

The contour map also shows a local minimum at the origin:

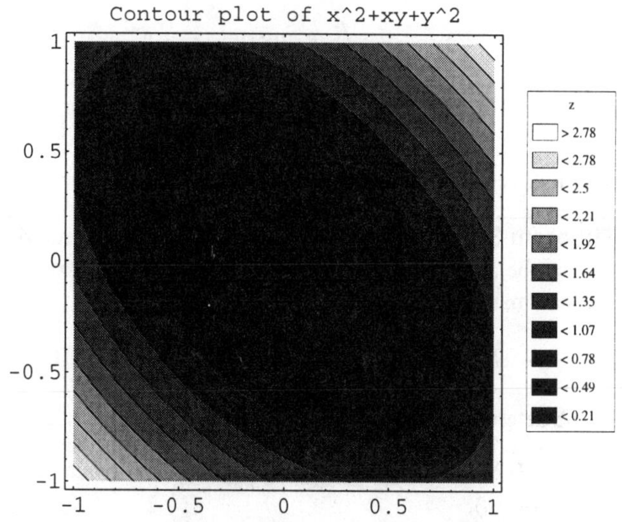

This time the Hessian matrix has more non-zero entries:

$$k''(0,0) = \begin{pmatrix} 2 & 1 \\ 1 & 2 \end{pmatrix}. \qquad \square$$

We could carry out the same calculation assuming that $B \neq 0$ or $C \neq 0$.

The general case. We'll handle the general case $f(x, y) = Ax^2 + Bxy + Cy^2$ as in the preceding example, by completing the square. We'll assume for convenience that $A \neq 0$. Here is the symbolic calculation:

$$\begin{aligned} f(x, y) &= Ax^2 + Bxy + Cy^2 = A\left(x^2 + \frac{B}{A}xy + \frac{C}{A}y^2\right) \\ &= A\left(\left(x + \frac{B}{2A}y\right)^2 + \left(\frac{C}{A} - \frac{B^2}{4A^2}\right)y^2\right). \end{aligned}$$

This shows that the type of stationary point depends on the sign of the coefficient of y^2. Notice that

$$\frac{C}{A} - \frac{B^2}{4A^2} \geq 0 \iff \frac{C}{A} \geq \frac{B^2}{4A^2} \iff 4AC - B^2 \geq 0.$$

We conclude:

- If $4AC - B^2 > 0$, then f has either a local maximum or local minimum at $(0, 0)$, depending on whether $A < 0$ or $A > 0$.
- If $4AC - B^2 < 0$, then f has a saddle point at $(0, 0)$.

(In the remaining case, where $4AC - B^2 = 0$, various things can happen, and we normally look for more information.)

In terms of derivatives. If we rewrite these conclusions using partial derivatives (recalling that $2A = f_{xx}(0, 0)$, $B = f_{xy}(0, 0) = f_{yx}(0, 0)$, and $2C = f_{yy}(0, 0)$), we see that

$$4AC - B^2 = f_{xx} f_{yy} - f_{xy}^2;$$

in other words, $4AC - B^2$ is the **determinant** of the Hessian matrix

$$f''(0, 0) = \begin{pmatrix} f_{xx}(0, 0) & f_{xy}(0, 0) \\ f_{yx}(0, 0) & f_{yy}(0, 0) \end{pmatrix}.$$

Let's rewrite what we've found as a general theorem. We assume, as above, that f has continuous second partial derivatives.

Theorem 3. (Stationary points and the Hessian matrix) Let (x_0, y_0) be a stationary point of a function f, let $f''(x_0, y_0)$ be the Hessian matrix of f, and let

$$D = f_{xx}(x_0, y_0) f_{yy}(x_0, y_0) - (f_{xy}(x_0, y_0))^2$$

be the determinant of $f''(x_0, y_0)$. Then

- If $D > 0$ and $f_{xx}(x_0, y_0) > 0$, then f has a local minimum at (x_0, y_0).
- If $D > 0$ and $f_{xx}(x_0, y_0) < 0$, then f has a local maximum at (x_0, y_0).
- If $D < 0$, then f has a saddle point at (x_0, y_0).
- If $D = 0$, then more information is needed.

The theorem makes many maximum-minimum calculations routine.

■ **Example 6.** The function $f(x, y) = xy - y - 2x + 2$ has one stationary point. Find it. What type of stationary point is it?

Solution: To find the stationary point, we solve

$$\nabla f(x, y) = (y - 2, x - 1) = (0, 0);$$

clearly, this occurs (only) at the point $(1, 2)$. At this point the Hessian matrix has the form

$$H_f(1, 2) = \begin{pmatrix} 0 & 1 \\ 1 & 0 \end{pmatrix}.$$

Therefore $D = -1$, and it follows from Theorem 3 that $(1, 2)$ is a saddle point. □

In higher dimensions

Similar ideas can be applied to give a second derivative test, based on the Hessian matrix, for functions of three or more variables. But the derivation and statement are considerably more complicated; we omit them. In practice, moreover, it's sometimes clear from other considerations or simple experimentation what type of stationary point a function has. Here's an example of such an *ad hoc* argument.

■ **Example 7.** Let $f(x, y, z)$ be a function of three variables, and suppose that f has a stationary point at the origin, that is, $\nabla f(0, 0, 0) = (0, 0, 0)$. Suppose also that $f_{xx}(0, 0, 0) > 0$ and $f_{yy}(0, 0, 0) < 0$. Show that f has neither a local maximum nor a local minimum at the origin.

Solution: Consider the function $g(t) = f(t, 0, 0)$. Then $g'(0) = f_x(0, 0, 0) = 0$, and $g''(0) = f_{xx}(0, 0, 0) > 0$. Thus $g(t)$ has a local *minimum* at $t = 0$. For similar reasons, the function $h(t) = f(0, t, 0)$ has a local *maximum* at $t = 0$. This difference means that f itself can't have either a local maximum or a minimum at $(0, 0, 0)$. □

Extremes on the boundary: optimization on closed regions

Recall what happens in elementary calculus for a differentiable function $f(x)$ defined on a closed interval $[a, b]$: f may assume its maximum and minimum values either at a stationary point, where $f'(x) = 0$, or at either of the endpoints $x = a$ and $x = b$.

The situation is similar for a function of two variables defined on a region, such as a rectangle or a circle, that has a definite "edge," or boundary: $f(x, y)$ may assume its maximum and minimum either at a stationary point or somewhere on the boundary of the region. We illustrate the situation, and one way to approach it, with a simple example.

■ **Example 8.** Where on the rectangle $R = [-1, 1] \times [1, 1]$ does $g(x, y)$ assume its minimum and maximum values?

Solution: We saw in Example 4, page 152, that g has only one stationary point—a saddle point—in the interior of R. Therefore, the maximum and minimum values of g must occur somewhere on the boundary of R. A look at the contour plot of g (notice the symmetry) shows that its enough to look along *any* boundary edge of R, such as the right edge. On this edge we have $x = 1$, so g behaves like a function of just one variable: $g(x, y) = g(1, y) = y$. Clearly, $g(1, y) = y$ is largest at $y = 1$ and smallest at $y = -1$. We conclude, therefore, that $g(1, 1) = 1$ and $g(1, -1) = -1$ are, respectively, maximum and minimum values of g on R. □

Exercises

1. Let $g(x, y) = xy$. (Its contour map is shown in Example 4.) To an ant walking along the surface $z = g(x, y)$ from lower left to upper right, the origin seems to be a low spot; another ant walking from upper left to lower right would experience the origin as a high spot.

   (a) An ant walks along the surface from $(0, -1)$ to $(0, 1)$. How does the ant's altitude change along the way?

   (b) Another ant walks along the surface from $(0.5, -1)$ to $(0.5, 1)$. How does the ant's altitude change along the way? Where is the ant highest? How high is the ant there?

2. See the contour map of $f(x, y) = \cos(x)\sin(y)$ on page 149.

   (a) The surface $z = f(x, y)$ resembles an egg carton. Where do the eggs go?

   (b) From the picture alone, estimate the coordinates of a local minimum point, a local maximum point, and a saddle point.

   (c) Use the formula $f(x, y) = \cos(x)\sin(y)$ to find (exactly) all the stationary points of f in the rectangle $R = [-3, 3] \times [-3, 3]$.

   (d) Find the maximum and minimum values of f in the rectangle $R = [-3, 3] \times [-3, 3]$.

3. Consider the function $f(x, y) = x(x - 2)\sin(y)$. Here's a contour map:

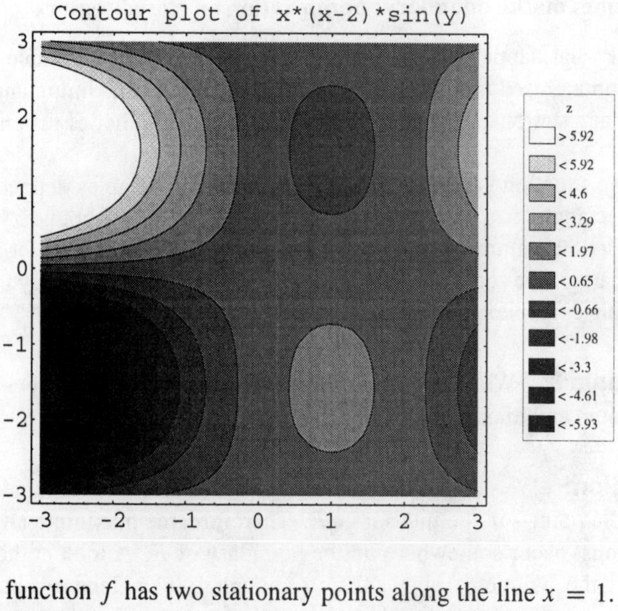

   (a) The function f has two stationary points along the line $x = 1$. Use the picture to estimate their coordinates. What type is each one?

(b) There are four stationary points inside the rectangle $R = [-3, 3] \times [-3, 3]$. Use the formula for f to find all four.

(c) The contour plot shows that f assumes its maximum and minimum values on $R = [-3, 3] \times [-3, 3]$ somewhere along the left boundary, i.e., where $x = -3$. Find these maximum and minimum values. [HINT: If $x = -3$, then $f(x, y) = f(-3, y) = 15\sin(y)$. This is a function of one variable, defined for $-3 \le y \le 3$.]

4. For each of the following functions, use the formula to find all stationary points. Then use technology (e.g., a properly chosen contour plot or surface plot) to decide what type of stationary point each one is.

(a) $f(x, y) = -x^2 - y^2$

(b) $f(x, y) = x^2 - y^2$

(c) $f(x, y) = 3x^2 + 2y^2$

(d) $f(x, y) = xy - y - 2x + 2$

5. Consider the linear function $L(x, y) = 1+2x+3y$. Does L have any stationary points? If so, what type are they? If not, why not?

6. Consider the linear function $L(x, y) = a + bx + cy$, where a, b, and c are any constants.

(a) The graph of L is a plane. Which planes have stationary points? For these planes, where are the stationary points?

(b) Under what conditions on a, b, and c will L have stationary points? In this case, where are the stationary points? Reconcile your answers with those in part (a).

7. Let $f(x, y) = x^2$. The graph of f is a cylinder, unrestricted in the y-direction.

(a) Use technology to plot the surface $z = f(x, y)$. Where in the xy-plane are the stationary points? What type are they? [HINT: There's a whole line of stationary points.]

(b) Use partial derivatives of f to find all the stationary points. Reconcile your answer with part (a).

8. Give an example as described in each part. [HINTS: (1) See Exercise 7 for ideas. (2) Check your answers by plotting.]

(a) A function $g(x, y)$ for which every point on the x-axis is a local minimum point

(b) A function $h(x, y)$ for which every point on the line $x = 1$ is a local maximum point

(c) A non-constant function $k(x, y)$ which has a local minimum at $(3, 4)$

9. Let $f(x, y) = x^2 + y^2 + kxy$.

(a) Show that the origin is a critical point of f regardless of the value of k.

(b) Show that if $k < -2$ or $k > 2$, the origin is a saddle point.

(c) Suppose that $-2 < k < 2$. Is the origin a local maximum, a local minimum, or a saddle point of f? Justify your answer.

(d) Suppose that $k = 2$. Is the origin a local maximum, a local minimum, or a saddle point of f? Justify your answer.

10. Let $f(x, y) = x^2+y^2-x^2y/2$. Find all the stationary points of f and indicate whether each is a local maximum, a local minimum, or a saddle point.

11. Let $f(x, y) = x^2y + xy^2 + x + y$.

    (a) Show that $(1, -1)$ is a stationary point of the function f.

    (b) Is the point $(1, -1)$ a local maximum, a local minimum, or a saddle point of f? Justify your answer.

12. (a) What does it mean for (x_0, y_0) to be a ***saddle point*** of the function $f(x, y)$?

    (b) Give an example of a function $g(x, y)$ that has a saddle point at $(2, 1)$.

13. Let $y = f(x)$ be a function of one variable, and suppose that both $f'(x_0) = 0$ and $f''(x_0) = 0$. Recall that in this case, "anything" can happen at x_0.

    (a) Give an example of a function f and a point x_0 as above for which x_0 is a local maximum point.

    (b) Give an example of a function f and a point x_0 as above for which x_0 is a local minimum point.

    (c) Give an example of a function f and a point x_0 as above for which x_0 is neither a local minimum point nor a local maximum point.

14. What (if anything) does the second derivative test say about local maxima and minima of each of the following functions?

    (a) $f(x) = \cos x$

    (b) $f(x) = e^x$

    (c) $f(x) = \frac{x^4}{4} - x^3 + x^2$

    (d) $f(x) = x^4$

15. Check that Theorem 3 gives the correct information for each of the functions f, g, h, j, and k in Examples 1–5.

16. Suppose that $f(x, y)$ has a stationary point at (x_0, y_0), and suppose that $f_{xx}(x_0, y_0)$ and $f_{yy}(x_0, y_0)$ have opposite signs. Show that (x_0, y_0) must be a saddle point.

17. For any nonzero constant a, the graph of $z = ax^2+ay^2$ is a circular paraboloid. This exercise explores the reasons for this terminology.

    (a) Explain why, for each real number c, the cross sections $z = c$ are circles. Describe the cross sections for $c = 0$, $c = 1$, and $c = 2$.

    (b) What is the shape of the cross sections $x = c$? Describe them for $c = 0$, $c = 1$, and $c = 2$.

    (c) What is the shape of the cross sections $y = c$? Describe them for $c = 0$, $c = 1$, and $c = 2$.

18. Repeat the previous exercise, but for the hyperbolic paraboloid $z = x^2 - y^2$.

19. Let $f(x, y) = 3x^2 - 6xy + 2y^3$. Find and classify all the stationary points of f.

20. Find and classify all the stationary points of $f(x, y) = x^3+y^3+3x^2-3y^2-8$. (Note: There are 4 stationary points in all.)

21. Find and classify all the stationary points of $f(x, y) = x^2 - xy - y^2$.

22. Find and classify all stationary points of $f(x, y) = x^4 + y^4$. Does Theorem 3 help?

23. Let $f(x, y) = x^2 + axy + by^2$, where a and b are real constants (positive, negative, or zero).

    (a) Explain why $(0, 0)$ is a critical point of f, *regardless* of the values of a and b.

    (b) Under what conditions on a and b will $(0, 0)$ be a local maximum? A local minimum? A saddle point? Explain your answers carefully, and give an example of each possibility.

    (c) Suppose (in this part only) that $b = a^2/4$. What sort of critical point occurs at $(0, 0)$? Why?

24. Let $f(x, y) = \sin(x) + \cos(2y)$. Show that $(\pi/2, 0)$ is a stationary point. What type is it?

25. Let $f(x, y) = x^2$; f is a cylindrical function.

    (a) Describe the shape of the f-graph near the stationary point $(0, 0)$.

    (b) Show from the formula that f has a local minimum at $(0, 0)$. (Note: The local minimum is not "strict," in the sense that $f(x, y) \geq f(0, 0)$ holds, but $f(x, y) > f(0, 0)$ may not.)

    (c) What does the second derivative test say about f at $(0, 0)$?

26. Consider functions of the form $f(x, y) = x^2 + bxy + y^2$, where b is any constant.

    (a) What values of b (if any) make $(0, 0)$ a local minimum? A local maximum? A saddle point?

    (b) If possible, use technology to plot level curves and a surface to illustrate each possibility in the preceding parts.

    (c) For what value of b is the Hessian determinant zero? How does f behave near the origin in this case? Draw (by hand) several level curves of f near the origin to see what's happening.

2.8 The chain rule

Note. *This section assumes some a basic familiarity with matrices and matrix multiplication. For a brief introduction or review, see Appendix C.*

Introduction. The chain rule, the product rule, and the quotient rule are all "combinatorial" results. They tell how to find derivatives of "new" functions that are formed by combining "old" functions in various ways. (They also assure us that, under appropriate conditions, these new derivatives exist.) The new derivatives are, of course, appropriate combinations of the old derivatives and functions.

But not necessarily the "obvious" combinations!

Composition: one variable vs. several

The chain rule—in any dimension—concerns derivatives of functions formed by composition. Composition is relatively simple in single-variable calculus: Composing two ordinary scalar-valued functions, say f and g, of one variable, produces another scalar-valued function of one variable, $f \circ g$, defined by $f \circ g(x) = f(g(x))$. If $f(x) = \sin x$ and $g(x) = x^2$, then $f \circ g(x) = \sin(x^2)$.

Remember: Composition isn't commutative.

Composition means exactly the same thing for functions of several variables—the output from one function is used as input to another. In multivariable calculus, however, both inputs and outputs may be either scalars or vectors, so it's important to keep careful track of each function's **domain space** and **image space**. The arrow notations in the next example can help in this regard. (The arrows point from the domain space to the image space.)

■ **Example 1.** Consider functions $f : \mathbb{R} \to \mathbb{R}$, $g : \mathbb{R}^2 \to \mathbb{R}$, and $\boldsymbol{h} : \mathbb{R} \to \mathbb{R}^2$, defined by

$$f(t) = t^3; \quad g(x, y) = x^2 + y^2; \quad \boldsymbol{h}(t) = (\cos t, \sin t).$$

Which compositions make sense? Which don't? Why?

Note the use of boldface for the vector-valued function.

Solution: The notations $f : \mathbb{R} \to \mathbb{R}$ and $g : \mathbb{R}^2 \to \mathbb{R}$ show (even before looking at function definitions) that $g \circ f$ can't make sense. Outputs from f are scalars; they have the wrong "type" to be inputs to g, which accepts 2-vectors. The composition $f \circ g$ does make sense; outputs from g are scalars, as are inputs to f. Symbolically, we have $f \circ g : \mathbb{R}^2 \to \mathbb{R} \to \mathbb{R}$ (or, simply, $f \circ g : \mathbb{R}^2 \to \mathbb{R}$), defined by

$$f \circ g(x, y) = f(x^2 + y^2) = (x^2 + y^2)^3.$$

Similarly, $g : \mathbb{R}^2 \to \mathbb{R}$ and $\boldsymbol{h} : \mathbb{R} \to \mathbb{R}^2$ can be composed to give the vector-valued function $\boldsymbol{h} \circ g : \mathbb{R}^2 \to \mathbb{R}^2$, defined by

$$\boldsymbol{h} \circ g(x, y) = \boldsymbol{h}(x^2 + y^2) = (\cos(x^2 + y^2), \sin(x^2 + y^2)).$$

Alternatively, g and $\boldsymbol{h}$ can be composed in the opposite order, to give an ordinary, calculus-style function $g \circ \boldsymbol{h} : \mathbb{R} \to \mathbb{R}$, defined by

$$g \circ \boldsymbol{h}(t) = g(\cos t, \sin t) = \cos^2 t + \sin^2 t = 1.$$

The composite function $\boldsymbol{h} \circ f : \mathbb{R} \to \mathbb{R}^2$ is defined by

$$\boldsymbol{h} \circ f(t) = \boldsymbol{h}(t^3) = (\cos(t^3), \sin(t^3));$$

it's a vector-valued function of one variable. □

The one-variable chain rule: multiply derivatives

Recall the one-variable chain rule:

> **Fact: (Chain rule in one variable)** Let f and g be differentiable functions, with a in the domain of g. Then
>
> $$(f \circ g)'(a) = f'(g(a)) \cdot g'(a).$$

There are other notations for saying the same thing. If we write $y = f(u)$ and $u = g(x)$, then the chain rule can be written as

We'll see several as we go along.

$$\frac{dy}{dx}(a) = \frac{dy}{du}(g(a)) \cdot \frac{du}{dx}(a).$$

In *any* symbolic form, the key idea is this:

> The derivative of the composition $f \circ g$ is a *product* of the derivatives f' and g'.

Where are the factors evaluated? What the chain rule says—multiply individual derivatives—sounds (and is) simple enough. Notice, however, that the two derivatives in the product $f'(g(a)) \cdot g'(a)$ are evaluated at different places. In the right factor, g' is evaluated at $x = a$; in the left factor, f' is is evaluated at $g(a)$. A function diagram—

Not at a.

$$a \overset{g}{\longmapsto} g(a) \overset{f}{\longmapsto} f(g(a))$$

shows why these choices make sense: The two derivatives are evaluated at the corresponding domain points.

Derivatives and linear functions: why the chain rule works. Why does the chain rule work? How would one ever guess such a formula?

The answer is surprisingly simple—and it's the same for any number of variables. The key idea has two parts:

- **The chain rule "works" for linear functions.** It's easy to show, right from scratch, that the chain rule holds for linear functions, i.e., if $f(x) = A + Bx$ and $g(x) = C + Dx$ for any constants A, B, C, and D. For linear functions, the derivatives are just the coefficients of x: In this case, $f'(x) = B$ and $g'(x) = D$.

  Using capital letters saves other letters for later.

  The composite $f \circ g(x)$ is another linear function:

  $$f \circ g(x) = f(C + Dx) = A + B(C + Dx) = A + BC + BDx.$$

  Thus $(f \circ g)'(x) = BD$, the product of f' and g'.

  As the chain rule predicts.

- **Differentiable functions are locally linear.** Most functions are not linear. However, every differentiable function g is *locally linear*, in the sense that near any point of its domain, g can be closely approximated by a linear approximation function, L_g. Similarly, f can be closely approximated near any point of *its* domain by another suitably-chosen linear function, L_f. Because $f \approx L_f$ and $g \approx L_g$, it follows that $f \circ g \approx L_f \circ L_g$. In particular, $(f \circ g)'(a) = (L_f \circ L_g)'(a)$. Because the chain rule "works" for $L_f \circ L_g$, it must work for $f \circ g$, too.

  We discussed it in detail in the preceding section.

  The previous paragraph says so.

Details and fine print appear in the next example.

■ **Example 2.** Let g and f be differentiable functions and a a domain point of g. Use linear approximations to derive the chain rule for $(f \circ g)'(a)$.

Solution: For (temporary) convenience, write $b = g(a)$. Then we have

$$a \overset{g}{\longmapsto} b \overset{f}{\longmapsto} f(b).$$

Check this carefully; details appear in the preceding section.

The linear approximation to g at $x = a$ is

$$L_g(x) = g(a) + g'(a)(x - a) = b + g'(a)(x - a);$$

notice that $L_g(a) = g(a)$ and $L'_g(a) = g'(a)$. Similarly, the linear approximation to f at $x = b$ is

$$L_f(x) = f(b) + f'(b)(x - b),$$

with $L_f(b) = f(b)$ and $L'_f(b) = f'(b)$. Composing L_g with L_f gives

$$a \overset{L_g}{\longmapsto} b \overset{L_f}{\longmapsto} f(b),$$

Check the second identity.

which resembles the earlier diagram. Working out the formula gives

$$L_f \circ L_g(x) = L_f(b + g'(a)(x - a)) = f(b) + f'(b)g'(a)(x - a).$$

Therefore, $L_f \circ L_g$ is a linear function of x, with derivative $f'(b) \cdot g'(a)$. This gives the chain rule:

$$(f \circ g)'(a) = (L_f \circ L_g)'(a) = f'(g(a)) \cdot g'(a).$$

(The first equality above *is* true, but a fully rigorous proof is a little beyond our scope.) □

The multivariable chain rule: multiply derivative matrices

The chain rule says the same thing in all dimensions: *The derivative of a composition $f \circ g$ is found by multiplying—in an appropriate sense—the derivatives of f and g.* The qualification is necessary because, for multivariable functions, derivatives are usually not single numbers, but vectors or matrices. In these cases, as we'll see, "multiplication" means matrix multiplication or the dot product. This difference aside, the multivariable chain rule is exactly the same as in the one-variable case.

The dot product can be thought of as a special case of matrix multiplication.

Derivatives as matrices. Functions of more than one variable and vector-valued functions generate a whole collection of derivatives and partial derivatives. It's convenient to store all this information in an array called the **derivative matrix** (or **Jacobian matrix**). Suppose, for instance, that $\boldsymbol{k} : \mathbb{R}^2 \to \mathbb{R}^2$ is defined by

$$\boldsymbol{k}(x, y) = (u(x, y), v(x, y)) = (x^2 + y, 2x - y^3).$$

Then the derivative matrix is

$$\boldsymbol{k}'(x, y) = \begin{pmatrix} \dfrac{\partial u}{\partial x} & \dfrac{\partial u}{\partial y} \\[2ex] \dfrac{\partial v}{\partial x} & \dfrac{\partial v}{\partial y} \end{pmatrix} = \begin{pmatrix} 2x & 1 \\ 2 & -3y^2 \end{pmatrix}.$$

The idea of a derivative matrix makes sense regardless of the dimensions of the domain and image spaces. We'll need dimensions no higher than three, but here is the general definition:

Including dimension 1.

Definition: (Derivative matrix) Let $\boldsymbol{f} : \mathbb{R}^n \to \mathbb{R}^m$ be a vector-valued function of n variables, with component functions $f_1(x_1, x_2, \ldots, x_n)$, $f_2(x_1, x_2, \ldots, x_n)$, $\ldots$, $f_m(x_1, x_2, \ldots, x_n)$. Let $\boldsymbol{X_0}$ be a point of the domain of $\boldsymbol{f}$. The derivative matrix of f at $\boldsymbol{X_0}$ is the $m \times n$ matrix

$$\boldsymbol{f}'(\boldsymbol{X_0}) = \begin{pmatrix} \frac{\partial f_1}{\partial x_1}(\boldsymbol{X_0}) & \frac{\partial f_1}{\partial x_2}(\boldsymbol{X_0}) & \cdots & \frac{\partial f_1}{\partial x_n}(\boldsymbol{X_0}) \\ \frac{\partial f_2}{\partial x_1}(\boldsymbol{X_0}) & \frac{\partial f_2}{\partial x_2}(\boldsymbol{X_0}) & \cdots & \frac{\partial f_2}{\partial x_n}(\boldsymbol{X_0}) \\ \vdots & \vdots & \vdots & \vdots \\ \frac{\partial f_m}{\partial x_1}(\boldsymbol{X_0}) & \frac{\partial f_m}{\partial x_2}(\boldsymbol{X_0}) & \cdots & \frac{\partial f_m}{\partial x_n}(\boldsymbol{X_0}) \end{pmatrix}.$$

Notice that each row of the matrix $\boldsymbol{f}'$ is the gradient vector of a component function. Indeed, the derivative matrix is sometimes written as

$$\boldsymbol{f}'(\boldsymbol{X_0}) = \begin{pmatrix} \nabla f_1(\boldsymbol{X_0}) \\ \nabla f_2(\boldsymbol{X_0}) \\ \vdots \\ \nabla f_m(\boldsymbol{X_0}) \end{pmatrix}.$$

The abstract definition may look formidable, but in specific cases the calculations are easy.

■ **Example 3.** Consider the functions

The first three are from Example 1, page 160.

$$f(t) = t^3; \quad g(x, y) = x^2 + y^2; \quad \boldsymbol{h}(t) = (\cos t, \sin t);$$

$$\boldsymbol{L}(x, y) = (1 + 2x + 3y, 4 + 5x + 6y).$$

Find their derivative matrices.

Solution: All the derivative calculations are easy. Here are the results:

But check them for yourself!

$$f'(t) = \left(3t^2\right); \quad g'(x, y) = (2x \quad 2y);$$

$$\boldsymbol{h}'(t) = \begin{pmatrix} -\sin t \\ \cos t \end{pmatrix}; \quad \boldsymbol{L}'(t) = \begin{pmatrix} 2 & 3 \\ 5 & 6 \end{pmatrix}.$$

Notice especially the size and shape of each matrix. In particular, $f'(t)$ is a 1×1 "matrix," i.e., a scalar; $g'(x, y)$ is simply the gradient ∇g, written as a row vector. □

Jacobi and the Jacobian. The Jacobian matrix and determinant are named for the German mathematician Karl Jacobi (1804–1851). Jacobi was not the first to organize derivative information in a matrix, but in his 1841 paper *De determinantibus functionalibus* Jacobi applied matrix theory to prove results about relations among sets of functions. Jacobi was also a popular teacher at the University of Königsberg, and is credited with having pioneered the seminar method of teaching. (For a wealth of historical information of this nature, see the History of Mathematics Archive (`http://www-groups.dcs.st-and.ac.uk/~history/`), a Web-based resource at the University of St. Andrews, in Scotland.)

Linear functions: a note on terminology. Linear functions, such as L in the preceding example, are those in which the variables appear in powers no greater than one. The name is reasonable because graphs and level curves of linear functions are lines, planes, or other "flat" objects. In particular, we allow linear functions to have nonzero constant terms (as L does, above).

For the record, the word "linear" is used a little differently in linear algebra courses. There, a "linear function" or "linear transformation" is usually required to have only zero constant terms. The difference is worth noting, but it shouldn't cause us trouble in context.

Linear functions, matrices, and derivatives. There's a close connection between linear functions (in our sense of the word) and matrices. For instance,

$$L(x, y) = (1 + 2x + 3y, 4 + 5x + 6y)$$

Convince yourself.

says exactly the same thing as the matrix equation

$$L(x, y) = \begin{pmatrix} 1 \\ 4 \end{pmatrix} + \begin{pmatrix} 2 & 3 \\ 5 & 6 \end{pmatrix} \cdot \begin{pmatrix} x \\ y \end{pmatrix},$$

where the dot stands for matrix multiplication.

More generally, *every* linear function L has the form

$$L(X) = C + M \cdot X,$$

In one variable, a linear function has the form $L(x) = c + mx$.

where M is a matrix (called the **coefficient matrix**), X is a vector input, C is a constant vector, and the dot represents matrix multiplication. Writing linear functions in matrix form has two main advantages for us:

Composition and matrix multiplication. Consider two linear functions given by matrix equations

$$L_1(X) = C_1 + M_1 \cdot X \quad \text{and} \quad L_2(X) = C_2 + M_2 \cdot X.$$

Then the composition $L_1 \circ L_2$ has matrix form

$$\begin{aligned} L_1 \circ L_2(X) &= L_1(C_2 + M_2 \cdot X) \\ &= C_1 + M_1 \cdot (C_2 + M_2 \cdot X) \\ &= C_1 + M_1 C_2 + (M_1 \cdot M_2) \cdot X. \end{aligned}$$

(The last equation follows from algebraic properties of matrix multiplication.) The first two summands above are constant vectors, so the composition has the general form $\boldsymbol{L}_1 \circ \boldsymbol{L}_2(\boldsymbol{X}) = (M_1 \cdot M_2) \cdot \boldsymbol{X} + \boldsymbol{C}$. In particular:

The coefficient matrix of $\boldsymbol{L}_1 \circ \boldsymbol{L}_2$ is the matrix product $M_1 \cdot M_2$.

Derivatives of linear functions. The preceding example➸ illustrates a simple but important property of linear functions and their matrices:

See the derivative of $\boldsymbol{L}$.

Fact: Let $\boldsymbol{L}$ be the linear function

$$\boldsymbol{L}(\boldsymbol{X}) = \boldsymbol{C}_0 + M \cdot \boldsymbol{X}.$$

Then the derivative matrix is $\boldsymbol{L}' = M$.

This fact, combined with the italicized sentence above, tells how to find the derivative matrix of a composition of linear functions:

The derivative matrix of the composition $\boldsymbol{L}_1 \circ \boldsymbol{L}_2$ is the matrix product $M_1 \cdot M_2$.

Linear approximations to functions. For a real-valued function f the linear approximation to f at $\boldsymbol{X}_0$ is the function

$$L(\boldsymbol{X}) = f(\boldsymbol{X}_0) + \nabla f(\boldsymbol{X}_0) \cdot (\boldsymbol{X} - \boldsymbol{X}_0);$$

the dot represents the dot product. If $\boldsymbol{f} = (f_1, f_2)$ (or (f_1, f_2, f_3)) is a vector-valued function, then the linear approximation to $\boldsymbol{f}$ at $\boldsymbol{X}_0$ is the matrix analogue of the preceding equation:

$$\boldsymbol{L}(\boldsymbol{X}) = \boldsymbol{f}(\boldsymbol{X}_0) + \boldsymbol{f}'(\boldsymbol{X}_0) \cdot (\boldsymbol{X} - \boldsymbol{X}_0),$$

where $\boldsymbol{f}'(\boldsymbol{X}_0)$ is now the derivative matrix,➸ and the dot now represents matrix multiplication. Notice that $\boldsymbol{L}$ is (like $\boldsymbol{f}$) a vector-valued function, such that

The rows are gradients.

$$\boldsymbol{L}(\boldsymbol{X}_0) = \boldsymbol{f}(\boldsymbol{X}_0) \quad \text{and} \quad \boldsymbol{L}'(\boldsymbol{X}_0) = \boldsymbol{f}'(\boldsymbol{X}_0).$$

A generic example. We've now assembled all the necessary ingredients to state the multivariable chain rule carefully. But first we present a "generic" example➸ (one that uses function names, not specific formulas) that suggests *why* it is that matrices and matrix multiplication prove to be natural bookkeeping devices for storing and organizing all the necessary information.➸ Linear approximation is the key idea.

The authors thank Reg Laursen for suggesting this calculation.

There's quite a lot of information, so organized storage is important!

■ **Example 4.** Let $z = f(x, y)$, where $x = g(s, t)$ and $y = h(s, t)$. (To avoid distractions, we'll suppose that the functions f, g, and h are all "well behaved" in the sense that all the necessary derivatives exist.) We can think of z as a composite function of s and t:

$$z = f(x, y) = f\left(g(s, t), h(s, t)\right).$$

Use linear approximation to estimate the partial derivatives $\partial z/\partial s$ and $\partial z/\partial t$ at the point (s_0, t_0). How is matrix multiplication involved?

Solution: First, some handy notation. We'll write

$$x_0 = g(s_0, t_0), \quad y_0 = h(s_0, t_0), \quad \text{and} \quad z_0 = f(x_0, y_0).$$

When it's convenient, we'll also use the vector forms

$$\boldsymbol{X_0} = (x_0, y_0), \quad \boldsymbol{S_0} = (s_0, t_0), \quad \text{and} \quad \boldsymbol{G} = (g, h).$$

(In the last case, $\boldsymbol{G}$ is the function $\boldsymbol{G} : \mathbb{R}^2 \to \mathbb{R}^2$; the coordinate functions g and h are scalar-valued functions of two variables.)

Consider the following linear approximation formulas for z, x, and y:

$$\begin{aligned} x &\approx x_0 + g_s(s_0, t_0)\,(s - s_0) + g_t(s_0, t_0)\,(t - t_0); \\ y &\approx y_0 + h_s(s_0, t_0)\,(s - s_0) + h_t(s_0, t_0)\,(t - t_0); \\ z &\approx z_0 + f_x(x_0, y_0)\,(x - x_0) + f_y(x_0, y_0)\,(y - y_0). \end{aligned}$$

We use some vector notation here to cut clutter.

The first two approximations give

$$x - x_0 \approx g_s(\boldsymbol{S_0})(s - s_0) + g_t(\boldsymbol{S_0})(t - t_0) \quad \text{and} \quad y - y_0 \approx h_s(\boldsymbol{S_0})(s - s_0) + h_t(\boldsymbol{S_0})(t - t_0).$$

Substituting these into our approximation for z gives

$$z \approx z_0 + f_x(\boldsymbol{X_0})\big[\, g_s(\boldsymbol{S_0})(s - s_0) + g_t(\boldsymbol{S_0})(t - t_0) \,\big] + f_y(\boldsymbol{X_0})\big[\, h_s(\boldsymbol{S_0})(s - s_0) + h_t(\boldsymbol{S_0})(t - t_0) \,\big]$$

Rearranging terms slightly gives

$$z \approx z_0 + \big[\, f_x(\boldsymbol{X_0}) g_s(\boldsymbol{S_0}) + f_y(\boldsymbol{X_0}) h_s(\boldsymbol{S_0}) \,\big]\,(s - s_0) + \big[\, f_x(\boldsymbol{X_0}) g_t(\boldsymbol{S_0}) + f_y(\boldsymbol{X_0}) h_t(\boldsymbol{S_0}) \,\big]\,(t$$

The line above carries the key insight: we've found a linear approximation to z in terms of the *new* variables s and t. This suggests that the quantities in square brackets must in fact be the sought-after partial derivatives of z with respect to s and t:

$$\begin{aligned} \frac{\partial z}{\partial s}(\boldsymbol{S_0}) &= \big[\, f_x(\boldsymbol{X_0}) g_s(\boldsymbol{S_0}) + f_y(\boldsymbol{X_0}) h_s(\boldsymbol{S_0}) \,\big]; \\ \frac{\partial z}{\partial t}(\boldsymbol{S_0}) &= \big[\, f_x(\boldsymbol{X_0}) g_t(\boldsymbol{S_0}) + f_y(\boldsymbol{X_0}) h_t(\boldsymbol{S_0}) \,\big]. \end{aligned}$$

We omitted input variable names for compactness.

Rewriting these facts in matrix form gives

$$\begin{bmatrix} z_s & z_t \end{bmatrix} = \begin{bmatrix} f_x & f_y \end{bmatrix} \cdot \begin{bmatrix} g_s & g_y \\ h_s & h_y \end{bmatrix}.$$

In Jacobian matrix notation, it's equivalent to write, simply,

$$(f(\boldsymbol{G}))' = f' \cdot \boldsymbol{G}',$$

where all symbols represent matrices, and the operation on the right is matrix multiplication. We've arrived. □

The multivariable chain rule. We can now state the chain rule in its general form.

> **Theorem 4. (The multivariable chain rule)** Let $\boldsymbol{f}$ and $\boldsymbol{g}$ be differentiable functions, with $\boldsymbol{X_0}$ in the domain of g. Then we have the matrix equation
>
> $$(\boldsymbol{f} \circ \boldsymbol{g})'(\boldsymbol{X_0}) = \boldsymbol{f}'(\boldsymbol{g}(\boldsymbol{X_0})) \cdot \boldsymbol{g}'(\boldsymbol{X_0});$$
>
> the dot represents matrix multiplication.

A proof sketch. The idea is exactly the same as in the single-variable case. We approximate f and g with appropriate linear functions L_f and L_g, for which the theorem "clearly" holds.➡ Then we conclude that the theorem holds in general.

That's the conclusion drawn just before the theorem.

For g we have the following linear approximation at X_0:

$$L_g(X) = g(X_0) + g'(X_0) \cdot (X - X_0).$$

Similarly, for f we have the following linear approximation at $g(X_0)$:

$$L_f(X) = f(g(X_0)) + f'(g(X_0)) \cdot (X - X_0).$$

The nature of the approximations $f \approx L_f$ and $g \approx L_g$ implies that $f \circ g \approx L_f \circ L_g$, and also that $(f \circ g)'(X_0) = (L_f \circ L_g)'(X_0)$. Finally, the last derivative, as we showed above, is the product of the corresponding derivative matrices. Thus,

$$(f \circ g)'(X_0) = (L_f \circ L_g)'(X_0) = f'(g(X_0)) \cdot g'(X_0). \qquad \square$$

■ **Example 5.** Consider the functions

$$f(u, v) = (uv, u - v) \quad \text{and} \quad g(x, y) = (x + y, x^2 + y^2).$$

What's $(f \circ g)'(x, y)$? What's $(f \circ g)'(3, 4)$?

Solution: The derivative matrices are

$$f'(u, v) = \begin{pmatrix} v & u \\ 1 & -1 \end{pmatrix} \quad \text{and} \quad g'(x, y) = \begin{pmatrix} 1 & 1 \\ 2x & 2y \end{pmatrix}.$$

Now the chain rule says:

$$(f \circ g)'(x, y) = \begin{pmatrix} v & u \\ 1 & -1 \end{pmatrix} \cdot \begin{pmatrix} 1 & 1 \\ 2x & 2y \end{pmatrix} = \begin{pmatrix} v + 2ux & v + 2uy \\ 1 - 2x & 1 - 2y \end{pmatrix}.$$

Substituting $u = x + y$ and $v = x^2 + y^2$ gives

$$(f \circ g)'(x, y) = \begin{pmatrix} (x^2 + y^2) + 2(x + y)x & (x^2 + y^2) + 2(x + y)y \\ 1 - 2x & 1 - 2y \end{pmatrix}.$$

To find $(f \circ g)'(3, 4)$ we could substitute above. Alternatively, we can observe that $g(3, 4) = (7, 25)$,

$$f'(7, 25) = \begin{pmatrix} 25 & 7 \\ 1 & -1 \end{pmatrix}, \quad \text{and} \quad g'(3, 4) = \begin{pmatrix} 1 & 1 \\ 6 & 8 \end{pmatrix}.$$

Thus, by the chain rule again,

$$(f \circ g)'(3, 4) = \begin{pmatrix} 25 & 7 \\ 1 & -1 \end{pmatrix} \cdot \begin{pmatrix} 1 & 1 \\ 6 & 8 \end{pmatrix} = \begin{pmatrix} 67 & 81 \\ -5 & -7 \end{pmatrix}. \qquad \square$$

■ **Example 6.** Sometimes composite expressions appear without explicit function names. For example, suppose that u is a function of x and y, while x and y are functions of s and t. Find the partial derivatives $\partial u/\partial s$ and $\partial u/\partial t$.

Solution: The chain rule still holds, but it looks a little different. First, write

$$u = u(x, y) \quad \text{and} \quad \boldsymbol{X}(s, t) = (x(s, t), y(s, t)).$$

Then

$$u'(x, y) = \begin{pmatrix} \dfrac{\partial u}{\partial x} & \dfrac{\partial u}{\partial y} \end{pmatrix} \quad \text{and} \quad \boldsymbol{X}'(s, t) = \begin{pmatrix} \dfrac{\partial x}{\partial s} & \dfrac{\partial x}{\partial t} \\ \dfrac{\partial y}{\partial s} & \dfrac{\partial y}{\partial t} \end{pmatrix}.$$

The chain rule says to take the product:

$$u'(s, t) = \begin{pmatrix} \dfrac{\partial u}{\partial x} & \dfrac{\partial u}{\partial y} \end{pmatrix} \cdot \begin{pmatrix} \dfrac{\partial x}{\partial s} & \dfrac{\partial x}{\partial t} \\ \dfrac{\partial y}{\partial s} & \dfrac{\partial y}{\partial t} \end{pmatrix}.$$

It follows that

$$\frac{\partial u}{\partial s} = \frac{\partial u}{\partial x}\frac{\partial x}{\partial s} + \frac{\partial u}{\partial y}\frac{\partial y}{\partial s}; \quad \frac{\partial u}{\partial t} = \frac{\partial u}{\partial x}\frac{\partial x}{\partial t} + \frac{\partial u}{\partial y}\frac{\partial y}{\partial t}.$$

(Notice the "symbolic cancellation" of partials.) □

Exercises

1. Let $f(x) = a + bx$, $g(x) = c + dx$, and $h(x) = x^2$ throughout this exercise; a, b, c, and d are constants.

   (a) Find values of a, b, c, and d such that $f \circ g(x) \neq g \circ f(x)$. (There are many possibilities.)

   (b) Find values of a, b, c, and d such that f and g are different functions, and $f \circ g(x) = g \circ f(x)$.

   (c) What conditions on a, b, c, and d guarantee that $f \circ g(x) = g \circ f(x)$.

   (d) Under what conditions on a and b is $f \circ h(x) = h \circ f(x)$?

2. Let $f(x) = ax^2$ and $g(x) = bx^3$, where a and b are nonzero constants. Under what conditions on a and b is $f \circ g(x) = g \circ f(x)$?

3. Write the derivative matrix for each of the following functions at the given point.

   (a) $\boldsymbol{f}(x, y) = (1x + 2y + 3, 4x + 5y + 6)$; $\boldsymbol{X_0} = (0, 0)$

   (b) $\boldsymbol{f}(x, y) = (1x + 2y + 3, 4x + 5y + 6)$; $\boldsymbol{X_0} = (1, 2)$

   (c) $\boldsymbol{g}(x, y, z) = (y + z, x + z, x + y)$; $\boldsymbol{X_0} = (1, 2, 3)$

   (d) $\boldsymbol{h}(t) = (\cos t, \sin t, t)$; $\boldsymbol{t_0} = 0$

   (e) $\boldsymbol{k}(s, t) = (1, 2, 3) + s(4, 5, 6) + t(7, 8, 9)$; $(s_0, t_0) = (1, 1)$

4. Let $\boldsymbol{f}$, $\boldsymbol{g}$, $\boldsymbol{h}$, and $\boldsymbol{k}$ be as in the preceding exercise. In each part below, decide whether the given composition makes sense. If it does, find a formula for the composite function, and use the chain rule to find the derivative matrix at the given point.

(a) $\boldsymbol{k} \circ \boldsymbol{f}$; $(x_0, y_0) = (0, 0)$
(b) $\boldsymbol{f} \circ \boldsymbol{g}$; $(x_0, y_0, z_0) = (1, 2, 3)$
(c) $\boldsymbol{g} \circ \boldsymbol{k}$; $(s_0, t_0) = (1, 1)$

5. In this exercise let $f(x) = x^2 + x$, $g(x) = \sin x$, and $x_0 = 0$.

   (a) Find L_g, the linear approximation to g at x_0.
   (b) Find L_f, the linear approximation to f at $g(x_0)$.
   (c) Find formulas for $f \circ g$ and $L_f \circ L_g$.
   (d) Show that $(f \circ g)'(x_0) = (L_f \circ L_g)'(x_0)$.
   (e) Plot $(f \circ g)$ and $(L_f \circ L_g)$ in the vicinity of $x = x_0$. Do the graphs "look right"?

6. Repeat the previous exercise, but let $f(x) = x^2 + x$, $g(x) = e^x$, and $x_0 = 0$.

7. Repeat the previous exercise, but let $f(t) = t^2 - 9t + 20$, $g(x, y) = x^2 + y^2$, and $\boldsymbol{X_0} = (2, 1)$. (In (e), plot the two functions as surfaces in xyz-space.)

8. Let f, g, $\boldsymbol{h}$, and $\boldsymbol{L}$ be the functions $f(t) = t^3$; $g(x, y) = x^2 + y^2$; $\boldsymbol{h}(t) = (\cos t, \sin t)$; $\boldsymbol{L}(x, y) = (1 + 2x + 3y, 4 + 5x + 6y)$. (Their derivative matrices are calculated in Example 3.)

   (a) Use the chain rule to find the derivative matrix $(f \circ g)'(x_0, y_0)$.
   (b) Use the chain rule to find the derivative matrix $(\boldsymbol{h} \circ g)'(x_0, y_0)$.
   (c) Use the chain rule to find the derivative matrix $(g \circ \boldsymbol{L})'(x_0, y_0)$. (Hint: To avoid clashing variable names, first rewrite $g(u, v) = u^2 + v^2$.)

9. Let g and $\boldsymbol{h}$ be as in the preceding exercise.

   (a) Use the chain rule to find the derivative matrix $(g \circ \boldsymbol{h})'(t)$.
   (b) Find a formula for $g \circ \boldsymbol{h}(t)$ in terms of t. Differentiate (without the chain rule!) to find $(g \circ \boldsymbol{h})'(t)$. Compare your answer to the previous part.

10. Let $f(t) = t^3$ and $\boldsymbol{h}(t) = (\cos t, \sin t)$.

    (a) Use the chain rule to find the derivative matrix $(\boldsymbol{h} \circ \boldsymbol{f})'(t)$.
    (b) Find a formula for $\boldsymbol{h} \circ f(t)$ in terms of t. Differentiate (without the chain rule!) to find $(\boldsymbol{h} \circ f)'(t)$. Compare your answer to the previous part.

11. Give an example of a function k such that k' is a 3×2 matrix.

12. Let $f(u, v) = \big(uv, u + v\big)$, $g(x, y) = \big(x^2 - y^2, x^2 + y^2\big)$, and $h(s, t) = (f \circ g)(s, t)$. Use the Chain Rule to find $h'(2, 1)$.

13. The voltage (V), current (I), and resistance (R) in an electrical circuit are related by Ohm's Law: $I = V/R = I(V, R)$. Suppose that the voltage produced by an aging battery decreases at a rate of 0.1 volts/hour and that the resistance in the circuit increases at the rate of 2 Ω/hour due to heating. At what rate is the current through the circuit changing when $R = 500\ \Omega$ and $V = 11$ volts?

14. Suppose that $f : \mathbb{R}^3 \to \mathbb{R}$ is a differentiable function and let $u(x, y, z) = f(x - y, y - z, z - x)$. Show that

$$\frac{\partial u}{\partial x} + \frac{\partial u}{\partial y} + \frac{\partial u}{\partial z} = 0.$$

15. Let $\boldsymbol{f}(x, y) = (x^2 y, x^3 + y^2, y - x)$. Evaluate $\boldsymbol{f}'(2, -3)$.

Chapter 3

Integrals

3.1 Multiple integrals and approximating sums

The two most important ideas of calculus—of one *or* several variables—are the derivative and the integral. In studying multivariable calculus up to now we've seen the derivative idea in many forms and settings, including partial derivatives, derivatives of vector-valued functions, gradients, directional derivatives, and linear and quadratic approximation.

Now we study integrals, the other main idea of our subject. Like derivatives, integrals have various meanings and interpretations in different multivariable settings. In this chapter our main focus is on **multiple integrals**, i.e., integrals of real-valued functions of two or three variables, integrated over regions in two- or three-dimensional space.◂ We'll study how multiple integrals are defined, how to calculate (or approximate) their values, and (last, but certainly not least) what they tell us in various circumstances.

Multiple integrals can also be defined in dimensions higher than three.

Integrals and approximating sums

All integrals—single, double, triple, or whatever—are defined to be certain limits of **approximating sums** (also known, sometimes, as **Riemann sums**). This important idea is always studied in single-variable calculus, but it may be quickly (and, perhaps, gratefully) forgotten. Readers whose memories are vague on this score have an excellent excuse: Although integrals are *defined* as limits of approximating sums, they are often *calculated* in an entirely different way, using antiderivatives. Here's a typical calculation:◂

There's not a Riemann sum in sight.

$$\int_0^1 x^2\,dx = \left.\frac{x^3}{3}\right]_0^1 = \frac{1}{3}.$$

This method of evaluating an integral—find an antiderivative for the integrand and plug in the endpoints—works just fine, thanks to the fundamental theorem of calculus.

So why bother at all with approximating sums? Here are two good reasons:

Antiderivative trouble. The antiderivative method depends on *finding* a convenient antiderivative of the integrand. Unfortunately, not every function, even in single-variable calculus, *has* an "elementary" antiderivative, i.e., an antiderivative with a symbolic formula built from the usual ingredients. The simple-looking function $f(x) = \sin(x^2)$ is an example—it has no elementary antiderivative.◂ The best we can do with the integral

Spend a moment looking for an antiderivative formula. Nothing works.

$$I = \int_0^1 \sin(x^2)\,dx,$$

therefore, is to approximate it, perhaps with some sort of sum, such as the left rule, the midpoint rule, or the trapezoid rule. (For the record, approximating I with a trapezoid-rule sum with 10 subdivisions gives $I \approx 0.311$.)

What integrals mean. The fundamental theorem often makes *calculating* integrals easy,➡ but approximating sums may illustrate more clearly what the answers *mean*. The following picture, for instance, illustrates the sense in which a midpoint-rule sum with four subdivisions approximates the area bounded by the curve $y = x^2$ from $x = 0$ to $x = 1$:

But not always—sometimes it's hard or impossible to find antiderivatives.

A midpoint sum estimate to $\int_0^1 x^2\,dx$: $M_4 = 63/192 \approx 0.328$

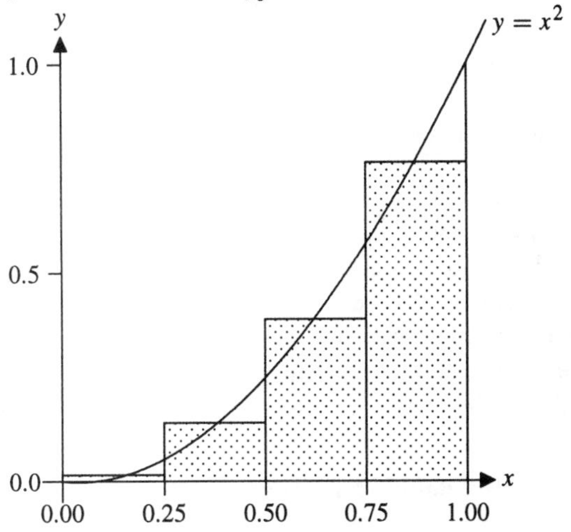

The basic idea of an integral as a limit of approximating sums is much the same for functions of two (or more) variables as for functions of one variable.

Integrals and sums: a review

Let's review the main single-variable ideas and notations that arise on the way to defining the integral of a function f over an interval $[a, b]$. The first step is to form Riemann sums:

> **Definition:** Let the interval $[a, b]$ be partitioned into n subintervals by any $n + 1$ points
>
> $$a = x_0 < x_1 < x_2 < \cdots < x_{n-1} < x_n = b;$$
>
> let $\Delta x_i = x_i - x_{i-1}$ denote the width of the i-th subinterval. Within each subinterval $[x_{i-1}, x_i]$, choose any point c_i. The sum
>
> $$\sum_{i=1}^{n} f(c_i)\Delta x_i = f(c_1)\Delta x_1 + f(c_2)\Delta x_2 + \cdots + f(c_n)\Delta x_n$$
>
> is called a **Riemann sum with n subdivisions** for f on $[a, b]$.

Left-rule, right-rule, and midpoint-rule approximating sums all fit this definition. Each of these sums is built from a **regular partition** of $[a, b]$ (i.e., one with subintervals of equal length) and some consistent scheme for choosing the **sampling points**

c_i. (For left, right, and midpoint sums, respectively, we choose each c_i as the left endpoint, the right endpoint, or the midpoint of the i-th subinterval.)

Graphical and numerical intuition suggests that, as the number n of subdivisions tends to infinity, all of these approximating sums should converge to some fixed number. Geometrically speaking, moreover, this number measures the signed area bounded by the f-graph from $x = a$ to $x = b$.

"Signed" means that areas under the x-axis are considered to be negative.

The formal definition of the integral makes these ideas precise:

Definition: Let the function f be defined on the interval $[a, b]$. The **integral of f over** $[a, b]$, denoted $\int_a^b f(x)\,dx$, is the number to which all Riemann sums S_n tend as n tends to infinity and as the widths of all subdivisions tend to zero. In symbols:

$$\int_a^b f(x)\,dx = \lim_{n\to\infty} S_n = \lim_{n\to\infty} \sum_{i=1}^{n} f(c_i)\,\Delta x_i\,,$$

if the limit exists.

Honesty dictates a brief admission: The limit in the definition, taken at face value, is a slippery customer. Understanding every ramification of permitting arbitrary partitions and sampling points, for example, can be tricky. Fortunately, these issues need not trouble us for the well-behaved functions (e.g., continuous functions) we typically meet in single-variable and multivariable calculus. For such functions, almost any respectable sort of approximating sum does what we expect—it approaches the true value of the integral as n tends to infinity.

Two variables: double integrals

Most of the differences between single-variable integrals and multivariable integrals are technical rather than theoretical. Indeed, the definitions of

$$\iint_R f(x, y)\,dA \quad \text{and} \quad \int_a^b f(x)\,dx$$

are almost identical. Here, f is a function of two variables, and R is a region—in the simplest case, a rectangle—in the xy-plane. On the other hand, the mechanics of evaluating these two types of integrals by antidifferentiation are quite different.

We'll get to that in the next section.

First, let's list the ingredients that go into defining the **double integral** $\iint_R f(x, y)\,dA$. Look, in each case, for points of similarity and difference with the one-variable situation.

R, the region of integration. In one variable the region of integration is *always* an interval $[a, b]$ in the domain of f; this is implicit in the notation $\int_a^b f(x)\,dx$. In two variables, by contrast, the region of integration, denoted by R, may be almost *any* two-dimensional subset of the plane. In simple cases R is a rectangle $[a, b] \times [c, d]$. In this case, we'll sometimes write

$$\int_a^b \int_c^d f(x, y)\,dy\,dx, \quad \text{not} \quad \iint_R f(x, y)\,dA.$$

(The first notation suggests a fact we'll see in the next section: that double integrals can sometimes be calculated by integrating "one variable at a time.")

What's dA? The symbol "dA" in the double integral resembles the "dx" that appears in single integrals. The "A" reminds us of "area."

Partitions. In one variable we partition an interval $[a, b]$ by cutting it (perhaps unevenly) into smaller intervals with endpoints $a = x_0 < x_1 < x_2 < \ldots < x_n = b$. The "size" of the i-th subinterval is simply its length, Δx_i.

In two variables we do much the same thing: We chop the plane region R into m smaller regions $R_1, R_2, R_3, \ldots, R_m$, perhaps of different sizes and shapes. The "size" of a subregion R_i is now taken to be its *area*, denoted by ΔA_i.

In practice—whatever the number of variables—it's usually convenient to choose the partition in some consistent way. In one variable, a **regular partition** (one with equal-length subintervals) is simplest. An analogous procedure in two variables, if R is a rectangle $[a, b] \times [c, d]$, is to cut R by an n-by-n grid in each direction, producing n^2 rectangular subregions in all. (This isn't the only possibility. Another alternative is to cut R into small *squares*.)

Approximating sums. In one variable, an approximating sum has the form

$$f(c_1)\Delta x_1 + f(c_2)\Delta x_2 + \cdots + f(c_n)\Delta x_n = \sum_{i=1}^{n} f(c_i)\Delta x_i,$$

where each c_i is a sample point chosen from the i-th subinterval.

A two-variable approximating sum is similar. From each subregion R_i we choose a sampling point $P_i = (x_i, y_i)$, and then form the approximating sum

$$S_m = f(P_1)\Delta A_1 + f(P_2)\Delta A_2 + \ldots + f(P_m)\Delta A_m = \sum_{i=1}^{m} f(P_i)\Delta A_i,$$

where ΔA_i is the area of R_i.

The following picture illustrates the idea, using the function $f(x, y) = x+y$ on the rectangle $R = [0, 4] \times [0, 4]$, with $m = n^2 = 16$ subdivisions. In each subrectangle, sampling is done at the corner closest to the origin:

The surface z=x+y and a Riemann approximation

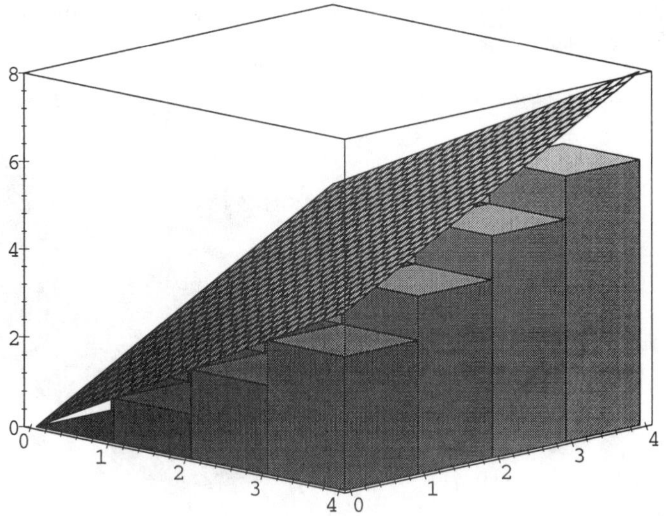

The picture shows that the approximating sum estimates the *volume* bounded on top by the surface $z = f(x, y)$ and whose base is the rectangle $R = [0, 4] \times [0, 4]$. This volume is the value of the integral $\iint_R (x + y)\, dA$. A close look and some back-of-the-envelope calculations show that the approximating sum adds up to 48. To summarize in symbols, we have

$$S_{16} = 48 \approx \iint_R (x + y)\, dA.$$

As the picture shows, all the approximating sums *underestimate* the true volume. Still, it's reasonable to expect these approximating sums S_m to converge to the "true" volume as m increases to infinity. A table of values lends credence to this expectation:

Calculated with help from technology, of course.

Approximating sums for various m	
number of subdivisions (m)	approximating sum (S_m)
2^2	32.00
4^2	48.00
8^2	56.00
12^2	58.67
16^2	60.00
20^2	60.80
24^2	61.33
28^2	61.71
32^2	62.00
100^2	63.36

The values of S_m seem to be creeping up toward a number around 64.

The next picture shows another approximating sum for the same function over the same interval. This time, sampling is done at the *midpoint* of each subrectangle:

```
The surface z=x+y and a midpoint Riemann approximation
```

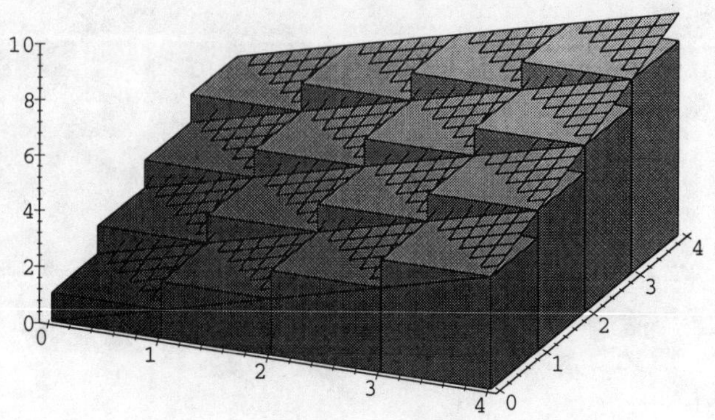

This time the approximating sum adds up to 64. In fact, 64 turns out to be the *exact* value of the integral. A close look at the picture suggests why the answer is exact: the overestimates and underestimates in the approximating sum cancel each other out. (We'll obtain the same result in the next section by other methods.)

■ **Example 1.** Consider the double integral $I = \iint_R f(x, y)\, dA$, where $f(x, y) = x + y$ and $R = [0, 4] \times [0, 4]$. (This is the same integral studied in the preceding paragraphs.) Calculate, this time by hand, an approximating sum S_4 with 4 equal subdivisions (2 in each direction). In each square subregion, evaluate f at the corner farthest from the origin. Does the sum underestimate or overestimate the integral?

Solution: All 4 subregions are squares with edge length 2 and therefore area 4; all corners have integer coordinates. The sampling points described above are these: $P_1 = (2, 2)$, $P_2 = (4, 2)$, $P_3 = (2, 4)$, $P_4 = (4, 4)$. The desired approximating sum, therefore, is

$$S_4 = \sum_{i=1}^{4} f(P_i)\Delta A_i = 4 \cdot 4 + 6 \cdot 4 + 6 \cdot 4 + 8 \cdot 4 = 96.$$

The sum overestimates the integral, because in each subrectangle the integrand function is evaluated where it takes its largest possible value. □

The integral as a limit

We defined the one-variable integral $\int_a^b f(x)\, dx$ above as the limit of approximating sums S_n, as n tends to infinity and the widths of all subdivisions tend to zero. The double integral $\iint_R f(x, y)\, dA$ is defined in a similar way.

Diameter: a technical subtlety. Before giving a detailed definition, we mention one technical subtlety.➥ It concerns the requirement that the *size* of subregions of R should go to zero as the *number* of subregions tends to infinity. If the subregions are squares, there's no problem—as the number of squares increases, the size of each square automatically decreases. Trouble could arise, however, if the subdivisions are extremely "oblong": this would happen if, say, we subdivided the rectangle $[0, 1] \times [0, 1]$ *only* in the x-direction, leaving the y-interval untouched. One technical "fix" that avoids such possible trouble is to define the **diameter** of a subregion to be the largest possible distance between any two points in the subregion. Requiring the diameter of all subregions to tend to zero—as we do in the following definition—automatically forces the subregions to be "small" in all directions.

This subtlety concerns only the formal definition of integrals, so it's unlikely to cause trouble in practice.

The following definition is adequate for the well-behaved functions $f(x, y)$ we'll study in this book:

> **Definition:** Let the function $f(x, y)$ be defined on the region R, and let S_m be an approximating sum with m subdivisions, as described above. Suppose that S_m tends to a number I as m tends to infinity and the diameter of all subdivisions tends to zero. Then I is the **double integral of f over R**, and we write
>
> $$I = \iint_R f(x, y)\, dA = \lim_{m\to\infty} \sum_{i=1}^{m} f(P_i)\Delta A_i.$$

As in the one-variable setting, the limit definition—although crucial to understanding integrals, and often useful for approximating them—almost never lends itself to calculating integrals *exactly*. For this purpose, fortunately, there are methods based on antidifferentiation. We'll see some soon.

Integrating constant functions. Constant functions are easy to integrate in one variable. The story is similar in multivariable calculus, thanks to the definition just given. Thus, let R be a region in the plane, let $f(x, y) = k$ be any constant function, and consider the integral $I = \iint_R f(x, y)\,dA$. Since $f(x, y) = k$ for all (x, y), *every* approximating sum for I has the same value:

$$S_m = \sum_{i=1}^{m} f(P_i)\Delta A_i = \sum_{i=1}^{m} k\,\Delta A_i = k\sum_{i=1}^{m}\Delta A_i = k \times \text{area of } R.$$

As the limit of approximating sums, the integral must have the same value. That is,

$$\iint_R f(x, y)\,dA = k \times \text{area of } R.$$

■ **Example 2.** Evaluate $\iint_R 3\,dA$, where R is the region inside the unit circle.

Solution: By the preceding observation, $\iint_R 3\,dA = 3 \times \text{area of } R = 3\pi$. □

Triple sums and triple integrals

The idea of integral can be extended to three (and even higher) dimensions. In this section we'll consider only the simplest case: the **triple integral** of a function $g(x, y, z)$ over a rectangular parallelepiped $R = [a, b] \times [c, d] \times [e, f]$, denoted by either

To put it more humbly, a "brick."

$$\iiint_R g(x, y, z)\,dV \quad \text{or} \quad \int_a^b \int_c^d \int_e^f g(x, y, z)\,dz\,dy\,dx.$$

(As for double integrals, the second notation suggests, correctly, that such integrals can sometimes be calculated one variable at a time.)

Triple integrals, just like single and double integrals, are defined formally as limits of approximating sums. An approximating sum in three dimensions is formed by subdividing a rectangular solid region R into m smaller rectangular subregions R_i, with *volume* ΔV_i, choosing a sampling point P_i in each subregion, and then evaluating the sum

$$\sum_{i=1}^{m} f(P_i)\Delta V_i.$$

The triple integral, finally, is defined as the limit of such sums as the number of subregions tends to infinity and the diameter of all subregions tends to zero.

See the remarks about diameter, above.

■ **Example 3.** Consider the triple integral $I = \iiint_R g(x, y, z)\,dV$, where $g(x, y, z) = x+y+z$ and $R = [0, 2] \times [0, 2] \times [0, 2]$. Calculate an approximating sum S_8 with 8 equal subdivisions (2 in each direction). In each subregion, evaluate g at the corner nearest to the origin.

Solution: All 8 subregions are cubes with edge length 1 and, therefore, volume 1. The sampling points are as follows—

$$P_1 = (0,0,0),\quad P_2 = (1,0,0),\quad P_3 = (0,1,0),\quad P_4 = (1,1,0),$$
$$P_5 = (0,0,1),\quad P_6 = (1,0,1),\quad P_7 = (0,1,1),\quad P_8 = (1,1,1).$$

and the approximating sum is

$$S_8 = \sum_{i=1}^{8} g(P_i)\Delta V_i = 0+1+1+2+1+2+2+3 = 12.$$

□

Interpreting multiple integrals

Integrals can be interpreted geometrically, physically, or in other ways. Following is a sampler of possibilities.

Double integrals and volume. The simplest geometric interpretation of a single-variable integral $\int_a^b f(x)\,dx$ is in terms of *area*: If $f(x) \geq 0$, then $\int_a^b f(x)\,dx$ measures the area of the region bounded above by the curve $y = f(x)$, below by the interval $[a, b]$ in the x-axis, and having vertical sides. In a similar vein, as we've already seen from pictures, if $f(x, y) \geq 0$ for (x, y) in R, then the double integral $\iint_R f(x, y)\,dA$ measures the *volume* of the three-dimensional solid bounded above by the surface $z = f(x, y)$, below by the region R in the xy-plane, with sides perpendicular to the xy-plane.

Double integrals and area. If R is a region in the xy-plane and g is the constant function $g(x, y) = 1$, then (as the previous paragraph says) the integral $\iint_R 1\,dA$ represents the volume of the solid S bounded below by R, having vertical sides and *constant height 1*. Recall, however, that the volume of any such "cylindrical" solid S is the *area* of the base times the height. Therefore, in this very special case, the volume of S and the area of R happen to have the *same* numerical value. In symbolic shorthand,

$$\iint_R 1\,dA = \text{area of } R,$$

for any plane region R. Perhaps surprisingly, this fact is often useful in calculating areas of plane regions. We'll see examples in later sections.

Triple integrals and volume. There isn't "room" in three-dimensional space even to plot a function $w = g(x, y, z)$—*four* variables would be needed. For this reason, interpreting triple integrals geometrically is, as a rule, difficult or impossible. There's one important exception to this rule.

For reasons similar to those explained for double integrals and area, integrating the *constant* function $g(x, y, z) = 1$ over a solid region R in xyz-space measures the *volume* of the region R. In symbols:

$$\iiint_R 1\,dV = \text{volume of } R.$$

Density and mass. Both double and triple integrals can often be interpreted physically, in the language of density and mass. (This view has the special advantage of making sense for both double and triple integrals.)

For a double integral $\iint_R f(x, y)\, dA$, one thinks of the plane region R as a flat plate with variable density—at any point (x, y), $f(x, y)$ gives the density, measured in appropriate units (e.g., grams per square centimeter). From this viewpoint, the double integral $\iint_R f(x, y)\, dA$ measures the total mass (in grams) of the plate R.

For a triple integral $\iiint_R g(x, y, z)\, dV$, one imagines a *solid* region R with variable density—at any point (x, y, z), the function value $g(x, y, z)$ is the solid's density, measured in appropriate units (e.g., grams per cubic centimeter). From this viewpoint, the triple integral $\iiint_R g(x, y, z)\, dV$ is the total mass (in grams) of the solid R.

Average value of a function. Let $f(x, y)$ be a function defined on a region R in the plane; suppose that R has total area $A(R)$. To form an approximating sum S_n for $I = \iint_R f(x, y)\, dA$, we chop R into n small subregions R_i, each with area ΔA_i; then we calculate the sum

$$S_n = f(P_1)\Delta A_1 + f(P_2)\Delta A_2 + \ldots f(P_n)\Delta A_n,$$

where the P_i are any convenient points inside R_i. Thus S_n can be thought of as a "weighted sum" of n output values of f, with the inputs P_i scattered around R, and each output value "weighted" by the size of the subregion it represents. Therefore, the quotient $S_n/A(R)$ can reasonably be thought of as an approximate **average value** of f over R. Taking this idea to the limit, and recalling that the area of R can also be calculated as an integral, leads to the definition.

An analogous definition holds for the average value of a one-variable function.

Definition: Let $f(x, y)$ be defined on a region R, with area $A(R)$. The **average value** of f over R is the ratio

$$\frac{\iint_R f(x, y)\, dA}{A(R)} = \frac{\iint_R f(x, y)\, dA}{\iint_R 1\, dA}.$$

Exactly the same reasoning applies to functions of three variables. If $g(x, y, z)$ is defined on a 3d-region R, with volume $V(R)$, then the average value of g over R is given by

$$\frac{\iiint_R g(x, y, z)\, dV}{V(R)} = \frac{\iiint_R g(x, y, z)\, dV}{\iiint_R 1\, dV}.$$

■ **Example 4.** Discuss average values in relation to the integrals of Example 1 and Example 3.

Solution: Example 1 and its preceding paragraphs are about the function $f(x, y) = x + y$ over the region $R = [0, 4] \times [0, 4]$, so $A(R) = 16$. We calculated several approximating sums for $I = \iint_R f(x, y)\, dA$; their values ranged from 32 to 96. These approximations produce, in turn, estimates for the average value of f over R, ranging from $32/16 = 2$ to $96/16 = 6$. As we remarked earlier, the exact value of I is 64, so the "correct" average value of f over R is 4.

Example 3 concerns the function $g(x, y, z) = x + y + z$ and the $R = [0, 2] \times [0, 2] \times [0, 2]$, which has volume 8. The (crude) approximating sum $S_8 = 12$ leads, therefore, to the estimate $12/8 = 1.5$ for the average value. In fact, the exact value of $I = \iiint_R g\, dA$ can be shown to be 24; assuming this, we get an average value of $24/8 = 3$ for g over R. □

We show this in the next section.

General advice. Multivariable integrals can be interpreted in many ways, depending on the situation, but is there any consistent theme?

One useful strategy for interpreting the information a given integral conveys is to think first about approximating sums. If approximating sums for an integral estimate a quantity, such as the mass of a solid object, than the integral measures the same quantity. We illustrate this line of reasoning in the next example.

■ **Example 5.** Depending on conditions, the audience at a public event (a night at the opera, say, or a rock concert) may either be evenly spread through the available space, or it may be tightly packed near the stage, and less crowded farther away.

Of course, people are counted in whole numbers, not fractions, so any effort to model the situation with calculus can only be approximate. (Despite this, calculus methods can and do produce useful estimates.) With this proviso, we proceed.

Suppose that at a certain event, the function $f(x, y)$ describes the "crowd density," in people per square meter, at the point x meters east and y meters north of some point on the stage, and let $R = [0, 50] \times [-15, 15]$. What information does the integral $I = \iint f(x, y)\, dA$ convey?

Solution: An approximating sum for the integral I has the form $f(P_1)\, \Delta A_1 + f(P_2)\, \Delta A_2 + \cdots + f(P_n)\, \Delta A_n$. Each summand is the product of a density (in people per unit area) and an area, so each summand approximates the number of people in a small area. Therefore, the approximating sum estimates the *total* number of people in a rectangular region extending 50 meters east, 15 meters north, and 15 meters south of the stage. The integral, in turn, can also be interpreted as gauging the total number of people in the region. (If the model is well designed, the integral—although still only an estimate—will improve on the approximating sums.) □

A word of caution. With so many possible interpretations for integrals, a final note of caution may be in order: The integral is defined as a *number*, say 5—not as a geometric or physical quantity. Although it's often useful and enlightening to *interpret* an integral in the language of area, volume, mass, or some other quantity, these are *only* interpretations, not intrinsic properties of an integral. In particular, there is nothing illogical about interpreting an integral $\iint_R f(x, y)\, dA = 5$ today as the volume of a certain solid, tomorrow as the mass of a plane region with variable density, the day after that in terms of average value, and next week as a crowd estimate.

Exercises

1. Calculate by hand (without technology) the midpoint sum with 4 subdivisions for $\int_0^1 x^2\, dx$. Then check your answer using technology. Finally, use technology to compute the midpoint sum with 100 subdivisions.

2. Calculate by hand (without technology) the double midpoint sum with $n = 3$ (i.e., 9 subdivisions in all) for the integral $\iint_R \sin(x)\sin(y)\, dA$ over the rectangle $R = [0, 1] \times [0, 1]$. Then check your answer using technology. Finally, use technology to compute the double midpoint sum with $n = 10$ (i.e., 100 subdivisions in all).

3. Calculate by hand (without technology) the triple midpoint sum with $n = 2$ (i.e., 8 subdivisions in all) for the triple integral $\iiint_R xyz\, dV$ over the cube

$R = [0, 4] \times [0, 4] \times [0, 4]$. Then check your answer using technology. Finally, use technology to compute the triple midpoint sum with $n = 4$, (i.e., $4^3 = 64$ subdivisions in all).

4. Let $f(x, y) = x + y$, let $R = [0, 4] \times [0, 4]$, and let $I = \iint_R f(x, y)\, dA$. Here is a contour plot of f:

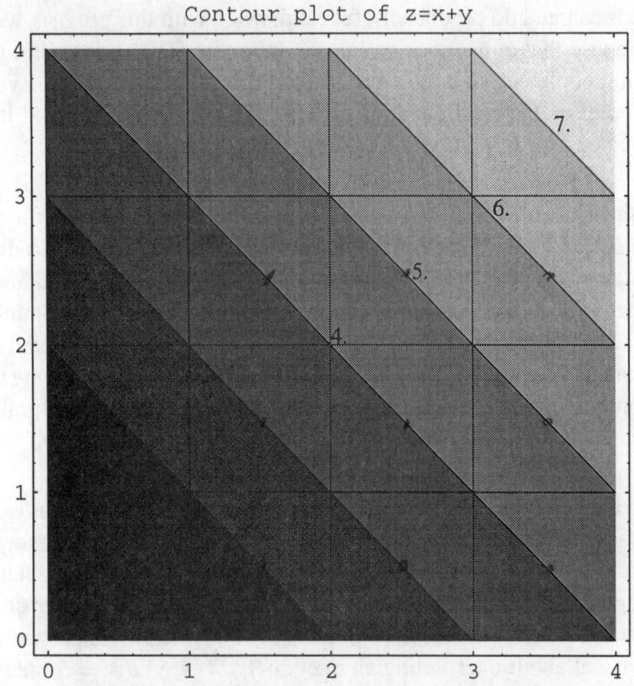

(a) Use the contour plot to evaluate a double midpoint sum for I with $n = 4$ (16 subdivisions in all).

(b) Use technology to check your answer to part (a).

(c) Your answer to part (a) is, in fact, the exact value of the integral I. How does the symmetry of the contour map show this?

5. Let $f(x, y) = x^2 + y^2$, let $R = [0, 4] \times [0, 4]$, and let $I = \iint_R f(x, y)\, dA$.

Here is a contour plot of f:

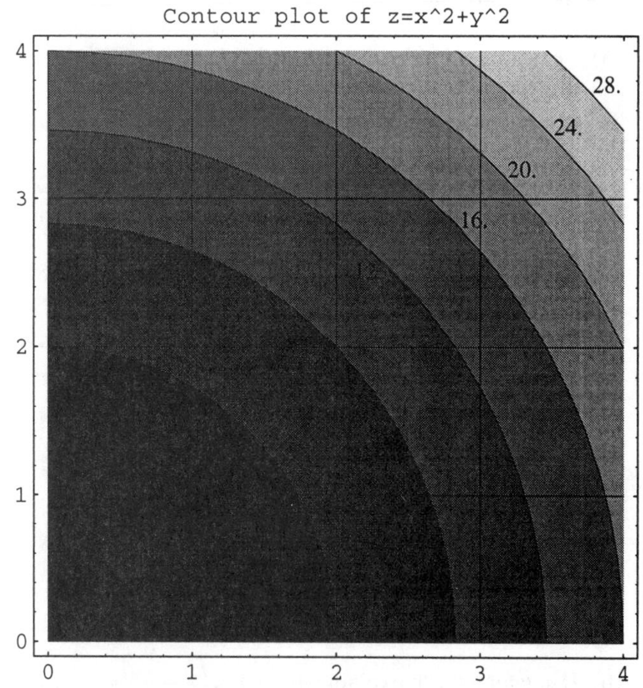

(a) Use the contour plot to evaluate a double midpoint sum for I with $n = 4$ (16 subdivisions in all).

(b) Use technology to check your answer to part (a).

(c) Would you expect your answer to part (a) to overestimate or underestimate the true value of I? How can you tell?

6. This problem is about the situation of Example 5, page 179; use the notation given there. You may want to sketch the area in question.

(a) Suppose the event is a free concert, the hall is full, the area in question is completely filled with seats, and that each seat in the hall occupies a rectangle 1 meter by 0.8 meters. Find a formula for $f(x, y)$ and calculate $I = \iint f(x, y)\,dA$. What does the answer mean in appropriate units?

(b) Suppose the event is a rock concert in an open (seatless) area, and that the crowd density decreases linearly from 3 people per square meter at the "front" of the hall (where $x = 0$) to 1 person per square meter at the "back" of the hall (where $x = 50$). Find a formula for $f(x, y)$; then estimate $I = \iint f(x, y)\,dA$ using a midpoint sum with 25 subdivisions (5 in each direction). (Use technology to find the answer.) What does the answer mean in appropriate units?

(c) Suppose the event is a sold-out opera, and that the hall has two classes of seats: high-priced seats in the "left half" of the hall (where $x \leq 25$) and low-priced seats in the "right" half of the hall (where $x \geq 25$). For their extra money, the higher-paying customers get larger seats, occupying 1.2 square meters; cheaper seats occupy only 0.8 square meters. Express the total number of people in the hall as the sum of two integrals over different rectangular domains. What *is* the total number of people?

7. In each case below interpret the given quantity in real-world language; be sure to include appropriate units.

   (a) Let $f(x, y)$ be the density at the point (x, y), in grams per square centimeter, of a metal plate lying in the xy-plane. (The unit of distance is centimeters.) Let R be a region in the plane. What physical quantity does $\iint_R f(x, y)\,dA$ measure?

   (b) Let $f(x, y)$ and R be as in the preceding part. Interpret $\dfrac{\iint_R f(x, y)\,dA}{\iint_R 1\,dA}$.

   (c) Let $g(x, y)$ be the depth (in inches) of snow at the point x miles east and y miles north of downtown Frostbite Falls, Minnesota. Let $R = [-50, 50] \times [-50, 50]$, and suppose that $\iint_R g(x, y)\,dA = 60{,}000$. What does the integral measure? What are the units?

   (d) Let $g(x, y)$ and R be as in the preceding part. Evaluate $\dfrac{\iint_R g(x, y)\,dA}{\iint_R 1\,dA}$.

   What does the answer mean about snow?

8. Let R denote the region of the xy-plane enclosed by the circle $x^2 + y^2 = 1$.

   (a) Evaluate $\displaystyle\iint_R 1\,dA$. Justify your answer.

   (b) Use geometry to explain why $\displaystyle\iint_R \sqrt{1 - x^2 - y^2}\,dA = \frac{2\pi}{3}$.

   [HINT: $x^2 + y^2 + z^2 = 1$ is the equation of a sphere in xyz-space.]

9. Calculate a midpoint approximating sum with four subdivisions (two in each direction) for the double integral $\displaystyle\iint_R (x + y^2)\,dA$, where R is the rectangle $[-4, 4] \times [3, 7]$.

10. Let R_1 denote the region $[0, 5] \times [-4, 4]$, R_2 the region $[0, 5] \times [0, 4]$, and R_3 the region $[-5, 0] \times [-4, 0]$. Suppose that $\iint_{R_2} f(x, y)\,dA = 10$, that $\iint_{R_3} f(x, y)\,dA = 24$, and that $f(x, y) = -f(-x, y)$. Evaluate $\iint_{R_1} f(x, y)\,dA$.

11. Suppose that the function $f : \mathbb{R}^2 \to \mathbb{R}$ has the values shown in the following table.

$y \backslash x$	1.5	2.0	2.5	3.0
3.0	4	6	9	6
2.5	6	9	7	5
2.0	4	8	6	4
1.5	3	5	5	7

Use the table to estimate $\displaystyle\int_2^3 \int_1^3 f(x, y)\,dx\,dy$.

3.2 Calculating integrals by iteration

The previous section was about *defining* multivariable integrals, as limits of approximating sums. This section is about *calculating* multivariable integrals, using antidifferentiation. Approximating sums are conceptually simple and (with technology) easy to calculate. But approximating sums are only approximate; to evaluate integrals *exactly*, we'd like an appropriate version of the single-variable antiderivative method, suitably modified for multivariable use.

It works thanks to the fundamental theorem of calculus.

Iteration: how it works

The key idea is to integrate a multivariable function *one variable at a time*, treating other variables as constants. The process is called **iterated integration.** To start, here's an example to show *how* it works; we'll see *why* it works in a moment.

In math-speak, "iterate" means "repeat."

■ **Example 1.** Let $f(x, y) = x + y$ and $R = [0, 4] \times [0, 4]$. Find $\iint_R f(x, y)\,dA$ by iterated integration. (We studied this integral in the preceding section; see the pictures and the table of values starting on page 173.)

Solution: We integrate first in x, treating y as a constant. Watch each step carefully:

Attaching variable names to the limits of integration is optional, but it can help remind us which variable is involved.

$$\begin{aligned}\iint_R f(x, y)\,dA &= \int_{y=0}^{y=4}\left(\int_{x=0}^{x=4}(x+y)\,dx\right)dy \\ &= \int_{y=0}^{y=4}\left(\frac{x^2}{2}+xy\bigg]_{x=0}^{x=4}\right)dy \\ &= \int_{y=0}^{y=4}(8+4y)\,dy \\ &= 8y+2y^2\bigg]_{y=0}^{y=4} = 64.\end{aligned}$$

Observe:

The same answer. The final answer, 64, should be familiar—it's what we estimated in the previous section using a midpoint approximation.

The answer as volume. Interpreted as a volume, the answer says that the solid bounded below by $R = [0, 4] \times [0, 4]$, above by the plane $z = x + y$, and having straight vertical walls has volume 64 cubic units. Here's a picture of the solid in question:

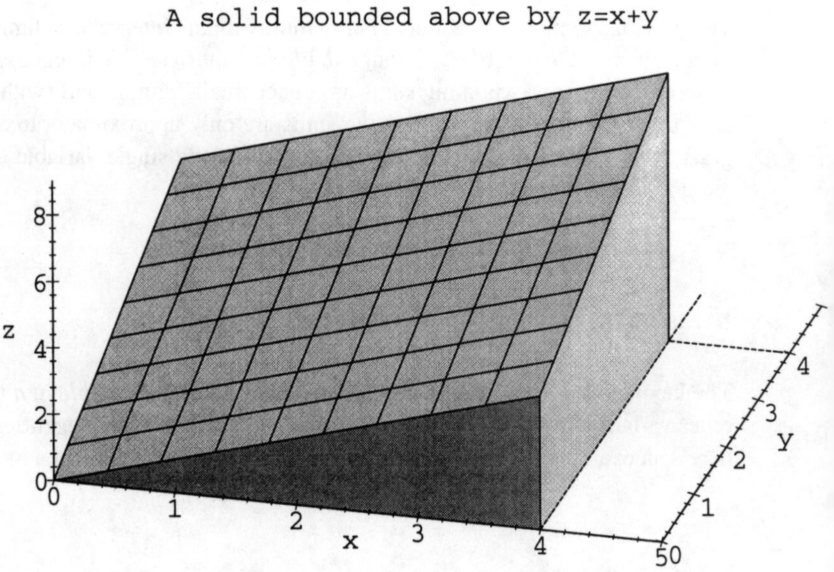

Checking answers with technology. Is the answer 64 geometrically reasonable? Is it symbolically correct? Technology (e.g., *Maple*, *Mathematica*, and their relatives) can help with both questions. Looking at a plot, like the one above, suggests that the answer 64 is at least in the ball park. As a further check, here is *Maple*'s version of the symbolic calculation above:

```
> int( int( x+y, x=0..4), y=0..4 );

                    64
```

Work from inside out. Iterated integrals are calculated from the inside out: The "inner" integral is found first.

The x-integral in the calculation above.

Either order works. There's nothing sacred about integrating first in x, and then in y. For well-behaved functions (those we meet *will* be well-behaved) we can integrate in either order—and both orders give the same answer. We'll return to this question below.

□

Cross-sectional area: an intermediate function. In the example above, the inner integral was calculated with respect to x, treating y as constant. The result was an intermediate function, g, of y alone; the formula is

$$g(y) = \int_0^4 (x+y)\,dx = 8 + 4y.$$

We integrated $g(y)$ with respect to y to get the final answer.

The function g has a nice geometric meaning: For any fixed y_0 in $[0, 4]$, $g(y_0)$ is the area under the curve $z = f(x, y_0)$ and above the xy-plane. In other words, $g(y)$ is the *area* of the cross section of the solid obtained by slicing with the plane $y = y_0$.

Here's the picture for $y = 2$:

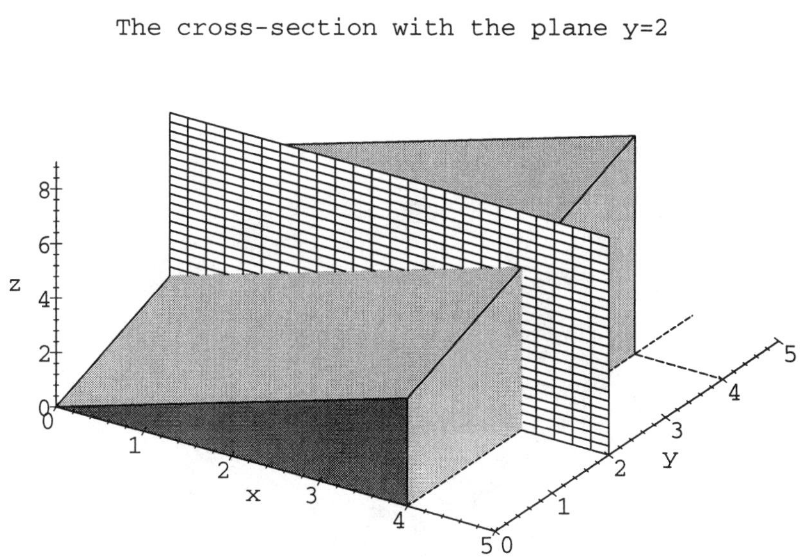

Since $g(y) = 8+4y$, we get $g(2) = 16$; this is the area of the part of the plane inside the the solid. As y runs from $y = 0$ to $y = 4$, $g(y)$ measures the area "swept out" by planes parallel to the one shown.

This area increases with y, as the picture shows.

As a matter of fact, the idea of a cross-sectional-area-function is not new; we used the same idea in single-variable calculus when we calculated volumes by integration. Back then, we put it like this:

In the example above, $g(y)$ plays the role of $A(x)$ below.

> **Fact:** Suppose that a solid lies with its base on the xy-plane, between the vertical planes $x = a$ and $x = b$. For all x in $[a, b]$, let $A(x)$ denote the area of the cross section at x, perpendicular to the x-axis. If $A(x)$ is a continuous function, then
>
> $$\text{volume} = \int_a^b A(x)\,dx.$$

Iteration: why it works

Why does the iteration method just demonstrated work? The Fact above gives some geometric feeling for the matter, at least when an integral can be thought of as the volume of a three-dimensional solid.

But not all integrals can or should be thought of in this way. A better reason why iteration works—a reason that makes sense in *any* dimension—is based on approximating sums. We'll describe the idea in two dimensions, but everything transfers readily to three (or even higher) dimensions. The main idea, in a nutshell, is that an approximating sum for a double integral can be grouped either along "rows" or along "columns." We'll explain first with an example and then give a more general argument.

■ **Example 2.** Let $f(x, y) = 2 - x^2 y$, $R = [0, 1] \times [0, 1]$, and $I = \iint_R f(x, y)\,dA$. Here's a picture of the solid whose volume is given by I:

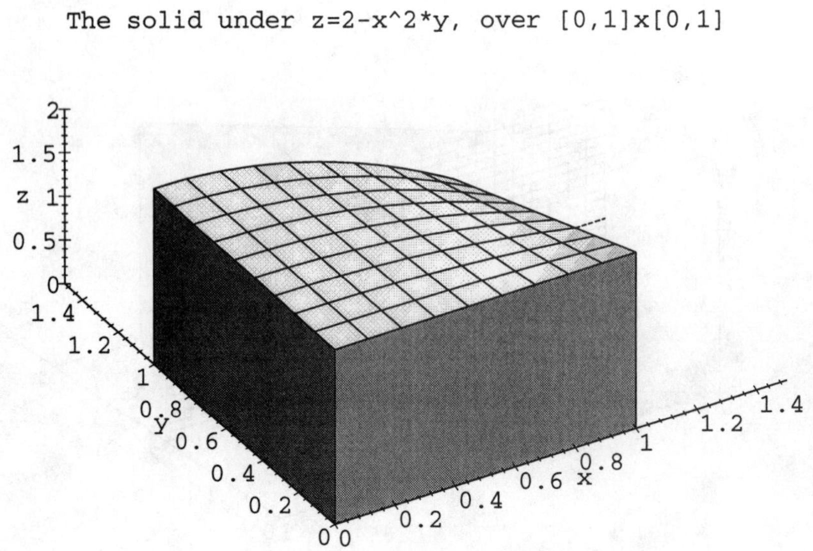

Compute a midpoint approximating sum with 100 equal subdivisions (10 in each direction) for the integral $I = \iint_R f(x, y)\,dA$. Compare it to the *exact* value of I, found by iteration.

But check our work.

Solution: It's easy to calculate I exactly, by iteration.[•] This time, we'll integrate first in y:

$$I = \int_0^1 \left(\int_0^1 (2 - x^2 y)\,dy \right) dx = \int_0^1 \left(\left. 2y - \frac{x^2 y^2}{2} \right]_0^1 \right) dx = \int_0^1 \left(2 - \frac{x^2}{2} \right) dx = \frac{11}{6}.$$

They're rounded to 1 decimal place.

To find an approximating sum, we'll first tabulate values (rounded to 2 decimals) of f at the midpoints of all 100 subrectangles of $[0, 1] \times [0, 1]$. For later use, column sums are at the bottom:[•]

Values of $f(x, y) = 2 - x^2 y$

$y \backslash x$	0.05	0.15	0.25	0.35	0.45	0.55	0.65	0.75	0.85	0.95
0.95	2.00	1.98	1.94	1.88	1.81	1.71	1.60	1.47	1.31	1.14
0.85	2.00	1.98	1.95	1.90	1.83	1.74	1.64	1.52	1.39	1.23
0.75	2.00	1.98	1.95	1.91	1.85	1.77	1.68	1.58	1.46	1.32
0.65	2.00	1.99	1.96	1.92	1.87	1.80	1.73	1.63	1.53	1.41
0.55	2.00	1.99	1.97	1.93	1.89	1.83	1.77	1.69	1.60	1.50
0.45	2.00	1.99	1.97	1.94	1.91	1.86	1.81	1.75	1.68	1.59
0.35	2.00	1.99	1.98	1.96	1.93	1.89	1.85	1.80	1.75	1.68
0.25	2.00	1.99	1.98	1.97	1.95	1.92	1.89	1.86	1.82	1.77
0.15	2.00	2.00	1.99	1.98	1.97	1.95	1.94	1.92	1.89	1.87
0.05	2.00	2.00	2.00	1.99	1.99	1.98	1.98	1.97	1.96	1.95
sum	20.0	19.9	19.7	19.3	19.0	18.4	17.9	17.2	16.3	15.4

The approximating sum is found by totaling all the function values and multiplying by $1/100$, the area of each subrectangle. (In this case, therefore, the approximating sum is the average of the 100 table entries.) The result is 1.83; this compares nicely to the exact answer, $11/6 \approx 1.8333$.

A computer helped!

Now consider any *column* in the table. The last column, for instance, contains numbers of the form $f(0.95, y)$, for 10 equally-spaced values of y, with $\Delta y = 0.1$. Therefore the column sum multiplied by 0.1 (the answer is 1.54) is a Riemann sum for the integral $\int_0^1 f(0.95, y)\,dy$. We can calculate this last integral directly:

$$\int_0^1 f(0.95, y)\,dy = \int_0^1 (2 - 0.95^2 y)\,dy = 1.54875;$$

that's not far from the Riemann sum. Tabulating results for the other columns shows the same pattern. The Riemann sum associated to each column closely approximates the corresponding y-integral:

Integrals are rounded to 2 decimals.

Column sums and y-integrals

x	0.05	0.15	0.25	0.35	0.45	0.55	0.65	0.75	0.85	0.95
column sum	20.0	19.9	19.7	19.3	19.0	18.4	17.9	17.2	16.3	15.4
$\int_0^1 f(x, y)\,dy$	2.00	1.99	1.97	1.94	1.90	1.85	1.79	1.72	1.64	1.55

The calculations are rather complicated, but the moral is quite simple: Whether adding up approximating sums or calculating integrals, it's OK to work with one variable at a time. □

A general argument. Let's summarize the ideas of the previous example in more general terms. To do so, suppose that we're given a rectangle $[a, b] \times [c, d]$ in the xy-plane, and a function f defined on R. We'll see why the integral $I = \iint_R f(x, y)\,dA$ can reasonably be calculated by iteration.

To begin, let's write an approximating sum for I, in the sense of the last section. First we subdivide both $[a, b]$ and $[c, d]$ into n equal subintervals. Their lengths, respectively, are $\Delta x = (b - a)/n$ and $\Delta y = (d - c)/n$. This produces a grid of n^2 subrectangles R_{ij}, where $1 \le i, j \le n$; each rectangle has area $\Delta x\,\Delta y$. Let (x_i, y_j) be the midpoint of the (i, j)-th rectangle; we'll use these midpoints as sampling points for an approximating sum S_{n^2} for I.

With all ingredients now in place, we can write the approximating sum:

$$S_{n^2} = \sum_{i,j=1}^{n} f(x_i, y_j)\,\Delta x\,\Delta y = f(x_1, y_1)\,\Delta x\,\Delta y + \cdots + f(x_n, y_n)\,\Delta x\,\Delta y.$$

The subscript on the "$\sum$" means that we sum over all possible values of *both* i and j from 1 to n.

There are n^2 summands in all.

We can group and add the summands in any order or pattern we like. Here's one convenient pattern:

It's OK to factor Δy out of each row, because Δy common to all summands.

$$\begin{aligned} S_{n^2} &= \left(f(x_1, y_1)\Delta x + f(x_2, y_1)\Delta x + \cdots + f(x_n, y_1)\Delta x \right) \Delta y \\ &+ \left(f(x_1, y_2)\Delta x + f(x_2, y_2)\Delta x + \cdots + f(x_n, y_2)\Delta x \right) \Delta y \\ &+ \ \ldots \\ &+ \left(f(x_1, y_j)\Delta x + f(x_2, y_j)\Delta x + \cdots + f(x_n, y_j)\Delta x \right) \Delta y \\ &+ \ \ldots \\ &+ \left(f(x_1, y_n)\Delta x + f(x_2, y_n)\Delta x + \cdots + f(x_n, y_n)\Delta x \right) \Delta y. \end{aligned}$$

Here's the first of two key points:

> *The sum inside parentheses on each line above is a Riemann sum with n subdivisions for a* single-variable *integral in x.*

Specifically, the sum on the first line is a Riemann sum for the integral $\int_a^b f(x, y_1)\, dx$; the sum on the second line approximates $\int_a^b f(x, y_2)\, dx$, and so on. Now, if n is large, then all of these sums are close to the integrals they approximate. Thus, for large n,

$$S_{n^2} \approx \int_a^b f(x, y_1)\, dx\ \Delta y + \int_a^b f(x, y_2)\, dx\ \Delta y + \cdots + \int_a^b f(x, y_n)\, dx\ \Delta y.$$

Here's the second key observation:

> *The sum on the right above is a Riemann sum with n subdivisions for the integral $\int_c^d g(y)\, dy$, where $g(y) = \int_a^b f(x, y)\, dx$.*

For large n this sum, too, is near the integral it approximates. In other words,

This should sound reasonable, and it is indeed true for the well-behaved functions we'll see in this book. But a rigorous proof is quite subtle; it depends on technical properties of the integrand function.

$$S_{n^2} \approx \sum_{j=1}^{n} g(y)\, \Delta y \approx \int_c^d g(y)\, dy = \int_c^d \left(\int_a^b f(x, y)\, dx \right) dy.$$

This shows (informally) what we hoped to show: For large n,

$$I \approx S_{n^2} \approx \int_c^d \left(\int_a^b f(x, y)\, dx \right) dy.$$

We conclude that we can, indeed, evaluate a double integral I over a rectangle R by integrating first in x and then in y. Nor does the *order* matter—we could just as well have reversed the roles of x and y throughout the argument above.

Iterated integrals in three variables

Iteration works exactly the same way for a triple integral defined on a cube in xyz-space.

■ **Example 3.** Let $f(x, y, z) = x + y + z$, $R = [0, 2] \times [0, 2] \times [0, 2]$, and $I = \iiint_R f(x, y, z)\, dV$. Calculate I exactly, by iteration. (Compare Example 3, page 176, where we calculated a crude approximating sum for I.) What does the answer *mean*?

Solution: We integrate in each of the three variables in turn:

$$\begin{aligned}
\iiint_R f(x,y,z)\,dV &= \int_0^2 \left(\int_0^2 \left(\int_0^2 (x+y+z)\,dx \right) dy \right) dz \\
&= \int_0^2 \left(\int_0^2 \left(\frac{x^2}{2} + xy + xz \Big]_0^2 \right) dy \right) dz \\
&= \int_0^2 \left(\int_0^2 (2+2y+2z)\, dy \right) dz \\
&= \int_0^2 \left(2y + y^2 + 2yz \Big]_0^2 \right) dz \\
&= \int_0^2 (8+4z)\, dz \\
&= 24.
\end{aligned}$$

What the answer means depends on our point of view. If we think of the integrand $f(x, y, z)$ as the density (in, say, grams per cubic centimeter) of the solid R at the point (x, y, z), then the integral tells the *mass* of the solid (in grams). □

Integrals over non-rectangular regions

Not every integral of interest is taken over a rectangular domain of integration. It's sometimes useful to integrate over regions with curved or slanted boundaries. Much the same process of iteration applies in this case as for integrals over rectangles and cubes, but some extra care is needed. We'll illustrate the process with an example.

■ **Example 4.** Find $I = \iint_R (2-x)\,dA$, where R is the plane region between the curves $y = 0$ and $y = x$, and $x = 1$.

It's essential to draw the domain; use an ordinary pair of xy-axes.

Solution: The integral gives the volume of this figure:

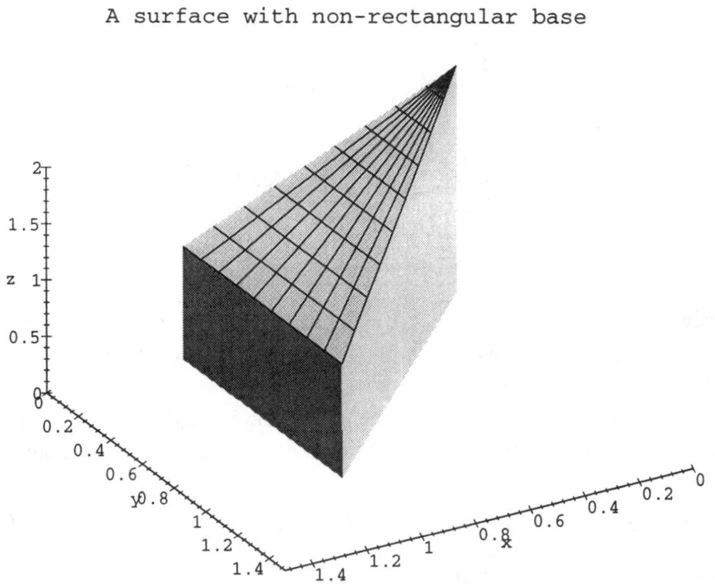

We can think of the domain R as bounded by the straight lines $x = 0$ and $x = 1$ on the left and right, and by the curves $y = 0$ and $y = x$ on the bottom and top. As in earlier cases, we can integrate by iteration. The *outer* integral, in x, runs from $x = 0$ to $x = 1$. The upper and lower limits of the *inner* integral, however, depend on x, as the following computation shows:

$$\begin{aligned} I &= \iint_R (2-x)\,dA = \int_{x=0}^{x=1} \left(\int_{y=0}^{y=x} (2-x)\,dy \right) dx \\ &= \int_{x=0}^{x=1} \left(2y - xy]_{y=0}^{y=x} \right) dx \\ &= \int_{x=0}^{x=1} \left(2x - x^2 \right) dx = \frac{2}{3}. \end{aligned}$$

In fact, this integral is not much different from those above, which were over rectangular domains. The variable limits of integration in the inner integral simply reflect the fact that the domain of integration is has varying "heights," depending on x. □

Changing the order of integration

For integrals over rectangles or cubes, we could integrate in any order we liked. A similar result holds in the present case, too—as long as the region has an appropriate shape. We'll illustrate by redoing the previous example, but integrating the variables in the opposite order.

■ **Example 5.** Redo the integral $I = \iint_R (2-x)\,dA$ of the previous example; but now integrate first in x and then in y.

Solution: We think of R, this time, as lying between the lines $y = 0$ and $y = 1$. For given y, R starts at the line $x = y$ and ends at the line $x = 1$. Now the calculation is similar to the one above:

$$\begin{aligned} I = \iint_R (2-x)\,dA &= \int_{y=0}^{y=1} \left(\int_{x=y}^{x=1} (2-x)\,dx \right) dy \\ &= \int_{y=0}^{y=1} \left(2x - \frac{x^2}{2} \Big]_{x=y}^{x=1} \right) dy \\ &= \int_{y=0}^{y=1} \left(\frac{3}{2} - 2y + \frac{y^2}{2} \right) dy = \frac{2}{3}. \end{aligned}$$ □

Not always so simple. Things aren't always quite so simple. Some domains of integration lend themselves more naturally to one order of integration than another. See the exercises for examples.

Three-variable integrals can also be more challenging. We'll return to them, and to some additional techniques for handling them, at the end of this chapter.

A theoretical application of integrals (optional)

We met the following theorem in Section 2.6:

> **Theorem 1. (Equality of cross partials)** Let $f(x, y)$ be a function; assume that the second partial derivatives f_{xy} and f_{yx} are defined and continuous on the domain of f. Then for all (x, y),
>
> $$f_{xy}(x, y) = f_{yx}(x, y).$$

Perhaps surprisingly, a relatively simple proof of the theorem can be given using an iterated double integrals.

The idea of the proof. We'll explain why $f_{xy}(0, 0) = f_{yx}(0, 0)$. (That will suffice, since there's nothing special about $(0, 0)$.) To do so, we'll show that for any small square $R = [0, h] \times [0, h]$,

$$\iint_R f_{xy}(x, y)\, dA = \iint_R f_{yx}(x, y)\, dA. \qquad (3.2.1)$$

First, let's think for a moment about how Equation 3.2.1 helps us. Suppose, for instance, that $f_{xy}(0, 0) > f_{yx}(0, 0)$. Then, since f_{xy} and f_{yx} are continuous functions,➸ we'd have $f_{xy}(x, y) > f_{yx}(x, y)$ for all (x, y) near $(0, 0)$. In particular, we'd have $f_{xy}(x, y) > f_{yx}(x, y)$ for all (x, y) on a small rectangle $R = [0, h] \times [0, h]$; in this case Equation 3.2.1 couldn't possibly hold.

Here's where the technical assumption comes in.

Therefore, it's enough to convince ourselves of Equation 3.2.1. To do so, we calculate both sides of Equation 3.2.1 as iterated integrals. Here's the left side (LHS):

$$\text{LHS} = \iint_R f_{xy}(x, y)\, dA = \int_0^h \left(\int_0^h f_{xy}(x, y)\, dy \right) dx.$$

Now notice that f_{xy} is (by definition!) the y-derivative of f_x, and f_x is the x-derivative of f. This means that we can calculate our integrals by antidifferentiation:

$$\begin{aligned} LHS &= \int_0^h \left(\int_0^h f_{xy}(x, y)\, dy \right) dx \\ &= \int_0^h \left(f_x(x, y)\big]_0^h \right) dx \\ &= \left(f(x, h) - f(x, 0) \right)\big]_0^h \\ &= f(h, h) - f(0, h) - f(h, 0) + f(0, 0). \end{aligned}$$

A similar calculation➸ shows that the right-hand integral in Equation 3.2.1 has the same value. □

See the exercises.

Exercises

1. Use iteration to calculate each of the following integrals by hand (without technology). Then check your answers symbolically, using *Maple*. In parts (a)–(c), plot a 3d-surface over an appropriate domain to see that your answer is reasonable. [NOTE: Most of these integrals appear among the exercises in Section 14.5.]

   (a) $\iint_R \sin(x) \sin(y)\, dA$; $R = [0, 1] \times [0, 1]$

(b) $\iint_R \sin(x+y)\,dA$; $R = [0,1] \times [0,1]$

(c) $\iint_R (x^2+y^2)\,dA$; $R = [0,4] \times [0,4]$

(d) $\iiint_R x\,dV$; $V = [0,1] \times [0,2] \times [0,3]$

(e) $\iiint_R y\,dV$; $V = [0,1] \times [0,2] \times [0,3]$

2. Use iteration to calculate each of the following nonrectangular integrals by hand (i.e., without technology). In each case, the inner integral should be in y and the outer integral in x. Check your answers symbolically, using *Maple*.

   (a) $\iint_R (x+y)\,dA$; R the region bounded by the curves $y = x$ and $y = x^2$

   (b) $\iint_R x\,dA$; R the region bounded by the curves $y = x^2$ and $y = \sqrt{x}$

   (c) $\iint_R 1\,dA$; R the first quadrant part of the circle $x^2 + y^2 \le 1$

3. Redo Exercise 2, but integrate first in x and then in y.

4. Consider the integral $I = \iint_R (x+y)\,dA$, where R is the region bounded by the curves $y = x^2$ and $y = 1$.

   (a) Calculate I by integrating first in y and then in x.

   (b) Calculate I by integrating first in x and then in y.

5. Let $f(x,y) = x$, and let R be the plane region bounded by the curves $y = e^x$, $y = 0$, $x = 0$, and $x = 1$.

   (a) Calculate $I = \iint_R f(x,y)\,dA$ by integrating first in y and then in x.

   (b) Calculate $I = \iint_R f(x,y)\,dA$ by integrating first in x and then in y. [HINT: First split the region R into two simpler pieces; each simpler piece should be bounded on the left by one curve and on the right by another.]

6. Let $y = f(x)$ be a function, with $f(x) \ge 0$ if $a \le x \le b$; let R be the plane region bounded by the curves $y = f(x)$, $y = 0$, $x = a$, and $x = b$.

   (a) What does *single-variable* calculus say about the area of R?

   (b) According to Section 14.5, the double integral $I = \iint_R 1\,dA$ gives the area of R. Use an iterated integral to reconcile this formula with the one in part (a).

7. Let $x = g(y)$ be a function with $g(y) \ge 0$ if $c \le y \le d$; let R be the plane region bounded by the curves $x = g(y)$, $x = 0$, $y = c$, and $y = d$.

   (a) What does *single-variable* calculus say about the area of R?

   (b) According to Section 14.5, the double integral $I = \iint_R 1\,dA$ gives the area of R. Use an iterated integral to reconcile this formula with the one in part (a).

8. Find the volume under the surface $z = xe^y$ and above the triangle with vertices at (0, 0), (2, 2), and (4, 0).

9. Evaluate the iterated integral $\int_0^1 \int_{2x}^2 x\sqrt{1+y^3}\,dy\,dx$ by reversing the order of integration.

10. Explain why $\int_0^3 \int_1^{e^x} \sin(x^2+y^2)\,dy\,dx = \int_1^{e^3} \int_{\ln y}^3 \sin(x^2+y^2)\,dx\,dy.$

11. Let T be the tetrahedron bounded by the planes, $x = 0$, $y = 0$, $z = 0$, and $x + 2y + 3z = 6$.

    (a) Express the volume of T as a triple iterated integral.

    (b) Express the volume of T as a double iterated integral.

    (c) Find the volume of T.

12. Evaluate the iterated integral $\int_0^1 \int_{x/2}^{1/2} e^{y^2}\,dy\,dx.$

13. We showed in this section that the left-hand integral in Equation 3.2.1 (page 191) has the value $f(h,h) - f(0,h) - f(h,0) + f(0,0)$. Mimic the calculation there to show that the right-hand integral has the same value.

3.3 Double integrals in polar coordinates

Easy integrals and hard integrals

What makes a double integral $I = \iint_R f(x,y)\,dA$ hard to calculate? Both f and R can play a role: if either one is complicated or messy to describe, then I may be correspondingly ugly. Let's see examples of both "good" and "bad" integrals.

■ **Example 1.** Discuss $I_1 = \iint_{R_1} x^2\,dA$ and $I_2 = \iint_{R_2} \sqrt{x^2+y^2}\,dA$, where R_1 is the rectangle $[0,1]\times[0,2\pi]$, and R_2 is the region inside the unit circle $x^2+y^2=1$.

Draw R_1 and R_2 for yourself.

Solution: The first integral is easy:

$$I_1 = \int_{x=0}^{x=1}\left(\int_{y=0}^{y=2\pi} x^2\,dy\right)dx = \int_0^1\left(x^2y\Big]_0^{2\pi}\right)dx = \int_0^1 2\pi x^2\,dx = \frac{2\pi}{3}.$$

The ingredients of I_2 are more complicated to describe. The circular region R_2 can be thought of as bounded by the curves $y=\sqrt{1-x^2}$ and $y=-\sqrt{1-x^2}$ on the top and bottom, and by the lines $x=-1$ and $x=1$ on the left and right. Now we can write I_2 in iterated form:

In xy-coordinates, at least. Polar coordinates will make things simpler.

$$I_2 = \int_{x=1}^{x=-1}\left(\int_{y=-\sqrt{1-x^2}}^{y=\sqrt{1-x^2}} \sqrt{x^2+y^2}\,dy\right)dx.$$

The integral looks—and is—complicated. Just to get started on the inner integral, we'd need this antiderivative formula

Even Maple has trouble with it.

From an integral table.

$$\int \sqrt{x^2+p^2}\,dx = \frac{1}{2}\left(x\sqrt{x^2+p^2} + p^2 \ln\left|x+\sqrt{x^2+p^2}\right|\right).$$

Faced with this prospect, we retreat. But only temporarily—we shall return to I_2 shortly. □

What went wrong, and what to do. The integral I_2 in Example 1 led to an unpleasant calculation in x and y for two reasons:

(i) the integrand, $\sqrt{x^2+y^2}$, has a complicated antiderivative in x or y;

No thanks to the square root.

(ii) the domain of integration, although geometrically simple, is messy to describe in rectangular coordinates.

No thanks, again, to square roots.

In *polar* coordinates, on the other hand, both the integrand and the domain have simple, uncluttered formulas. The integrand is

$$f(x,y) = \sqrt{x^2+y^2} = r.$$

The domain of integration is, in polar language, a lot like a rectangle; it's defined by the inequalities

$$0 \le r \le 1 \quad\text{and}\quad 0 \le \theta \le 2\pi.$$

In this case, apparently, both the integrand f and the domain R "deserve" to be described in polar coordinates, not in rectangular coordinates. It seems reasonable, therefore, that the double integral I_2 should also be calculated using polar rather than rectangular coordinates.

That hunch is correct. We'll see in this section how to calculate double integrals in polar form, using r and θ, as opposed to Cartesian➽ form, using x and y. Integrals such as I_2, in which the integrand, the domain of integration, or both are simplest in polar form, are natural candidates for polar treatment.

Or rectangular.

Polar "rectangles"

A rectangle in Cartesian coordinates is defined by two inequalities of the form

$$a \le x \le b; \quad c \le y \le d;$$

each of the coordinates x and y ranges through an interval. A **polar rectangle** is defined by two similar inequalities:

$$a \le r \le b; \quad \alpha \le \theta \le \beta;$$

again, each of the coordinates r and θ ranges through an interval. Here are generic pictures of both types of rectangle:

A Cartesian rectangle

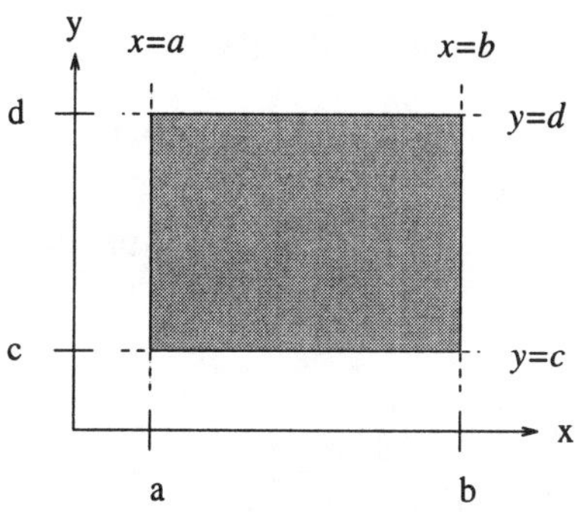

A polar rectangle

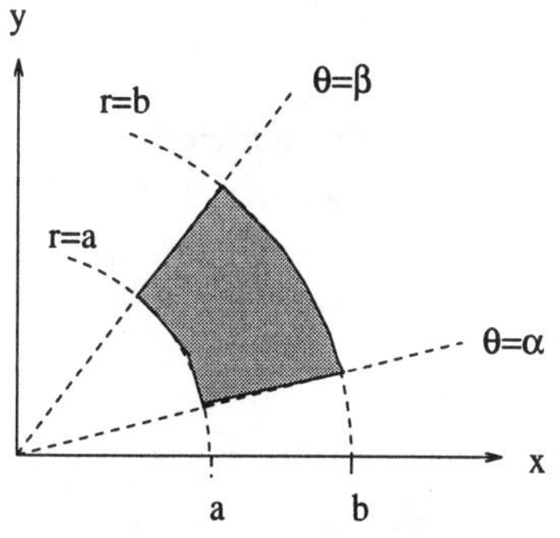

For polar integrals, as for rectangular ones, rectangles (in the appropriate sense) are the simplest regions over which to integrate.

Polar integration—how it works

A double integral $I = \iint_R f(x, y)\, dA$ in rectangular coordinates, where $R = [a, b] \times [c, d]$ is an ordinary rectangle, is written in iterated form➽ as

We'll integrate first in y, this time.

$$\iint_R f(x, y)\, dA = \int_a^b \int_c^d f(x, y)\, dy\, dx.$$

(Here and below we omit some parentheses—it's always understood that the inner integral is done first.)

Now suppose we're given a double integral $I = \iint_R g(r, \theta)\, dA$ in polar coordinates, where R is a *polar* rectangle, defined by inequalities

$$a \le r \le b \quad \text{and} \quad \alpha \le \theta \le \beta,$$

and $g(r, \theta)$ is a function defined on R. The following Fact gives the appropriate integral formula:

Fact: (Double integrals in polar coordinates) Let g and R be as above. Then

$$\iint_R g(r,\theta)\,dA = \int_{\theta=\alpha}^{\theta=\beta}\int_{r=a}^{r=b} g(r,\theta)\,r\,dr\,d\theta.$$

The formula raises several important observations:

Trading x and y for r and θ. Any function $f(x, y)$ can be "traded" for an equivalent function $g(r, \theta)$, using the relations

$$x = r\cos\theta \quad\text{and}\quad y = r\sin\theta.$$

The same method works for *equations* in x and y. The equation $x = y$, for example, says in polar coordinates that $r\cos\theta = r\sin\theta$, or, equivalently, that $\tan\theta = 1$. This polar equation describes the same line as the original Cartesian equation. We'll use these principles below when evaluating polar integrals.

A useful mnemonic. Compare the formulas above for integrating in rectangular and polar coordinates. An important difference between the two has to do with the "dA" expression. The full mathematical story is much deeper, but as a first quick aid to memory, the following formulas are very handy:

We won't go into great depth, but we'll give some informal justification soon.

$$\begin{aligned} dA &= dx\,dy \quad \text{for Cartesian coordinates;}\\ dA &= r\,dr\,d\theta \quad \text{for polar coordinates.}\end{aligned}$$

That extra factor of r . . . . What's that mysterious extra r doing in the polar formula $dA = r\,dr\,d\theta$? Why not just $dA = dr\,d\theta$?

We certainly owe the reader an explanation. We'll honor that debt in a moment, when we discuss *why* the formula works. But first let's see *that* it works.

■ **Example 2.** Let R_2 be the region inside the unit circle. Use polar coordinates to calculate that troublesome integral $I_2 = \iint_{R_2}\sqrt{x^2+y^2}\,dA$ from Example 1, page 194.

This is always possible.

Solution: First we write all the data in polar form. For the integrand, we have $f(x, y) = \sqrt{x^2+y^2} = r = g(r, \theta)$. For the domain of integration, we translate the Cartesian equation $x^2 + y^2 = 1$ into its (simpler!) polar form, $r = 1$. The rest is easy:

Watch each step; work from inside out.

$$\begin{aligned}\iint_{R_2}\sqrt{x^2+y^2}\,dA &= \int_{\theta=0}^{\theta=2\pi}\int_{r=0}^{r=1} r\,dA\\ &= \int_{\theta=0}^{\theta=2\pi}\int_{r=0}^{r=1} r^2\,dr\,d\theta\\ &= \int_{\theta=0}^{\theta=2\pi}\left.\frac{r^3}{3}\right]_0^1 d\theta\\ &= \int_{\theta=0}^{\theta=2\pi}\frac{1}{3}\,d\theta = \frac{2\pi}{3}.\end{aligned}$$

Notice, especially, the similarity to the integral I_1 of Example 1—I_1 and I_2 turned out to have the same value. This is no accident, of course. After rewriting in polar coordinates, I_2 turned out to be the same integral in r and θ as I_1 is in x and y. □

Polar integrals—what they mean

Polar integrals have exactly the same interpretations as any other double integrals.➸ Depending on the situation and on our point of view, an integral might represent a volume, the area of a plane region, the mass of a thin plate, or many other things.

We discussed several possible interpretations of integrals in Section 3.1.

For example, both the integrals I_1 and I_2 of Example 1, page 194, can be interpreted as volumes of solids. For I_2, the solid lies above the unit disk and below the surface $z = \sqrt{x^2 + y^2}$, as shown:

A polar solid

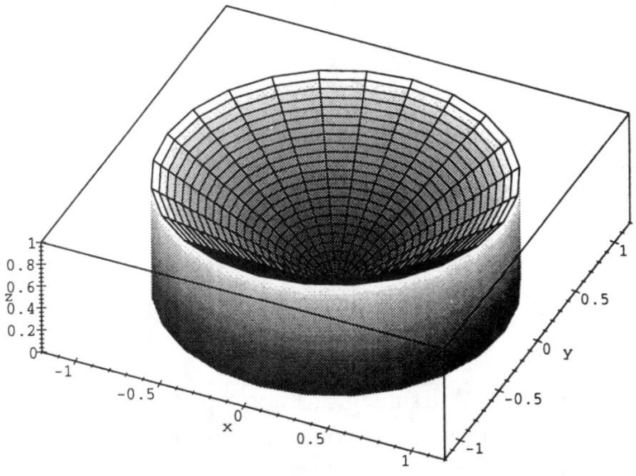

As we just calculated, this figure has volume $2\pi/3 \approx 2.094$ cubic units.➸

Does this seem reasonable from the picture, given the general size of units?

Polar integration—why it works

Why does the polar integration formula work? Where, especially, does the r in $dA = r\,dr\,d\theta$ come from?

All properties of integrals—whether in Cartesian, polar, or any other form—stem ultimately from properties of the approximating sums that are used to define integrals. For any function f defined on a region R, we have

$$\iint_R f\,dA = \lim_{m\to\infty} \sum_{i=1}^{m} f(P_i)\Delta A_i,$$

where ΔA_i is the area of the i-th subregion of R and P_i is a sampling point chosen inside this subregion.

If $R = [a,b] \times [c,d]$ is a Cartesian rectangle, then it's natural to subdivide R into smaller rectangles, each with sides Δx and Δy. Any such rectangle has area $\Delta A_i = \Delta x\,\Delta y$. In the limit that defines the integral, therefore, $dA = dx\,dy$.

If R is a polar rectangle, the picture is a little different. In this case, a "polar grid" is the natural way to subdivide R. Here's the picture; each bulleted point represents a "centered" sampling point in its respective subdivision.

A polar grid on a polar rectangle

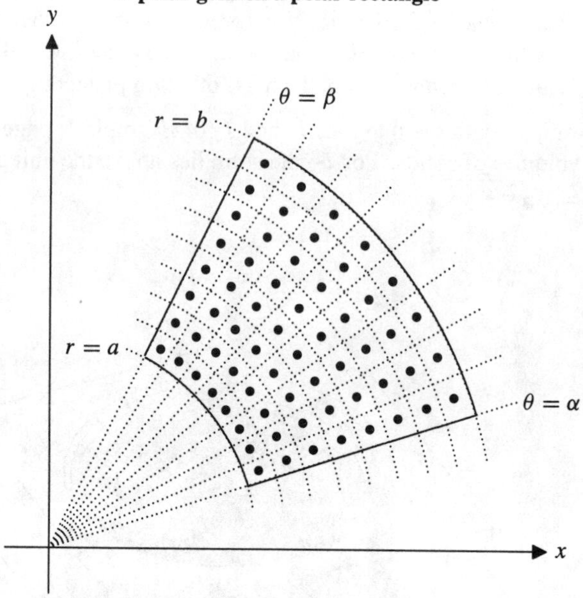

Notice:

Similar subregions. All subregions correspond to the same $\Delta\theta$ (the angle between any two adjacent dotted radial lines) and the same Δr (the radial distance from one arc to the next).

But not identical. Similar as they are, the subregions shown are not identical. Here's the key point:

> *In a polar grid, the subregions have different sizes. The area depends not only on $\Delta\theta$ and Δr, but also on r—larger values of r produce larger subregions.*

This fact explains the difference between polar and Cartesian integrals—and hints at why that extra r is needed.

The area of one subregion. Each subregion above is a small polar rectangle, with polar dimensions $\Delta\theta$ and Δr, inner radius r, and outer radius $r + \Delta r$. It's an important fact that for such a polar rectangle,

$$\text{area} = \Delta A_i = \frac{r + r + \Delta r}{2} \Delta r \, \Delta\theta.$$

(We leave verification of this straightforward fact to the exercises.) The first factor on the right is crucial—it represents the *average* radius of the given subregion, i.e., the r-coordinate of the i-th bulleted midpoint (r_i, θ_i) shown above. Therefore,

$$\Delta A_i = r_i \, \Delta r \, \Delta\theta.$$

This is just what we've been waiting for. It shows that for a polar rectangular region, a midpoint approximating sum has the form

$$\sum_{i=1}^{m} f(r_i, \theta_i) \, \Delta A_i = \sum_{i=1}^{m} f(r_i, \theta_i) \, r_i \, \Delta r \, \Delta\theta.$$

The integral itself, therefore, has the limiting form $\int_R f(r, \theta) \, r \, dr \, d\theta$.

Polar integrals over non-rectangular regions

Polar integrals, like Cartesian integrals, can be taken over non-rectangular regions. The method is similar, too.

> **Fact:** Let R be the region bounded by the radial lines $\theta = \alpha$ and $\theta = \beta$, by an "inner" curve $r = r_1(\theta)$, and by an "outer" curve $r = r_2(\theta)$. ("Inner" and "outer" are understood relative to the origin.) Let $g(r, \theta)$ be a function defined on R. Then
>
> $$\iint_R g\,dA = \int_{\theta=\alpha}^{\theta=\beta} \int_{r=r_1(\theta)}^{r=r_2(\theta)} g(r, \theta)\, r\, dr\, d\theta.$$

We illustrate with an example.

■ **Example 3.** Use a polar integral to find the area of the region R inside the cardioid $r = 1 + \cos\theta$. (See the picture in Appendix B, or draw one for yourself.)

Solution: We'll use the familiar principle➻ that

It applies equally well in polar coordinates!

$$\text{area of } R = \iint_R 1\,dA,$$

but we'll calculate the integral in polar form, as follows:

$$\begin{aligned}\iint_R 1\,dA &= \int_{\theta=0}^{\theta=2\pi} \int_{r=0}^{r=1+\cos\theta} 1r\,dr\,d\theta \\ &= \int_{\theta=0}^{\theta=2\pi} \left.\frac{r^2}{2}\right]_0^{1+\cos\theta} d\theta \\ &= \int_{\theta=0}^{\theta=2\pi} \frac{(1+\cos\theta)^2}{2}\,d\theta.\end{aligned}$$

The last integral takes a little doing by hand.➻ But *Maple* has no trouble:

But it's not really difficult.

```
> int( (1+cos(t))^2, t=0 .. 2*Pi );

                    3*Pi/2
```

□

Exercises

1. Let R be the polar rectangle defined by $a \le r \le b$ and $\alpha \le \theta \le \beta$.

   (a) Show that the area of R is $\dfrac{a+b}{2}(b-a)(\beta-\alpha)$. [HINT: Appendix B discusses the area of a circular sector, or "wedge."]

   (b) Use part (a) to show that a polar rectangle with dimensions Δr and $\Delta\theta$ and inner radius r has area $\dfrac{r+r+\Delta r}{2}\Delta r\,\Delta\theta$.

2. Let $f(x, y) = y$, let R be the upper half of the region inside the unit circle $x^2 + y^2 = 1$, and let $I = \iint_R f\,dA$.

(a) Calculate I as an iterated integral in rectangular coordinates, with the inner integral in y.

(b) Calculate I as an iterated integral in rectangular coordinates, with the inner integral in x.

(c) Calculate I as an iterated integral in polar coordinates.

3. In this section, we used polar coordinates to calculate that

$$I_2 = \iint_{R_2} \sqrt{x^2 + y^2}\, dA = 2\pi/3,$$

where R_2 is the region inside the unit circle $r = 1$. (See the picture on page 197.) Use formulas for the volumes of cones and cylinders to find the same answer by elementary means.

4. This exercise is about the integral I_1 of Example 1 (page 194).

(a) Draw (use technology if necessary, but try without it) the solid whose volume is given by I_1.

(b) Evaluate I_1 again but with the inner integral in x, not y.

5. Use a polar double integral in each of the following parts. (Draw each region first.)

(a) Find the area of the region inside the cardioid $r = 1 + \sin\theta$.

(b) Find the area of the region bounded by $y = x$, $y = 0$, and $x = 1$. Could you find the answer another way? [HINT: First write the boundary equations in polar form.]

(c) Find the area of the region bounded by the circle of radius 1/2, centered at (0, 1/2). Could you find the answer another way? [HINT: One approach is to first write a Cartesian equation for the circle, then change it to polar form.]

6. Use polar coordinates to calculate each of the following quantities.

(a) Find $\iint_R \frac{1}{\sqrt{x^2 + y^2}}\, dA$, where R is the region inside the cardioid $r = 1 + \sin\theta$ and above the x-axis.

(b) Find the volume of the solid under the surface $z = 1 - x^2 - y^2$ and above the xy-plane. [HINT: First decide where the surface hits the xy-plane.]

(c) Find the volume of the conical solid under the surface $z = 1 - \sqrt{x^2 + y^2}$ and above the xy-plane. [HINT: First decide where the surface hits the xy-plane.]

7. Use a double integral to find the area of the region inside the cardioid $r = 1 + \cos\theta$ and above the x-axis.

3.4 More triple integrals; cylindrical and spherical coordinates

The definition revisited. Triple integrals, as we saw briefly in Section 3.1, are defined in the same way as single and double integrals. For a function $f(x, y, z)$, defined on a solid region S in xyz-space, the integral is defined to be a limit of approximating sums:

$$\iiint_S f\,dV = \lim_{m\to\infty} \sum_{i=1}^{m} f(P_i)\Delta V_i.$$

The sum on the right is formed by subdividing S into m small subregions S_i, $i = 1, 2, \ldots, m$. The ith subregion has volume ΔV_i; within each subregion a sampling point P_i is chosen. In the approximating sum, therefore, the contribution of each value $f(P_i)$ is "weighted" by the volume of its respective subdivision.

Pros and cons of the definition. Approximating sums aren't hard to calculate (with help from computing)—if the number of subdivisions is modest. But that's a big "if" for functions of three variables. We saw in Section 3.1 that subdividing a three-dimensional cube rapidly generates thousands of subregions.➻ Approximating sums, moreover, do no more than their name implies: They approximate an integral rather than evaluate it exactly.

Dividing each edge into n pieces produces n^3 subdivisions.

Therefore it's useful, for both practical and theoretical reasons, to have exact, antiderivative-based methods. In this section we survey several useful techniques for evaluating triple integrals by antidifferentiation. We'll introduce two new coordinate systems for $\mathbb{R}^3$; an important theme will be choosing the "right" coordinate system to simplify a given integral.

Triple integrals, plain and fancy

Triple integrals are easiest to calculate if the region of integration S is a rectangular solid, or **parallelepiped**, of the form $S = [a, b] \times [c, d] \times [e, f]$.➻ ("Brick" is a less pretentious alternative.) In this case we can integrate f by iteration, one variable at a time—without any fuss over variable limits of integration. (See Example 3, page 176 for a straightforward example.)

Ordinary rectangles are nicest for double integrals.

Integrals get stickier if the domain of integration is not rectangular, as, indeed, it often is not.➻ Very irregular domains may be beyond our powers, but it turns out that many useful domains (in two and three dimensions) fit a standard pattern that we *can* readily handle. To save space we'll call such domains **basic**. (This is just a useful shorthand, not a formal definition.)

This is so for 2-d integrals, too.

Basic domains in two variables. The simplest regions in the plane are rectangles, in either the Cartesian sense or the polar sense.➻ The next simplest plane regions—the ones we'll call basic—are types like these:

See the previous section for more on polar rectangles.

- the region bounded *above* by a curve $y = f_2(x)$, bounded *below* by a curve $y = f_1(x)$, and on the left and right by $x = a$ and $x = b$;
- the region bounded on the right by a curve $x = f_2(y)$, on the left by a curve $x = f_1(y)$, and on the bottom and top by $y = c$ and $y = d$;
- the polar region bounded on the outside by a curve $r = f_2(\theta)$, on the inside by a curve $r = f_1(\theta)$, and by the lines $\theta = \alpha$ and $\theta = \beta$.

For any such domain R we can readily integrate a function $f(x, y)$ by iteration, with an integral of one of these types:

$$\int_a^b \int_{y=f_1(x)}^{y=f_2(x)} f(x, y)\, dy\, dx; \qquad \int_c^d \int_{x=f_1(y)}^{x=f_2(y)} f(x, y)\, dx\, dy;$$

$$\int_\alpha^\beta \int_{r=f_1(\theta)}^{x=f_2(\theta)} g(r, \theta)\, r\, dr\, d\theta.$$

(In the last case, $f(x, y)$ must first be changed to the form $g(r, \theta)$.) The inner integrals all have variable limits, but with due care taken, that's no problem. Technology, too, can often help with routine calculations.

Draw your own picture of R.

■ **Example 1.** Let R be the region in the first quadrant bounded on the outside by the cardioid $r = 1 + \cos\theta$ and on the inside by the circle $r = 1$. Let $f(x, y) = 1/\sqrt{x^2 + y^2}$. Find $\iint_R f(x, y)\, dA$.

Solution: The integral can be done by iteration. Note that

$$f(x, y) = 1/\sqrt{x^2 + y^2} = 1/r = g(r, \theta),$$

and that the region lies between $\theta = 0$ and $\theta = \pi/2$. Therefore

$$\iint_R f(x, y)\, dA = \int_0^{\pi/2} \int_{r=1}^{r=1+\cos\theta} dr\, d\theta = \int_0^{\pi/2} \cos\theta\, d\theta = 1. \qquad \square$$

Basic domains in three variables. Various sorts of domains in three variables could be considered "almost as nice" as ordinary rectangular solids. For simplicity, we'll concentrate on two main types:

- the solid S bounded above by a surface $z = f_2(x, y)$, below by a surface $z = f_1(x, y)$, and lying above a region R in the xy-plane;
- a "rectangular solid" in either cylindrical or spherical coordinates (we'll give more details later).

Notice that in the first case, R can be thought of as the "shadow" S casts in the xy-plane if a light shines straight downward, parallel to the z-axis. In the same way, the first of the three basic 2-dimensional regions mentioned earlier has $[a, b]$ as its "shadow" in the x-axis, and $[a, b]$ is the domain of the outer integral. A very similar fact is true—for the same reason—for triple integrals:

> **Fact:** Let S be a region of the first type above, and let $f(x, y, z)$ be defined on S. Then
>
> $$\iiint_S f(x, y, z)\, dV = \iint_R \int_{z=f_1(x,y)}^{z=f_2(x,y)} f(x, y, z)\, dz\, dA.$$

The Fact says that if we integrate first in z, we can trade the original triple integral (over S) for a double integral (over the "shadow" region R). That's often a good trade, because we have sharp tools for handling double integrals. Sometimes (in three dimensions as in two) the shadow region needs to be calculated from the given data.

■ **Example 2.** Let S be the region bounded below by the surface $z = x^2 + y^2$, and above by the surface $z = 4$; let $f(x, y, z) = 2z$. Calculate $\iiint_S f(x, y, z)\, dz$.

Solution: The domain of integration S is shown below; it's the part between the surfaces:

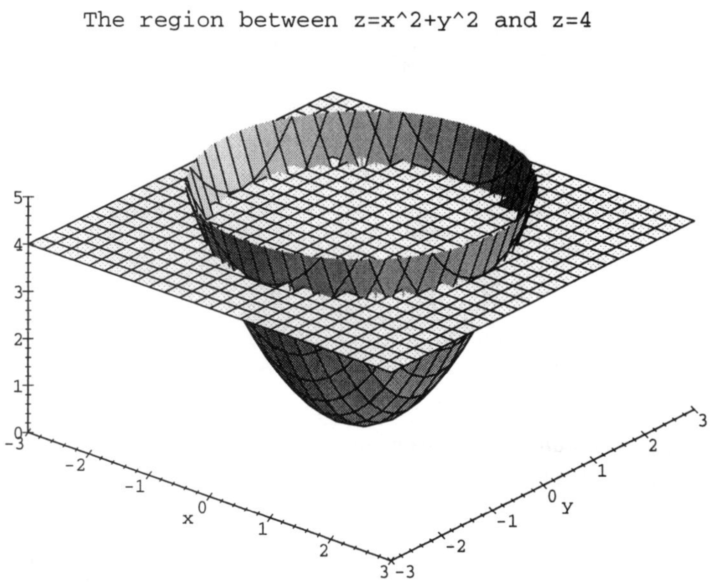

To find the shadow region R, we need to find where the surfaces intersect. We do so by setting the surface equations equal, i.e., $x^2 + y^2 = 4$. Thus the surfaces intersect above the circle $x^2 + y^2 = 4$ in the xy-plane, and the shadow region R lies inside this circle. Now we can use the Fact above:

Check the steps.

$$\iiint_S 2z\, dV = \iint_R \int_{z=x^2+y^2}^{z=4} 2z\, dz\, dA = \iint_R \left(16 - \left(x^2 + y^2\right)^2\right) dA.$$

The last integral naturally deserves polar treatment:

Check the result, perhaps with technology.

$$\int_0^{2\pi} \int_0^2 (16 - r^4) r\, dr\, d\theta = \frac{128\pi}{3} \approx 134. \qquad \square$$

Cylindrical coordinates

As we saw earlier, polar coordinates in the plane radically simplify the formulas used to describe certain regions and functions. **Cylindrical coordinates** represent a natural extension of the same idea to $\mathbb{R}^3$. The idea is simple: Use r, θ, and z (rather than x, y, and z) to describe points in three-dimensional space. Here's the picture:

Cylindrical coordinates

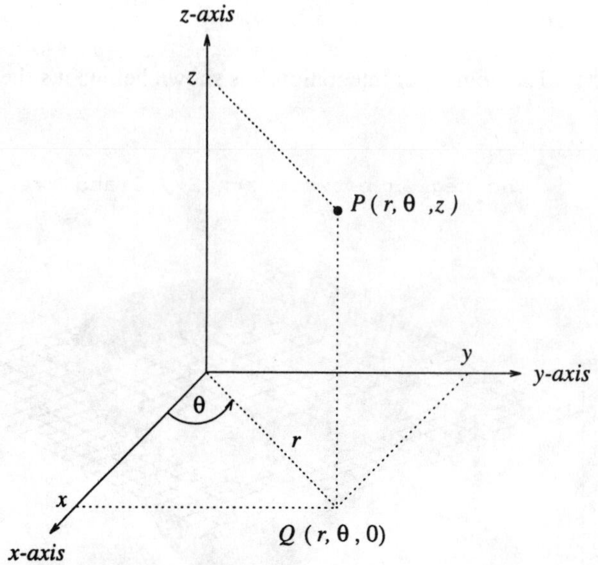

The various coordinates are related to each other just as for polar coordinates. Look for these relations in the picture:

$$x = r\cos\theta; \quad y = r\sin\theta; \quad \sqrt{x^2 + y^2} = r.$$

Why the name? Simple shapes. Cylindrical coordinates deserve the name because they're so well suited to describing circular cylinders and related shapes. Here are some samples:

Cylinders. In cylindrical coordinates the graph of the equation $r = a$, for any fixed $a > 0$, is an infinite cylinder of radius a, centered along the z-axis.

Horizontal planes. The graph of the equation $z = a$ is the horizontal plane at height a above (or below) the xy-plane.

Vertical planes. The graph of the equation $\theta = a$ is a vertical plane, perpendicular to the xy-axis, making angle a radians with the x-axis.

Cones. The graph of the equation $z = mr$ is a cone centered on the z-axis, with vertex at the origin; m measures the slope of the sides of the cone.

Integration in cylindrical coordinates. Recall the formula for double integrals in polar coordinates. For a polar rectangle R, we've seen that

$$\iint_R f(x, y)\, dA = \iint_R g(r, \theta)\, r\, dr\, d\theta.$$

For short, $dA = r\, dr\, d\theta$; dA is called the **area element in polar coordinates.**

The corresponding formula for cylindrical coordinates is similar. If S is a region in cylindrical form,

$$\iiint_S f(x, y, z)\, dV = \iiint_S g(r, \theta, z)\, r\, dr\, d\theta\, dz.$$

For short, $dV = r\,dr\,d\theta\,dz$ is the **volume element in cylindrical coordinates**.

The form of the volume element should not be too surprising, given what we've seen for polar coordinates. For cylindrical coordinates, the idea is to subdivide the region of integration into small "cylindrical rectangles," and use these to construct approximating sums. Here's a picture of one cylindrical subdivision:

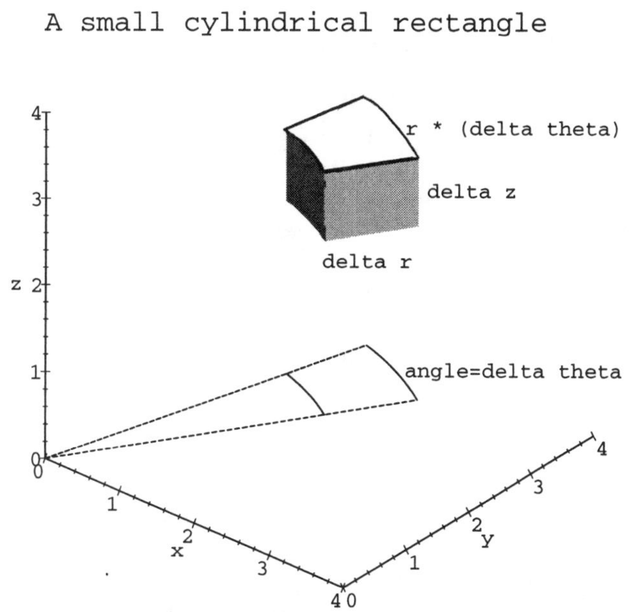

We saw in the previous section that one such rectangle has base area approximately $r\,\Delta r\,\Delta\theta$. Since the thickness is Δz, it follows that

$$\text{volume of one box} \approx r\,\Delta r\,\Delta\theta\,\Delta z.$$

Thus any approximating sum in cylindrical coordinates has the form

$$\sum_{i=1}^{m} g(r_i, \theta_i, z_i)\,\Delta V_i \approx \sum_{i=1}^{m} g(r_i, \theta_i, z_i) r_i\,\Delta r\,\Delta\theta\,\Delta z.$$

In the limit, therefore,

$$\iiint_S f(x, y, z)\,dV = \iiint_S g(r, \theta, z)\,r\,dr\,d\theta\,dz.$$

Cylindrical coordinates are the natural choice when the region of integration or the integrand (or both) have certain types of symmetry.

■ **Example 3.** Derive the classical formula for the volume of a cone with height h and radius a.

Solution: The desired cone C has cylindrical equation $z = rh/a$.➻ The part we want lies below $z = h$ and above $z = rh/a$. Let's calculate the volume in cylindrical coordinates:

Convince yourself of this; note that $z = h$ when $r = a$.

$$\text{volume} = \iiint_C 1\,dV = \int_0^{2\pi} \int_{r=0}^{r=a} \int_{z=rh/a}^{z=h} r\,dz\,dr\,d\theta.$$

It's a routine matter to calculate the iterated integral on the right; the answer is $\pi a^2 h/3$. □

Spherical coordinates

Spherical coordinates offer yet another way of describing three-dimensional space. Spherical coordinates are denoted ρ, θ, and ϕ. The first coordinate measures three-dimensional distance from the origin and the other two coordinates are angular (θ is the same angle as in the polar and cylindrical case). Here's the general picture:

Spherical coordinates

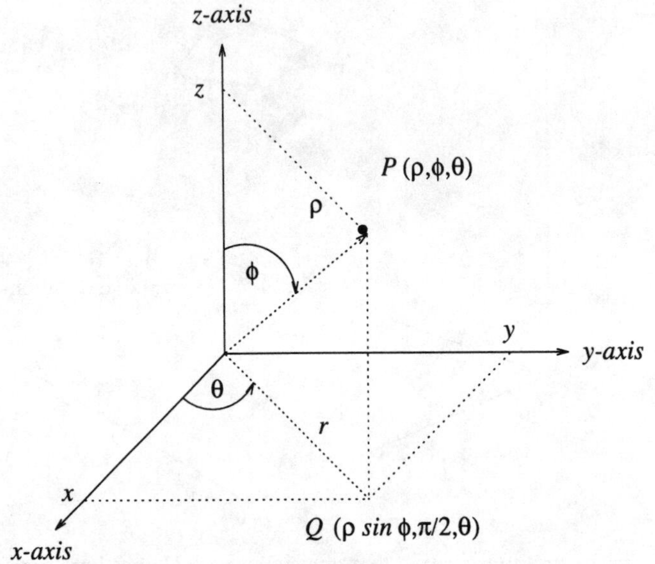

A close and careful look at the picture reveals various relations among spherical and other coordinates. These are especially important:

Do you see why they hold?

$$z = \rho\cos\phi; \quad r = \rho\sin\phi.$$

Combining these relations with those already known from polar coordinates gives

$$x = r\cos\theta = \rho\sin\phi\cos\theta; \quad y = r\sin\theta = \rho\sin\phi\sin\theta; \quad z = \rho\cos\phi.$$

These relations let us convert any function $f(x, y, z)$, in Cartesian coordinates, into a new function $g(\rho, \theta, \phi)$ in spherical coordinates.

Integration in spherical coordinates. Triple integrals in spherical coordinates have their own special form. If S is a spherical region and $f(x, y, z)$ is defined on S, then

$$\iiint_S f(x, y, z)\, dV = \iiint_S g(\rho, \theta, \phi)\, \rho^2 \sin\phi \, d\rho\, d\theta\, d\phi.$$

In short, $dV = \rho^2 \sin\phi \, d\rho\, d\theta\, d\phi$ is the **volume element in spherical coordinates.**

Explaining why the volume element has precisely this form is similar to the argument given above for cylindrical coordinates. We omit details, but the following figure contains all the necessary ingredients.

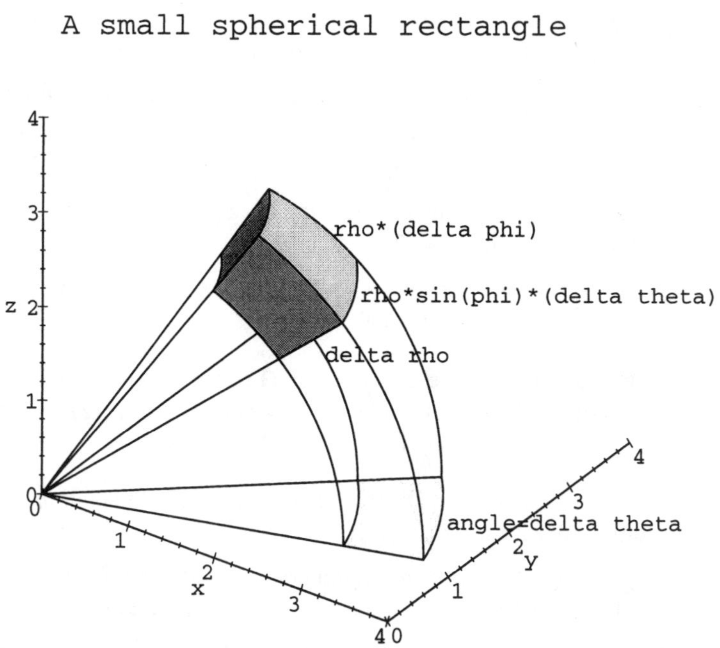

(The lengths of three various edges are marked; their product gives the volume element.) It gives (after some very careful examination) a good estimate to the volume of one small spherical rectangle:

$$\text{volume of one box} \approx \rho\,\Delta\phi\,\rho\,\sin\phi\,\Delta\theta\,\Delta\rho = \rho^2\sin\phi\,\Delta\rho\,\Delta\theta\,\Delta\phi.$$

The integral formula follows as a result.

■ **Example 4.** Use a spherical integral to verify the classical formula for the volume of a solid sphere of radius a.

Solution: Let S be the solid sphere; it's defined by the inequalities

$$0 \le \phi \le \pi; \quad 0 \le \theta \le 2\pi; \quad 0 \le \rho \le a.$$

Thus the sphere is, in effect, a "spherical rectangle," and the integral calculation is straightforward (though a little complicated):

$$\text{volume} = \iiint_S 1\,dV = \int_0^{\pi}\int_0^{2\pi}\int_0^{a} \rho^2\sin\phi\,d\rho\,d\theta\,d\phi.$$

The answer, after some scratchwork, is $4\pi a^3/3$. □

Exercises

1. Find the volume of each region below. (Note: Setting up the integrals is most important; it's OK to check calculations with technology.)

(a) The volume of a cylinder of radius a and height h. (Use cylindrical coordinates, of course.)

(b) The volume of the region bounded above by the plane $z = x + y$, below by the xy-plane, and on the sides by the cylinder $r = 1$.

(c) The volume of a sphere of radius a, using cylindrical coordinates.

(d) The volume of the region bounded below by the surface $z = x^2 + y^2$ and above by the plane $z = x + y$.

(e) The volume of the region under the cone $z = r$, above the xy-plane, and inside the cylinder $r = 2$.

(f) The region above the cone $z = \sqrt{x^2 + y^2}$ and below the sphere $x^2 + y^2 + z^2 = 1$. (Use spherical coordinates)

2. Calculate each integral below. (Notes: (1) Setting up the integrals is most important; it's OK to check calculations with technology. (2) The regions are the same as those in the previous exercise.)

   (a) $\iiint_C z\,dV$, where C is the cylinder of radius a and height h. (Use cylindrical coordinates, of course.)

   (b) $\iiint_S x\,dV$, where S is the region bounded above by the plane $z = x + y$, below by the xy-plane, and on the sides by the cylinder $r = 1$.

   (c) $\iiint_S x\,dV$, where S is the region under the cone $z = r$, above the xy-plane, and inside the cylinder $r = 2$. [HINT: Make a guess first, based on the symmetry of the region.]

   (d) $\iiint_S z\,dV$, where S is the region under the cone $z = r$, above the xy-plane, and inside the cylinder $r = 2$.

   (e) $\iiint_S x\,dV$, where S is a sphere of radius a. (Note: Make a guess first.)

3. By hand, verify the calculation in Example 3.

4. By hand, verify the calculation in Example 4.

5. Sketch each surface described below in cylindrical coordinates.

   (a) $z = -r$

   (b) $\theta = 1$

   (c) $r = \theta$, for $0 \le \theta < 2\pi$

   (d) $z = 1 - r$

   (e) $z = 1 - r^2$

6. Write a formula in cylindrical coordinates for each surface described below.

   (a) The surface with Cartesian equation $x = 1$.

   (b) A cylinder of radius 3, centered on the z-axis.

   (c) A sphere of radius 3, centered at the origin.

   (d) The surface with Cartesian equation $x^2 - x + y^2 - y = 0$. Describe the surface in words.

7. Describe in words the surface associated with each of the following equations in spherical coordinates. Then give a Cartesian equation for each surface.

(a) $\rho = a$, for any $a > 0$

(b) $\phi = \pi/4$

(c) $\theta = \pi/4$

8. Give a geometric description of the solid whose volume is given by the triple integral $\int_0^{2\pi}\int_1^2\int_0^5 r\,dz\,dr\,d\theta$.

9. Rewrite the triple integral $\int_0^{2\pi}\int_0^1\int_0^{\sqrt{4-r^2}} r^2\,dz\,dr\,d\theta$ in

   (a) rectangular coordinates;

   (b) spherical coordinates.

10. The volume of a solid is $\int_0^2\int_0^{\sqrt{2x-x^2}}\int_{-\sqrt{4-x^2-y^2}}^{\sqrt{4-x^2-y^2}} dz\,dy\,dx$.

    (a) Describe the solid by giving equations for the surfaces that form its boundary.

    (b) Rewrite the integral in cylindrical coordinates.

11. A hemispherical bowl of radius 5 cm is filled with water to within 3 cm of the top. Set up an integral for the volume of water in the bowl.

12. A hemisphere of radius six inches is truncated at the bottom to make a bowl of volume 99π cubic inches. Find the height of the bowl.

13. Give an equation in Cartesian coordinates for the surface described in spherical coordinates by the equation $\rho = 3$.

14. (a) Express the point $(x, y, z) = (2, 3, 1)$ in cylindrical coordinates.

    (b) Express the point $(x, y, z) = (2, 3, 1)$ in spherical coordinates.

    (c) Express the point $(r, \theta, z) = (2, 3, 1)$ in Cartesian coordinates.

    (d) Express the point $(\rho, \theta, \phi) = (2, 3, 1)$ in Cartesian coordinates.

15. Let R be the "ice cream cone" region enclosed by a sphere of radius 2 centered at the origin and the cone $z = \sqrt{x^2/3 + y^2/3}$. Write the triple integral $\int_R y\,dV$ as an iterated integral in

    (a) rectangular coordinates

    (b) cylindrical coordinates

    (c) spherical coordinates

3.5 Multiple integrals overviewed; change of variables

In this chapter so far we have encountered several approaches to multiple integrals. In this final section, we first compare these ideas of the integral, by reviewing some of the approaches and methods we've seen. Then we link some apparently different approaches to the integral by presenting a general change-of-variables formula that permits us to transform one integral to another.

Integrals reviewed

Different types of integrals. In elementary calculus there is only one main type of integral. For a function $f : \mathbb{R} \to \mathbb{R}$ and an interval $[a, b]$, the familiar integral $I = \int_a^b f(x)\,dx$ is a certain real number, defined as a limit of approximating sums. Although I has various possible geometric, physical, and numerical interpretations (as area under the f-graph, as displacement if f represents velocity, as a weighted average of f-values, and so on), they are nothing more than different ways of looking at the same mathematical object.

In multivariable calculus the possibilities for integrals are much broader. Integrands may be functions of one or more variables, and they may be either scalar- or vector-valued functions. The sets over which functions are integrated may also be of various types, including ordinary intervals, 1-dimensional curves in $\mathbb{R}^2$ or $\mathbb{R}^3$, planar regions in $\mathbb{R}^2$, solid regions in $\mathbb{R}^3$, and 2-dimensional surfaces in $\mathbb{R}^3$. We've seen several such possibilities, with symbolic forms such as these:

$$\int_a^b \boldsymbol{v}(t)\,dt; \quad \iint_R f(x, y)\,dA; \quad \iint_R g(r, \theta)\,r\,dr\,d\theta; \quad \iiint_S h(r, \theta, z)\,r\,dr\,d\theta\,dz.$$

In Chapter 5 we'll see still more integrals, including **line integrals** and **surface integrals**, in which the domains of integration are, respectively, either curves or surfaces in $\mathbb{R}^2$ or $\mathbb{R}^3$.

Derivatives, integrals, and fundamental theorems. A first course in calculus usually culminates with the **fundamental theorem of calculus**, which relates derivatives and integrals. In its simplest◂ form, the fundamental theorem says that under appropriate conditions,

But most important.

$$\int_a^b f'(x)\,dx = f(b) - f(a).$$

The theorem says that, in a certain sense, single-variable integration "undoes" single-variable differentiation, and vice versa.

As one might guess, the ideas behind the fundamental theorem can be extended and re-interpreted to fit the multivariable setting. For instance, a version of the **fundamental theorem for line integrals** says that

$$\int_\gamma \nabla f \cdot d\boldsymbol{X} = f(\boldsymbol{b}) - f(\boldsymbol{a}), \tag{3.5.1}$$

where $f(x, y)$ is a function of two variables, and γ is a curve joining $\boldsymbol{a}$ to $\boldsymbol{b}$ in $\mathbb{R}^2$. Much remains to be explained, of course, including the precise meaning of the line integral on the left. But the general similarity to the ordinary fundamental theorem should be apparent. In each case, some sort of integral is applied to some sort of derivative of f,◂ and the answer involves f itself.

The gradient ∇f is a type of derivative of f.

Integrating functions of one variable Our first goal is to review the various types of integrals encountered so far. The simplest involve only one input variable. Single-variable integrals like this one—

$$\int_0^1 x^2 dx = \left.\frac{x^3}{3}\right]_0^1 = \frac{1}{3},$$

are the stock in trade of elementary calculus; the integrand is a scalar-valued function of one variable and the domain of integration is an interval. Integrals of this basic type do arise in multivariable calculus, but often as "byproducts" to some higher-dimensional problem. Arclength calculations are one good example.➽

To review arclength, see Section 1.5.

■ **Example 1.** The position function

$$\boldsymbol{r}(t) = (\cos t,\ \sin t,\ t)$$

describes a helix (spiral) in $\mathbb{R}^3$. Find the arclength of one "turn"; think of t as time.

Solution: Position is a *vector*-valued function of time t, but the speed at time t, given by

$$|\boldsymbol{r}'(t)| = |(-\sin t, \cos t, 1)| = \sqrt{\sin^2 t + \cos^2 t + 1} = \sqrt{2}$$

is *scalar-valued.* (In this case, the speed function happens to be constant.) One turn of the helix takes 2π units of time, so the arclength is given by the ordinary integral

$$\int_0^{2\pi} |\boldsymbol{r}'(t)|\, dt = \int_0^{2\pi} \sqrt{2}\, dt = 2\sqrt{2}\pi \approx 8.89$$

units of distance. □

We have also seen how to integrate *vector-valued* functions of one variable over an interval in the domain. An important use of such integrals is in modeling phenomena of motion.

■ **Example 2.** A particle moves in the xy-plane. At time t, its velocity vector is

$$\boldsymbol{v}(t) = (5, -32t).$$

Find $\int_0^6 \boldsymbol{v}(t)\, dt$. What does the answer mean about the particle's motion?

Solution: The calculation is routine;➽ we integrate in each coordinate separately:

But check our work.

$$\int_0^6 \boldsymbol{v}(t)\, dt = \left(\int_0^6 5\, dt, \int_0^6 -32t\, dt \right) = (30, -576).$$

More interesting than the numerical answer is what it tells: the particle's *displacement* over the time interval $0 \le t \le 6$. If $\boldsymbol{p}(t)$ is the particle's position at time t, then $\boldsymbol{p}'(t) = \boldsymbol{v}(t)$, and

$$\int_0^6 \boldsymbol{v}(t)\, dt = \boldsymbol{p}(6) - \boldsymbol{p}(0).$$

Thus if $\boldsymbol{v}(t)$ represents velocity in units of feet per second, the calculation means that over the 6-second interval, the object moves 30 feet to the right and 576 feet downward. □

Multiple integrals. We have also studied **multiple integrals**—in which the integrand is a scalar-valued function of two or three variables, and the domain of integration is a subset of $\mathbb{R}^2$ or $\mathbb{R}^3$. In the simplest case, the integrand is a function of two variables and the domain of integration is a rectangle. In this happy event, the integral can be calculated by "iteration," i.e., by integrating in one variable and then the other.

The name "multiple integral" is standard, but a little vague; it doesn't say whether inputs, outputs, or both are "multiple."

■ **Example 3.** Let $f(x, y) = 2 - x^2 y$, $R = [0, 1] \times [0, 1]$, and $I = \iint_R f(x, y)\, dA$. Calculate I as an iterated integral. What does the answer mean?

Solution: We'll integrate first in x and then in y:

$$\begin{aligned} I &= \int_0^1 \left(\int_0^1 (2 - x^2 y)\, dx \right) dy = \int_0^1 \left(2x - \frac{x^3 y}{3} \Big]_0^1 \right) dy \\ &= \int_0^1 \left(2 - \frac{y}{3} \right) dy = \frac{11}{6}. \end{aligned}$$

The other order of integration is possible, too; that calculation appears, along with a useful picture, in Example 2, Section 3.2.

And gives the same answer!

Geometrically, the integral gives the volume of the solid bounded below by the domain of integration, R, and above by the surface $z = f(x, y)$. Alternatively, the integral can be thought of as the *average value* of f over the rectangle $[0, 1] \times [0, 1]$, which has area 1. (The answer 11/6, a little less than two, accords well with the picture and with the table of representative values of f.) □

Techniques like those just shown can be used to integrate a function $f(x, y, z)$ over a rectangular solid region in $\mathbb{R}^3$.

Integrals over non-rectangles; polar coordinates. If a domain of integration is not rectangular, then various strategies are used, depending on the situation. One possibility—using variable limits of integration on the inner integral—is illustrated in Examples 4 and 5, Section 3.2.

Another possibility arises when the domain of integration is more conveniently represented in polar coordinates r and θ. Things work best when the integrand also lends itself to polar form.

■ **Example 4.** Let R be the region inside the unit circle, and let $f(x, y) = x^2 + y^2$. Use polar coordinates to calculate $I = \iint_R f(x, y)\, dA$. What does the answer mean geometrically?

Solution: First we write everything in polar form, using the following change-of-variable equations:

$$x = r\cos\theta; \quad y = r\sin\theta; \quad x^2 + y^2 = r^2; \quad dA = r\, dr\, d\theta.$$

(The last identity is called the **area element** in polar coordinates.) For the integrand, we have $f(x, y) = x^2 + y^2 = r^2$. The domain of integration is a "polar rectangle," described by the inequalities

$$0 \le r \le 1; \quad 0 \le \theta \le 2\pi.$$

Thus, rewritten in polar coordinates, the integral I becomes

$$I = \int_0^{2\pi} \int_0^1 r^2\, r\, dr\, d\theta = \int_0^{2\pi} \int_0^1 r^3 dr\, d\theta.$$

Integrating in r and then θ gives $I = \pi/2 \approx 1.57$. Geometrically, the integral measures the volume of the region under the paraboloid $z = x^2 + y^2$ and above the unit disk $x^2 + y^2 \leq 1$. □

Cylindrical and spherical coordinates. Similar change-of-variable strategies work for triple integrals, when the domain of integration, the integrand, or both are best described in cylindrical coordinates (r, θ, z) or spherical coordinates (ρ, θ, ϕ). In each case, a **volume element** formula applies. The volume elements are, respectively,

$$dV = r\, dr\, d\theta\, dz \quad \text{for cylindrical coordinates,}$$

and

$$dV = \rho^2 \sin\phi\, d\rho\, d\theta\, d\phi \quad \text{for spherical coordinates.}$$

Here's a typical calculation in spherical coordinates.

■ **Example 5.** Calculate $I = \iiint_P \frac{1}{x^2 + y^2 + z^2}\, dV$, where P is the part of the unit ball $x^2 + y^2 + z^2$ that lies in the first octant.

Solution: The first octant in $\mathbb{R}^3$ is the set of points whose spherical coordinates satisfy $0 \leq \theta \leq \pi/2$ (this assures that $x \geq 0$ and $y \geq 0$) and $0 \leq \phi \leq \pi/2$ (this assures that $z \geq 0$). Points that are also inside the unit ball satisfy $0 \leq \rho \leq 1$. Thus P is—from the spherical coordinate point of view—a rectangular solid. The integrand is also best converted to spherical coordinates:

$$\frac{1}{x^2 + y^2 + z^2} = \frac{1}{\rho^2}.$$

Now everything is ready:

$$\begin{aligned} I &= \iiint_P \frac{1}{x^2 + y^2 + z^2}\, dV \\ &= \int_0^{\pi/2} \int_0^{\pi/2} \int_0^1 \frac{1}{\rho^2} \rho^2 \sin\phi\, d\rho\, d\theta\, d\phi \\ &= \frac{\pi}{2} \int_0^{\pi/2} \sin\phi\, d\phi = \frac{\pi}{2}. \end{aligned}$$ □

The general change-of-variables formula

The most mysterious ingredients in the foregoing examples, in which we changed from Cartesian to other coordinates, are the area and volume elements in polar, cylindrical, and spherical coordinates:

$$dA = r\, dr\, d\theta, \quad dV = r\, dr\, d\theta\, dz, \quad \text{and} \quad dV = \rho^2 \sin\phi\, d\rho\, d\theta\, d\phi.$$

In fact, all three of these formulas are instances of a more general principle, known as the change-of-variables formula. It describes how integrals change when one set of coordinate variables (such as x and y) are exchanged for another (such as r and θ).

Changing one variable. The change-of-variables formula is best understood as the multivariable version of integration by substitution in elementary calculus. An example will help fix the ideas and notation. Notice that although our use of the symbols u and x may seem "reversed" compared to what's usual in elementary calculus, the reversal is *only* in nomenclature—the ideas below are exactly the ones familiar from elementary calculus.

■ **Example 6.** Use substitution to calculate $\int_0^{\pi/2} 3\sin^2 u \cos u \, du$.

Solution: The substitutions $x = \sin u$ and $dx = \cos u \, du$ seem natural. Indeed, they work just fine:

$$\int_{u=0}^{u=\pi/2} 3\sin^2 u \cos u \, du = \int_{x=0}^{x=1} 3x^2 \, dx = 1.$$

Notice the change in limits of integration: Since $x = \sin u$, the u-interval $I_u = [0, 2\pi]$ corresponds to the x-interval $I_x = [0, 1]$.

The first equality is the main point. If we write $f(x) = 3x^2$ and $x = x(u) = \sin u$, then the first equality has the form

$$\int_{I_u} f(x(u)) \frac{dx}{du} \, du = \int_{I_x} f(x) \, dx. \qquad \square$$

A magnification factor. Notice especially the "extra" factor dx/du on the left above. In fact, the whole substitution method can be summarized succinctly by the formal equation

Perhaps even too succinctly—it packs a lot into a small space.

$$\frac{dx}{du} \, du = dx.$$

The main point is that, because of the relationship between u and x, a tiny change du in u produces a corresponding change dx in x, and that du and dx are related as the equation indicates. In effect, dx/du plays the role of a "magnification factor," either enlarging or diminishing the effect of small changes in u.

Changing several variables. The change-of-variables recipe for multiple integrals uses similar ingredients. For double integrals, the usual setup is to start with an integral $\iint_{R_{xy}} f(x, y) \, dA$, where f is a function and R_{xy} is a region in the xy-plane. In addition, we're given (or somehow find) two functions relating x and y (the "old" coordinates) to u and v (the "new" coordinates):

$$x = x(u, v) \quad \text{and} \quad y = y(u, v).$$

We assume that these functions "map" the uv-plane (or part of it) into the xy-plane in such a way that some region R_{uv} in the uv-plane corresponds to R_{xy}, the original domain of integration in the xy-plane. (Ideally, the uv-region R_{uv} is simpler or more familiar than the xy-region R_{xy}.)

So far, the one- and two-variable change-of-variable setups are similar. An important difference, however, is that instead of the ordinary *derivative* dx/du, we now use the **Jacobian determinant**, i.e., the determinant of the derivative matrix of the mapping $(u, v) \to (x, y)$:

$$\frac{\partial(x, y)}{\partial(u, v)} = \det \begin{pmatrix} x_u & x_v \\ y_u & y_v \end{pmatrix}.$$

(The peculiar-looking "fraction" on the left is just a mnemonic shorthand.) It turns out that the determinant above plays the same role for the mapping $(u, v) \to (x, y)$ as does the derivative dx/du in the one-variable setting: For any input (u, v), the absolute value $|\partial(x, y)/\partial(u, v)|$ is the factor by which the mapping *magnifies areas* near (u, v). In particular, the mapping transforms a tiny rectangle at (u, v), with area $\Delta u \, \Delta v$, to another tiny near-rectangle at (x, y), with area approximately

$$\Delta x \, \Delta y \approx \left| \frac{\partial(x, y)}{\partial(u, v)} \right| \Delta u \, \Delta v.$$

We can now state the general theorem. Some technical hypotheses are added, but the final equation is the main point:

> **Theorem 2 (Change of variables in double integrals).** Let coordinates (x, y) and (u, v) be related as above. We assume that all derivatives in question exist and are continuous, that the mapping $(u, v) \to (x, y)$ is one-to-one, and that the regions R_{uv} and R_{xy} correspond to each other under the mapping. Then
>
> $$\begin{aligned} \iint_{R_{xy}} f(x, y) \, dA_{xy} &= \iint_{R_{uv}} f(x(u, v), y(u, v)) \left| \frac{\partial(x, y)}{\partial(u, v)} \right| dA_{uv} \\ &= \iint_{R_{uv}} f(x(u, v), y(u, v)) \left| \frac{\partial(x, y)}{\partial(u, v)} \right| du \, dv. \end{aligned}$$

■ **Example 7.** What does the theorem say about polar coordinates in double integrals?

Solution: Using r and θ in place of u and v gives $x = r \cos\theta$ and $y = r \sin\theta$. Thus the Jacobian matrix is

The names aren't sacred.

$$\begin{pmatrix} x_r & x_\theta \\ y_r & y_\theta \end{pmatrix} = \begin{pmatrix} \cos\theta & -r\sin\theta \\ \sin\theta & r\cos\theta \end{pmatrix};$$

it has determinant is r. Thus the theorem says that

$$\iint_{R_{xy}} f(x, y) \, dA = \iint_{R_{r\theta}} f(x(r, \theta), y(r, \theta)) \, r \, dr \, d\theta;$$

this is the familiar area element for polar coordinates. □

In three variables. The same formula holds for triple integrals in three variables, except that the Jacobian determinant has the form $\partial(x, y, z)/\partial(u, v, w)$. The work is potentially quite messy, but in many cases of interest the change of coordinates is simple.

The Jacobian matrix is 3×3.

■ **Example 8.** Let a, b, and c be positive numbers. Find the volume of the ellipsoid $\dfrac{x^2}{a^2} + \dfrac{y^2}{b^2} + \dfrac{z^2}{c^2} \le 1$ in $\mathbb{R}^3$.

Solution: If we use the "new" coordinates

$$u = \frac{x}{a}, \quad v = \frac{y}{b}, \quad \text{and} \quad w = \frac{z}{c},$$

then the ellipsoid E in xyz-space corresponds to the *unit ball* B in uvw-space; it has formula $u^2 + v^2 + w^2 \le 1$ and (known) volume $4\pi/3$. To use the change-of-variable formula, we need the Jacobian determinant of the mapping $(u, v, w) \to (x, y, z)$. In this case it's easy to find. The formulas above give

$$x(u, v, w) = au; \quad y(u, v, w) = bv; \quad z(u, v, w) = cz.$$

Check the easy calculation.

Thus the Jacobian matrix is diagonal, with constant determinant

$$\frac{\partial(x, y, z)}{\partial(u, v, w)} = abc.$$

The change-of-variable formula now finishes our problem:

$$\begin{aligned} \text{volume of E} &= \iiint_E 1\, dV_{xyz} = \iiint_B 1 \left| \frac{\partial(x, y, z)}{\partial(u, v, w)} \right| dV_{uvw} \\ &= abc \iiint_B 1\, dV_{uvw} = \frac{4\pi}{3} abc. \end{aligned}$$

Notice, finally, that the result has a very pleasing geometric meaning. At every point of uvw-space, the mapping $(u, v, w) \to (x, y, z) = (au, bv, cw)$ "magnifies" distances by factors of a, b, and c respectively in the three coordinate directions; the corresponding effect on volumes is to magnify by the factor abc. □

■ **Example 9.** This example problem explores the fact that the Jacobian determinant of a mapping $(u, v) \to (x, y)$ describes an area magnification factor. In each case below, let R_{uv} be the unit square $0 \le u \le 1$; $0 \le v \le 1$; it has area 1. Let R_{xy} be the image of R_{uv} in the xy-plane, under the mapping $(u, v) \to (x, y)$. In each case following: (i) describe R_{xy}; (ii) find the area of R_{xy}; (iii) find the Jacobian determinant of the given mapping. (In each case, R_{xy} is a parallelogram, so it's enough to find its corners.)

$$\text{(a)} \quad x = 2u,\ y = v; \qquad \text{(b)} \quad x = 1 + 2u,\ y = 3 + 4v.$$

Solution: We answer in words—but recommend strongly that readers draw their own pictures.

In case (a), the image region R_{xy} is the rectangle $[0, 2] \times [0, 1]$, so the magnification factor is 2. The derivative matrix is $\left[\begin{smallmatrix} 2 & 0 \\ 0 & 1 \end{smallmatrix}\right]$; its Jacobian determinant is 2.

In case (b), the image region R_{xy} is the rectangle $[1, 3] \times [3, 7]$, with area 8—this is also the magnification factor. The derivative matrix is $\left[\begin{smallmatrix} 2 & 0 \\ 0 & 4 \end{smallmatrix}\right]$; its Jacobian determinant is 8. □

The idea of the proof. Approximating sums are the idea behind the proof of the change-of-variables theorem. Recall that any double integral $\iint_{R_{uv}} g(u, v)\, dA$ is defined as a limit of approximating sums of the form

$$\lim_{n\to\infty} \sum_{i=1}^{n} g(u_i, v_i)\, A(R_i),$$

where $A(R_i)$ is the area of a small uv-rectangle containing (u_i, v_i).

In the case at hand, suppose that

$$g(u, v) = f(x(u, v), y(u, v)) \left| \frac{\partial(x, y)}{\partial(u, v)} \right| .$$

Then the approximating sum above becomes

$$\lim_{n\to\infty} \sum_{i=1}^{n} f(x(u_i, v_i), y(u_i, v_i)) \left| \frac{\partial(x, y)}{\partial(u, v)}(u_i, v_i) \right| A(R_i)$$

By the earlier remark about area magnification, this sum is approximately equal to

$$\lim_{n\to\infty} \sum_{i=1}^{n} f(x_i, y_i)\, A(R'_i)$$

where $A(R'_i)$ is the area of a small xy-rectangle R'_i, containing (x_i, y_i). But this last sum approximates the integral $\iint_{R_{xy}} f(x, y)\, dA$. We've arrived at last.

Exercises

1. In Example 1 (page 211) we calculated the length of one turn of a certain spiral in $\mathbb{R}^3$. The same curve can be parametrized by

$$\boldsymbol{r}(t) = (\cos(t^2), \sin(t^2), t^2); \quad 0 \le t \le \sqrt{2\pi}.$$

Use this parametrization to recalculate the arclength.

2. This exercise is about the curve with position vector $\boldsymbol{r}(t) = (t + \sin t, 1 + \cos t)$, a cycloid.

   (a) Plot the curve either by hand or using technology.

   (b) Find (exactly, by antidifferentiation) the length of one arch of the curve. (Hint: $2 + 2\cos t = 4\cos^2(t/2)$. To simplify the computation, find the length from $t = 0$ to $t = \pi$ and double the value.)

3. Redo the previous exercise, but use the curve $\boldsymbol{r}(t) = (t/2 + \sin t, 1 + \cos t)$. (This curve represents the path traced by a point on a wheel that slips as it rolls.) In this case you'll need to estimate the final integral; use the midpoint rule with 10 subdivisions.

4. Consider the situation described in Example 2 (page 211). Assume that the units of time and distance are seconds and feet, respectively.

   (a) Write an integral for the distance traveled by the particle over the 6-second interval. Either calculate the integral exactly (use technology or a table of integrals) or approximate its value using the midpoint rule.

   (b) Find the vector-valued acceleration function $\boldsymbol{a}(t)$. What is the appropriate unit of measurement?

   (c) Calculate $\int_0^6 \boldsymbol{a}(t)\, dt$. What does the answer mean about the particle's motion?

5. Consider the integral $I = \iint_R x^2 y \, dA$, where R is the region bounded by the curves $y = x^2$ and $y = 1$.

   (a) Calculate I by integrating first in y, then in x.

   (b) Calculate I by integrating first in x, then in y.

6. Let $f(x, y) = y$, and let R be the plane region bounded by the curves $y = \ln x$, $y = 0$, and $x = e$.

   (a) Calculate $I = \iint_R f(x, y) \, dA$ by integrating first in y, then in x.

   (b) Calculate $I = \iint_R f(x, y) \, dA$ by integrating first in x, then in y.

7. Consider the integral I of Example 4 (page 212). We said there that the integral gives the volume of the solid region (call is S) under the paraboloid $z = x^2+y^2$ and above the unit disk $x^2 + y^2 \leq 1$.

   (a) The solid region S described above is a solid of revolution in the sense of Calculus II. Explain why—i.e., tell which plane region must be revolved around which axis to produce S.

   (b) Use the disk method of Calculus II to find the volume of S as an integral in one variable.

8. Consider the ellipse E in the xy-plane defined by $\dfrac{x^2}{a^2} + \dfrac{y^2}{b^2} \leq 1$, where a and b are positive numbers.

   (a) Set up an elementary calculus-style integral for the area of E. (It's laborious to solve by hand; try technology, if available.)

   (b) Use the idea of Example 8 (page 215) to find the area of E using an appropriate change of variable.

   (c) How does the area calculated in the previous part behave as $a \to b$?

9. (For students with determinant experience.) As in Example 7, find an appropriate determinant to explain the volume element formula $dA = r \, dr \, d\theta \, dz$ for cylindrical coordinates.

10. (For students with determinant experience.) As in Example 7, find an appropriate determinant to explain the volume element formula $dA = \rho^2 \sin\phi \, d\rho \, d\theta \, d\phi$ for spherical coordinates.

11. This exercise resembles Example 9, page 216. It explores the fact that the Jacobian determinant of a mapping $(u, v) \to (x, y)$ describes an area magnification factor. In each case below, let R_{uv} be the unit square $0 \leq u \leq 1$; $0 \leq v \leq 1$; it has area 1. Let R_{xy} be the image of R_{uv} in the xy-plane, under the mapping $(u, v) \to (x, y)$. In each part following: (i) sketch R_{xy}; (ii) find the area of R_{xy}; (iii) find the Jacobian determinant of the given mapping. (Note: In each case, R_{xy} is a parallelogram, so it's enough to find its corners.)

    (a) $x = u$, $y = 2v$

    (b) $x = 2 + 3u$, $y = 4 + 5v$

    (c) $x = 1 + 2u + 3v$, $y = 4 + 5u + 6v$ (Hint: Use the cross product to find the area of the parallelogram.)

12. Let D be the region in xyz-space defined by the inequalities $1 \le x \le 2$, $0 \le xy \le 2$, and $0 \le z \le 1$. Write the triple integral $\iiint_D (x^2y + 3xyz)\, dx\, dy\, dz$ as an iterated integral in the coordinates $u = x$, $v = xy$, and $w = 3z$.

13. Let $g(x, y) = (u, v) = (-x + y, x + 2y)$, let R be a region in the xy-plane, and let T be the image of R under the mapping g. Furthermore, suppose that $f(x, y) > 0$ for all $(x, y) \in \mathbb{R}^2$. Is the statement

$$\iint_R f(x, y)\, dA < \iint_T f\big(x(u, v), y(u, v)\big)\, dA$$

true? Justify your answer.

Chapter 4

Other Topics

4.1 Linear, circular, and combined motion

We saw in the preceding section how the calculus of vectors and vector-valued functions can be used to model phenomena of motion.

Motions that arise in practical applications can often be thought of as combinations of simpler, more basic motions. A point on a moving bicycle wheel, for instance, follows a path that's determined by two simpler types of motion: the rotation of the wheel and the linear motion of the bicycle itself.

We'll model this situation in detail later in this section.

To model such motions, therefore, it makes sense first to study some simpler, elementary motions, and then see how to combine them—using vector algebra—to form more complex, "compound" motions.

Modeling compound motions

A two-part industrial robot arm might appear as below:

A robot arm

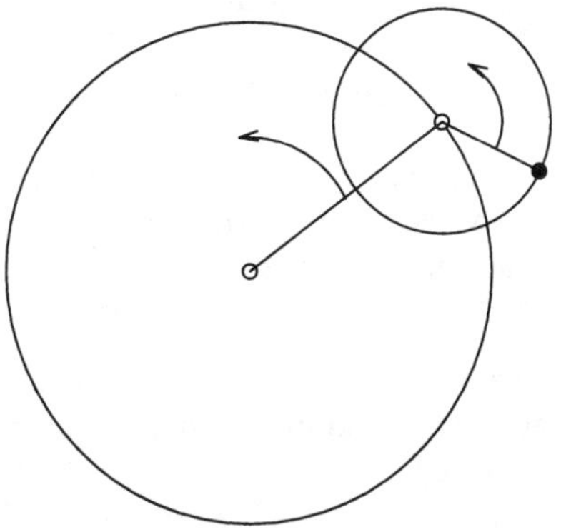

The long arm is pinned at the center of the large circle, the short section is pinned to the end of the long arm, and each arm rotates separately (driven by a motor, perhaps) about its fixed end. What path does the black dot (which might represent a tool at the end of the compound arm) follow as the two arms rotate? What are the velocity and acceleration of the black dot at any given time? (Controlling acceleration is practically important, because acceleration is proportional to force; excessive acceleration could damage the machinery.)

At first blush these questions might seem very difficult. The answers depend, after all, on several variables: the lengths of the two arms and the speed and direction of each rotation. As we'll see, however, the calculus of vector-valued functions makes modeling such motions quite straightforward. Our strategy will be to break the compound arm's motion into two simpler motions, analyze each separately, and then, quite literally, reassemble the parts.

As a visual teaser, here are two paths the black dot might follow, depending on the data:

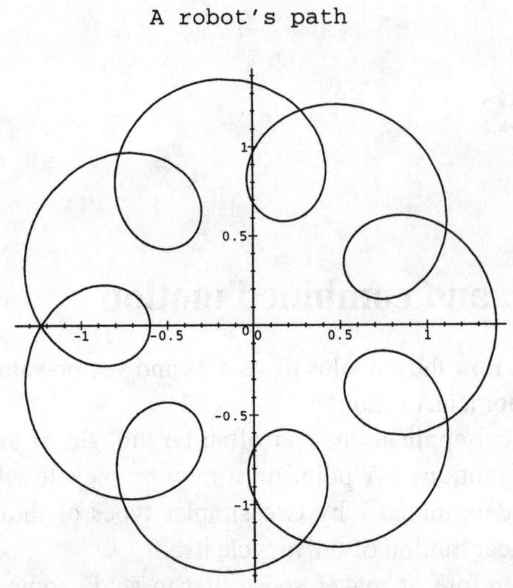

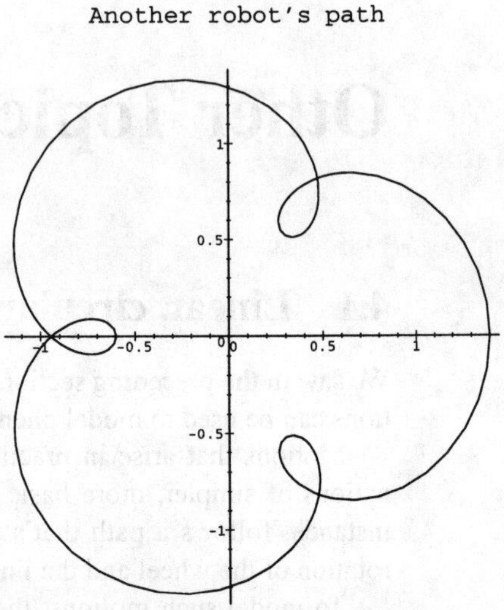

We'll explain soon how the paths are found.

Modeling linear motion

Motion in a straight line, with constant speed, is called **uniform linear motion**. This type of motion is easily modeled, using linear position functions. As we saw in the preceding section, the linear position function

$$p(t) = (x_0, y_0) + t(a, b)$$

describes a particle that (i) starts at position (x_0, y_0) at time $t = 0$; (ii) moves in the direction of the velocity vector (a, b); (iii) has constant speed $\sqrt{a^2 + b^2}$. Choosing the constants judiciously lets us model any instance of uniform linear motion.

■ **Example 1.** Suppose that a particle moves linearly, at constant speed, from $(1, 2)$ at $t = 0$ to $(5, 6)$ at $t = 3$. Find the position, velocity, and acceleration functions. What is the particle's speed?

Solution: The vector $v = (4, 4) = (5, 6) - (1, 2)$ points in the right direction; it joins the point $(1, 2)$ to the point $(5, 6)$. To arrange that the trip takes 3 seconds, we use velocity vector $(4, 4)/3$. The desired position function, therefore, is

$$p(t) = (1, 2) + t\left(\frac{4}{3}, \frac{4}{3}\right);$$

the velocity and acceleration are

$$\boldsymbol{v}(t) = \left(\frac{4}{3}, \frac{4}{3}\right) \quad \text{and} \quad \boldsymbol{a}(t) = (0, 0).$$

The speed is $\sqrt{4^2 + 4^2}/3 = \sqrt{32}/3 \approx 1.89$ units per second. □

Modeling circular motion

Constant-speed motion along a circular path is called **uniform circular motion**. The following two special cases will be very useful:

> **Fact:** For any center point (x_0, y_0) and any radius $R > 0$, the position function
>
> $$\boldsymbol{p}(t) = (x_0, y_0) + (R\cos t, R\sin t)$$
>
> models *counterclockwise* uniform circular motion about (x_0, y_0) with radius R, at constant speed R. The position function
>
> $$\boldsymbol{p}(t) = (x_0, y_0) + (R\cos t, -R\sin t)$$
>
> models *clockwise* uniform circular motion at constant speed R.

All parts of the Fact are easy to verify; see the exercises. Using the Fact—and a few handy tricks—we can model all sorts of uniform circular motion. We illustrate several by example.

■ **Example 2.** **(Different speeds.)** A particle has position function

$$\boldsymbol{p}(t) = (x_0, y_0) + (R\cos(at), R\sin(at)),$$

where $a > 0$ is a positive constant. What difference does the a make?

Solution: The a makes no difference to the particle's path—the particle still travels at constant speed along the circle of radius R about (x_0, y_0). Velocity and acceleration are affected, however. Differentiation gives

$$\boldsymbol{v}(t) = a(-R\sin(at), R\cos(at));$$

it follows that $|\boldsymbol{v}(t)| = aR$. Similarly,

$$\boldsymbol{a}(t) = \boldsymbol{v}'(t) = a^2(-R\cos(at), -R\sin(at)).$$

Thus the effect of a is to multiply the speed by a and the magnitude of acceleration by a^2. □

■ **Example 3.** **(Changing speed.)** A particle moves counterclockwise, with constant speed s, around the circle of radius R, centered at (x_0, y_0). Find the position function.

Or just check it for yourself; it's easy.

Solution: By the Fact, the position function

$$p(t) = (x_0, y_0) + (R\cos t, R\sin t)$$

has constant speed R. The preceding example shows that to change the speed from R to s, we multiply t by s/R, to get the new position function

$$p(t) = (x_0, y_0) + (R\cos(t\,s/R), R\sin(t\,s/R)).$$

Now it's easy to see that $p(t)$ has speed s, as desired. □

Draw a picture!

■ **Example 4.** A particle starts at $(2, 3)$ at $t = 0$. It moves at constant speed 1 around the circle centered at $(2, 0)$, until it reaches the point $(-1, 0)$. Find the position function.

Solution: The circle has radius 3, so $p(t) = (2, 0) + (3\cos t, 3\sin t)$ describes a particle starting from the east pole $(5, 0)$ at $t = 0$. To start at angle $\pi/2$, we could use the position function $p(t) = (2, 0) + (3\cos(t + \pi/2), 3\sin(t + \pi/2))$. But this has speed 3; to arrange speed 1 we divide t by 3, to get $p(t) = (2, 0) + (3\cos(t/3 + \pi/2), 3\sin(t/3 + \pi/2))$. The particle arrives at $(-1, 0)$ when $t = 3\pi/2$. □

Circular motion, acceleration, and centripetal force

Suppose that a particle moves with uniform speed s about a circle of radius R, centered at the origin. Then, by the work above,

$$p(t) = (R\cos(st/R), R\sin(st/R)).$$

Carefully!

Differentiating shows that the acceleration function is

$$a(t) = \frac{s^2}{R^2}(-R\cos(st/R), -R\sin(st/R)) = -\frac{s^2}{R}(\cos(st/R), \sin(st/R)).$$

This harmless-looking calculation has an important physical meaning:

> *For uniform circular motion with speed s, around a circle with radius R, the acceleration vector always points toward the center of the circle, and has magnitude s^2/R.*

See Section 2.5

It may seem, at first, slightly surprising that constant-speed circular motion generates any acceleration at all. But by Newton's second law, acceleration is proportional to force—and we've all felt the force (called **centripetal force**) that's generated when we swing an object around on a string, or when we drive around a tight curve in a car. The italicized remark says, moreover, that this force is inversely proportional to the radius and directly proportional to the square of the speed. The automotive lesson is clear: Taking tight corners at high speed is a poor idea.

Combining motions means adding vector-valued functions

Many interesting motions—such as that of the robot-arm with which we began this section—are simple combinations of uniform linear and circular motion. It's a pleasant physical fact that these combinations of simpler motions can be modeled by *adding* (in the vector sense) the vector-valued functions that represent the simpler motions.

■ **Example 5.** Consider the jointed-arm robot mentioned above. Assume: (i) the long arm has length 1; (ii) the short arm has length 0.4; (iii) the long arm rotates uniformly counterclockwise, making 1 rotation in 2π seconds; (iv) the short arm rotates uniformly counterclockwise, making 8 rotations in 2π seconds. Find the position function for the black dot.

Solution: The end of the long arm undergoes uniform circular motion, with radius 1, center $(0, 0)$, and unit speed, so it has position function

$$\boldsymbol{p}_{\text{long}}(t) = (\cos t, \sin t).$$

The short arm, if it were pinned at the origin, would have position function

$$\boldsymbol{p}_{\text{short}}(t) = (0.4\cos(8t), 0.4\sin(8t)).$$

The robot combines the two motions by moving the center of the short arm's rotation to the end of the long arm. This amounts simply to adding the position functions as vectors, i.e.,

$$\boldsymbol{p}(t) = \boldsymbol{p}_{\text{short}}(t) + \boldsymbol{p}_{\text{long}}(t) = (\cos(t) + 0.4\cos(8t), \sin(t) + 0.4\sin(8t)).$$

Plotting this function in parametric form for $0 \le t \le 2\pi$, produced the left-hand curve at the start of this section. □

Similar results can be had by combining uniform linear and circular motions.

■ **Example 6.** Combine the uniform linear motion

$$\boldsymbol{p}_{\text{linear}}(t) = (t, 1)$$

and the uniform circular motion

$$\boldsymbol{p}_{\text{circular}}(t) = (\cos t, -\sin t).$$

What does the combination $\boldsymbol{p}(t) = (t + \cos t, 1 - \sin t)$ represent?

Solution: Here's the picture for $0 \le t \le 4\pi$:

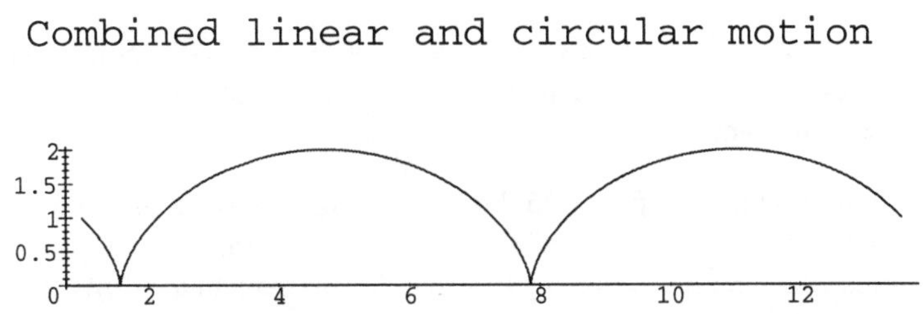

The curve is called a **cycloid**—it represents the path traced by a point on the rim of a rolling wheel. (We'll revisit the cycloid in exercises.) □

Exercises

1. In each part below, find a position function that models the given uniform linear motion. Then use technology to plot the particle's path as a parametric curve.

   (a) The particle starts at $(0, 0)$ at $t = 0$ and travels to $(1, 2)$ at constant speed 1.

   (b) The particle starts at $(1, 2)$ at $t = 0$ and travels to $(0, 0)$ at constant speed 1.

   (c) The particle starts at $(1, 2)$ at $t = 0$ and travels to $(5, 6)$ at constant speed 100.

   (d) The particle is at $(1, 2)$ at time $t = 1$ and at $(5, 6)$ at time $t = 10$.

2. This exercise confirms the Fact on page 223. Let $\boldsymbol{p}(t) = (x_0, y_0) + (R\cos t, R\sin t)$.

   (a) Show that $|\boldsymbol{p}(t) - (x_0, y_0)| = R$ for all t.

   (b) Show that $\boldsymbol{p}$ has constant speed R.

   (c) Find $\boldsymbol{v}(0)$. How does the answer reflect the counterclockwise direction?

3. In this exercise let $\boldsymbol{p}(t) = (1, 2) + (3\cos t, -3\sin t)$ describe a particle's position for $-\pi/2 \le t \le \pi/2$.

   (a) Describe the particle's path geometrically. In which direction does it travel?

   (b) Show that $\boldsymbol{p}$ has constant speed 3.

   (c) Find $\boldsymbol{v}(0)$. What does the answer say about the direction of travel?

4. In each part below, give a position function that models the uniform circular motion described. Use technology to plot the position curve.

   (a) with center $(0, 0)$, radius 3, speed 1, and counterclockwise direction; one full rotation starting from the east pole

   (b) with center $(0, 0)$, radius 3, speed 2, and clockwise direction, one half rotation, starting from the east pole

   (c) with center $(1, 2)$, passing through $(4, 5)$, speed 1, counterclockwise direction, one full rotation starting from $(4, 5)$

5. Redo Example 5, page 225, but assume that the small arm rotates counterclockwise 4 times (rather than 8 times) as fast as the larger arm. Find a position function; plot it for $0 \le t \le 2\pi$.

6. Redo Example 5, page 225, but assume that the small arm rotates *clockwise* (rather than counterclockwise) 4 times as fast as the larger arm. Find the position function; plot it for $0 \le t \le 2\pi$. Try to explain the different appearance from the previous problem.

7. Redo Example 5, page 225, but this time assume that the large arm rotates counterclockwise 4 times as fast as the smaller arm. Find a position function; plot it for $0 \le t \le 2\pi$.

8. Consider a robot like that of Example 5, page 225, but assume (i) the long arm has length 2; (ii) the short arm has length 1; (iii) the long arm rotates uniformly counterclockwise, making 1 rotation in 2π seconds; (iv) the short arm rotates uniformly counterclockwise, making 4 rotations in 2π seconds.

   (a) Find the position function for the black dot.

   (b) Find the velocity function for the black dot.

   (c) Find the speed function for the black dot. Use it to find or estimate the length of one "cycle." (Note: You will probably have to use a numerical method to estimate the arclengh integral.)

   (d) Find the acceleration function for the black dot.

9. This exercise is about Example 6, page 225.

   (a) Find the velocity function $\boldsymbol{v}(t)$. Use it to find points at which the velocity is $(0, 0)$. How do these points appear on the graph?

   (b) Find the speed function $s(t)$. Use it to find (exactly!) the length of one arch of the curve.

   (c) Show that the acceleration has constant magnitude.

10. (a) Suppose that a bicycle with tires of radius a moves at a constant speed v_0 along a straight road. Assume that the road is the x-axis and that the bicycle moves to the right. Explain why the point on the tire which is at the origin at time $t = 0$ is at the point $\boldsymbol{p}(t) = \big(v_0 t - a\sin(v_0 t/a), a - a\cos(v_0 t/a)\big)$ at time $t > 0$. [HINT: Think of the motion of the point on the tire as a combination of sliding and spinning.]

   (b) Suppose that a bicycle with tires of radius a moves at a constant speed v_0 along a straight road. At time $t = 0$ an insect starts crawling from the center of the wheel along a spoke at constant linear speed s_0. Find an expression for the location of the insect at time t. (Assume that the road is the x-axis and that the bicycle moves to the right.)

   (c) Suppose that a bicycle with tires of radius a moves at a constant speed v_0 up a hill. Assume that the road is the line $y = mx$ and that the bicycle moves to the right. Find an expression for the location at time $t > 0$ of the point on the tire which is at the origin at time $t = 0$.

Project: Cycloids and Epicycloids

About cycloids

A cycloid is the path traced by a point on the rim of a wheel (we'll assume it has radius 1) as it rolls, without slipping, along a line (we'll take it to be the y-axis). A cycloid can be modeled as the vector sum of a uniform linear motion (of the center of the wheel) and a uniform circular motion (of the point on the rim, around its center). Thus the cycloid has a vector-valued position function of the form

$$\boldsymbol{p}(t) = \text{linear} + \text{circular} = (at, 1) + (\cos(bt), -\sin(bt));$$

a and b are constants.

Problems on cycloids

1. Use technology to plot some curves of the form above, for several values of a and b. How do the values of a and b affect the shape of the curve?

2. If the wheel rolls to the right, without slipping, at a constant rate of 1 unit per second, what's the formula for $\boldsymbol{p}(t)$? In particular, why does $-\sin$ appear rather than sin?

3. What situation does the formula $\boldsymbol{p}(t) = (t, 1)+(\cos(2t), -\sin(2t))$ represent? (Hint: The answer involves slipping; 2 could be called the slipping coefficient.) Draw the curve. At what points on the curve, if any, does the point on the rim of the wheel move to the *left*? What is its speed then?

4. Explain why the wheel does not slip if $a > 0$ and $\boldsymbol{p}(t) = (at, 1)+(\cos(at), -\sin(at))$

About epicycloids

An **epicycloid** (sometimes called a **roulette**) is the curve traced out by a point P on the rim of wheel as the wheel rolls without slipping around *another* wheel, say of radius R. Let's assume for simplicity that the outer wheel has radius 1. Here's a picture:

Tracing an epicycloid

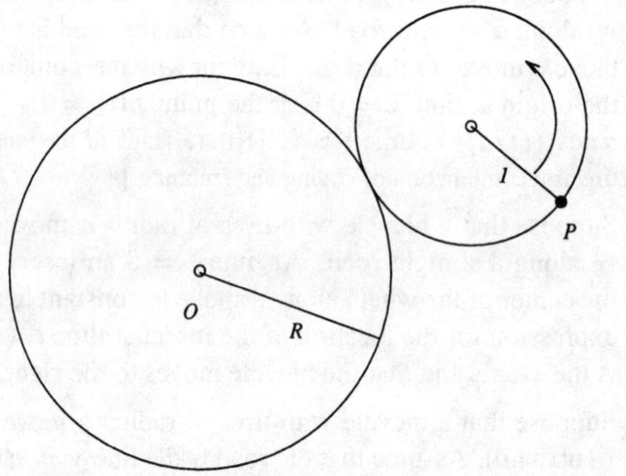

It should be clear that the motion of P is the vector sum of two uniform circular motions. The following formula describes the motion:

$$\boldsymbol{p}(t) = \text{first arm} + \text{second arm} = (R+1)(\cos(at), \sin(at)) + (\cos(bt), \sin(bt)).$$

Problems on epicycloids

1. Explain why the motion formula above "works." Why does $R+1$ appear in it?

2. Suppose that $a = 1$ and $R = 3$ and $b = 2$. Find all the cusps—i.e., the places where the point P has zero velocity. How do these points appear on the epicycloid?

3. Give a formula for, and draw with technology, an epicycloid with 5 cusps.

4. Suppose that $a = 1$. Under what conditions on R and b will the wheel roll without slipping? (Hint: The answer is $b = R - 1$. Try to explain why.)

4.2 Using the dot product: more on curves

In this section we use vector tools—projections, length, the dot product, perpendicularity, etc.—to handle plane curves more efficiently than we could do otherwise. In particular, we'll use vector tools to start the process of building interesting and beautiful new curves, such as rotated sinusoids and conchoids,➡ out of old ones. Some of these "new" curves are, in fact, quite ancient. The Conchoid of Nicomedes, for instance, is named for a Greek mathematician of the third century B.C., who used it in studying such classical problems as trisecting angles and duplicating a cube.

We'll define the term later in this section.

Our basic objects: curves in vector form

Throughout this section our basic object of study will be a parametrized plane curve C. We'll describe such a curve either in vector form

$$\boldsymbol{r}(t) = (x(t), y(t)); \qquad a \le t \le b,$$

or in parametric form

$$x = x(t); \qquad y = y(t); \qquad a \le t \le b.$$

The differences between the vector and parametric forms are mainly cosmetic, and we'll toggle freely back and forth between them. When using the vector version, we'll think of $\boldsymbol{r}(t)$ as the position vector with tail at the origin and head at $(x(t), y(t))$. As t varies, the arrowhead "sweeps out" the curve C, as shown:

Position (radius) vectors sweep out a curve

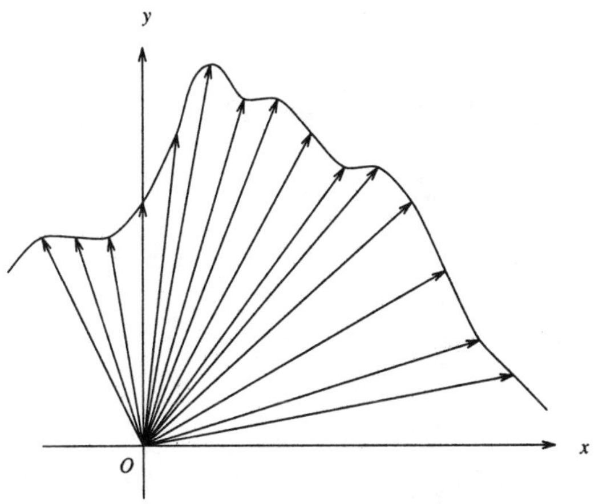

The notation $\boldsymbol{r}(t)$ suggests "radius"—as it should, because all the position vectors emanate "radially" from the origin.

In this section we'll combine the dot product with the vector representation of curves to produce new curves from old.

Rotations

The dot product and its connection to angles can be used to rotate curves in the plane.

Rotating vectors. It's often useful, given a vector $\boldsymbol{v} = (a, b)$, to find the new vector $\boldsymbol{w}$ formed by rotating $\boldsymbol{v}$ counterclockwise through an angle α. The following Fact tells exactly how to do this.

Fact: **(How to spin a vector.)** Let $\boldsymbol{v} = (a, b)$ be any non-zero vector, and let α be an angle, in radians. The vector

$$\boldsymbol{w} = (a\cos\alpha - b\sin\alpha, a\sin\alpha + b\cos\alpha)$$

is the result of rotating $\boldsymbol{v}$ counterclockwise through α radians.

How was the formula for $\boldsymbol{w}$ found? We'll leave that good question to the exercises, but it's easy to check that $\boldsymbol{w}$ does what it's claimed to do. Now let θ be the angle between $\boldsymbol{v}$ and $\boldsymbol{w}$; the Fact claims that $\theta = \alpha$. Taking some dot products is the key to seeing why. Straight calculations show:

$$\boldsymbol{v}\cdot\boldsymbol{w} = (a^2+b^2)\cos\alpha; \quad \boldsymbol{v}\cdot\boldsymbol{v} = a^2+b^2; \quad \boldsymbol{w}\cdot\boldsymbol{w} = a^2+b^2.$$

We know, too, that $\boldsymbol{v}\cdot\boldsymbol{w} = |\boldsymbol{v}||\boldsymbol{w}|\cos\theta$. Combining all our dot products now gives

$$(a^2+b^2)\cos\alpha = \boldsymbol{v}\cdot\boldsymbol{w} = |\boldsymbol{v}||\boldsymbol{w}|\cos\theta = \sqrt{a^2+b^2}\sqrt{a^2+b^2}\cos\theta.$$

Thus $\cos\alpha = \cos\theta$, as we'd hoped.

Rotating curves. The preceding Fact tells how to rotate any given vector through any given angle. Rotating $\boldsymbol{v} = (1, 2)$ through $\pi/3$ radians, for example, gives the vector

$$\begin{aligned}\boldsymbol{w} &= (1\cos(\pi/3) - 2\sin(\pi/3), 1\sin(\pi/3) + 2\cos(\pi/3))\\ &= \left(\frac{1}{2} - \sqrt{3}, \frac{\sqrt{3}}{2} + 1\right) \approx (-1.23, 1.86).\end{aligned}$$

Something much more interesting happens if we apply the Fact to the vector formula for a curve.

■ **Example 1.** Consider the curve C given in vector form by

$$\boldsymbol{r}(t) = (t, 3+\sin t); \qquad -3.5 \le t \le 3.5.$$

Using $\alpha = \pi/3$, apply the rotation formula to the position vector $\boldsymbol{r}(t)$, to form a new curve $\boldsymbol{r}_\alpha(t)$. What happens?

Solution: If $\alpha = \pi/3$, then $\cos\alpha = 1/2$ and $\sin\alpha = \sqrt{3}/2$. Applying the rotation formula gives

After a little simplification.

$$\boldsymbol{r}_\alpha(t) = \frac{1}{2}(t - (3+\sin t)\sqrt{3}, t\sqrt{3} + (3+\sin t)); \qquad -3.5 \le t \le 3.5.$$

Plotting $\boldsymbol{r}$ and $\boldsymbol{r}_\alpha$ on the same axes gives this result:

Rotating a curve through pi/3

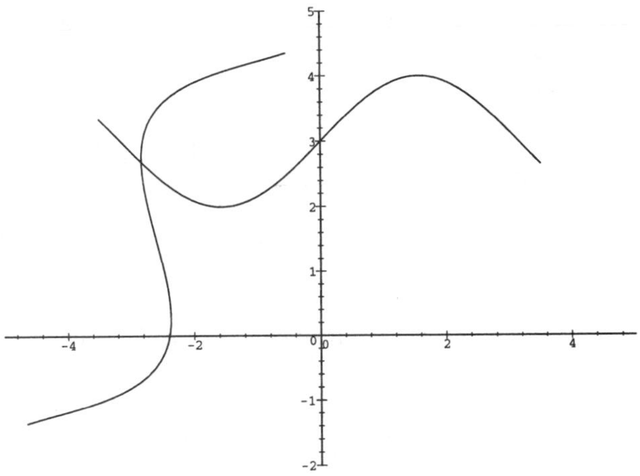

The new curve is the result of rotating the original curve $\pi/3$ radians counterclockwise, with center at the origin. Upon reflection this should not be too surprising—rotating *every* position vector through $\pi/3$ has exactly the same effect on the curve itself. □

New curves from old: conchoids

There are many ways to construct new curves from old. Following is one possibility, based on the dot product.

A **conchoid** is a curve constructed in the following way.

> We start with a given curve C, a point P_0, and a fixed positive number k. For each point P on C, another point Q is determined by moving k units of distance "outward" along the line joining P_0 and P. The set of all such points Q is the conchoid.

The following picture illustrates the idea; it shows several corresponding pairs P and Q.

Constructing a conchoid

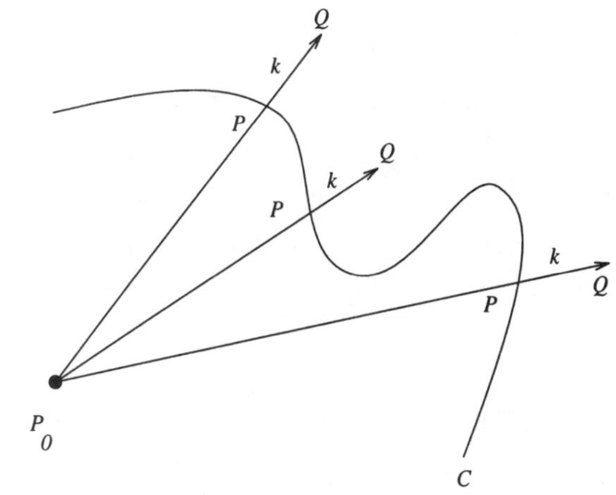

As P moves along the original curve C, the corresponding points Q sweep out the conchoid. (The conchoid curve is not shown.)

For simplicity we'll assume that $P_0 = (0, 0)$, $k = 1$, and C is given in vector form as $\boldsymbol{p}(t) = (x(t), y(t))$, for $a \le t \le b$. Then, as the picture and description show, the conchoid curve can be written by extending each of the vectors $\boldsymbol{p}(t)$ by one unit of length, i.e., by adding the unit vector $\boldsymbol{p}(t)/|\boldsymbol{p}(t)|$. This gives the following vector equation for the conchoid:

$$q(t) = \boldsymbol{p}(t) + \frac{\boldsymbol{p}(t)}{|\boldsymbol{p}(t)|} = \boldsymbol{p}(t)\left(1 + \frac{1}{|\boldsymbol{p}(t)|}\right).$$

For instance, $\boldsymbol{p}(t)$ describes a circle centered at the origin, then $|\boldsymbol{p}(t)|$ is constant, so the right-hand factor above is a positive constant, and the conchoid is simply a larger circle.

If $\boldsymbol{p}(t)$ is a straight line, the resulting curve is a conchoid of Nicomedes. See the exercises for examples.

Exercises

1. This exercise is about the Fact on page 230. Use the same notations as given there for $\boldsymbol{v}$ and $\boldsymbol{w}$.

   (a) Show that $\boldsymbol{v} \cdot \boldsymbol{w} = (a^2 + b^2) \cos\alpha$.

   (b) Show that $|\boldsymbol{w}| = |\boldsymbol{v}|$.

   (c) Let $\boldsymbol{u} = (a\cos\alpha + b\sin\alpha, -a\sin\alpha + b\cos\alpha)$. Show that $|\boldsymbol{u}| = |\boldsymbol{v}|$, and that the angle between $\boldsymbol{u}$ and $\boldsymbol{v}$ is α.

   (d) How are $\boldsymbol{u}$ and $\boldsymbol{w}$ related to each other, and to $\boldsymbol{v}$?

2. Use rotation and other such tricks to find vector formulas for each type of curve following.

   (a) The parabola $y = x^2$, but rotated $\pi/4$ radians counterclockwise.

   (b) A sine curve, but rotated so as to wiggle from northwest to southeast.

   (c) The cardioid $r = 1 + \cos\theta$, but rotated 45 degrees clockwise. (Hint: First write the cardioid in parametric form.)

3. Find the distance from the point P to the line ℓ in each case following.

   (a) from $P(2, 3)$ to the line ℓ through $(0, 0)$ with direction vector $\boldsymbol{i}$

   (b) from $P(2, 3)$ to the line ℓ through $(1, 1)$ with direction vector $(-1, 1)$

   (c) from $P(2, 3)$ to the line ℓ through $(1, 1)$ and $(0, 2)$

   (d) from the origin to the line with Cartesian equation $y = mx + b$; assume that $b \neq 0$. (Could you do this without vectors?)

4. Consider the curve C given by

$$\boldsymbol{r}(t) = (3\cos t, \sin t); \quad 0 \le t \le 2\pi.$$

   (a) Give an equation in x and y for C.

(b) Find and plot the new curve C_1 that's obtained by rotating C $\pi/4$ radians about the origin.

(c) Find and plot the new curve C_2 that's obtained by rotating C $\pi/2$ radians about the origin. Give an equation in x and y for C_2.

5. Use the general conchoid formula given in this section to find (i.e., give a vector equation for) and plot each conchoid below. In all parts, let $P_0 = (0, 0)$ and $k = 1$.

   (a) The conchoid based on $\boldsymbol{p}(t) = (2\sin t, 2\cos t)$, $0 \le t \le 2\pi$.

   (b) The conchoid based on the straight line $x = 1$. (Hint: First write a vector equation for this line.)

   (c) The conchoid based on the spiral $x = t\cos t$, $y = t\sin t$, $0 \le t \le 4\pi$.

Project: Constructing Pedal Curves

There are many ways to create new curves from old. This project explores one of the classical ways of doing so.

Given a smooth curve C and a fixed point O, the **pedal curve** of C with respect to O is defined as follows. (We'll always take O to be the origin, $(0, 0)$.) For a given point P on C, we draw the tangent line ℓ to C at P. The associated point Q is the point on ℓ such that $\overline{OQ}$ is perpendicular to ℓ. (If ℓ passes through O, then $Q = O$.) As P moves along C, the associated point Q traces out the pedal curve. The left-hand picture below shows several points P and their associated points Q. The right-hand picture shows a curve (the circle) and its pedal curve. (Can you guess what type of curve this pedal curve is?)

Constructing a pedal curve

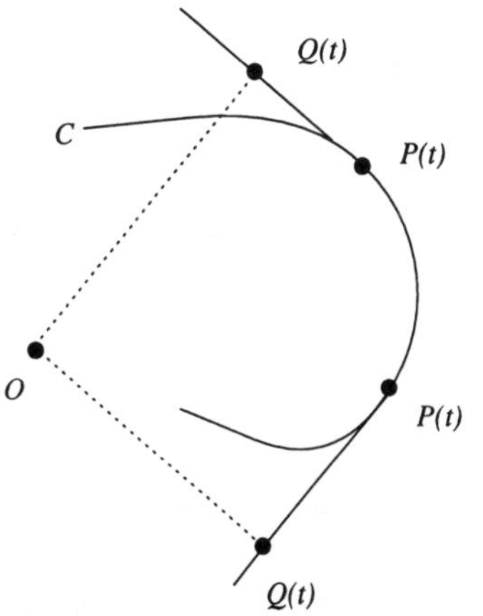

A circle and its pedal curve

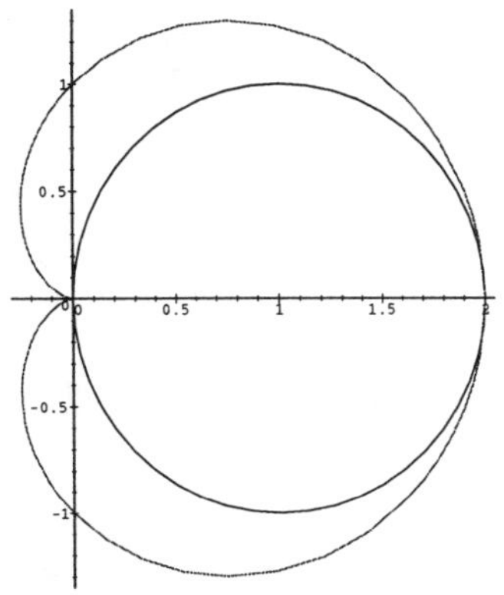

Suppose that the curve C is defined by the position vector $\boldsymbol{p}(t)$, with velocity vector $\boldsymbol{p}'(t)$. Then (some hints for deriving the formula are given below) the pedal curve is defined by the position vector

$$\boldsymbol{q}(t) = \boldsymbol{p}(t) - \frac{\boldsymbol{p}(t)\cdot\boldsymbol{p}'(t)}{|\boldsymbol{p}'(t)|^2}\boldsymbol{p}'(t).$$

Problems

1. What is the pedal curve of a circle centered at the origin? First give a geometric explanation. Then show how the formula above agrees with your answer.

2. What is the pedal curve of a straight line? (Assume that the line misses the origin.) First give a geometric explanation for your answer. Then show how the formula above agrees with your answer.

3. Study at the picture at upper right; the pedal curve is the outside curve. Can you guess the formula for this pedal curve? (Hint: Put your heart into it.) Then find a formula for the circle, and use the formula above to show that your guess is correct. (Another hint: Your guess should be a polar curve; you'll need to change it to parametric form.)

4. Look at the left-hand picture above; observe that the vector joining $\mathbf{P}(t)$ to $\mathbf{Q}(t)$ is the component of $P(t)$ in the direction of the velocity vector at $P(t)$. Using this fact, carefully derive the formula above.

5. For each curve below, find a vector formula for the pedal curve. Then use technology to plot both the curve and its pedal on the same axes.

   (a) the parabola $y = x^2$ (the pedal is called a **cissoid of Diocles**, named for an ancient Greek mathematician; Diocles (ca. 150 B.C.) was first to recognize the special reflection properties of parabolic mirrors).

   (b) the parabola $y = x^2 - 1/4$

   (c) the ellipse $x = 2\cos t + \sqrt{3}$, $y = \sin t$, $0 \le t \le 2\pi$

4.3 Curvature

Anyone who's driven a winding road, even on flat terrain and at moderate speeds, knows the strain that puts on the engine, the brakes, and the passengers. The tightest curves, with the smallest "turning radius," are the worst. Passing through them spills coffee, shifts cargo, and throws passengers into each others' laps.

The idea

Curvature, which we'll define and calculate in this section, is a mathematical measure of how fast a plane curve➽ "turns" near a given point. The automotive analogy shows that there are good practical reasons for measuring curvature, such as designing roads and tracks of various kinds.

The curve might be a road.

For an arbitrary curve, such as the parabola $y = x^2$, we'll see that curvature varies from point to point: It's largest at the vertex and becomes smaller far away from the vertex, where the parabola becomes more and more like a straight line. On a circle, by contrast, the curvature is the same at all points—*the smaller the radius, the larger the curvature.*➽ On a straight line, the curvature is everywhere zero. Our first task (a non-trivial one) will be to define curvature, using vector ingredients, in a way that captures these intuitive properties.

Think about the italicized phrase carefully. Is it reasonable?

Defining curvature

Let a curve C be given by a position function $\boldsymbol{p}(t) = (x(t), y(t))$. We'll define the curvature of C at the point $\boldsymbol{p}(t_0) = (x(t_0), y(t_0))$. First, recall that for any t, the velocity vector $\boldsymbol{v}(t) = (x'(t), y'(t))$ is *tangent* to C at the point $\boldsymbol{p}(t)$. Our definition of curvature, therefore, will measure how fast the velocity vector turns with respect to distance traveled along the curve. For this purpose, the *length* of the velocity vector (i.e., the speed of the curve) is immaterial—we're interested in the *direction*. Therefore we'll write the velocity vector in the form

$$\boldsymbol{v}(t) = s(t)\boldsymbol{T}(t) = \sqrt{x'(t)^2 + y'(t)^2}\frac{\big(x'(t), y'(t)\big)}{\sqrt{x'(t)^2 + y'(t)^2}},$$

where $s(t)$ is the speed at time t and $\boldsymbol{T}(t)$ is called the **unit tangent vector** to C at t. The name is logical because $\boldsymbol{T}(t)$ is a unit vector, and points in the direction of the curve at t.

The curvature at $t = t_0$ tells how fast the tangent direction (given by $\boldsymbol{T}(t)$) varies "with respect to arclength," i.e., with respect to distance traveled along the curve. Finding $|\boldsymbol{T}'(t_0)|$, the magnitude of the derivative $\boldsymbol{T}'(t_0)$, is a good first step. It tells how much the tangent direction changes per unit of *time*. To find the rate with respect to *arclength*, we divide by the speed of the curve at $t = t_0$.➽ Here, at last, is the formal definition:

Does this seem reasonable? Convince yourself that it is.

> **Definition:** **(Curvature.)** Let the curve C be defined by the position function $\boldsymbol{p}(t) = (x(t), y(t))$; assume that the speed $s(t_0) \neq 0$. The curvature of C at the point $\boldsymbol{p}(t_0)$ is defined by
>
> $$\frac{|\boldsymbol{T}'(t_0)|}{s(t_0)}.$$

We'll see by examples that the definition makes good sense.

Lines. For a linear position function, the tangent direction $\boldsymbol{T}(t)$ is a *constant* vector, so $\boldsymbol{T}'(t) = (0, 0)$. It follows that (as we'd expect) a line has constant curvature zero.

Circles. If C is a circle of radius R, then C has a position function of the form

$$\boldsymbol{p}(t) = (x_0, y_0) + R(\cos t, \sin t).$$

But check the work!

It follows that for all t,

$$\boldsymbol{v}(t) = R(-\sin t, \cos t), \quad s(t) = R, \quad \text{and} \quad \boldsymbol{T}(t) = (-\sin t, \cos t).$$

Thus, for every t,

$$\text{curvature} = \frac{|\boldsymbol{T}'(t)|}{s(t)} = \frac{|(-\cos t, -\sin t)|}{R} = \frac{1}{R}.$$

As expected, the curvature is the same at every point on a circle, and the larger the radius, the smaller the curvature.

Radius of curvature. If C has curvature K at a point P, then we say that C has **radius of curvature** $1/K$ at P. The preceding calculation shows that this definition makes sense for circles. We'll see below what it means geometrically.

Different speeds, same curvature. Curvature, by rights, should be intrinsic to the curve itself; it should not depend on the speed of a particular parametrization. The denominator in the definition above takes care of this. To see how, suppose that the circle of radius R mentioned above is parametrized at *twice* the speed, with the parametrization

A road's curviness doesn't depend on how fast cars drive.

$$\boldsymbol{p}(t) = (x_0, y_0) + R(\cos(2t), \sin(2t)).$$

The new, speedier parametrization gives

$$\boldsymbol{v}(t) = 2R(-\sin(2t), \cos(t)), \quad s(t) = 2R, \quad \text{and} \quad \boldsymbol{T}(t) = (-\sin(2t), \cos(2t)).$$

For all t,

$$\text{curvature} = \frac{|\boldsymbol{T}'(t)|}{s(t)} = \frac{|2(-\cos(2t), -\sin(2t))|}{2R} = \frac{1}{R},$$

just as we found earlier.

Calculating curvature

The definition above explains nicely what curvature *is*, but the formula is difficult to calculate with. For all but the simplest curves, calculations quickly become very messy. The following Fact gives a much more convenient formula for finding curvature.

Fact: Let the curve C be defined by the position function $\boldsymbol{p}(t) = (x(t), y(t))$. If the speed $s(t) \neq 0$, then

$$\text{curvature} = \frac{|x'(t)y''(t) - y'(t)x''(t)|}{(\sqrt{x'(t)^2 + y'(t)^2})^3}.$$

More succinctly, curvature $= |x'y'' - y'x''|/s^3$.

The Fact can be proved via a straightforward but long and tedious calculation, starting from the definition. A shorter but slightly more subtle calculation is outlined in the exercises. Our main concern is with *using* the formula, so we omit details here.

■ **Example 1.** Discuss the curvature of the parabola $y = x^2$. Where is the curvature most and least? How does the curvature behave as $x \to \infty$?

Solution: We can parametrize the parabola as $\boldsymbol{p}(t) = (t, t^2)$. Then $x' = 1$, $x'' = 0$, $y' = 2t$, and $y'' = 2$, so applying the formula gives

$$\text{curvature} = \frac{2}{(\sqrt{1+4t^2})^3}.$$

Thus the curvature has its largest value, 2, at the origin, where $t = 0$; the radius of curvature there is $1/2$. At the point $(1, 1)$, where $t = 1$, the curvature is $2/5\sqrt{5} \approx 0.18$; the radius of curvature there is the reciprocal $5\sqrt{5}/2 \approx 5.59$. The following picture shows what the radii of curvature mean, geometrically:

Radii of curvature at two points on y=x^2

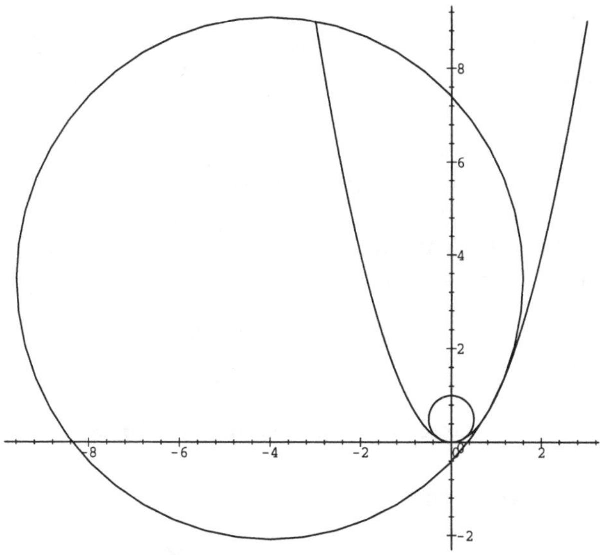

At each point P in question, a circle is drawn. The circle is (i) tangent to the curve at P; and (ii) has radius equal to the radius of curvature at P. (The center of each circle is on the line perpendicular to C at P.) Such a circle is called the **osculating circle** to C at P, because it "kisses" the curve at the given point.➻ The curvature formula shows, too, that the curvature tends to zero as $t \to \infty$. Equivalently, the osculating circles get larger and larger as $t \to \infty$. □

Who said mathematicians can't employ vivid metaphors?

Exercises

1. Show that the line ℓ through $(1, 2)$ in the direction of $(3, 4)$ has constant zero curvature. (Hint: First write a position function for ℓ; then use the definition of curvature.)

2. Suppose that C is the graph of a function $y = f(x)$. Show that in this case, the curvature of C at any point $(x, f(x))$ has the form

$$\text{curvature} = \frac{|f''(x)|}{(\sqrt{1+f'(x)^2})^3}.$$

3. Explain geometrically why the parabola $y = x^2$ should have the same curvature at the symmetric points (t, t^2) and $(-t, t^2)$. Does the formula guarantee this? How?

4. Use technology to draw a picture like that in Example 1, page 237. (Note: The real point of this exercise is to find suitable parametrizations for both osculating circles.)

5. Find the curvature of the curve $y = x^3$ at the point $(0, 0)$. Explain your answer geometrically.

6. Find the curvature of the curve $y = x^3$ at the point $(1, 1)$. Use technology to draw both the curve and the osculating circle there.

7. Here are steps that lead (through considerable symbolic calculation) to the curvature formula in the Fact on page 236. (Alas, there is no really easy way to derive the formula.)

   Let C be given by $\mathbf{p}(t) = \big(x(t), y(t)\big)$. To find the curvature K at a point P, suppose that the tangent vector at P makes angle ϕ with the positive x-axis. Let $l(t)$ be the arclength along C, measured from some fixed point P_0 (it doesn't really matter which point is fixed) to $\mathbf{p}(t)$. By definition, the curvature is the absolute rate of change of ϕ, with respect to arclength l. That is,

$$K = \left|\frac{d\phi}{dl}\right|.$$

   We'll show in steps how to calculate this derivative.

   (a) For all t with $dl/dt \neq 0$,

$$\frac{d\phi}{dl} = \frac{d\phi/dt}{dl/dt}.$$

   Explain why.

   (b) Explain why $dl/dt = s(t) = \sqrt{x'(t)^2 + y'(t)^2}$.

   (c) If $x'(t) \neq 0$, then

$$\tan(\phi(t)) = \frac{y'(t)}{x'(t)}.$$

   Explain why this equation holds. (Hint: Draw a typical tangent vector $(x'(t), y'(t))$.)

   (d) Differentiate both sides of the preceding equation to show that

$$\sec^2(\phi(t))\,\frac{d\phi}{dt} = \frac{x'(t)y''(t) - y'(t)x''(t)}{x'(t)^2}.$$

(e) Use the preceding equation to show that

$$\frac{d\phi}{dt} = \frac{x'(t)y''(t) - y'(t)x''(t)}{x'(t)^2 + y'(t)^2}.$$

(Hint: Use the facts (i) $\sec^2(\phi(t)) = 1 + \tan^2(\phi(t))$; and (ii) $\tan(\phi(t)) = y'(t)/x'(t)$.)

(f) Show that the Fact's formula is true. (Hint: Use the formulas just derived for $d\phi/dt$ and dl/dt.)

4.4 Lagrange multipliers and constrained optimization

Let $f(x, y)$ be a function defined on a domain in $\mathbb{R}^2$; suppose we want to find maximum and minimum values of f. We've seen in earlier sections where to look. Stationary points of f are the candidates, and the second derivative test helps us sort out the possibilities. If, say, $f(x, y) = x^2 - 4x + 2y^2$, then $\nabla f(x, y) = (2x - 4, 4y)$, so $(2, 0)$ is the only stationary point, and it's easy to see that f has a local minimum there.

Sometimes it's of interest to find maximum or minimum values of a function subject to some "constraint" on legal inputs. For instance, we might ask for the largest and smallest values of $f(x, y)$ when (x, y) is constrained to lie on the circle $x^2 + y^2 = 9$. Solving the gradient equation $\nabla f = (0, 0)$, as above, will do us no good—the constraint circle doesn't pass through the one stationary point $(2, 0)$. As we'll see, gradients are *still* the right tools—but they need to be used in a way appropriate to the present problem.

Naming the problem, and a graphical example

A little attention paid to jargon now will save trouble later. The general problem at hand is called **constrained optimization**; the function to be maximized or minimized is called the **objective function**; the restriction on inputs is described by a **constraint equation**. (Sometimes constraints are given by inequalities, or by more than one equation.) Looking at the situation graphically will suggest the role that gradients play.

That is, optimized.

■ **Example 1.** Optimize $f(x, y) = x^2 - 4x + 2y^2$ subject to the constraint $x^2 + y^2 = 9$.

Solution: The constraint describes a circle C in the xy-plane; to "see" the objective function in the same picture, it's natural to look at level curves of f near C:

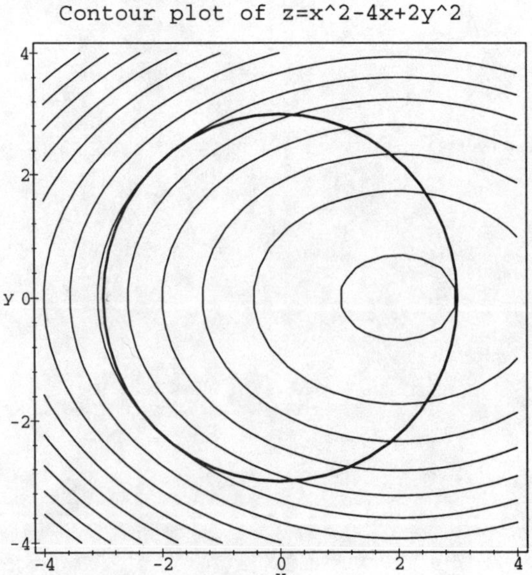

The picture deserves a close look:

The objective function Level curves of f are ellipses; they're centered at $(2, 0)$, where f has a local minimum. Larger ellipses, therefore, represent larger values of f.

The constraint Inputs (x, y) that satisfy the constraint are points that lie on the circle C. If we imagine an ant walking along the surface $z = f(x, y)$ above the curve C, our problem is to decide where the ant's "altitude"—described by the level curves—is largest and smallest.

A graphical answer A close look along C suggests special attention to four points on the circle: $(-3, 0)$ and $(3, 0)$, where $y = 0$; and $(-2, \sqrt{5})$ and $(-2, -\sqrt{5})$, where $x = -2$. At these points, our wandering ant appears to experience highs and lows of altitude; *between* these points, the ant travels either uphill or down. Checking numerical values of f at these points supports these impressions:

Do you agree? Look especially carefully near the west pole of the circle.

$$f(3, 0) = -3; \quad f(-3, 0) = 21; \quad f(-2, \sqrt{5}) = f(-2, -\sqrt{5}) = 22.$$

We'd estimate, then, that f assumes its (constrained) maximum value, 22, at the points $(-2, \pm\sqrt{5})$, and its minimum value, -3, at $(3, 0)$.

A useful observation At each special point P just mentioned, a crucial property holds:

The level curve of f through P is tangent to the constraint curve.

□

The condition just mentioned—suitably interpreted in terms of gradients—is the main idea of this section. We'll see how how to use this property, which we'll call the **Lagrange condition,** to solve constrained optimization problems symbolically. First, however, let's acknowledge that other approaches are possible for the problem above; one follows.

After the Italian-French mathematician Joseph-Louis Lagrange (1736–1813).

The exercises contain a third technique.

■ **Example 2.** Redo Example 1, this time by parametrizing the constraint curve.

Solution: The circle $x^2 + y^2 = 9$ is easily parametrized; one way is to let

$$\mathbf{X}(t) = (x(t), y(t)) = (3\cos t, 3\sin t); \quad 0 \le t \le 2\pi.$$

$\vec{X}'(t) = (-3\sin t, 3\cos t)$

Constraining (x, y) to lie on the circle means, in effect, optimizing the composite function

Check the algebra.

$$h(t) = f(\mathbf{X}(t)) = 9\cos^2 t - 12\cos t + 18\sin^2 t = 9\sin^2 t - 12\cos t,$$

for t in $[0, 2\pi]$. This reduces the problem to an old familiar form; the rest is routine. Since

$$h'(t) = 18\sin t\cos t + 12\sin t = \sin t\,(18\cos t + 12),$$

we have $h'(t) = 0$ if either $\sin t = 0$ (i.e., $y = 0$) or $\cos t = -2/3$ (i.e., $x = -2$); these are the same candidate points as we found earlier.

Observe, too, that by the chain rule, the key derivative is a dot product:

$$h'(t) = \nabla f(\mathbf{X}(t)) \cdot \mathbf{X}'(t).$$

$\vec{\nabla} f(\vec{X}(t)) = (6\cos t - 4, 12\sin t)$

$\vec{\nabla} f \cdot X'(t) = -18\sin t\cos t + 12\sin t + 36\cos t\sin t = 18\sin t\cos t + 12\sin t$

At any maximum or minimum point we must have $h'(t) = 0$; this means, geometrically, that at all such points the gradient ∇f is *perpendicular* to $\mathbf{X}'$, the tangent vector to the constraint curve. This is another way of stating the Lagrange condition mentioned earlier, because at any point $P(x, y)$, the gradient of f is perpendicular to the level curve of f through P. (We saw this important fact in Section 2.4, on gradients. □

See the first few pictures there.

Gradients and the Lagrange condition

The main point so far, illustrated by the last two examples, is as follows:

> *At a maximum or minimum point for a constrained optimization problem, the gradient of the objective function must be perpendicular to the constraint set.*

This complicated-sounding condition is easiest to understand and use with the help of level curves. The key fact, which links all the main ideas, is the connection between gradients and level sets:

For more details, see the beginning of Section 2.4.

> **Fact: (Gradients and level sets)** Let $g(x, y)$ be a differentiable function, with (x_0, y_0) a point in the domain of g. Let C be the level curve of g that passes through (x_0, y_0). If $\nabla g(x_0, y_0) \neq (0, 0)$, then $\nabla g(x_0, y_0)$ is perpendicular to C at (x_0, y_0).

Proof. The proof is a nice application of the chain rule. Suppose that the curve C is parametrized by a vector-valued function $\boldsymbol{X}(t)$, with $\boldsymbol{X}(t_0) = (x_0, y_0)$. Then $\boldsymbol{X}'(t_0)$ is a vector tangent to C at (x_0, y_0). (We're assuming the slightly more technical fact that such a parametrization must exist, because of our assumption that $\nabla g(x_0, y_0) \neq (0, 0)$.) Because g is constant on the curve C, the composite function $g(\boldsymbol{X}(t))$ is constant in t. Therefore, by the chain rule,

Be sure you agree!

$$0 = \frac{d}{dt}(g(\boldsymbol{X}(t))(t_0) = \nabla g(x_0, y_0) \cdot \boldsymbol{X}'(t_0).$$

This means that $\nabla g(x_0, y_0)$ is perpendicular to $\boldsymbol{X}'(t_0)$, and therefore also to C. □

In higher dimensions. Though we'll work mainly in two variables, the Fact above holds just as well for functions of three (or even more) variables, except that in these cases, the level set is a surface, not a curve. For example, the level set $g(x, y, z) = x^2 + y^2 + z^2 = 1$ is a sphere in $\mathbb{R}^3$. The Fact says (and it's not hard to "see") that at any point (x, y, z) on the sphere, the gradient vector $(2x, 2y, 2z)$ is perpendicular to the sphere.

The gradient of the constraint function. The constraint in an optimization problem is usually described by an equation. If we write the constraint equation in the form $g(x, y) = 0$, where $g(x, y)$ is a *function*, then the constraint curve becomes the level curve $g(x, y) = 0$. The preceding Fact says, moreover, that at any point (x, y) on this level curve, the gradient vector $\nabla g(x, y)$ of the constraint function is either the zero vector or is perpendicular to the level curve.

The gradient of the objective function. Let (x_0, y_0) be a point on the curve $g(x, y) = 0$, and suppose that the objective function $f(x, y)$ assumes either a local maximum or a local minimum value at (x_0, y_0) (compared with nearby points on the constraint curve). Then the following useful fact holds:

> *The gradient $\nabla f(x_0, y_0)$ is perpendicular to the constraint curve at (x_0, y_0).*

The picture in Example 1 illustrates this fact. At all four "candidate" points, the level curves of f are *parallel* to the constraint curve, so the gradient ∇f must be *perpendicular* to the constraint curve. The calculation at the end of Example 2 explains why this fact holds.

Lagrange multipliers

The facts above, taken together, say something striking and useful: If (x_0, y_0) is a local maximum or local minimum point for the constrained optimization problem, then *both* vectors $\nabla f(x_0, y_0)$ and $\nabla g(x_0, y_0)$ are perpendicular to the level curve $g(x, y) = 0$. It follows, therefore, that these vectors must be parallel to each other, i.e., scalar multiples. The formal result follows; we state it for two variables, but it holds in any dimension.

Theorem 1. (Lagrange multipliers) Let $f(x, y)$ and $g(x, y)$ be functions, and consider the problem of optimizing $f(x, y)$ subject to the constraint $g(x, y) = 0$. If f assumes a constrained local maximum or local minimum value at (x_0, y_0), then for some scalar λ,

$$\nabla f(x_0, y_0) = \lambda \nabla g(x_0, y_0).$$

(The scalar λ is called a **Lagrange multiplier**.)

In simple cases, the theorem makes quick work of finding constrained maxima and minima.

■ **Example 3.** Optimize $f(x, y) = x + y$, subject to the constraint $x^2 + y^2 = 9$.

Solution: We'll write $g(x, y) = x^2+y^2-9$; then the constraint becomes $g(x, y) = 0$, as in the theorem. (This trick always works; notice that the constant 9 is "absorbed" into the function g.) Then $\nabla f(x, y) = (1, 1)$ and $\nabla g(x, y) = (2x, 2y)$.

The theorem says that a constrained maximum or minimum occurs, if at all, at a point (x, y) for which

$$\nabla f(x, y) = \lambda \nabla g(x, y)$$

holds for some scalar λ. In addition, we have the constraint equation $g(x, y) = 0$. In our setting, this means

$$(1, 1) = \lambda(2x, 2y) \quad \text{and} \quad x^2 + y^2 = 9.$$

This adds up to three equations➧ in the three unknowns x, y, and λ.

The gradient equation counts for two!

There are many ways to solve these equations. One is to observe first that

$$(1, 1) = \lambda(2x, 2y) \implies x = y;$$

plugging this into the constraint equation gives $2x^2 = 9$, or $x = \pm 3/\sqrt{2}$. (We needn't bother to solve for λ; it's x and y we care about.) Our candidate points, therefore, are $(3/\sqrt{2}, 3/\sqrt{2})$ and $(-3/\sqrt{2}, -3/\sqrt{2})$. Checking values of f gives $f(3/\sqrt{2}, 3/\sqrt{2}) = 3\sqrt{2}$, a constrained maximum, and $f(-3/\sqrt{2}, -3/\sqrt{2}) = -3\sqrt{2}$, a constrained minimum.

A picture of the situation suggests the same conclusion:

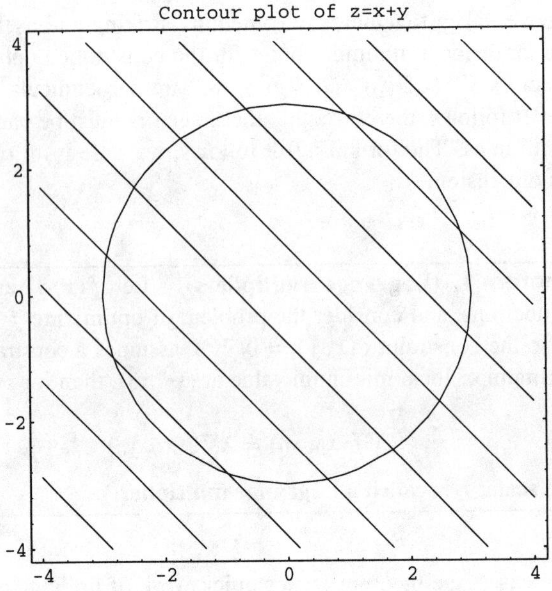

As in Example 1, contour lines of f are parallel to the constraint curve at the maximum and minimum points. □

Caveats

The theorem is often useful, but several remarks and cautions are in order. It's especially important to understand what the theorem *doesn't* say.

A necessary condition, but not sufficient. The theorem says that the Lagrange condition $\nabla f = \lambda \nabla g$ is *necessary* for (x_0, y_0) to be a constrained optimum point—but not, in itself, sufficient. (In the same spirit, $f'(x_0) = 0$ is necessary, but not always sufficient, for x_0 to be a local maximum or minimum point.) At a point that satisfies the condition, f might assume a constrained maximum, a constrained minimum, or *neither*. Deciding among these possibilities can be hard; pictures may help.

There may be no solution. Not every constrained optimization problem has a solution. Even in this case, however, the Lagrange condition may be of use.

■ **Example 4.** Optimize $f(x, y) = x + y$ subject to the constraint $g(x, y) = y = 0$.

Solution: It's clear that $f(x, 0) = x$ can be arbitrarily large positive or negative, so f has neither a constrained maximum nor a constrained minimum.

Notice, though, that here $\nabla f(x, y) = (1, 1)$ and $\nabla g(x, y) = (0, 1)$. Thus the Lagrange condition says $(1, 1) = \lambda(0, 1)$, which is clearly impossible. Therefore, by the theorem, no constrained maxima or minima exist. □

When does a solution exist? The optimization problem in the last example had no solution. The difficulty there was that the constraint set $y = 0$ is unbounded, so $f(x, y)$ was free to blow up in magnitude.

General theory (a little beyond our scope in this book) guarantees, however, that if f and g are differentiable functions, and the constraint set $g(x, y) = 0$ is bounded, then f does indeed assume (finite) constrained maximum and minimum values. In this case, the theorem guarantees that these values must occur where the Lagrange condition is satisfied.

If the constraint set is unbounded, as in the preceding example, then the objective function *may or may not* assume constrained maximum and/or minimum values—it depends on the objective function, and there's no simple rule for deciding which is the case.

Solving may be difficult. For a function $f(x, y)$ and a constraint $g(x, y) = 0$, the Lagrange condition and the constraint equation produce a total of three equations—not necessarily linear equations—in three unknowns. For functions of three variables, four unknowns are involved. Solving such systems can be very difficult, or even impossible. Fortunately, many problems of interest lead to relatively simple systems of equations. The fact that the particular value of λ usually doesn't matter may also save some effort.

The extra unknown is λ.

Example 5. Optimize $f(x, y, z) = x+y+z$ subject to the constraint $g(x, y, z) = x^2 + y^2 + z^2 - 3 = 0$.

Solution: Here the constraint set—a sphere of radius $\sqrt{3}$—is finite, so we know that constrained maximum and minimum values do exist. With these data, the Lagrange condition says

$$\nabla f = (1, 1, 1) = \lambda \nabla g = \lambda(2x, 2y, 2z).$$

It follows (as in a previous calculation) that $x = y = z$. Plugging this into the constraint equation gives

$$x^2 + y^2 + z^2 - 3 = 0 \implies 3x^2 = 3 \implies x = \pm 1.$$

Thus the candidate points are $(1, 1, 1)$ and $(-1, -1, -1)$, a constrained maximum point and a constrained minimum point, respectively. □

Exercises

1. In each part below, first use the Lagrange multiplier method to find constrained maximum and minimum values, if they exist. Then redo the exercise by elementary calculus methods (i.e., use the constraint equation to rewrite the objective function as a one-variable function).

   (a) $f(x, y) = xy$, subject to $g(x, y) = x + y - 1 = 0$

   (b) $f(x, y) = x + y$, subject to $g(x, y) = xy - 1 = 0$

2. Use the Lagrange multiplier method to find (if they exist) constrained maximum and minimum values in each part below. If no such maxima or minima exist, explain why not. At each constrained maximum or minimum point, calculate both ∇f and ∇g.

   (a) $f(x, y) = x - y$, subject to $g(x, y) = x^2 + y^2 - 1 = 0$

(b) $f(x, y) = xy$, subject to $g(x, y) = x^2 + y^2 - 1 = 0$

(c) $f(x, y) = x^2 + y^2$, subject to $g(x, y) = x + y - 2 = 0$

(d) $f(x, y, z) = 2x + y + z$, subject to $g(x, y, z) = x^2 + y^2 + z^2 - 6 = 0$

3. Here's yet another way to optimize $f(x, y) = x^2 - 4x + 2y^2$, subject to the constraint $x^2 + y^2 = 9$ (the exercise of Example 1, page 240). Carry out the details as described below.

(a) Substitute $y^2 = 9 - x^2$ into the objective function, to produce a new function h of one variable, x.

(b) On what x-interval should $h(x)$ be optimized? Why?

(c) Use single-variable methods to maximize and minimize $h(x)$ on the appropriate x-interval. Check that your results agree with those of Example 1.

4. The following picture shows level curves of $f(x, y) = x + 2y$, and also the curve $g(x, y) = 4x^2 + 9y^2 - 36 = 0$. This exercise is about optimizing f subject to the constraint $g(x, y) = 0$.

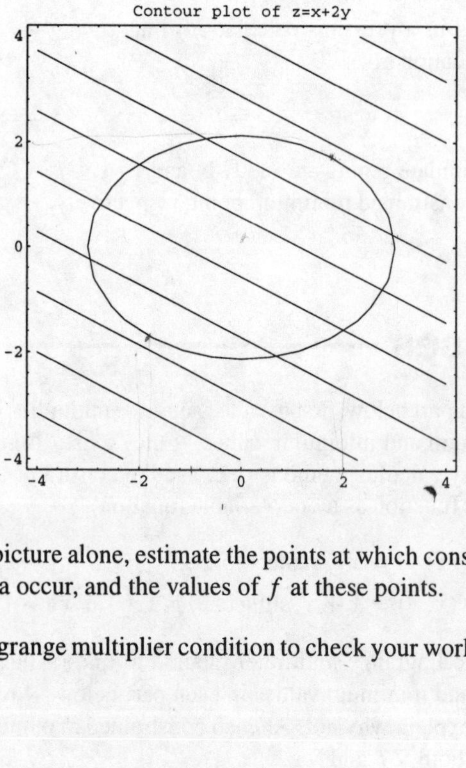

(a) Using the picture alone, estimate the points at which constrained maxima and minima occur, and the values of f at these points.

(b) Use the Lagrange multiplier condition to check your work in the previous part.

5. The picture below shows several level curves of $f(x, y) = x + 2y$. This exercise is about maximizing and minimizing f subject to the constraint $g(x, y) =$

$x^2 + y^2 - 9 = 0$.

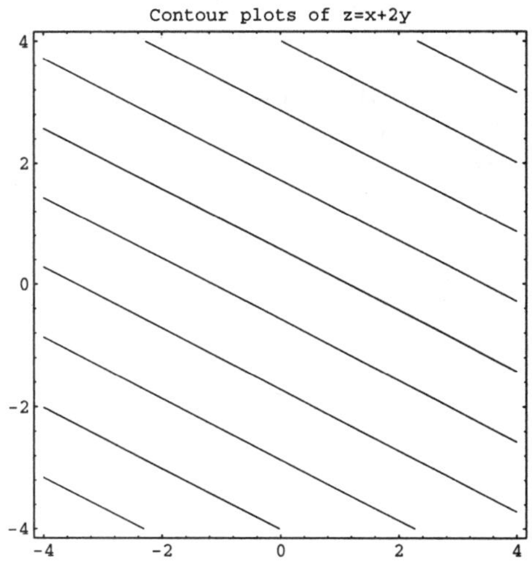

(a) Carefully draw the constraint set $g(x, y) = 0$ into the picture. Label some contour lines with their z-values.

(b) Using the picture alone, estimate the points at which constrained maxima and minima occur, and the values of f at these points.

(c) Use the Lagrange multiplier condition to check your work in the previous part.

6. The picture below shows several level curves of $f(x, y) = x^2 + xy + y^2$. This exercise is about minimizing f subject to the constraint $g(x, y) = x + y - 2 = 0$. (There is no constrained maximum.)

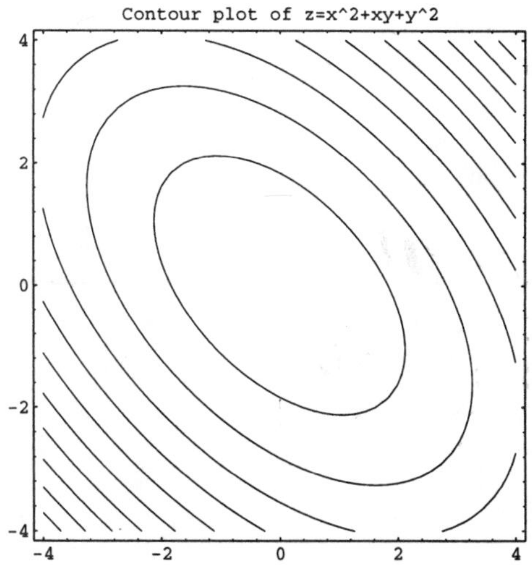

(a) Carefully draw the constraint set $g(x, y) = 0$ into the picture. Label some contour lines with their z-values.

(b) Using the picture alone, estimate the points at which the constrained minimum occurs, and the values of f at this point.

(c) Use the Lagrange multiplier condition to check your work in the previous part.

7. Farmer Brown has 100 feet of fence, and wants to enclose as much area as possible in a rectangular pig pen. Help Farmer Brown. (Use the Lagrange multiplier to reinterpret and solve this tired old standard.)

8. Farmer Jones has 100 feet of fence, and wants to enclose as much area as possible in (for some reason) a right-triangular pig pen. Use the Lagrange method to assist this eccentric agriculturalist.

Chapter 5

Vector Calculus

5.1 Line integrals

A peek ahead

In work so far we have met many variants on the basic single-variable definite integral $\int_a^b f(x)\,dx$. Now we meet yet another variant: the **line integral**

$$\int_\gamma f(X) \cdot dX,$$

where $f : \mathbb{R}^2 \to \mathbb{R}^2$ is a vector-valued function, and γ is a directed curve in $\mathbb{R}^2$. (Higher-dimensional versions exist, too.) As we'll see a little later in this chapter, line integrals have a close connection to double integrals—a remarkable fact, given the apparent differences between line integrals and double integrals. This important result is known as Green's theorem.➻

About which more soon.

Line integrals are also essential tools in physics. They come with the territory whenever vector-valued functions are used, as when physicists model such phenomena as forces, fluid flow, electricity, and magnetism. The mathematical theory of line integrals was developed in the early 1800's partly to solve physical problems. Green's theorem, for example, can be understood as a quantitative property of fluid flow.

But we're getting far ahead of ourselves. Our first goal is to introduce line integrals themselves, starting from their basic ingredients.

Vector fields

The integrand in a line integral➻ is a vector-valued function $f : \mathbb{R}^2 \to \mathbb{R}^2$. The function f has the general form

For now we'll stick to line integrals in $\mathbb{R}^2$.

$$f(x, y) = (\, P(x, y), Q(x, y)\,),$$

where $P : \mathbb{R}^2 \to \mathbb{R}$ and $Q : \mathbb{R}^2 \to \mathbb{R}$ are real-valued functions of two variables. In effect, f is a *pair* of scalar functions.

How might we visualize f, which accepts two input variables and produces two output variables?➻ The best approach for us will be to think of f as a **vector field** in $\mathbb{R}^2$: To any input point (x, y) in $\mathbb{R}^2$, f assigns the 2-dimensional vector $(P(x, y), Q(x, y))$, which we can represent as an arrow based at (x, y). By drawing (or having a machine draw) many of these arrows we can get a sense of how f behaves.

That's four in all—too many to draw.

■ **Example 1.** Consider the vector-valued function $f(x, y) = (x - y, x + y)$ as a vector field. Plot f on the input domain $[-3, 3] \times [-3, 3]$. Discuss the picture. Where are the arrows vertical and horizontal? Why?

Solution: Here is *Maple*'s picture:

The vector field f(x,y)=(x-y,x+y)

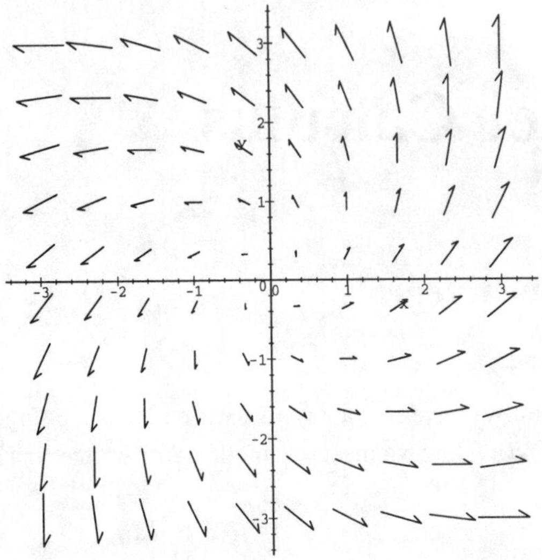

There's a lot to see:

Only a sample. As with any plot, this one only samples values of f. The picture shows 100 output arrows.

Arrows are scaled down. By definition, $f(1, 1) = (0, 2)$—a vertical arrow of length two. In the picture, however, the arrow at $(1, 1)$ is much shorter. This scaling is usually done (by plotting programs and by humans) to prevent arrows overlapping with their neighbors. Although some information is lost in this process, it seldom matters much. In any event, arrows that *look* longer *are* longer.

Horizontal and vertical arrows. An arrow is vertical if its first coordinate (given here by $P(x, y) = x - y$) is zero. This occurs along the line $y = x$; the picture agrees. Arrows are horizontal where $Q(x, y) = x + y = 0$, i.e., along the line $y = -x$.

Vanishing arrows. Since $f(0, 0) = (0, 0)$, the arrow at the origin has zero length, and so "vanishes." (The origin is the *only* such point.)

In general, the vector picture suggests something like a "reverse whirlpool," with water welling slowly up at the center, and then spinning around counterclockwise, faster and faster. □

■ **Example 2.** Plot $g(x, y) = (y, x)$ as a vector field. Can you "see" a connection to the scalar function $h(x, y) = xy$?

Solution: Here is *Maple*'s picture, this time in $[-5, 5] \times [-5, 5]$:

The vector field g(x,y)=(y,x)

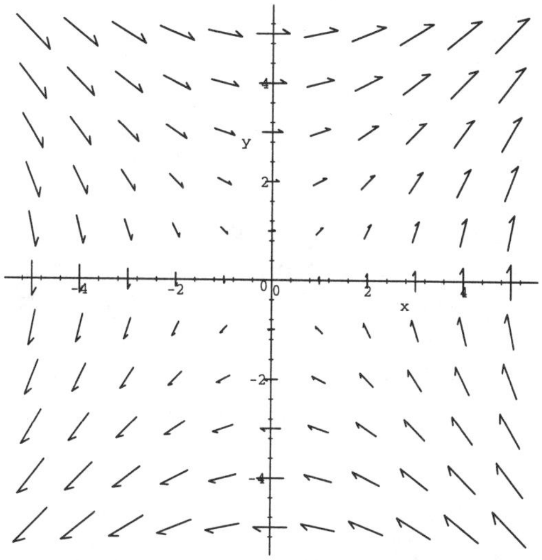

Here, $P(x, y) = y$ vanishes along the x-axis; as the picture shows, arrows there are vertical. Similarly, $Q(x, y) = x = 0$ along the y-axis; there, arrows are horizontal. As in the preceding example, $\mathbf{g}(x, y) = (0, 0)$ *only* at the origin. Notice too that the general appearance of the picture seems reminiscent of the contour map of $h(x, y) = xy$. This connection to $h(x, y)$ is no accident, of course. In fact,

$$\mathbf{g}(x, y) = (y, x) = \nabla h(x, y).$$

We'll pursue this connection carefully in later sections (and briefly in this section's exercises). □

Oriented curves

The other main ingredient in a line integral is the **oriented curve** over which the integral is taken. An oriented curve is, as the name suggests, a curve with a "direction of travel" specified, from starting point to ending point. The curve may be either in the plane or in three-dimensional space. The upper half of the unit circle, for example, can be traversed counterclockwise, from $(1, 0)$ to $(-1, 0)$; the "reversed" curve goes clockwise, from $(-1, 0)$ to $(1, 0)$.

For curves, "orientation" is just math-speak for "direction." In other settings, orientation refers to something a little more complicated.

It's traditional to denote an oriented curve by γ (the Greek letter gamma), and the reversed curve by $-\gamma$. (Notice, however, that γ and $-\gamma$ are exactly the *same set of points*; only the direction is different.) This sign convention turns out to be a convenient aid to memory. We'll see soon that in line integrals, reversing a curve's orientation changes the *sign* of the integral.

To parametrize a curve γ means to specify parametric equations and a parametrizing interval, as in

We discussed this at length in Chapter 1.

$$x = x(t); \quad y = y(t); \quad a \le t \le b.$$

(To avoid technical troubles, we always assume that $x'(t)$ and $y'(t)$ exist and aren't simultaneously zero, except perhaps at the endpoints. Otherwise the parametrization might "stop" or change direction.) Notice that such a parametrization also automatically specifies an orientation: The curve starts at $(x(a), y(a))$ and ends at

$(x(b), y(b))$. Therefore, if we want to parametrize a curve in a given direction, we may need to choose our parametrization to assure that this comes about. An example follows; Section 1.2 describes a technique for reversing a curve's direction.

■ **Example 3.** Let γ be the curve with parametrization

$$x = \cos t; \quad y = \sin t; \quad 0 \le t \le \pi.$$

Describe γ and $-\gamma$. How would $-\gamma$ be parametrized?

Solution: The curve γ is the upper half of the unit circle, parametrized counterclockwise, so $-\gamma$ is the same curve, parametrized clockwise. There are many ways to parametrize $-\gamma$. One possibility is to think of $-\gamma$ as the graph of a function:

$$x = t; \quad y = \sqrt{1-t^2}; \quad -1 \le t \le 1. \qquad \square$$

Calculating line integrals

With all ingredients at hand, we're ready to calculate line integrals. We'll start with a simple example. Watch both the calculations and the useful shorthands.

■ **Example 4.** Let γ be the upper half-circle mentioned above, oriented counterclockwise, and let $f(x, y) = (x - y, x + y)$ be the vector field shown in the first example. Evaluate the line integral

$$\int_\gamma f(X) \cdot dX.$$

Solution: To evaluate any line integral, the idea is to use a parametrization of γ to write everything in terms of one variable, say t.

To parametrize this γ, we'll use the obvious choice:

$$X(t) = (x(t), y(t)) = (\cos t, \sin t); \quad 0 \le t \le \pi.$$

Along the curve γ—where the integral is taken—we can write

$$f(X) = f(X(t)) = (x - y, x + y) = (\cos t - \sin t, \cos t + \sin t);$$

this expresses the integrand as a (vector-valued) function of t alone.

Now for the dX factor. It follow from our parametrization that $dx/dt = -\sin t$ and $dy/dt = \cos t$; equivalently,

$$dx = -\sin t\, dt \quad \text{and} \quad dy = \cos t\, dt.$$

We can write the same information in vector form, as

$$dX = (dx, dy) = (-\sin t\, dt, \cos t\, dt) = X'(t)\, dt.$$

Now we've written both $f(X)$ and dX as vector functions of t, so the dot product in the integral makes sense:

Check the algebra in the last step.

$$\begin{aligned} f(X) \cdot dX &= f(X(t)) \cdot X'(t)\, dt \\ &= (\cos t - \sin t, \cos t + \sin t) \cdot (-\sin t, \cos t)\, dt = 1\, dt. \end{aligned}$$

The result is a scalar function of t; all that's left is to integrate over the parameter interval. Our conclusion finally boils down to this:

$$\int_\gamma f(X)\cdot dX = \int_0^\pi f(X(t))\cdot X'(t)\,dt = \int_0^\pi 1\,dt = \pi. \qquad \square$$

The definition puts the technique just illustrated in general terms:

Definition: (Line integral) Let γ be an oriented curve in $\mathbb{R}^2$, and let $f:\mathbb{R}^2\to\mathbb{R}^2$ be a vector field defined on and near γ. Let $X(t)$ be a differentiable parametrization of γ, with $a\le t\le b$. The line integral of f along γ, denoted by

$$\int_\gamma f(X)\cdot dX,$$

is defined by

$$\int_a^b f(X(t))\cdot X'(t)\,dt.$$

An equivalent notation. If we write $f(x,y) = (P(x,y), Q(x,y))$ and $X = (x,y)$, then the line integral can be written in any of the alternative forms

$$\int_\gamma f(X)\cdot dX = \int_\gamma (P(x,y), Q(x,y))\cdot(dx,dy) = \int_\gamma P dx + Q dy.$$

(The last form is often used in print, perhaps for its typographical simplicity.)➻

For some reason, the letters P and Q are chosen quite consistently.

Understanding line integrals; force and work

Let's rewrite the line integral formula in yet another way, as

$$\int_a^b f(X(t))\cdot\frac{X'(t)}{|X'(t)|}\,|X'(t)|\,dt.$$

This version leads to a useful➻ physical interpretation of the line integral in terms of work.

And intuitively helpful.

To see how, notice first that the vector to the right of the dot product is a unit vector. Let's think of f as describing a **force field**. Then, for each t, the dot product

$$f(X(t))\cdot\frac{X'(t)}{|X'(t)|}$$

is the component of the force vector $f(X(t))$ in the direction of $X'(t)$, i.e., in the direction *tangent* to the curve γ at the point $X(t)$. Therefore, this dot product measures the *work* done per unit of distance by the force $f(X(t))$.➻ The last factor in the integral, $|X'(t)|$, tells the *speed* of the curve at the point $X(t)$, i.e., the distance traveled per unit of time. Over a short time interval Δt, the force remains essentially constant. Over this interval, therefore, the work done by f along γ is approximately

We related work to the dot product in Chapter 1.

$$f(X(t))\cdot\frac{X'(t)}{|X'(t)|}\,|X'(t)|\,\Delta t.$$

From this expression our conclusion follows:

The line integral $\int_\gamma f(X) \cdot dX$ tells the work done by the force f along the oriented curve γ.

Here, for example, are the vector field and the curve in the preceding calculation:

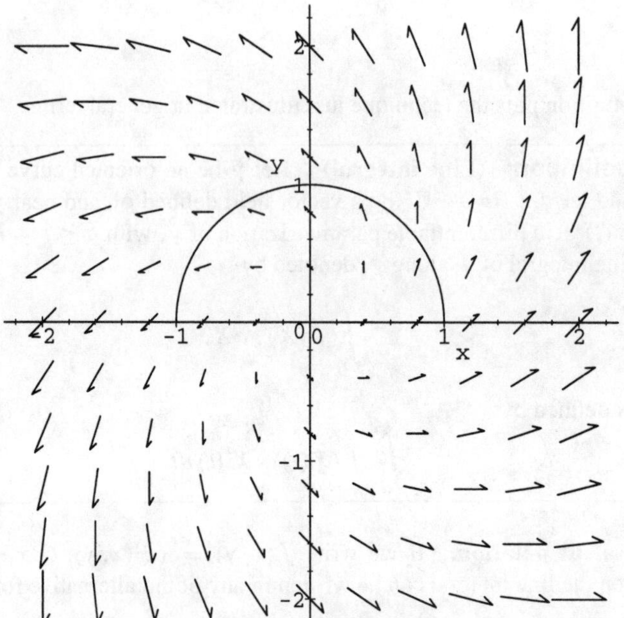

The curve runs counterclockwise—as does the force field. Therefore the work done is positive, as we calculated.

Exercises

1. In each part below, calculate the line integral $\int_\gamma f(X) \cdot dX$ for the given data. Do these by hand; it's OK to check answers using technology.
   (a) $f(x, y) = (x, y)$; γ is the upper half circle, parametrized counterclockwise
   (b) $f(x, y) = (x, y)$; γ is the upper half circle, parametrized clockwise
   (c) $f(x, y) = (x, y)$; γ is the line segment joining $(-1, 0)$ to $(1, 0)$
   (d) $f(x, y) = \nabla g(x, y)$, where $g(x, y) = x^2 - y$, and γ is the unit circle, parametrized counterclockwise
   (e) $f(x, y) = (-y, x)$; γ is the unit circle, parametrized counterclockwise
   (f) $f(x, y) = (-y, x)$; γ is the unit circle, parametrized clockwise
   (g) $f(x, y) = (x, 0)$; γ is the line segment from $(0, 0)$ to (a, b)
   (h) $f(x, y) = (y, 0)$; γ is the line segment from $(0, 0)$ to (a, b)
   (i) $f(x, y) = (x, y)$; γ is the line segment from (a, b) to (c, d)

2. Let $f(x, y)$ be any vector field and γ any curve. Suppose that γ is parametrized by $X(t)$, for $0 \le t \le 1$.

(a) Explain why $-\gamma$ is parametrized by $\boldsymbol{X}(1-t)$, for $0 \le t \le 1$.

(b) Use the previous part to show that

$$\int_{-\gamma} \boldsymbol{f}(\boldsymbol{X}) \cdot d\boldsymbol{X} = -\int_{\gamma} \boldsymbol{f}(\boldsymbol{X}) \cdot d\boldsymbol{X}.$$

(In other words, turning the curve around changes the sign of the answer.)

3. In each part below, do three things: (i) use technology (see the examples above) to draw both the vector field $\boldsymbol{f}$ and the curve γ in the rectangle $[-2, 2] \times [-2, 2]$; (ii) Use the picture to guess whether the line integral $\int_{\gamma} \boldsymbol{f}(\boldsymbol{X}) \cdot d\boldsymbol{X}$ is positive, negative, or zero. (iii) Use the definition (and technology) to calculate the line integral.

   (a) $\boldsymbol{f}(x, y) = (x/(x^2 + y^2), y/(x^2 + y^2))$; γ is the unit circle

   (b) $\boldsymbol{f}(x, y) = (-y/(x^2 + y^2), x/(x^2 + y^2))$; γ is the unit circle

   (c) $\boldsymbol{f}(x, y) = (x, 0)$ γ is the upper half of the unit circle, counterclockwise

   (d) $\boldsymbol{f}(x, y) = \nabla g(x, y)$, where $g(x, y) = x^2 + y^2$, and γ is the upper half of the unit circle

4. This exercise is related to Example 2, page 250. Use the functions and notation defined there.

   (a) In Section 2.4 we mentioned an important connection between the gradient of a function $h(x, y)$ and the level curves of h. What is that property?

   (b) Sketch (by hand) the level curves $z = 0$, $z = \pm 4$, and $z = \pm 9$ of the function $h(x, y) = xy$ into the vector field picture (or a copy of it) in Example 2. Does the property mentioned in (a) show up in your picture?

5. Let $h(x, y) = x + 2y$, and let $\boldsymbol{g}(x, y) = \nabla h(x, y)$.

   (a) Calculate $\boldsymbol{g}(x, y)$. Then, by hand, plot $\boldsymbol{g}(x, y)$ as a vector field in the rectangle $[0, 3] \times [0, 3]$; draw arrows at all points with integer coordinates. (The arrows will need to be scaled down to fit. There are 16 arrows in all.)

   (b) Plot the level curves $h(x, y) = 1$, $h(x, y) = 2$, and $h(x, y) = 3$. What relation is there between level curves and the vector field of (a)?

6. Repeat the previous exercise, but use $h(x, y) = y - x^2$. (This time, use technology to plot the vector field, but check a few entries by hand.)

5.2 More on line integrals; a fundamental theorem

In the preceding section we introduced the idea of the line integral $\int_\gamma f(X) \cdot dX$, or, equivalently, $\int_\gamma P\,dx + Q\,dy$, where $f(x, y) = (P(x, y), Q(x, y))$ is a vector field in the plane, and γ is an oriented curve in the domain of f.

We'll use both notations in this section.

This section continues the story. We add some basic properties of line integrals, most of them closely analogous to properties of ordinary, single-variable integrals. The analogy culminates in a fundamental theorem for line integrals, which (like the "ordinary" fundamental theorem of elementary calculus) relates derivatives to integrals.

Building intuition: force and flow—but not area

Line integrals, unlike ordinary integrals, are not especially easy to visualize *geometrically*. In particular, the value of a line integral $\int_\gamma f(X) \cdot dX$ is *not* in any helpful sense the "area under a graph." This is not to say that pictures are useless—just that they need to be interpreted quite differently from the standard pictures of elementary calculus.

Tempting as it is to think so.

Force. Line integrals were developed around problems in physics; not surprisingly, therefore, physical intuition is often best for understanding what line integrals say. We gave one such approach in the preceding section: If we interpret f (or (P, Q), in the other notation) as a force field in the plane, then the line integral becomes the work done by the force on an object that moves along the curve, from start to finish. This viewpoint helps explain, for instance, why reversing the orientation of a curve changes the sign of a line integral: The work done in moving an object along a curve in one direction is opposite to the work required in the other direction.

Fluid flow. Another physical way of thinking about line integrals is in terms of fluid flow. From this point of view, a vector field f describes, at each point, the velocity of fluid moving across the xy-plane. For an oriented curve γ, the line integral $\int_\gamma f \cdot dX$ is called a **flow integral**; it measures the tendency of fluid to flow along the curve. If γ is a **closed curve**, with the same starting and ending point, then the line integral is called the **circulation** of f along γ.

I.e., a loop.

■ **Example 1.** Here are two flow fields, each with a curve γ (the unit circle, oriented counterclockwise):

The flow f(x,y)=(x-y,x+y)

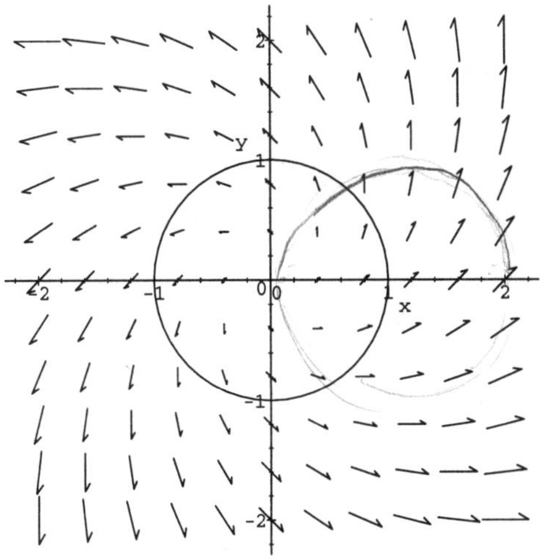

The flow f(x,y)=(x,y)

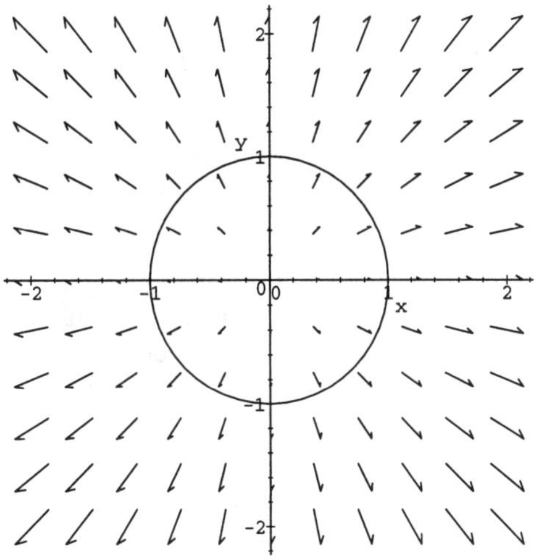

What do the pictures suggest about the line integrals (i.e., flow integrals) $\int_\gamma \boldsymbol{f} \cdot d\boldsymbol{X}$?

Solution: In the left-hand picture, the fluid appears to be swirling counterclockwise—in the same direction as γ—so we'd expect the flow integral to be *positive*. Indeed it is; a routine calculation shows that

Details left to the exercises.

$$\int_\gamma \boldsymbol{f} \cdot d\boldsymbol{X} = \int_\gamma (x - y)\,dx + (x + y)\,dy = 2\pi.$$

In the right-hand picture, the flow is everywhere *perpendicular* to the curve γ, so we'd expect zero circulation. Again, the calculation agrees:

$$\int_\gamma \boldsymbol{f} \cdot d\boldsymbol{X} = \int_\gamma x\,dx + y\,dy = 0. \qquad \square$$

New line integrals from old

Line integrals have algebraic properties similar to those of ordinary integrals. The following theorem collects several of the most useful properties.

Theorem 1. (Algebra with line integrals) Let $f = (P, Q)$ and $g = (R, S)$ be vector fields in $\mathbb{R}^2$, let c be a constant, and let γ be an oriented curve. Then

- $\int_\gamma (f \pm g) \cdot dX = \int_\gamma f \cdot dX \pm \int_\gamma g \cdot dX.$
- $\int_\gamma (cf) \cdot dX = c \int_\gamma f \cdot dX.$
- Let $-\gamma$ be the same curve as γ, but with orientation reversed. Then
$$\int_{-\gamma} f \cdot dX = -\int_\gamma f \cdot dX.$$
- Suppose that γ is the union of two oriented curves γ_1 and γ_2. Then
$$\int_\gamma f \cdot dX = \int_{\gamma_1} f \cdot dX + \int_{\gamma_2} f \cdot dX.$$

(The last part is mainly of interest when the two curves meet end to end. The upper and lower halves of the unit circle do so, for example.)

All parts of the theorem should seem reasonable; they're straightforward consequences of the definition of line integral and of the corresponding properties of ordinary integrals. We omit proofs.

The theorem lets us combine old line integral results to find new ones.

■ **Example 2.** Let $f(x, y) = (x - y, x + y)$ and $g(x, y) = (x, y)$. Use "known" results to calculate $\int_\gamma -y\,dx + x\,dy$, where γ is the unit circle, oriented counterclockwise.

We've intentionally varied the notation.

Solution: Notice first that $f - g = (x - y, x + y) - (x, y) = (-y, x) = h(x, y)$. Therefore,

$$\int_\gamma h \cdot dX = \int_\gamma -y\,dx + x\,dy$$

is the integral we seek. But we've already calculated line integrals of f and g; combining them produces our answer:

$$\int_\gamma h \cdot dX = \int_\gamma f \cdot dX - \int_\gamma g \cdot dX = 2\pi - 0 = 2\pi.$$

A picture of $\boldsymbol{h}$ near γ suggests that the integral is indeed positive:

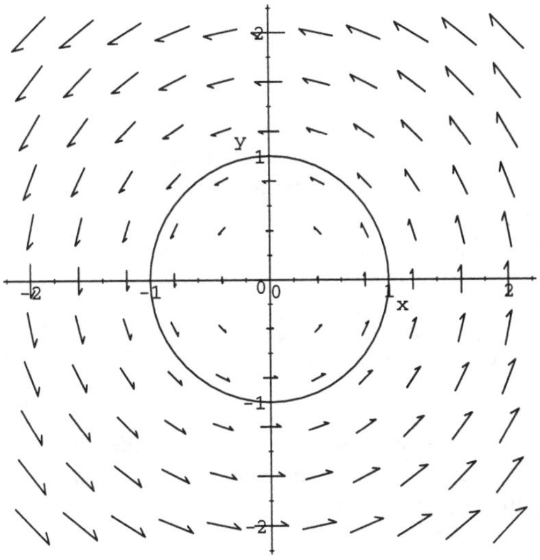

□

The curve matters, not the parametrization. Another important property of line integrals is a little more subtle than those above. As the notation $\int_\gamma \boldsymbol{f} \cdot d\boldsymbol{X}$ suggests, a line integral certainly depends on both the vector field $\boldsymbol{f}$ and the curve γ. But any curve can be parametrized in many ways, and a line integral should *not* depend on the particular parametrization chosen for γ. Without this property, called **independence of parametrization**, our definition of the line integral wouldn't make sense.

Fortunately, line integrals *do* enjoy this property. The next example illustrates why.

■ **Example 3.** Let γ be the upper half of the unit circle, parametrized counterclockwise. Consider these two parametrizations of γ:

$$\boldsymbol{X}(s) = (\cos s, \sin s); \quad 0 \le s \le \pi;$$

$$\boldsymbol{X}(t) = (\cos(t^2), \sin(t^2)); \quad 0 \le t \le \sqrt{\pi}.$$

Show that both parametrizations of the curve γ give the same value for the line integral $\int_\gamma -y\,dx + x\,dy$.

Solution: The first parametrization gives

$$x = \cos s; \quad dx = -\sin s\,ds; \quad y = \sin s; \quad dy = \cos s\,ds.$$

Therefore,

$$\int_\gamma -y\,dx + x\,dy = \int_0^\pi (\sin^2 s + \cos^2 s)\,ds = \pi.$$

The calculation is similar with the second parametrization:

$$x = \cos(t^2); \quad dx = -2t\sin(t^2)\,dt; \quad y = \sin(t^2); \quad dy = 2t\cos(t^2)\,dt.$$

This gives➻

Check the last (easy) step.

$$\int_\gamma -y\,dx + x\,dy = \int_0^{\sqrt{\pi}} (\sin^2(t^2) + \cos^2(t^2))\, 2t\, dt = \pi.$$

The results are the same, as they should be. □

The general argument for independence of parametrization is similar. Suppose we're given *any* two differentiable parametrizations $\boldsymbol{X}(s) = (x(s), y(s))$ and $\boldsymbol{X}(t) = (x(t), y(t))$ for a curve γ, where $a \le s \le b$ and $c \le t \le d$. For the line integral $\int_\gamma P\,dx + Q\,dy$, these parametrizations lead, respectively, to the two integrals

$$\int_a^b \left(P(\boldsymbol{X}(s))\frac{dx}{ds} + Q(\boldsymbol{X}(s))\frac{dy}{ds} \right) ds \quad \text{and} \quad \int_c^d \left(P(\boldsymbol{X}(t))\frac{dx}{dt} + Q(\boldsymbol{X}(t))\frac{dy}{dt} \right) dt.$$

For details on change of variables, see Section 3.5.

Fortunately, these integrals have the same value. In fact, it can be shown that s is a differentiable function of t, i.e., that $s = s(t)$, where $s(c) = a$ and $s(d) = b$. Then, making the change of variable $s = s(t)$ in the first integral gives the second, as follows:

$$\int_a^b \left(P(\boldsymbol{X}(s))\frac{dx}{ds} + Q(\boldsymbol{X}(s))\frac{dy}{ds} \right) ds = \int_c^d \left(P(\boldsymbol{X}(s(t)))\frac{dx}{ds} + Q(\boldsymbol{X}(s(t)))\frac{dy}{ds} \right) \frac{ds}{dt}\, dt$$

$$= \int_c^d \left(P(\boldsymbol{X}(s(t)))\frac{dx}{dt} + Q(\boldsymbol{X}(t))\frac{dy}{dt} \right) dt.$$

(We used the ordinary chain rule in the last equation.)

The fundamental theorem for line integrals

There are other equivalent versions; this one is best for present purposes.

The fundamental theorem of elementary calculus relates the integral and the derivative: If a function f and its derivative f' are continuous on $[a, b]$, then

$$\int_a^b f'(x)\,dx = f(b) - f(a).$$

A similar result holds for line integrals. To assure that all the ingredients exist, we assume that the function $h(x, y)$ has continuous partial derivatives on and near the curve γ, and that γ is smooth, except perhaps at its endpoints.

Theorem 2. (Fundamental theorem for line integrals) Let $h(x, y) : \mathbb{R}^2 \to \mathbb{R}$ be a function. Let γ be an oriented curve, starting at $\boldsymbol{X}_0 = (x_0, y_0)$ and ending at $\boldsymbol{X}_1 = (x_1, y_1)$. Then

$$\int_\gamma \nabla h \cdot d\boldsymbol{X} = h(\boldsymbol{X}_1) - h(\boldsymbol{X}_0).$$

If γ is a closed curve (i.e., $\boldsymbol{X}_1 = \boldsymbol{X}_0$), then

$$\int_\gamma \nabla h \cdot d\boldsymbol{X} = 0.$$

Proof. The proof is a pleasing and straightforward exercise with the chain rule. Let the curve γ be parametrized, as usual, by a function

$$\boldsymbol{X}(t) = (x(t), y(t)); \quad a \le t \le b.$$

Then the line integral has the form

$$I = \int_\gamma \nabla h \cdot d\boldsymbol{X} = \int_a^b \left(h_x(\boldsymbol{X}(t)), h_y(\boldsymbol{X}(t)) \right) \cdot \boldsymbol{X}'(t)\, dt.$$

Now the composite $h(\boldsymbol{X}(t))$ is a new function of t, and, by the chain rule in several variables,

$$\frac{d}{dt}(h(\boldsymbol{X}(t))) = \nabla h(\boldsymbol{X}(t)) \cdot \boldsymbol{X}'(t).$$

In other words, the integrand in the last integral above is the t-derivative of $h(\boldsymbol{X}(t))$. Therefore, by the ordinary fundamental theorem of calculus,

$$I = \int_a^b \frac{d}{dt}(h(\boldsymbol{X}(t)))\, dt = h(\boldsymbol{X}(b)) - h(\boldsymbol{X}(a)) = h(x_1, y_1) - h(x_0, y_0),$$

as claimed.

The gradient advantage. The fundamental theorem implies a remarkable property of any vector field that happens to be a gradient. For such fields, a line integral depends only on the endpoints of the curve γ—not on the curve itself. (All bets are off, of course, if the curve wanders outside the domain in which the gradient is defined.) Line integrals with this property are said to be **independent of path**.

The fundamental theorem makes short work of evaluating line integrals when—but only when—we can find a function whose gradient is the vector field in the integrand. If $\nabla h = \boldsymbol{f}$, then h is called a **potential function** for the vector field $\boldsymbol{f}$.

■ **Example 4.** Let $\boldsymbol{g}(x, y) = (y, x)$, let $\boldsymbol{f}(x, y) = (x - y, x + y)$, and let γ be the line segment from $(0, 0)$ to $(2, 1)$. Use the theorem (if possible), to find $\int_\gamma \boldsymbol{g} \cdot d\boldsymbol{X}$ and $\int_\gamma \boldsymbol{f} \cdot d\boldsymbol{X}$.

Solution: Notice first that if $h(x, y) = xy$, then

$$\nabla h(x, y) = (y, x) = \boldsymbol{g}(x, y).$$

(We found this potential function h purely by guessing; we'll give a more systematic approach in a moment.) We *could* parametrize γ, but the theorem makes that work unnecessary. By the theorem,

$$\int_\gamma \boldsymbol{g} \cdot d\boldsymbol{X} = h(2, 1) - h(0, 0) = 2.$$

Now consider the other line integral. Guessing a potential function h for which $\nabla h = (x - y, x + y)$ seems difficult—in fact, it's impossible. (We'll explain why just below.) Therefore the theorem is no help to us in evaluating $\int_\gamma \boldsymbol{f} \cdot d\boldsymbol{X}$.➻ □

We could always evaluate the integral by parametrizing the curve, of course.

When is a vector field a gradient? Why can't $f(X)$ in the preceding example be a gradient? The reason is that if a potential function h existed, then we'd have

$$h_x(x, y) = x - y \quad \text{and} \quad h_y(x, y) = x + y.$$

From this it would follow, in turn, that

$$h_{xy}(x, y) = -1 \quad \text{and} \quad h_{yx} = 1.$$

This, however, is impossible, because the "cross partial derivatives" of a well-behaved function must be equal.[«] This example illustrates a useful important general fact—in effect, a test for a vector field *not* being a gradient.

We proved this in Section 3.2.

> **Fact:** (**Gradients and cross partials**) A vector field $f(x, y) = (P(x, y), Q(x, y))$ may or may not be a gradient in the vicinity of a curve γ. If $P_y \neq Q_x$ along γ, then $f(x, y)$ is not a gradient field.

Finding potential functions

If a vector field passes the test just mentioned, a potential function can often be found by antidifferentiating in x or y, separately. We illustrate by example.

■ **Example 5.** Find a potential function for the vector field

$$(P, Q) = (y\cos(xy) + 1, x\cos(xy)).$$

Use it to calculate $\int_\gamma P\,dx + Q\,dy$, where γ is the upper half of the unit circle, oriented counterclockwise.

Solution: A quick check shows that

$$P_y = \cos(xy) - xy\sin(xy) = Q_x,$$

so it's worth continuing our search.

Note that if $h(x, y)$ is any potential function, then

$$h_x = y\cos(xy) + 1 \quad \text{and} \quad h_y = x\cos(xy).$$

Therefore, antidifferentiating in x (treating y as a constant) gives

$$h(x, y) = \int (y\cos(xy) + 1)\,dx = \int y\cos(xy)\,dx + \int 1\,dx.$$

Let's make the u-substitution $u = xy$ (and $du = y\,dx$) in the first integral on the right.[«] This gives

Remember—we're treating y as a constant.

$$\int y\cos(xy)\,dx = \int \cos u\,du = \sin(xy) + C;$$

Putting the pieces together gives

$$h(x, y) = \int y\cos(xy)\,dx + \int 1\,dx = \sin(xy) + x + C.$$

Now notice that C may legally depend on y, since our integration was only in the variable x. Therefore, we can write

$$h(x, y) = \int y\cos(xy)\,dx + \int 1\,dx = \sin(xy) + x + C(y),$$

where $C(y)$ is some still-to-be-chosen function of y. To help choose $C(y)$, we differentiate our newly-produced function $h(x, y)$ with respect to y:

Check details carefully.

$$h_y(x, y) = \frac{d}{dy}\left(\sin(xy) + x + C(y)\right) = x\cos(xy) + C'(y).$$

The requirement above is that $h_y = x\cos(xy)$. Thus $C'(y) = 0$, so C can be taken as any real number constant. We conclude (and it's easy to check) that for any constant C, $h(x, y) = \sin(xy) + x + C$ is a potential function for (P, Q).

All our work in calculating h makes finding the line integral a snap. By the fundamental theorem, the answer depends only on the endpoints of γ:

$$\int_\gamma P\,dx + Q\,dy = h(-1, 0) - h(1, 0) = -2.$$

(Calculating the integral by parametrization would be unpleasant.) □

Exercises

Notes. Technology will be quite helpful in some exercises following; see sample commands in the preceding section's exercises. It will save time and effort, for example, to use technology to evaluate standard integrals.

1. Throughout this exercise, let $f(x, y) = (x, y)$; think of f as the velocity of a fluid flow. A picture is shown in Example 1 of this section.) In each part below, calculate the flow integral $\int_\gamma f(X) \cdot dX$ for the given curve γ. Then explain briefly how the sign of the answer (positive, negative, or zero) could have been predicted from the picture. (Example 1 should help with (a).)

   (a) γ is the unit circle, oriented counterclockwise

   (b) γ is the upper half of the circle of radius 1, centered at $(1, 0)$

   (c) γ is the lower half of the circle of radius 1, centered at $(1, 0)$

   (d) γ is the line segment from $(0, 1)$ to $(1, 0)$.

   (e) γ is the line segment from $(0, 2)$ to $(1, 0)$.

2. Throughout this exercise, let $f(x, y) = (x - y, x + y)$; think of f as the velocity of a fluid flow. A picture is shown in Example 1 of this section. In each part below, calculate the flow integral $\int_\gamma f(X) \cdot dX$ for the given curve γ. Then explain briefly how the sign of the answer (positive, negative, or zero) could have been predicted from the picture. (Example 1 should help with (a).)

   (a) γ is the unit circle, oriented counterclockwise

   (b) γ is the full circle of radius 1, centered at $(1, 0)$ (Hint: First parametrize the circle. Section 2.1 showed how.)

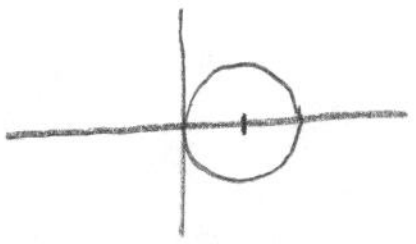

   (c) γ is the upper half of the circle of radius 1, centered at $(1, 0)$

3. Throughout this exercise, let $f(x, y) = (x, 0)$.

   (a) By hand or with technology, draw the vector field f in the rectangle $[-2, 2] \times [-2, 2]$. (No written answer required; the point is to see the picture.)

(b) Let γ be the unit circle, parametrized counterclockwise. Find $\int_\gamma \boldsymbol{f}(\boldsymbol{X}) \cdot d\boldsymbol{X}$. How does the vector field picture predict the sign of the answer?

(c) Let γ be the circle of radius 1 with center at $(1, 0)$. Find $\int_\gamma \boldsymbol{f}(\boldsymbol{X}) \cdot d\boldsymbol{X}$. How does the vector field picture predict the sign of the answer?

(d) Let γ be the circle with center (a, b) and radius c. Find $\int_\gamma \boldsymbol{f}(\boldsymbol{X}) \cdot d\boldsymbol{X}$.

4. Repeat the previous exercise, but let $\boldsymbol{f}(x, y) = (0, x)$.

5. In each part following, apply the test of the Fact on page 262. If the vector field passes the test, mimic the technique of Example 5 to find a potential function h.

(a) $(P, Q) = (x, y)$

(b) $(P, Q) = (y, x)$

(c) $(P, Q) = (-y, x)$

(d) $(P, Q) = (1, \sin x)$

(e) $(P, Q) = (\sin x, 1)$

(f) $(P, Q) = (x/(x^2 + y^2), y/(x^2 + y^2))$

6. In each part of the previous exercise, find $\int_\gamma P\,dx + Q\,dy$, where γ is the line segment from $(1, 1)$ to $(2, 2)$

7. In each part below, calculate the line integral $\displaystyle\int_\gamma P\,dx + Q\,dy$.

(a) $(P, Q) = (3xy, x - y^2)$; γ is the boundary (oriented counterclockwise) of the rectangle with vertices $(0, 0)$, $(2, 0)$, $(2, 1)$, and $(0, 1)$.

(b) $(P, Q) = (3xy, x - y^2)$; γ is the line segment from $(0, 0)$ to $(2, 1)$.

(c) $(P, Q) = (e^x \sin y, e^x \cos y)$; γ is the unit circle, oriented counterclockwise.

(d) $(P, Q) = (e^x \sin y, e^x \cos y)$; γ is the line from the origin to the point $(1, \pi/2)$.

(e) $(P, Q) = (-y, x)$; γ is the cardioid with polar equation $r = 1 + \cos\theta$, oriented counterclockwise.

8. Is the function $\boldsymbol{f}(x, y) = \left(3x^2 \cos(x^3 + y), \cos(x^3 + y)\right)$ the gradient of a function? If so, of what function? If not, explain how you know that it is not.

5.3 Relating line and area integrals: Green's theorem

The fundamental theorem for line integrals says that if a vector field f happens to be a gradient—i.e., if $f = \nabla h$ for some function h—then

We met it in the preceding section.

$$\int_\gamma f \cdot dX = \int_\gamma \nabla h \cdot dX = h(X_1) - h(X_0), \tag{5.3.1}$$

where X_0 and X_1 are the starting and ending points of γ. Two consequences of this identity are especially important:

Independence of path. Equation 5.3.1 means that the line integral of a gradient field depends only on the endpoints X_0 and X_1—not on the particular curve joining them. If γ' is another curve joining X_0 to X_1, and γ' is also in the domain of h, then

$$\int_\gamma f \cdot dX = \int_{\gamma'} f \cdot dX.$$

Closed curves, zero integrals, and conservation. A curve is called **closed** if it starts and ends at the same point. The fundamental theorem says that integrating a gradient field around any closed curve must give zero:

$$\int_\gamma \nabla h \cdot dX = h(X_0) - h(X_0) = 0.$$

In physical terms, this result means that if a force field f happens to be a gradient, then the force does zero total work in moving a particle around *any* closed curve. Physicists call force fields with this property **conservative**: in such fields, work and energy are "conserved." The fundamental theorem says (among other things) that every gradient field is conservative.

■ **Example 1.** The figure below shows several curves joining $(-1, 1)$ to $(1, 1)$; the vector field $f = (P, Q) = (y^2, 2xy)$ is superimposed.

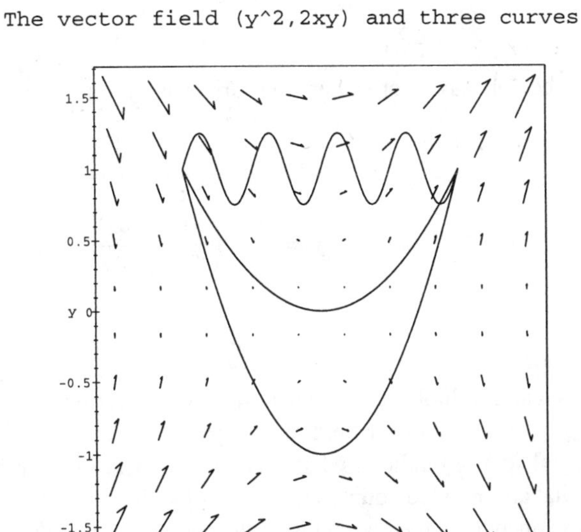

Explain why the line integral has the same value for all three curves. What *is* this value?

Solution: It would be unpleasant to parametrize all three curves. Fortunately, we don't have to: the vector field $(y^2, 2xy)$ is a gradient. We can tell (either by guesswork or by the method outlined in the previous section) that $h(x, y) = xy^2$ is a suitable potential function, i.e., $(y^2, 2xy) = \nabla(xy^2)$. Therefore, by the fundamental theorem, the line integral—for all three curves—is

$$\int_\gamma \nabla h \cdot dX = h(1, 1) - h(-1, 1) = 2.$$

The answer *looks* reasonable, too; the curves are oriented generally with rather than against the flow, so we expect a positive result.

Take a close look at the picture to convince yourself.

Another analogue of the fundamental theorem

Green's theorem is another result in the same spirit as other fundamental theorems we've seen. While the fundamental theorem in Equation 5.3.1 relates a line integral over a curve to values of a potential function at the *ends* of the curve, Green's theorem, by contrast, relates a double integral over a region in the plane to a line integral over the *boundary* of the region.

The endpoints of a curve are, in a sense, the curve's "boundary."

Green's theorem says that, under appropriate assumptions,

We'll state them carefully soon.

$$\oint_\gamma P\,dx + Q\,dy = \iint_R \left(\frac{\partial Q}{\partial x} - \frac{\partial P}{\partial y} \right) dA. \tag{5.3.2}$$

Let's take a careful look at the equation and its various parts. On the left, $(P(x, y), Q(x, y))$ is a vector field on $\mathbb{R}^2$, γ is a **simple closed curve,** and the special integral sign ($\oint$) indicates that γ is the boundary of a region. (We'll always use curves that are oriented counterclockwise.) On the right, the domain of integration, R, is assumed to be the region inside the curve γ. (In mathematical parlance, γ is the **boundary** of the region R.) Notice that the integrand, $Q_x(x, y) - P_y(x, y)$, is a *scalar*-valued function of two variables—the appropriate object to integrate over R.

We'll explain "simple" below.

Geometrically, the relation between γ and R is typically as in the examples below:

Six regions and their boundary curves

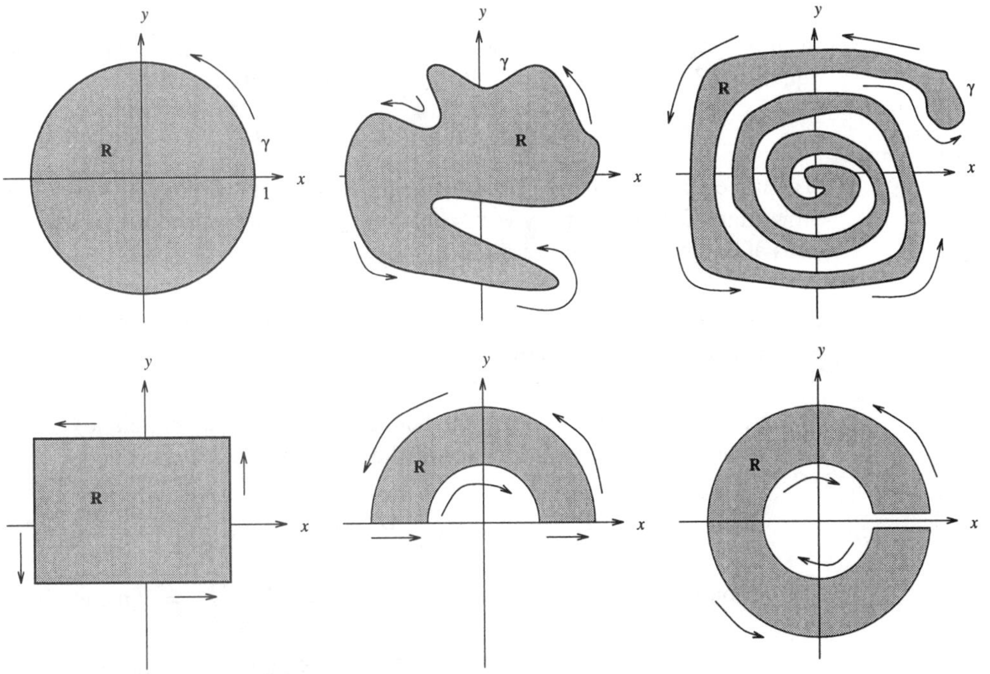

The first region is the **unit disk**; it's defined by the inequality $x^2 + y^2 \leq 1$, or, in polar coordinates, simply $r \leq 1$.

Notice that in each picture, the boundary curve γ is oriented so that the region R, shown shaded, remains on the *left* as one traverses γ. The boundary curves γ, moreover, are all **simple** (they have no self-intersections, as a figure-8 does), and they are either **smooth** or made up of a few smooth pieces. In full technical regalia, curves like those above, which bound reasonable areas, are called **piecewise-smooth Jordan curves**.

Jordan curves. Curves in the plane, even closed curves that don't intersect themselves, can be very complicated. Though it seems reasonable that such a curve should divide the plane into two regions, one inside and one outside, it's surprisingly difficult to prove this rigorously. The first proof was given by the French mathematician Camille Jordan (1838–1922). In his honor, curves that bound area are known as Jordan curves.

Before stating the theorem carefully, let's see what it says in some simple cases.

■ **Example 2.** Suppose the vector field (P, Q) is a gradient ∇h, i.e.,

$$(P(x, y), Q(x, y)) = \nabla h(x, y)$$

for some function h defined on and near a region R with boundary curve γ. Does Equation 5.3.2 hold?

Solution: Yes. By the fundamental theorem of line integrals,

$$\oint_\gamma P\,dx + Q\,dy = \oint_\gamma \nabla h \cdot dX = 0,$$

since γ is a closed curve. Now consider the double integral on the right side of Equation 5.3.2. By assumption, $(P, Q) = (h_x, h_y)$. Thus, for all (x, y),

$$Q_x - P_y = h_{yx} - h_{xy} = 0.$$

(The last equation holds because of the equality of cross partial derivatives.) We've shown, therefore, that for any vector field (P, Q) that happens to be a gradient, both sides of Equation 5.3.2 are zero. □

■ **Example 3.** Let $(P, Q) = (x - y, x + y)$ and let R be the unit disk. What does Equation 5.3.2 say in this case? Is it true?

Solution: The vector field and the boundary curve γ (the unit circle) are shown together in Example 1, page 256, and the line integral is calculated:

$$\oint_\gamma P\,dx + Q\,dy = 2\pi.$$

Check it!

For the double integral in Equation 5.3.2, we have $Q_x - P_y = 2$; therefore,

$$\iint_R (Q_x - P_y)\,dA = \iint_R 2\,dA = 2\,\text{area}\,(R) = 2\pi.$$

Again, Equation 5.3.2 holds true. □

A third example is a little more subtle.

■ **Example 4.** Consider the vector field

$$(P, Q) = \left(\frac{-y}{x^2 + y^2}, \frac{x}{x^2 + y^2}\right)$$

on the unit disk. What does Equation 5.3.2 say now? Is it true?

We leave the important calculation to you; see the exercises.

Solution: To find the line integral, we use the "usual" parametrization $X(t) = (\cos t, \sin t)$, for $0 \le t \le 2\pi$. The result is

$$\oint_\gamma P\,dx + Q\,dy = \int_0^{2\pi} \left(\sin^2 t + \cos^2 t\right)\,dt = 2\pi.$$

The calculation is worth a careful check.

For the double integral, we get

$$Q_x = P_y = \frac{y^2 - x^2}{(x^2 + y^2)^2},$$

so $Q_x - P_y = 0$. Therefore—apparently—$\iint_R (Q_x - P_y)\,dA = 0$. In this case, it seems, Equation 5.3.2 fails. Is something wrong?

Actually, no; there's a simple explanation. The vector field (P, Q) is not defined at the point $(0, 0)$. Neither, therefore, are Q_x and P_y, so there is no guarantee that the integral $\iint_R (Q_x - P_y)\,dA$ even exists. The moral, therefore, is that we need to take care with domains. We do so in the formal statement of the theorem. □

Theorem 3. (Green's theorem) Let R be a region in $\mathbb{R}^2$ whose boundary is a piecewise-smooth Jordan curve γ. Let P and Q be scalar-valued functions, having continuous partial derivatives on and near R. Then

$$\oint_\gamma P\,dx + Q\,dy = \iint_R \left(\frac{\partial Q}{\partial x} - \frac{\partial P}{\partial y} \right) dA.$$

Older student makes good. George Green (1793–1841), a miller from Nottingham, England, spent only two years in elementary school. Despite this possible handicap, he published 10 mathematical papers, including several on potential theory. He entered Cambridge University as an undergraduate at age 40, only 8 years before his death. Green's theorem itself, despite the name, probably predates George Green.

The idea of the proof. A fully rigorous proof of Green's theorem is quite subtle. It would require, among other things, a rigorous definition of "counterclockwise," and a careful treatment of domains that can be quite irregular. However, the basic idea of the proof boils down to several careful applications of the ordinary fundamental theorem of calculus. To see how, suppose first that R happens to be a region of the special sort shown below:

A special region

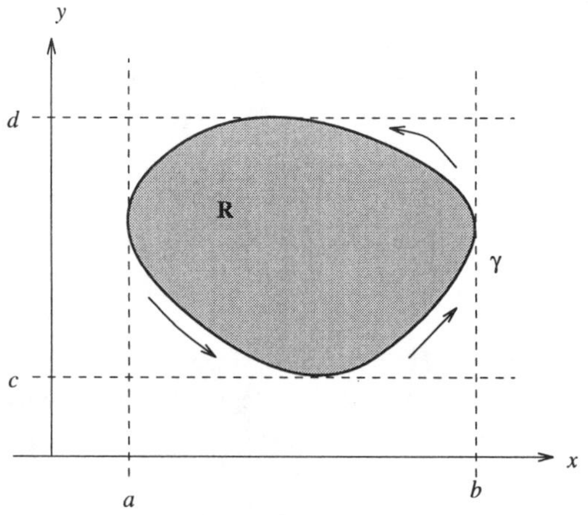

The region is special in that every horizontal line and every vertical line intersects γ at most twice. To prove the theorem, we'll show two identities:

$$-\iint_R P_y\,dA = \oint_\gamma P\,dx \quad \text{and} \quad \iint_R Q_x\,dA = \oint_\gamma Q\,dy. \tag{5.3.3}$$

For the first identity, we'll think of R as shown—

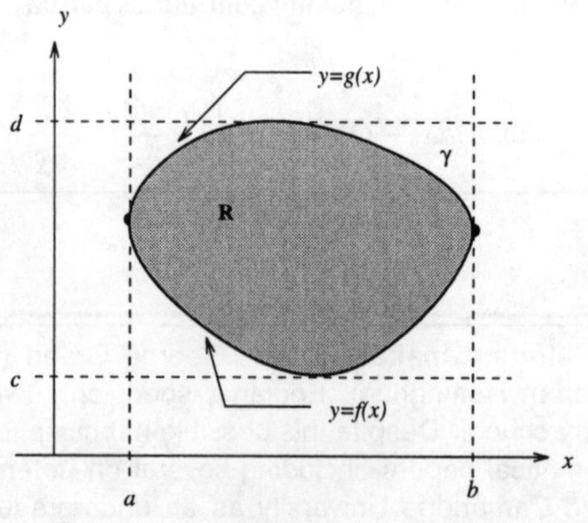

the region is bounded by upper and lower boundary curves $y = g(x)$ and $y = f(x)$ respectively, with $a \le x \le b$. Then the double integral becomes

$$-\iint_R P_y(x, y)\, dA = -\int_{x=a}^{x=b} \left(\int_{y=f(x)}^{y=g(x)} P_y(x, y)\, dy \right) dx$$
$$= -\int_{x=a}^{x=b} \left(P(x, g(x)) - P(x, f(x)) \right) dx.$$

This is the key step.

(We used the ordinary fundamental theorem, in the variable y, in the last step.

Now consider the the line integral $\oint_\gamma P(x, y)\, dx$. Now γ consists of upper and lower boundary curves; we can parametrize both in the same way. For the lower curve, we'll use

$$x = t; \quad y = f(t); \quad a \le t \le b.$$

For the upper curve, we'll do almost the same thing:

$$x = t; \quad y = g(t); \quad a \le t \le b.$$

Notice that in both cases, $dx = dt$. Also, since the upper curve is oriented from right to left, we'll attach a minus sign to the integral. Using these parametrizations, we get

$$\oint_\gamma P(x, y)\, dx = \int_{\gamma_{\text{bottom}}} - \int_{\gamma_{\text{top}}} = \int_a^b P(t, f(t))\, dt - \int_a^b P(t, g(t))\, dt.$$

See the exercises.

The last expression looks familiar: Except for the difference in variable names, this expression is exactly what we obtained for $\iint P_y\, dA$. Thus the first identity in Equation 5.3.3 holds as claimed. The second identity is proved in a similar way.

These arguments show that Green's theorem holds for regions of the special type shown above. To see that it holds on more general regions, the trick is to break R

into several smaller regions, each of the special type just mentioned, as shown:

Breaking up a region

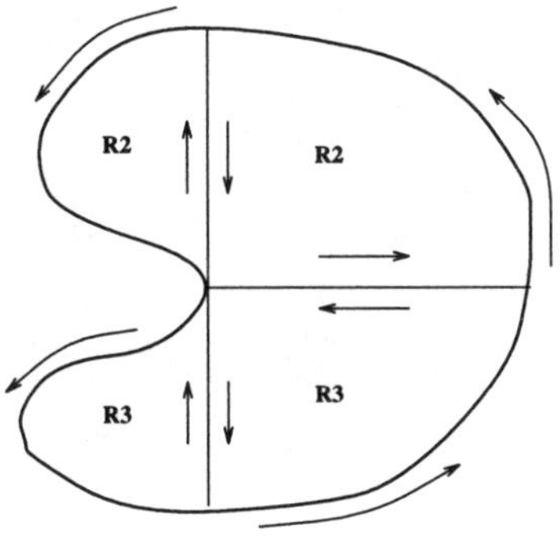

Notice that each "inner" boundary piece is traversed twice, once in each direction. We've shown that the theorem holds on each subregion R_i, with boundary γ_i, i.e.,

$$\iint_{R_i} (Q_x - P_y)\, dA = \oint_{\gamma_i} P\, dx + Q\, dy.$$

Adding these results together for $i = 1 \ldots 4$ gives $\iint_R (Q_x - P_y)\, dA$ on the left, and $\oint_\gamma P\, dx + Q\, dy$ on the right (since the line integrals on inside boundary edges cancel each other out). This completes the proof of Green's theorem. □

The integral sometimes saves trouble, by trading a messy line integral for a simpler area integral.➽

Sometimes the trade goes the other way.

■ **Example 5.** Find the work done by the force field $(P, Q) = (x - y, x + y)$ in moving an object around the square S with corners at $(-1, -1)$, $(1, -1)$, $(1, 1)$, and $(-1, 1)$.

Solution: Integrating around the square would take four separate parametrizations. Using Green's theorem to reduce to a double integral makes things much simpler. Since $Q_x - P_y = 2$, Green's theorem says

$$\oint_\gamma P\, dx + Q\, dy = \iint_S 2\, dA = 2 \operatorname{area}(S) = 8.$$

□

Green's theorem on a region with holes

Green's theorem, as stated above, applies to regions with only one boundary curve. However, a simple trick shows that it applies, in slightly different form, to regions like that shown below:

Cutting a donut

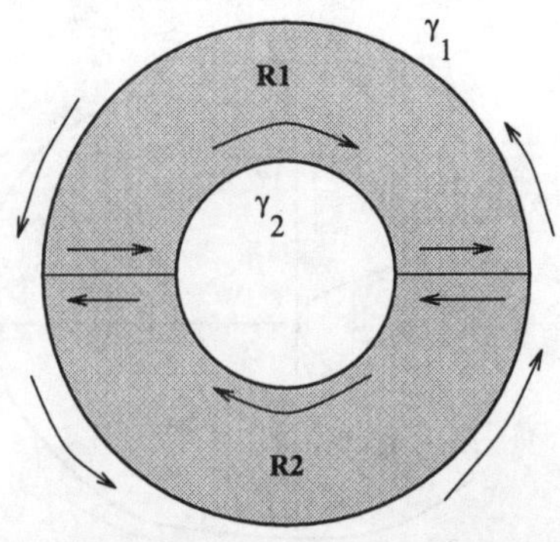

If R is the entire donut-shaped region, then we can imagine R as the union of the two simpler regions R_1 (the upper half) and R_2 (the lower half); we'll call their boundary curves C_1 and C_2. (In the picture, γ_1 and γ_2 are the outer and inner circles, respectively. Each of C_1 and C_2 consists of two semicircles and two line segments.) Green's theorem *does* apply to R_1 and R_2, to give

$$\iint_{R_1} (Q_x - P_y)\,dA = \oint_{C_1} P\,dx + Q\,dy$$

and

$$\iint_{R_2} (Q_x - P_y)\,dA = \oint_{C_2} P\,dx + Q\,dy.$$

Adding these equations together (and keeping track of cancellations along the line segments, which are traversed twice in opposite directions) gives

$$\iint_{R} (Q_x - P_y)\,dA = \oint_{\gamma_1} P\,dx + Q\,dy - \oint_{\gamma_2} P\,dx + Q\,dy,$$

The minus sign is there because the inner circle is traversed clockwise.

where, now, both line integrals are taken in the counterclockwise sense.

Here is the general principle:

> **Fact: (Green's theorem for a region with holes)** Let the situation be as in Green's theorem, but assume that R has outer boundary curve γ_1 and inner boundary curve γ_2. Then
>
> $$\iint_{R} (Q_x - P_y)\,dA = \oint_{\gamma_1} P\,dx + Q\,dy - \oint_{\gamma_2} P\,dx + Q\,dy.$$

■ **Example 6.** Let

$$(P, Q) = \left(-\frac{y}{x^2+y^2}, \frac{x}{x^2+y^2}\right),$$

and let S be a square of side 1, centered at the origin, oriented counterclockwise. Find $\oint_S P\,dx + Q\,dy$.

Solution: The line integral would be difficult to work by parametrization, but with Green's theorem (as extended) the calculation becomes simple. We'll let R be the region between S and the unit circle γ. We showed in Example 4, page 268, that $Q_x - P_y = 0$ for all (x, y) in R. Therefore, by the preceding Fact,

$$\begin{aligned} 0 &= \iint_R (Q_x - P_y)\, dA = \oint_\gamma P\,dx + Q\,dy - \oint_S P\,dx + Q\,dy \\ &\Longrightarrow \oint_\gamma P\,dx + Q\,dy = \oint_S P\,dx + Q\,dy. \end{aligned}$$

But we also showed in Example 4 that $\oint_\gamma P\,dx + Q\,dy = 2\pi$; this, too, is the value of the desired line integral. □

Exercises

1. In each part following, use Green's theorem to evaluate the line integral $\oint_\gamma P\,dx + Q\,dy$. (Trade each line integral for a double integral.) All curves are traversed counterclockwise.

   (a) $(P, Q) = (y^2 + x, x + y)$; γ is the square with vertices at $(0, 0)$, $(1, 0)$, $(1, 1)$, and $(0, 1)$.

   (b) $(P, Q) = (x - y, x + y)$; γ is the square with vertices at $(0, 0)$, $(1, 0)$, $(1, 1)$, and $(0, 1)$.

   (c) $(P, Q) = (y^2 + x, x + y)$; γ is the circle of radius 1, with center $(0, 0)$.

   (d) $(P, Q) = (x - y, x + y)$; γ is the circle of radius 1, with center $(1, 0)$.

   (e) $(P, Q) = (y^2 + x, x + y)$; γ is the boundary of the region $0 \le r \le 1$; $0 \le \theta \le \pi/2$.

   (f) $(P, Q) = (x - y, x + y)$; γ is the boundary of the region $0 \le r \le 1$; $0 \le \theta \le \pi/2$.

2. In this section we proved the first identity in Equation 5.3.3, page 269, assuming that the region had a certain special shape. Mimic the proof given there to prove the remaining identity: $\iint_R Q_x\,dA = \oint_\gamma Q\,dy$. (Hint: Because of its special shape, R can be assumed to have *left* boundary $x = h(y)$ and *right* boundary $x = k(y)$.)

3. Consider the curves and the vector field shown in Example 1 (page 265). The vector field is $(P, Q) = (y^2, 2xy)$.

   (a) The line integral $\int_\gamma P\,dx + Q\,dy$ has the same value for *any* curve starting at $(0, 0)$ and ending at $(1, 1)$. Explain why.

   (b) Find the common value mentioned in the previous part.

   (c) Let (a, b) be any point in the xy-plane, and let γ be any curve from $(0, 0)$ to (a, b). Find $\int_\gamma P\,dx + Q\,dy$.

4. Repeat the previous exercise, but use the vector field

$$(P, Q) = \left(\frac{2x}{1 + x^2 + y^2}, \frac{2y}{1 + x^2 + y^2} \right).$$

   (Hint: Look for a potential function.)

5. This exercise and the next one explore the fact that the force of gravity is conservative (in the sense discussed in this section).

   Near the surface of the earth, we can think of the force of gravity on an object as essentially constant, and pointing straight down. We can model this situation in the xy-plane by letting $(P, Q) = (0, -k)$, where k is a positive constant. Show that (P, Q) is a conservative force field by finding a potential function. What does the answer mean physically?

6. For motion on a large scale (e.g., spacecraft flight), gravity is better modeled as a force whose direction is always toward the center of the earth and whose magnitude is inversely proportional to the square of the distance to the center of the earth. This can be modeled in the xy-plane by a force of the form

$$(P, Q) = k\left(\frac{-x}{(x^2+y^2)^{3/2}}, \frac{-y}{(x^2+y^2)^{3/2}}\right),$$

   where k is any positive constant.

   (a) Show that (P, Q) is conservative by finding a potential function.

   (b) Find the work done by this force in moving an object from the point $(1, 1)$ to $(3, 4)$. Is the answer positive or negative?

7. Prove Green's theorem under the special assumption that R is the rectangle $[a, b] \times [c, d]$. Use these steps:

   (a) By parametrizing all four sides of the rectangle, show that

$$\oint_\gamma P\,dx+Q\,dy = \int_a^b P(t, c)dt - \int_a^b P(t, d)dt + \int_c^d Q(b, t)dt - \int_c^d Q(a, t$$

   (b) Explain why $\iint_R (Q_x - P_y)\,dA = \iint_R Q_x\,dA - \iint_R P_y\,dA$.

   (c) Show, explaining all steps, that

$$\begin{aligned}\iint_R Q_x\,dA &= \int_c^d \int_a^b Q_x(x, y)\,dx\,dy\\ &= \int_c^d (Q(b, y) - Q(a, y))\,dy.\end{aligned}$$

   (d) Show, explaining all steps, that

$$-\iint_R P_y\,dA = \int_a^b (P(x, c) - P(x, d))\,dx.$$

8. Consider the vector field

$$(P, Q) = \left(-\frac{y}{x^2+y^2}, \frac{x}{x^2+y^2}\right);$$

   let γ be the unit circle, oriented counterclockwise.

   (a) Parametrize γ to find $\int_\gamma P\,dx + Q\,dy$.

   (b) Is (P, Q) a gradient field? Why or why not?

(c) Find $\int_S P\,dx + Q\,dy$, where S is any square that contains the unit circle. (Hint: Mimic the last example of this section.)

(d) Find $\int_C P\,dx + Q\,dy$, where C is any circle that does *not* contain the origin. (Hint: Does the ordinary version of Green's theorem apply now?)

9. Consider the vector field

$$(P, Q) = \left(\frac{x}{x^2 + y^2}, \frac{y}{x^2 + y^2} \right);$$

let γ be the unit circle, oriented counterclockwise.

(a) Parametrize γ to find $\int_\gamma P\,dx + Q\,dy$.

(b) Carefully check that $Q_x(x, y) = P_y(x, y)$ for all $(x, y) \neq (0, 0)$.

(c) Does Green's theorem apply to (P, Q) on the unit disk? If so, what does it say? If not, why not?

10. Let $(x_1, y_1), (x_2, y_2), \ldots, (x_n, y_n), (x_{n+1}, y_{n+1}) = (x_1, y_1)$ be points on a counterclockwise path around the boundary of a polygonal region in the xy-plane (i.e., the boundary of the region is made up of line segments connecting adjacent points on the list). Show that the area of the region is

$$\text{area} = \frac{1}{2} \sum_{k=1}^{n} \big((x_{k+1} + x_k)(y_{k+1} - y_k) \big).$$

11. Let f be the field $f(x, y) = (2xy + x, xy - y)$ and γ be the perimeter of the square bounded by the lines $x = 0$, $x = 1$, $y = 0$, and $y = 1$. Use Green's Theorem to evaluate

(a) the counterclockwise circulation of f around γ;

(b) the outward flux of f through γ.

12. Use Green's Theorem to evaluate the line integral $\displaystyle\oint_\gamma y^2\,dx + x^2\,dy$ where γ is the circle $x^2 + y^2 = 4$.

$$\frac{-\sin t - b}{\cos^2 t + 2a\cos t + a^2 + \sin^2 t + 2b\sin t + b^2} = \frac{-\sin t - b}{2a\cos t + a^2 + 2b\sin t + b^2}, \frac{\cos t + a}{2a\cos t + a^2 + 2b\sin t + b^2} \left(-\sin t, \cos t\right)$$

$$\frac{+\sin^2 t + b\sin t + \cos^2 t + a\cos t}{2a\cos t + a^2 + 2b\sin t + b^2} = \frac{b\sin t + a\cos t}{2a\cos t + a^2 + 2b\sin t + b^2}$$

5.4 Surfaces and their parametrizations

Calculating line integrals has offered us plenty of practice with parametrizing curves in the xy-plane. Parametrizations were key, in particular, to proving two fundamental theorems for line integrals.

The next few sections introduce *surface integrals*, in which the domain of integration is a two-dimensional surface in three-dimensional space, such as a plane, the surface of a sphere or ellipsoid, or the graph of a function $z = f(x, y)$. Surface integrals are higher-dimensional versions of line integrals; we'll stress this connection whenever possible. In particular, surface integrals satisfy their own versions of the fundamental theorems we've studied for line integrals. Our goal in the rest of this chapter is to state and understand two such fundamental theorems, which relate various types of integrals and derivatives.

Essential to calculating such integrals, as one might expect, is the ability to parametrize surfaces in space. This brief section illustrates some useful ideas and techniques.

Curves, surfaces, and dimensions

The general setup for parametrizing a curve looks something like this:

Parametrizing a curve in the plane

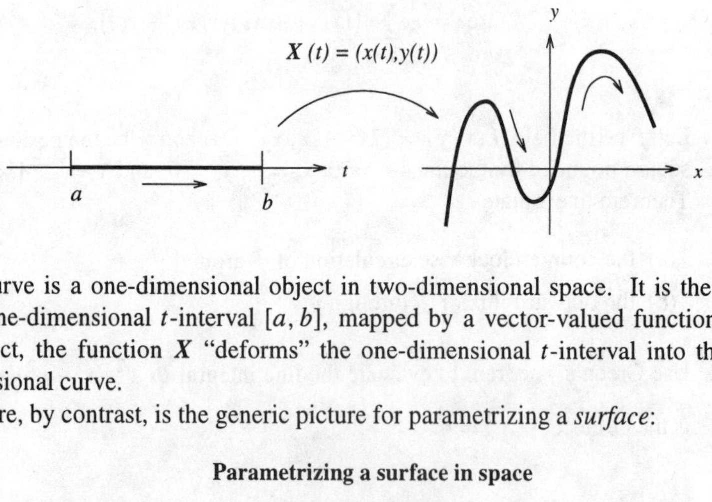

The curve is a one-dimensional object in two-dimensional space. It is the image of a one-dimensional t-interval $[a, b]$, mapped by a vector-valued function $\boldsymbol{X}(t)$. In effect, the function $\boldsymbol{X}$ "deforms" the one-dimensional t-interval into the one-dimensional curve.

Here, by contrast, is the generic picture for parametrizing a *surface*:

Parametrizing a surface in space

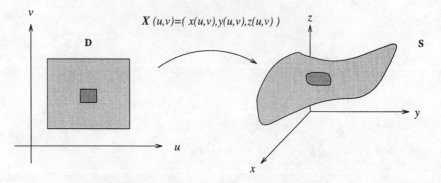

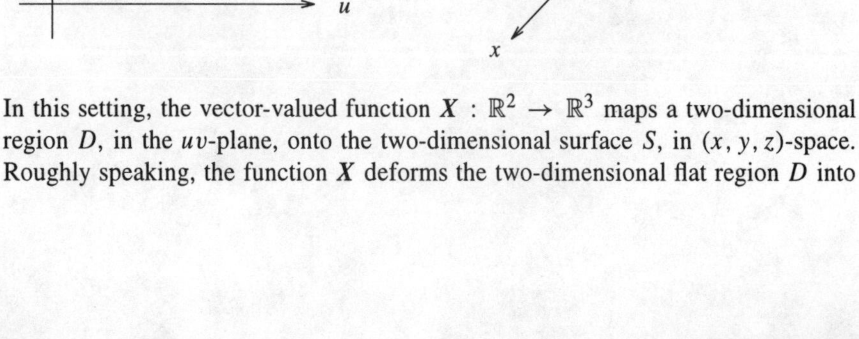

In this setting, the vector-valued function $\boldsymbol{X} : \mathbb{R}^2 \to \mathbb{R}^3$ maps a two-dimensional region D, in the uv-plane, onto the two-dimensional surface S, in (x, y, z)-space. Roughly speaking, the function $\boldsymbol{X}$ deforms the two-dimensional flat region D into the two-dimensional surface S. (The small darker rectangle in the domain is mapped to a "near-rectangle" in the surface.)

It *can* happen, by the way, that a function $\boldsymbol{X} : \mathbb{R}^2 \to \mathbb{R}^3$ maps a two-dimensional set in the domain into a smaller-dimensional set in xyz-space. For example, the constant function defined by $\boldsymbol{X}(u, v) = (1, 2, 3)$ maps all of $\mathbb{R}^2$ into a single point. All of the functions we'll use to parametrize surfaces, however, will "respect" dimensions, mapping two-dimensional sets to two-dimensional surfaces.

Surface parametrizations: a sampler

Any given surface in space can be parametrized in many various ways (as can any curve in the plane), using various domain sets. Some examples will give a sense of the possibilities.

Function graphs. The easiest surfaces to parametrize are graphs of functions $z = f(x, y)$—or, more often, parts of such graphs. Parametrizing a particular part of a graph may require fiddling with domains.

■ **Example 1.** Parametrize S_1, the part of the graph of $z = x^2 + y^2$ that lies above the square $[0, 2] \times [0, 2]$ in the xy-plane.

Solution: Points on the surface satisfy the equation $z = x^2 + y^2$, so we can set

$$\boldsymbol{X}(u, v) = (u, v, u^2 + v^2); \quad 0 \le u \le 2; \quad 0 \le v \le 2.$$

Plotting this surface with technology➸ is an excellent way to see that our parametrization is correct:

Maple, in this case.

Part of the surface z=x^2+y^2

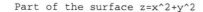

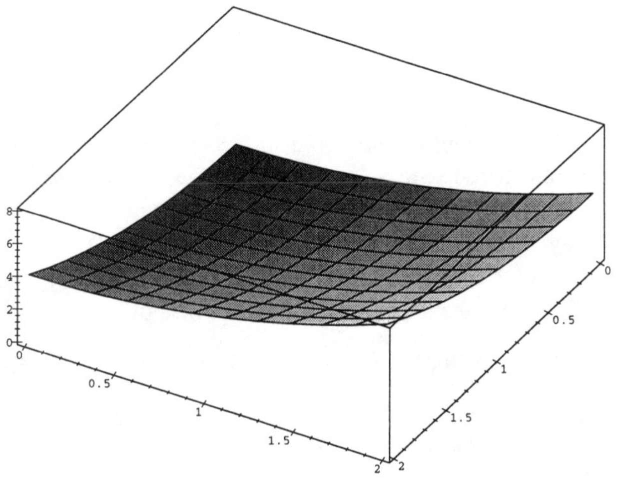

□

The surface shown is only a small part of the full graph $z = x^2 + y^2$. Let's look at another surface carved from the same graph.

■ **Example 2.** Parametrize S_2, the part of the graph of $z = x^2 + y^2$ that lies above the unit disk in the xy-plane.

Solution: Exactly the same parametrization works as in the preceding example, except that the domain D is now the unit disk.

If we prefer to parametrize S_2 using a rectangle as the domain set D, we can work in polar coordinates. With respect to r and θ, the unit disk is the rectangle described by $0 \le r \le 1$ and $0 \le \theta \le 2\pi$. In these coordinates, moreover, $x = r\cos\theta$ and $y = r\sin\theta$, and $x^2 + y^2 = r^2$. For consistency of notation, we'll write $u = r$ and $v = \theta$. Now our desired parametrization becomes

$$X(u, v) = (u\cos v, u\sin v, u^2); \quad 0 \le u \le 1; \quad 0 \le v \le 2\pi;$$

the domain set is a rectangle. Plotting the data gives what we expect:

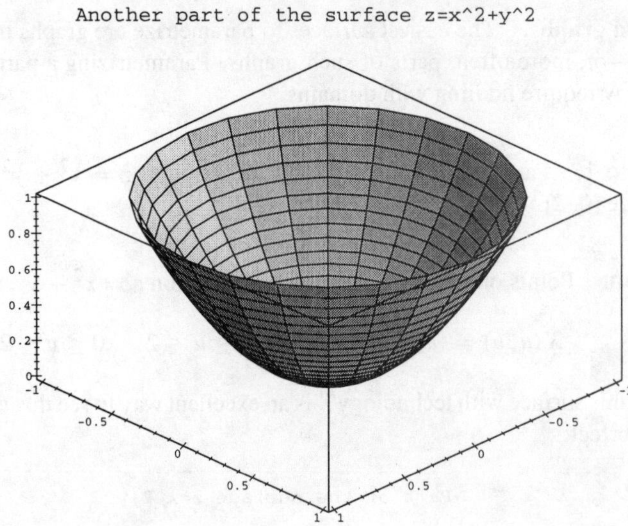

□

Planes and area magnification. If a surface S is a plane, or part of a plane, then S can be parametrized with a *linear* parametrization function. The next example illustrates. Most important, the example shows how areas in the uv-domain and on the surface are related.

■ **Example 3.** Describe the surface S parametrized by

$$X(u, v) = (x_0 + Au + Bv, y_0 + Cu + Dv, z_0 + Eu + Fv); \quad 0 \le u \le \Delta u; \quad 0 \le v \le \Delta v.$$

(All letters except u and v denote constants.)

The uv-domain D has area $\Delta u\, \Delta v$. What is the area of the corresponding surface S?

Solution: The parametrization function can be rewritten in the form

$$X(u, v) = (x_0, y_0, z_0) + u(A, C, E) + v(B, D, F).$$

Therefore, the surface S consists of all points of this form, where u and v range from 0 to Δu and Δv, respectively. From this it follows:

The surface S is the parallelogram spanned by the vectors $\Delta u\,(A, C, E)$ and $\Delta v\,(B, D, F)$; one corner is at (x_0, y_0, z_0).

For brevity, we'll write

$$\boldsymbol{X}_u = (A, C, E) \quad \text{and} \quad \boldsymbol{X}_v = (B, D, F).$$

Now recall that the area of the parallelogram spanned by any two vectors $\boldsymbol{X}_u$ and $\boldsymbol{X}_v$ is the absolute value of their cross product. (We discussed this property of the cross product in Section 1.8.) In the present case, therefore, the area of S is

$$\text{area } S = |(\Delta u\, \boldsymbol{X}_u) \times (\Delta v\, \boldsymbol{X}_v)| = \Delta u\, \Delta v\, |\boldsymbol{X}_u \times \boldsymbol{X}_v| . \qquad \square$$

Area magnification for any parametrization. The notation in the preceding example is no accident: The vector $\boldsymbol{X}_u$ is the partial derivative with respect to u of the parametrization function:

$$\boldsymbol{X}_u = \frac{\partial}{\partial u}(x_0 + Au + Bv, y_0 + Cu + Dv, z_0 + Eu + Fv) = (A, C, E).$$

Similarly,

$$\boldsymbol{X}_v = \frac{\partial}{\partial v}(x_0 + Au + Bv, y_0 + Cu + Dv, z_0 + Eu + Fv) = (B, D, E).$$

The same notation is convenient for *any* parametrization function $\boldsymbol{X}(u, v) = (x(u, v), y(u, v), z(u, v))$. Thus, we'll write

$$\boldsymbol{X}_u = \left(\frac{\partial}{\partial u}x(u, v), \frac{\partial}{\partial u}y(u, v), \frac{\partial}{\partial u}z(u, v)\right),$$

and

$$\boldsymbol{X}_v = \left(\frac{\partial}{\partial v}x(u, v), \frac{\partial}{\partial v}y(u, v), \frac{\partial}{\partial v}z(u, v)\right).$$

We showed earlier that for a linear parametrization function, the cross product $|\boldsymbol{X}_u \times \boldsymbol{X}_v|$ describes the factor by which $\boldsymbol{X}$ increases areas.

A similar result holds for *any* differentiable parametrization function $\boldsymbol{X}$. Near a specific domain point (u_0, v_0), each coordinate function of $\boldsymbol{X}$ is closely approximated by its linear approximation function, which has the same derivatives $\boldsymbol{X}_u(u_0, v_0)$ and $\boldsymbol{X}_v(u_0, v_0)$ as does $\boldsymbol{X}$. An important result follows:

Near any point (u_0, v_0) in the domain, the parametrization $\boldsymbol{X}$ magnifies areas by a factor of approximately $|\boldsymbol{X}_u(u_0, v_0) \times \boldsymbol{X}_v(u_0, v_0)|$.

We'll pursue this result further in the next section.

Exercises

1. In each part below, find a parametrization of the given surface S on a rectangular domain D in uv-space. If at all possible, use technology to plot the surface over the given domain (and thus see whether the result is correct).

(a) S is the part of the cone $z = \sqrt{x^2 + y^2}$ that lies above the square $[-1, 1] \times [-1, 1]$ in the xy-plane.

(b) S is the part of the cone $z = \sqrt{x^2 + y^2}$ that lies above the unit disk in the xy-plane.

(c) S is the part of the plane $z = 2x + 3y + 4$ that lies above the square $[0, 1] \times [0, 1]$ in the xy-plane.

(d) S is the part of the plane $z = 2x + 3y + 4$ that lies above the unit disk in the xy-plane.

2. In each part of the preceding exercise, calculate the magnification factor $|\boldsymbol{X}_u(u_0, v_0) \times \boldsymbol{X}_v(u_0, v_0)|$ for $(u_0, v_0) = (1, 0)$.

3. A surface S has the parametrization $\boldsymbol{X}(u, v) = (\sin v \cos u, \sin v \sin u, \cos v)$, with $0 \le u \le 2\pi$ and $0 \le v \le \pi$.

   (a) Use technology to plot the surface S. What is the surface?

   (b) Explain the link to spherical coordinates.

4. Let S be the *upper half* of the unit sphere $x^2 + y^2 + z^2 = 1$.

   (a) Parametrize S; let D be the unit disk $u^2 + v^2 \le 1$ in the uv-plane.

   (b) Parametrize S; let D be a rectangle in the uv-plane. (Hint: See the previous exercise.)

5.5 Surface integrals

Surface integrals differ from line integrals in that, for the former, the domain of integration is a surface in space, not a curve in the plane. Line and surface integrals are similar, on the other hand, in that calculating both types of integrals begins with parametrizing the curve or surface in a convenient way. Once this is done, line and surface integrals reduce (albeit in somewhat different ways, as we'll see) to "ordinary" integrals in one or two variables.

Line and surface integrals are also similar in that both help answer natural physical questions about vector phenomena. If $\boldsymbol{f}$ is a vector field in the plane, representing the velocity of a flow near a closed oriented curve γ, then we've seen that the line integral $\int_\gamma \boldsymbol{f} \cdot d\boldsymbol{X}$ measures the circulation of the flow, i.e., the tendency of the fluid to flow *around* γ with (or against) the direction of orientation. In a similar spirit, we'll see that if a vector field $\boldsymbol{f}$ represents a three-dimensional flow near a surface S, then we can use a special type of surface integral, called a **flux integral**, to measure the flow through the surface S. We'll see flux integrals in the next section.

We think of the surface as permeable, like a fish net, so that fluid flows freely through it.

Defining the surface integral

Let S be a surface in $\mathbb{R}^3$ and let $f(x, y, z)$ be a scalar-valued function defined on S. How might we sensibly define

$$\iint_S f\,dS,$$

the surface integral of f on S? A "good" definition should, among other things, allow us to calculate the surface area of a surface as the integral $\iint_S 1\,dS$. (The mnemonic symbol dS is analogous to dA for area integrals and dV for volume integrals.)

Parametrization is the key. Suppose that S is parametrized by a well-behaved function

$$\boldsymbol{X}(u, v) = (\,x(u, v), y(u, v), z(u, v)\,),$$

defined on a convenient domain D (a rectangle, say, or a disk) in the uv-plane. Composing f with $\boldsymbol{X}$ gives

$$f(\boldsymbol{X}(u, v)) = f(\,x(u, v), y(u, v), z(u, v)\,),$$

a function of u and v defined on D. It's tempting, perhaps, simply to integrate this function over D and call the result the surface integral. But that's a little too naive. After all, it's possible to parametrize a surface in various ways, using uv-domains of various sizes and shapes. A "good" definition of surface integral, therefore, must somehow take account of how $\boldsymbol{X}$ maps D onto S.

Area magnification—the factor by which the parametrization $\boldsymbol{X}$ shrinks or stretches areas in mapping D onto S—turns out to be the important idea.

■ **Example 1.** In the following picture, S is the part of the surface $z = 5 - x^2 - y^2$ that lies above the rectangle $[-1.5, 1.5]\times[-1.5, 1.5]$ in the xy-plane. (This rectangle is shown, too, as are some vectors we'll discuss in a moment.)

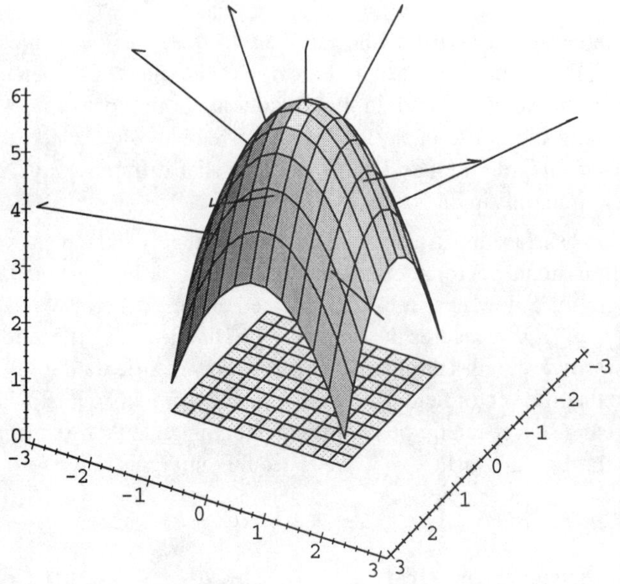

The surface is parametrized by

$$X(u, v) = (x, y, z) = (u, v, 5 - u^2 - v^2),$$

for (u, v) in the rectangle $[-1.5, 1.5] \times [-1.5, 1.5]$ in the uv-plane. (In this case, we can consider the uv-plane and the xy-plane as the same, but that's only because the present surface happens to be a graph.) Discuss how X magnifies areas.

Solution: The picture shows both the domain and the range of the parametrization X. The domain D is the flat rectangle in the xy-plane (or the uv-plane, as we're thinking of it); the range S is the curved surface.

Notice especially the grids on both D and S. The rectangular grid on D is mapped by X to the grid of curves on S—each grid line in D is "lifted" straight up to a corresponding curve on S. In the same way, X lifts each small grid rectangle in D to a slightly curved, parallelogram-shaped grid element in S.

Now compare the relative areas of grid elements in D and S. As the picture shows, the degree of area magnification varies from place to place. The magnification is least at the vertex of S, where the surface is essentially horizontal; it's greatest at the "corners" of S, where the surface is steepest. Everywhere on this surface, however, the magnification factor appears to be greater than one. □

Normal vectors. How can we calculate the magnification factor of a parametrizing function at various points (u, v) in the domain? We found the answer symbolically in the preceding section. We observed that for any domain point (u, v), the 3-vectors

$$X_u = \frac{d}{du}X(u, v) \quad \text{and} \quad X_v = \frac{d}{dv}X(u, v)$$

are both tangent to S at $X(u, v)$, so their cross product $X_u \times X_v$ is *perpendicular* to S at the same point. The magnitude $|X_u \times X_v|$, moreover, gives the area magnification factor at (u, v).

Let's see how this works for the surface shown earlier. First,

$$X(u, v) = (u, v, 5 - u^2 - v^2) \implies X_u = (1, 0, -2u) \quad \text{and} \quad X_v = (0, 1, -2v).$$

Next, routine calculations show that

$$X_u \times X_v = (2u, 2v, 1) \quad \text{and} \quad |X_u \times X_v| = \sqrt{4u^2 + 4v^2 + 1}.$$

The last result agrees with what we observed earlier: The magnification factor is never less than one, and is least at $(u, v) = (0, 0)$, i.e., at the vertex of the paraboloid. As u and v increase, so does the magnification factor. At $(u, v) = (1, 1)$, for instance, the magnification factor is 3—a plausible result, judging from the picture.➥

Find the point on the surface corresponding to $(u, v) = (1, 1)$. Does it seem plausible that the area magnification is 3 there?

The calculation also explains the vectors that appear in the picture. We calculated the perpendicular vector $X_u \times X_v = (2u, 2v, 1)$ at each (u, v) with integer coordinates; the vectors are shown based at $X(u, v)$. As expected, each vector looks perpendicular to the surface at the given point.

The definition There's still more to be seen in the picture in Example 1. Given a grid of subdivisions in the uv-domain D,➥ the mapping X produces a corresponding grid of subdivisions on S itself.➥

The square grid at the bottom.

The curved grid on the surface.

Let's use such a grid on S to define the surface integral. Given a function $f(x, y, z)$ defined on S, the surface integral is approximated by a sum of the form

$$\sum_{i=1}^{n} f(x_i, y_i, z_i) \cdot \text{area}(S_i),$$

where (x_i, y_i, z_i) is a point in the ith subdivision S_i. But thanks to the parametrization, $(x_i, y_i, z_i) = X(u_i, v_i)$, for some point (u_i, v_i), and

$$\text{area}(S_i) \approx |X_u(u_i, v_i) \times X_v(u_i, v_i)| \cdot \text{area}(D_i),$$

where D_i is the subdivision of D that corresponds to S_i. Therefore,

$$\sum_{i=1}^{n} f(x_i, y_i, z_i) \cdot \text{area}(S_i) \approx \sum_{i=1}^{n} f(X(u_i, v_i))\, |X_u(u_i, v_i) \times X_v(u_i, v_i)| \cdot \text{area}(D_i).$$

Now the right side is an approximating sum for the integral

$$\iint_D f(X(u, v))\, |X_u \times X_v|\, du\, dv.$$

As the grid becomes finer and finer, the approximating sums tend to their respective integrals.

This analysis motivates the following definition:

> **Definition: Surface integral of a function.** Let S be a surface in xyz-space, parametrized by a function $X(u, v)$ defined on a domain D in the uv-plane. Let $f(x, y, z)$ be a function defined on S. The surface integral of f on S is defined by
>
> $$\iint_S f\, dS = \iint_D f(X(u, v))\, |X_u \times X_v|\, du\, dv,$$
>
> if the integral exists.

An important special case occurs when the integrand is the constant function $f(x, y, z) = 1$:

> **Definition: Surface area.** Let S, D, and X be as above. The **surface area** of S is defined as
>
> $$\iint_S 1\, dS = \iint_D |X_u \times X_v|\, du\, dv,$$

Surface area of a graph. If a surface has the form $z = f(x, y)$, for (x, y) in a domain D, then the preceding definition has an especially simple form. Here we can use the parametrization function

$$\mathbf{X}(u, v) = (u, v, f(u, v)),$$

Check the calculation.

so $\mathbf{X}_u = (1, 0, f_u)$ and $\mathbf{X}_v = (0, 1, f_v)$. Taking the cross product gives

$$\mathbf{X}_u \times \mathbf{X}_v = (-f_u, -f_v, 1) \quad \text{and} \quad |\mathbf{X}_u \times \mathbf{X}_v| = \sqrt{1 + f_u^2(u, v) + f_v^2(u, v)}.$$

Therefore the area of S is given by the integral

$$\text{area} = \iint_D \sqrt{1 + f_u^2(u, v) + f_v^2(u, v)}\, du\, dv.$$

Observe the resemblance to the arclength formula for a one-dimensional curve $y = f(x)$, from $x = a$ to $x = b$:

$$\text{length} = \int_a^b \sqrt{1 + f'(x)^2}\, dx,$$

Notice, finally, that in the simplest case of all, in which $f(x, y)$ is constant, the surface S is parallel to D, and the surface area formula reduces simply to

$$\text{area}(S) = \iint_D \sqrt{1 + f_u^2(u, v) + f_v^2(u, v)}\, du\, dv = \iint_D 1\, du\, dv = \text{area}(D).$$

■ **Example 2.** Find the area of the part of the paraboloid $z = x^2 + y^2$ that lies above the unit disk $x^2 + y^2 \le 1$.

Solution: By the formula just found, the area is $\iint_D \sqrt{1 + 4x^2 + 4x^2}\, dx\, dy$, where D is the unit disk. This integral is best handled in polar coordinates. With

We'll use x and y, not u and v, because still another set of coordinates is coming.

$$x = r\cos\theta, \quad y = r\sin\theta, \quad \text{and} \quad dx\, dy = r\, dr\, d\theta,$$

we get

$$\text{area} = \int_0^{2\pi} \int_0^1 \sqrt{1 + 4r^2} r\, dr\, d\theta = \frac{\pi\left(5\sqrt{5} - 1\right)}{6} \approx 5.33.$$

□

Surface area for non-graphs. The same formula works—sometimes with a bit more mess—for surfaces that are given parametrically rather than as graphs of functions.

■ **Example 3.** The sphere of radius a is not the graph of a function $z = f(x, y)$. Find its area anyway, using the spherical coordinate parametrization

$$\mathbf{X}(u, v) = (a \sin u \cos v, a \sin u \sin v, a \cos u),$$

for $0 \le u \le \pi$ and $0 \le v \le 2\pi$.

Solution: In this case we get

$$X_u = a(\cos u \cos v, \cos u \sin v, -\sin u); \quad X_v = a(-\sin u \sin v, \sin u \cos v, 0).$$

Straightforward but slightly messy calculations now show that

$$\boldsymbol{X_u} \times \boldsymbol{X_v} = a^2(\sin^2 u \cos v, \sin^2 u \sin v, \sin u \cos u)$$

and that

$$|\boldsymbol{X_u} \times \boldsymbol{X_v}| = a^2 \sin u.$$

Now the area formula gives

$$\text{area} = a^2 \int_{v=0}^{v=2\pi} \int_{u=0}^{u=\pi} \sin u \, du \, dv = 4\pi a^2,$$

as the classical formula says. □

Non-constant integrands: mass and center of mass. In surface area integrals the integrand function is constant. This is not always the case, of course. For example, a surface S might have variable density (i.e., mass per unit area) $\rho(x, y, z)$ at different points (x, y, z). In this case, the **mass** of the surface is given by

$$\iint_S \rho(x, y, z) \, dS.$$

The surface's **center of mass** is the point $(\bar{x}, \bar{y}, \bar{z})$, with coordinates given by

$$\bar{x} = \frac{\iint_S x \, \rho(x, y, z) \, dS}{\text{mass}}; \quad \bar{y} = \frac{\iint_S y \, \rho(x, y, z) \, dS}{\text{mass}}; \quad \bar{z} = \frac{\iint_S z \, \rho(x, y, z) \, dS}{\text{mass}}.$$

(Each coordinate of the center of mass is the "weighted average" of that coordinate over the surface.)

■ **Example 4.** Suppose that the paraboloid of Example 2 has constant density $\rho(x, y, z) = 1$. Find the mass and the center of mass.

We're being informal about units; density would be measured in units of mass per unit of area.

Solution: Because the density is one, the mass is found from the same integral as in Example 2; we got

$$\frac{\pi\left(5\sqrt{5} - 1\right)}{6} \approx 5.33.$$

Symmetry considerations suggest that the center of mass lies somewhere along the z-axis, so we'll search only for the z-coordinate. By definition,

$$\bar{z} = \frac{\iint_S z \, \rho(x, y, z) \, dS}{\text{mass}}.$$

Applying the definition to the integral in the numerator leads (by the same process as in Example 2) to the uv-integral

$$\iint_D (u^2 + v^2)\sqrt{1 + 4u^2} \, du \, dv,$$

and from there, using polar coordinates, to

$$\int_0^{2\pi} \int_0^1 r^2 \sqrt{1 + 4r^2} \, dr \, d\theta.$$

This last integral can be done by standard symbolic methods, or even left to technology. The answer turns out to be

$$\int_0^{2\pi} \int_0^1 r^2\sqrt{1+4r^2}\, dr\, d\theta = 2\pi \left(\frac{5\sqrt{5}}{24} + \frac{1}{120}\right) \approx 2.98.$$

Therefore the z-coordinate of the center of mass is

$$\bar{z} = \frac{\iint_S z\,\rho(x, y, z)\, dS}{\text{mass}} \approx \frac{2.98}{5.33} \approx 0.56.$$

Thus the center of mass is a little above the geometric center of the paraboloid, as we might expect, given its shape. □

Exercises

1. We observed in this section that if a surface S is parametrized by $\boldsymbol{X}(u, v)$, then the vector $\boldsymbol{X_u} \times \boldsymbol{X_v}$ has two important properties: (i) it's perpendicular to S at $\boldsymbol{X}(u, v)$; (ii) its magnitude gives the area magnification factor of $\boldsymbol{X}$ at (u, v). In each part below, calculate $\boldsymbol{X_u} \times \boldsymbol{X_v}$ and $|\boldsymbol{X_u} \times \boldsymbol{X_v}|$ for the given parametrization $\boldsymbol{X}$. (Answers will involve u and v.)

   (a) $\boldsymbol{X}(u, v) = (u, v, u^2 + v^2)$
   (b) $\boldsymbol{X}(u, v) = (u, v, \sqrt{u^2 + v^2})$
   (c) $\boldsymbol{X}(u, v) = (u \cos v, u \sin v, u^2)$
   (d) $\boldsymbol{X}(u, v) = (u \cos v, u \sin v, u)$
   (e) $\boldsymbol{X}(u, v) = (\sin u \cos v, \sin u \sin v, \cos u)$

2. Let a nonzero vector $\boldsymbol{n}$ be perpendicular to a smooth surface S at a point (x_0, y_0, z_0). Then the plane through (x_0, y_0, z_0) with normal vector $\boldsymbol{n}$ is the tangent plane to S at (x_0, y_0, z_0). Use this fact and the vector $\boldsymbol{X_u} \times \boldsymbol{X_v}$ to find the tangent plane to each of the following surfaces at the given point $\boldsymbol{X}(u_0, v_0)$. If possible, use technology to plot both the surface and the tangent plane in an appropriate window.

   (a) $\boldsymbol{X}(u, v) = (u, v, u^2 + v^2)$ at $(u_0, v_0) = (1, 1)$
   (b) $\boldsymbol{X}(u, v) = (u, v, \sqrt{u^2 + v^2})$ at $(u_0, v_0) = (1, 1)$
   (c) $\boldsymbol{X}(u, v) = (u \cos v, u \sin v, u^2)$ at $(u_0, v_0) = (\sqrt{2}, \pi/4)$
   (d) $\boldsymbol{X}(u, v) = (\sin u \cos v, \sin u \sin v, \cos u)$ at $(u_0, v_0) = (\pi/2, 0)$

3. In each part below, use the integral formula to find the surface area both of the given surface and of the region it lies above (or below).

   (a) The part of the plane $z = 3$ that lies above the unit disk $0 \le r \le 1$.
   (b) The part of the plane $z = 2x + 3y + 4$ that lies above the unit disk $0 \le r \le 1$.
   (c) The part of the plane $z = 2x + 3y + 4$ that lies above the unit square $[0, 1] \times [0, 1]$.

(d) The part of the surface $z = x^2 - y^2$ that lies above (or below) the disk $x^2 + y^2 \leq 1$.

(e) The part of the surface $z = x^2 - y^2$ that lies above (or below) the disk $x^2 + y^2 \leq a^2$.

4. Use cylindrical coordinates r, θ, and z to find the area of the part of the cone $z = r$ that lies between $z = 1$ and $z = 2$. (Hint: Parametrize the cone using $\boldsymbol{X}(u, v) = (u \cos v, u \sin v, u)$. What is the uv-domain D?)

5. Repeat the previous exercise, but instead of the cone $z = r$, use the paraboloid $z = r^2$. (Hint: Parametrize the paraboloid using $\boldsymbol{X}(u, v) = (u \cos v, u \sin v, u^2)$. What is the uv-domain D?)

6. It's a fact that if S is a graph, parametrized by $\boldsymbol{X}(u, v) = (u, v, f(u, v))$, for (u, v) in D, then

$$\text{area}(S) = \iint_D \sec \alpha \, du \, dv,$$

where α is the angle between the normal vector $\boldsymbol{X_u} \times \boldsymbol{X_v}$ and the vertical vector $\boldsymbol{k}$. Show this fact. (This fact gives some more insight into how and why area magnification changes with the "steepness" of the surface.)

5.6 Derivatives and integrals of vector fields

The next section presents our last two analogues of the fundamental theorem of calculus: the **divergence theorem** and **Stokes' theorem**. Whereas the elementary fundamental theorem relates derivatives and integrals of scalar-valued *functions*, these higher-dimensional theorems involve certain derivatives and integrals of *vector fields*.

We saw earlier in this chapter how to integrate a vector field $\boldsymbol{f}$ in $\mathbb{R}^2$ along an oriented curve γ, using the line integral $\int_\gamma \boldsymbol{f} \cdot d\boldsymbol{X}$. We interpreted the result, physically, either as work done along γ (if $\boldsymbol{f}$ is thought of as a force) or as circulation around γ (if $\boldsymbol{f}$ is thought of as the velocity field of a flow). In this section we'll meet another type of vector integral. The **flux integral** of a vector field $\boldsymbol{f}$ over a surface S measures how much fluid flows across (i.e., perpendicular to) the surface in unit time. We'll also see two ways of *differentiating* a vector field in space; each type of derivative has its own geometric and physical significance. In short, this section introduces the objects and operations needed to state our final theorems.

The idea is exactly the same in $\mathbb{R}^3$.

Flux integrals

Let $\boldsymbol{f}(x, y, z) = (P(x, y, z), Q(x, y, z), R(x, y, z))$ be a vector field in $\mathbb{R}^3$; we'll think of $\boldsymbol{f}$ as the velocity field of a moving fluid. Let S be a surface in $\mathbb{R}^3$; imagine S as a permeable membrane suspended within the flow, like a fish net in a moving stream. Consider the problem of measuring the **flux** across S, i.e., the rate of flow per unit of time across S. (In fish net terms, the question is how much water flows *through* the net per unit of time.)

It's clear from physical intuition that the answer depends on the angle at which the surface S meets the flow. The flux will be greatest (in absolute value) if the surface is perpendicular to the flow, and least if the surface is parallel to the flow. This means, in other words, that the flux at any point (x, y, z) on the surface is the component of the flow vector (P, Q, R) in the direction *perpendicular* to the surface. If $\boldsymbol{n} = \boldsymbol{n}(x, y, z)$ is a unit vector perpendicular to the surface at (x, y, z), then the dot product

$$\boldsymbol{n} \cdot \boldsymbol{f} = \boldsymbol{n} \cdot (P, Q, R)$$

gives the component in question. Integrating this component over the surface gives the flux we're aiming for:

Recall: taking the dot product with a unit vector gives the component in that direction.

> **Definition: (Flux integral)** Let $\boldsymbol{f}$ be a vector field and S a surface in $\mathbb{R}^3$, and let $\boldsymbol{n}(x, y, z)$ denote a unit vector, normal to S at each point (x, y, z). The surface integral
>
> $$\iint_S \boldsymbol{f} \cdot \boldsymbol{n} \, dS,$$
>
> called the flux integral, measures the flow per unit time across S in the direction of $\boldsymbol{n}$.

Observe:

Not really different. The flux integral involves two vectors ($\boldsymbol{f}$ and $\boldsymbol{n}$), but taking their dot product produces a scalar-valued function. Thus the flux integral is really just a particular form of the surface integral we studied in the last section.

Two choices of normal vector. A two-dimensional surface in $\mathbb{R}^3$ has *two* normal directions at any point; they're opposite to each other. That's why the flux integral definition includes the proviso about flow in the direction of the normal vector. (In practice, the two directions are often easy to sort out.)

For some surfaces, however, there *is* no consistent choice of normal direction. (The Möbius strip is the simplest example.) Surfaces with this property are called non-orientable. We won't need to worry about this problem in this book. To be fully rigorous, however, the definition above would need to require that S be orientable.

Easy to calculate. Flux integrals are calculated exactly like any other surface integral. Recall that if the surface S is parametrized by a function $\boldsymbol{X}(u, v)$, defined on a domain D in uv-space, then the vector

$$\boldsymbol{X}_u \times \boldsymbol{X}_v$$

is normal to S at $\boldsymbol{X}(u, v)$. For a *unit* normal vector $\boldsymbol{n}$, therefore, we may as well use

$$\boldsymbol{n} = \frac{\boldsymbol{X}_u \times \boldsymbol{X}_v}{|\boldsymbol{X}_u \times \boldsymbol{X}_v|}.$$

This may look formidable, but a pleasant surprise is in store. By our definition of the surface integral,

Recall it from the preceding section.

$$\begin{aligned}
\iint_S \boldsymbol{f} \cdot \boldsymbol{n}\, dS &= \iint_S \boldsymbol{f} \cdot \frac{\boldsymbol{X}_u \times \boldsymbol{X}_v}{|\boldsymbol{X}_u \times \boldsymbol{X}_v|}\, dS \\
&= \iint_D \boldsymbol{f}(\boldsymbol{X}(u, v)) \cdot \frac{\boldsymbol{X}_u \times \boldsymbol{X}_v}{|\boldsymbol{X}_u \times \boldsymbol{X}_v|} |\boldsymbol{X}_u \times \boldsymbol{X}_v|\, du\, dv \\
&= \iint_D \boldsymbol{f}(\boldsymbol{X}(u, v)) \cdot (\boldsymbol{X}_u \times \boldsymbol{X}_v)\, du\, dv.
\end{aligned}$$

The bottom line, then, is that the flux integral may be even easier to calculate than a surface integral.

■ **Example 1.** Let S be the part of the surface $z = x^2 + y^2$ above the unit disk, and let $\boldsymbol{f} = (x, y, z)$. Find the flux integral. In which direction does the normal vector $\boldsymbol{n}$ point?

Solution: We can parametrize S as a graph, using $\boldsymbol{X}(u, v) = (u, v, u^2 + v^2)$, with (u, v) in the unit disk D. We showed in the preceding section that for this parametrization,

$$\boldsymbol{X}_u = (1, 0, 2u), \quad \boldsymbol{X}_v = (0, 1, 2v), \quad \text{and} \quad \boldsymbol{X}_u \times \boldsymbol{X}_v = (-2u, -2v, 1).$$

Therefore the flux integral is

$$\begin{aligned}
\iint_S \boldsymbol{f} \cdot \boldsymbol{n}\, dS &= \iint_D (u, v, u^2 + v^2) \cdot (-2u, -2v, 1)\, du\, dv \\
&= \iint_D (-u^2 - v^2)\, du\, dv = -\frac{\pi}{2}.
\end{aligned}$$

(The last integral is easily done in polar coordinates.)

Convince yourself that the answer is right.

Notice that our choice of normal vector n has positive z-coordinate, so it points upward (or toward the inside of the paraboloid bowl). The sign of the answer means that for the given flow field, more fluid flows "out of the bowl" than into it. □

Divergence and curl: derivatives of a vector field

A vector field $f(x, y, z) = (P(x, y, z), Q(x, y, z), R(x, y, z))$ can be thought of as a function $f : \mathbb{R}^3 \to \mathbb{R}^3$, with Jacobian matrix

$$\begin{pmatrix} P_x & P_y & P_z \\ Q_x & Q_y & Q_z \\ R_x & R_y & R_z \end{pmatrix}.$$

Many possible combinations of derivatives can be formed from all these data. The two following combinations turn out to have special physical interest:

Definition: (Divergence and curl) Let $f = (P, Q, R)$ be a vector field on $\mathbb{R}^3$. The **divergence** of f is the *scalar function* defined by

$$\operatorname{div} f = P_x + Q_y + R_z.$$

The **curl** of f is the *vector field* defined by

$$\operatorname{curl} f = (R_y - Q_z, P_z - R_x, Q_x - P_y).$$

Notice:

From the Jacobian matrix. Both the divergence and the curl are formed in regular ways from the entries of the Jacobian matrix above: The divergence is the sum of the diagonal entries (called the **trace** of the matrix). Each component of the curl field is the difference of two Jacobian entries that are symmetric with respect to the diagonal.

Easy to calculate. Calculating the divergence and curl of a vector field is a mechanical matter—indeed, *Maple* and other such programs have commands that do so. More interesting is to see what the answers mean.

What div means. If we think of f as a flow, then at any given point (x, y, z) in $\mathbb{R}^3$, the divergence $P_x + Q_y + R_z$ measures the total tendency of fluid to flow *away* from the point.

To get some feeling for why this is so, notice first that $P(x, y, z)$ describes the flow's velocity in the x-direction. Thus P_x is the acceleration in the x-direction: If $P_x(x, y, z) > 0$, the fluid is speeding up in the x-direction at (x, y, z), and so tends to "diverge" from (x, y, z). If $P_x(x, y, z) < 0$, the fluid is "slowing down," and so tends to "converge," or pile up, at (x, y, z). Adding up the similar contributions in the y- and z-directions gives the bottom line on whether the fluid diverges or converges at (x, y, z).

Curl and grad. If f happens to be a gradient field, i.e., if $f = \nabla h = (h_x, h_y, h_z)$ for some function $h(x, y, z)$, then a simple but important calculation (left to the exercises) shows that

$$\operatorname{curl} \nabla h = (0, 0, 0).$$

In words:

Every gradient field has curl $(0, 0, 0)$.

Thus the curl of a vector field f measures, in some sense, the extent to which f *differs* from being a gradient field. As we've seen in earlier sections, vector fields that aren't gradients do appear to "curl" around certain points.

Picturing divergence and curl. We defined divergence and curl for vector fields in $\mathbb{R}^3$. Unfortunately, three-dimensional vector fields are quite difficult to draw accurately on a flat page—too much detail is lost in projecting three dimensions onto two. In the following pictures therefore, we'll think mainly of the two-dimensional versions of divergence and curl. The idea is that a 2-dimensional vector field can, when convenient, be thought of as a 3-dimensional field that happens to be independent of z. The vector field $\mathbf{f}(x, y) = (x - y, x + y)$, for instance, can be thought of as the "slice" at $z = 0$ of the 3-dimensional field $\mathbf{f}(x, y, z) = (x - y, x + y, 0)$. It's natural, therefore, to define the divergence and curl of a two-dimensional field $\mathbf{f}(x, y) = (P(x, y), Q(x, y))$ as follows:➽

Check that this definition of curl is consistent with the earlier one.

$$\operatorname{div} \mathbf{f} = P_x + Q_y; \quad \operatorname{curl} \mathbf{f} = \left(0, 0, Q_x - P_y\right).$$

Notice a property of the curl vector: It points in the z-direction, *perpendicular* to the xy-plane. The curl vector, in effect, is a perpendicular axis *around* which the flow "curls."

■ **Example 2.** Discuss the divergence and curl of the vector fields

$$\mathbf{f}(x, y) = (x - y, x) \quad \text{and} \quad \mathbf{g}(x, y) = (x^2, 2y),$$

shown following:

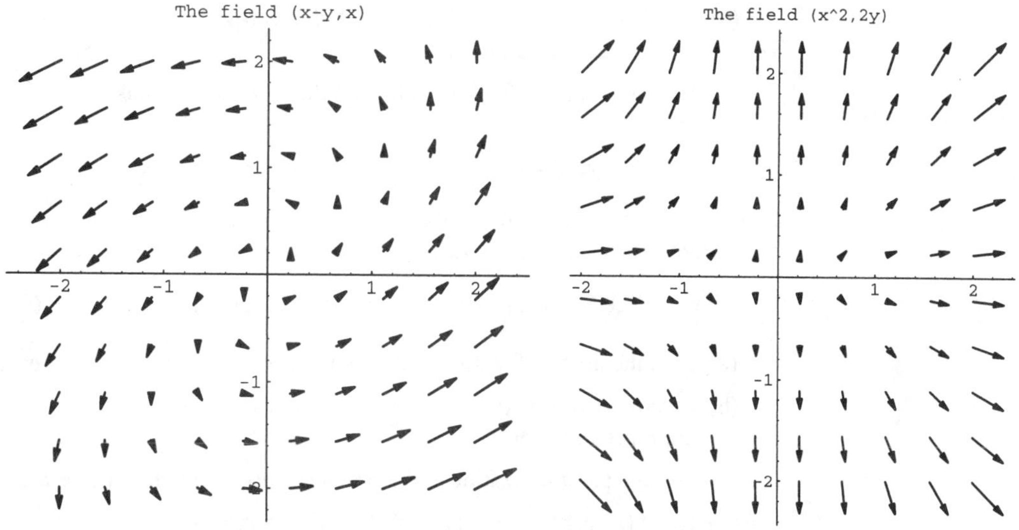

Solution: All the symbolic calculations are easy.➽ Let's start with the divergence; calculating $P_x + Q_y$ gives the scalar-valued functions

But check them for yourself.

$$\operatorname{div} \mathbf{f} = 1; \quad \operatorname{div} \mathbf{g} = 2x + 2.$$

Both results can be seen, at least qualitatively, in the pictures. Look first at $\mathbf{f}$. At *every* point (x, y), incoming arrows are shorter than outgoing arrows. Thus the divergence—which measures *outflow*—is everywhere positive for $\mathbf{f}$. For the field $\mathbf{g}$, the formula $\operatorname{div} \mathbf{g} = 2x + 2$ means that the divergence changes sign at $x = -1$. This can be seen in the picture, too: Where $x < -1$, comparing lengths of incoming and outgoing arrows reveals a net *inflow*, or negative divergence. Where $x > -1$, the same feature shows a net *outflow*, or positive divergence.

Calculating the curl is easy, too. Here are the results:

$$\text{curl}\,\boldsymbol{f} = (0, 0, 1); \quad \text{curl}\,\boldsymbol{g} = (0, 0, 0).$$

These results also appear (qualitatively) in the pictures. The field $\boldsymbol{f}$ *does* appear to curl, counterclockwise, around a vertical axis. (A clockwise curl would produce a negative z-coordinate.) The field $\boldsymbol{g}$, by contrast, does not seem to curl around on itself, so the curl vector appears to be zero. There's a good reason for this: $\boldsymbol{g}$ is a gradient field. Specifically,

$$\boldsymbol{g} = (x^2, 2y) = \nabla\left(\frac{x^3}{3} + y^2\right).$$

As we observed earlier, *every* gradient field has zero curl. □

Exercises

1. Throughout this exercise, let S be the part of the surface $z = x^2 + y^2$ that lies above the unit disk, as in Example 1. (Use the same parametrization as was used there.)

   (a) Let $\boldsymbol{f}(x, y, z) = (x, 0, 0)$. Find the flux across S; discuss the sign of the answer (as in Example 1).

   (b) Let $\boldsymbol{f}(x, y, z) = (0, 1, 0)$. Find the flux across S; discuss the sign of the answer.

   (c) Let $\boldsymbol{f}(x, y, z) = (0, 0, z)$. Find the flux across S.

2. Redo all three parts of the previous exercise, but use the cylindrical coordinate parametrization $\boldsymbol{X}(u, v) = (u \cos v, u \sin v, u^2)$.

3. Let the surface S be the triangle with corners at $(1, 0, 0)$, $(0, 1, 0)$, and $(0, 0, 1)$.

   (a) Find the area of S without integrating; use an appropriate cross product.

   (b) Parametrize S as the graph of a function $z = f(x, y)$, for (x, y) in an appropriate region D.

   (c) Use the parametrization of the previous part to find the surface area of S.

   (d) Let $\boldsymbol{f}(x, y, z) = (x, y, z)$. Find the flux across S.

   (e) Let $\boldsymbol{f}(x, y, z) = (a, b, c)$; a, b, and c are all constants. Find the flux across S. Under what conditions on a, b, and c is the flux zero?

4. Let S be the part of the cylinder $x^2 + y^2 = 1$ from $z = 0$ to $z = 1$.

   (a) Find the surface area of S by elementary means. (Imagine cutting the cylinder and unrolling it.)

   (b) Parametrize S using cylindrical coordinates. Use the result to calculate the surface area of S by integration.

   (c) Let $\boldsymbol{f}(x, y, z) = (1, 0, 0)$. Find the flux across S.

   (d) Let $\boldsymbol{f}(x, y, z) = (0, 0, R(x, y, z))$. Show that the flux across S is zero regardless of the function $R(x, y, z)$.

5. Suppose that $f = \nabla h = (h_x, h_y, h_z)$ for any smooth function $h(x, y, z)$. Explain why

$$\text{curl } \nabla h = (0, 0, 0).$$

6. For each vector field below, find the divergence and the curl.

   (a) $f = (x, y, z)$
   (b) $f = (y, z, x)$
   (c) $f = (-y, x, z)$
   (d) $f = (-y/(x^2 + y^2), x/(x^2 + y^2), 0)$

7. Consider the vector field $f(x, y) = (\sin(xy), \cos(x))$, as shown:

The field (sin(xy),cos(x))

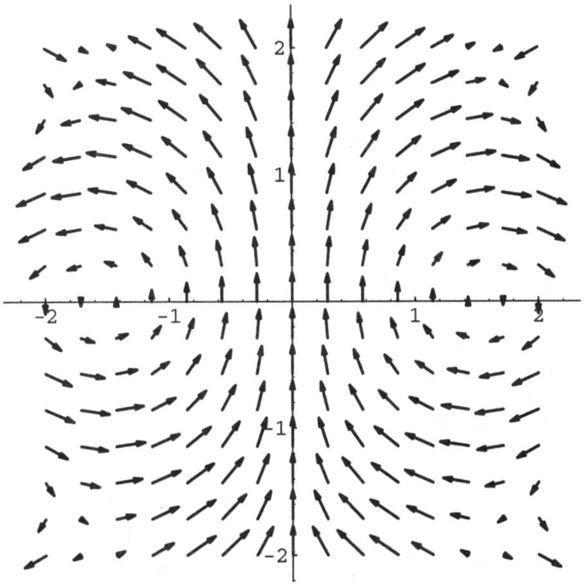

   (a) Find formulas for the divergence and curl of f.
   (b) Show that the divergence is zero everywhere along the x-axis. How does this appear in the picture?
   (c) Find the divergence at $(1, 1)$, $(-1, 1)$, $(-1, -1)$, and $(1, -1)$. How do the signs of the answers appear in the picture?
   (d) Find the curl at $(\pi/2, 0)$ and at $(-\pi/2, 0)$. Relate the sign difference to the direction of curl.

8. In each part below, use technology to plot the given vector field in the xy-plane, in the rectangle $[-2, 2] \times [-2, 2]$. (When plotting each field, simply ignore the z-coordinate.) In each case, try to "see" where the divergence is positive and and negative, and look for the presence or absence of curl.) Then calculate the divergence and the curl.

   (a) $f = (1, 0, 0)$
   (b) $f = (x, 0, 0)$
   (c) $f = (x^2, 0, 0)$
   (d) $f = (x - y, x + y, 0)$

9. Show that for any differentiable vector field f, div (curl f) $= 0$.

5.7 Back to fundamentals: Stokes' theorem and the divergence theorem

It's traditional to end a calculus course with a "fundamental theorem," one that links the main concepts of differentiation and integration. We'll stick to that tradition. In fact, we'll raise the ante substantially, by stating *five* fundamental theorems. We've developed all the necessary objects and processes, and it's time to assemble the pieces.

Five fundamental theorems

We'll collect all five theorems for comparison. Notation is as follows (bold symbols indicate vector quantities):

γ: an oriented curve in $\mathbb{R}^2$ or $\mathbb{R}^3$

D: a region in $\mathbb{R}^2$

S: a 2-dimensional surface in $\mathbb{R}^3$

V: a 3-dimensional solid in $\mathbb{R}^3$

$\boldsymbol{n}$: a unit vector field, normal to a surface S at each point of S

$\boldsymbol{f}$: a vector field in $\mathbb{R}^2$ or $\mathbb{R}^3$

f: a scalar-valued function of one or more variables

a, b: fixed points in $\mathbb{R}$

$\boldsymbol{a}, \boldsymbol{b}$: fixed points in $\mathbb{R}^2$ or $\mathbb{R}^3$

Technical hypotheses. To avoid sidetracks, we'll make several technical assumptions. We state these assumptions mainly for the record—all are satisfied automatically in typical simple examples. We should add, however, that the assumptions are genuinely important; in their absence there's no guarantee that the objects in question even exist. For a non-orientable surface,[◂] for instance, there *is* no suitable choice of normal vector, so the surface integrals in question don't make sense.

We discussed orientability briefly in the preceding section.

We'll assume, then, that all functions and derivatives mentioned in the following theorems exist and are continuous; this assures, in turn, that the integrals exist. Curves are assumed to be either smooth or piecewise-smooth (i.e., the union of several smooth curves or surfaces, with "kinks" only where the pieces join). Surfaces are orientable, and smooth except perhaps along edges where smooth pieces join.

Five theorems. With these provisos, here are the theorems; the last two are new.

Theorem 4. (Fundamental theorem of calculus)

$$\int_a^b f'(x)\,dx = f(b) - f(a).$$

Theorem 5. (Fundamental theorem for line integrals) If γ starts at $\boldsymbol{a}$ and ends at $\boldsymbol{b}$, then

$$\int_\gamma \nabla f \cdot d\boldsymbol{X} = f(\boldsymbol{b}) - f(\boldsymbol{a}).$$

Theorem 6. **(Green's theorem)** Let $f = (P, Q)$ be a vector field in $\mathbb{R}^2$, γ a closed curve (oriented counterclockwise), and D the region inside γ. Then

$$\iint_D (Q_x - P_y)\, dA = \int_\gamma P\, dx + Q\, dy.$$

Theorem 7. **(Stokes' theorem)** Let $f = (P, Q, R)$ be a vector field in $\mathbb{R}^3$. Let S be a surface in $\mathbb{R}^3$, bounded by a closed curve γ, with unit normal $\boldsymbol{n}$. Then

$$\iint_S (\text{curl}\, \boldsymbol{f}) \cdot \boldsymbol{n}\, dS = \pm \oint_\gamma \boldsymbol{f} \cdot d\boldsymbol{X}.$$

(The sign depends on the direction of $\boldsymbol{n}$.)

Theorem 8. **(Divergence theorem)** Let $f = (P, Q, R)$ be a vector field in $\mathbb{R}^3$. Let V be a solid region in $\mathbb{R}^3$, bounded by a surface S, with outward unit normal $\boldsymbol{n}$. Then

$$\iiint_V \text{div}\, \boldsymbol{f}\, dV = \iint_S \boldsymbol{f} \cdot \boldsymbol{n}\, dS.$$

All five theorems have the same theme: A function or vector field $\boldsymbol{f}$ is given. On the left side of each equation, some sort of derivative of $\boldsymbol{f}$ is integrated over some domain in $\mathbb{R}$, $\mathbb{R}^2$, or $\mathbb{R}^3$. On the right side of each equation, the expression involves $\boldsymbol{f}$ itself, evaluated on a lower-dimensional set—the *boundary* of the original domain.

More on Stokes' theorem

Let's see first what the theorem says, by example.

■ **Example 1.** Let S be the part of the surface $z = x^2 + y^2$ above the unit disk, and let $\boldsymbol{f} = (P, Q, R) = (-y, x, z)$. What does Stokes' theorem say in this case? Is it true?

Solution: We'll calculate the integrals on both sides of Stokes' theorem and see that they're equal.

Notice first that the boundary of S is a circle of radius 1 in $\mathbb{R}^3$—the intersection of the paraboloid and the plane $z = 1$. We can parametrize this boundary using

$$\boldsymbol{X}(t) = (\cos t, \sin t, 1); \quad 0 \le t \le 2\pi.$$

The line integral $\int_\gamma \boldsymbol{f} \cdot d\boldsymbol{X}$ is now easy to calculate:➸

But check details.

$$\int_\gamma P\, dx + Q\, dy + R\, dz = \int_0^{2\pi} \left(\sin^2 t + \cos^2 t\right) dt = 2\pi.$$

To handle the surface integral, recall that we used the same surface in Example 1, page 289. There, we used the parametrization $\boldsymbol{X}(u, v) = (u, v, u^2 + v^2)$, with (u, v) in the unit disk D; we found the normal vector $\boldsymbol{X}_u \times \boldsymbol{X}_v = (-2u, -2v, 1)$.

For the vector field $\boldsymbol{f} = (-y, x, z)$, calculation gives➸ curl $\boldsymbol{f} = (0, 0, 2)$.

Check it!

Therefore the flux integral is

$$\iint_S (\text{curl}\, \boldsymbol{f}) \cdot \boldsymbol{n}\, dS = \iint_D (0, 0, 2) \cdot (-2u, -2v, 1)\, du\, dv = \iint_D 2\, du\, dv.$$

The last integral is twice the area of the unit disk, or 2π.

Thus the line and surface integrals are the same in this case; Stokes' theorem holds. (With the opposite choice of normal vector, the flux integral would change sign; but Stokes' theorem permits that.) □

From Stokes to Green. Stokes' theorem is, in a natural sense, an extension to $\mathbb{R}^3$ of Green's theorem, which "lives" in $\mathbb{R}^2$. (Indeed, Stokes' theorem is sometimes known as "Green's theorem for surfaces.") We won't prove Stokes' theorem, since its proof involves the same main ideas (appropriately translated) as that of Green's theorem.

Instead of proving Stokes' theorem, let's see exactly how Green's theorem is a special case. First, recall that a vector field $\boldsymbol{f} = (P(x, y), Q(x, y))$ in $\mathbb{R}^2$ can be thought of, when convenient, as a special 3-dimensional vector field $(P, Q, 0)$, which depends only on x and y, and has zero for the last coordinate. For this new vector field, an easy calculation shows that

We discussed this in the preceding section.

$$\text{curl}\,(P, Q, R) = (R_y - Q_z, P_z - R_x, Q_x - P_y) = (0, 0, Q_x - P_y).$$

Thus the third coordinate of curl $\boldsymbol{f}$ turns out to be the area integrand in Green's theorem.

The second step is to think of the domain D in Green's theorem as a very simple surface in $\mathbb{R}^3$. Since D lies flat in the xy-plane, the upward-pointing vector $\boldsymbol{k}$ can be taken as a unit normal to D. (The downward-pointing vector $-\boldsymbol{k}$ could also have been chosen; doing so would change the sign of the surface integral.)

Taken together, these remarks mean that in the present situation,

$$\iint_D (\text{curl}\, \boldsymbol{f}) \cdot \boldsymbol{n}\, dS = \iint_D (0, 0, Q_x - P_y) \cdot \boldsymbol{k}\, dS = \iint_D (Q_x - P_y)\, dA.$$

In other words, the flux integral in Stokes' theorem boils down, in this special case, to the area integral of Green's theorem.

In a similar way, the boundary curve γ—which lies in the plane $z = 0$—can can be parametrized either as a curve in $\mathbb{R}^2$ or as a curve in $\mathbb{R}^3$. If the function

$$X(t) = (x(t), y(t)); \quad a \le t \le b$$

parametrizes γ in $\mathbb{R}^2$, then

$$X(t) = (x(t), y(t), 0); \quad a \le t \le b$$

parametrizes γ in $\mathbb{R}^3$. In the latter case, the line integral becomes

$$\int_\gamma P\, dx + Q\, dy + R\, dz = \int_\gamma P\, dx + Q\, dy,$$

because $dz = 0$. Thus the line integrals in Stokes' theorem and Green's theorem turn out, in this special case, to be identical.

We've shown what we wanted to show: For domains in the plane, each side of Stokes' identity reduces to the corresponding side of Green's identity.

Curl, flux, and circulation: a physical interpretation. The mathematical theory of line and surface integrals grew up around physical problems, and the terms "curl," "flux," and "circulation" are all borrowed from physics. Naturally enough, therefore, Stoke's theorem can be interpreted, at least intuitively, in the physical language of flow.

We start by imagining a 3-dimensional fluid flow f, defined on and near a surface S. At each point (x, y, z) on the surface S, the derivative curl f is a new vector, which measures the tendency of the flow to "curl" at (x, y, z). (The vector curl f acts as an axis around which the curling occurs.) It follows that if $\boldsymbol{n}$ is a unit normal to S at (x, y, z), then the dot product $(\text{curl } f) \cdot \boldsymbol{n}$—the left-hand integrand in Stokes' theorem—tells how much of the fluid's rotation occurs along the surface (rather than, say, perpendicular to it). Therefore, the surface integral

$$\iint_S (\text{curl } f) \cdot \boldsymbol{n}\, dS,$$

measures, in some sense, the *total* rotation of the flow along the surface.

The reasoning here is rough and ready, but it can be made mathematically precise.

The line integral $\pm \oint_\gamma f \cdot d\boldsymbol{X}$ is easier to interpret: It measures the fluid's circulation around the boundary of S. Stokes' theorem asserts, then, that two quantities are equal: (i) the fluid's circulation around the boundary of S; and (ii) the total rotation of fluid along S itself. That (i) and (ii) are equal seems physically believable, since each phenomenon can be thought of as causing the other.

More on the divergence theorem

The divergence theorem asserts that, under appropriate hypotheses,

$$\iiint_V \text{div } f\, dV = \iint_S f \cdot \boldsymbol{n}\, dS. \tag{5.7.1}$$

Let's pick the equation apart, thinking in physical terms.

On the left: a triple integral. The left side of Equation 5.7.1 is a triple integral over the solid region V; the integrand is the scalar-valued function div f. If f represents a flow, then we've seen that at any domain point (x, y, z), the function div f measures the net flow *away* from (x, y, z) per unit time. Integrating div f over V, therefore, gives the total net flow out of V per unit time, i.e., the net rate at which fluid "leaves" V.

On the right: a flux integral. The right side of Equation 5.7.1 is a flux integral of the type described in the previous section; it measures the rate at which fluid crosses the boundary surface S. Because the unit normal $\boldsymbol{n}$ is chosen to point *outward* from V, the flux integral measures the flow across S in the *outward* direction.

Divergence and flux: why they're equal. The last two paragraphs say, in effect, that the integrals on both sides of Equation 5.7.1 measure the same thing: the flow out of V. From this point of view, the divergence theorem should sound physically reasonable—two integrals that measure the same quantity should have the same value.

■ **Example 2.** Let V be the solid in $\mathbb{R}^3$ bounded above by the plane $z = 1$ and below by the paraboloid $z = x^2 + y^2$. Let f be the vector field $(P, Q, R) = (x, y, z)$. The solid and the field appear as follows:

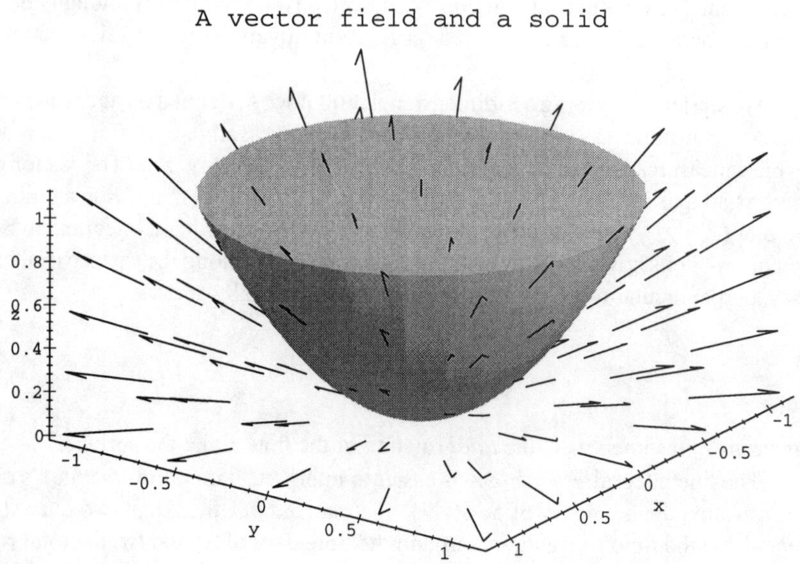

What does the divergence theorem say in this case? Is it true?

Solution: In this case the boundary surface S has two parts: (i) S_1, the part of the paraboloid $z = x^2 + y^2$ below the plane $z = 1$; and (ii) S_2, the part of the plane $z = 1$ for which $x^2 + y^2 \leq 1$.

The drum.

The drumhead.

Let's calculate both sides of Equation 5.7.1, starting on the left. It's easy to see, first, that

$$\operatorname{div} \boldsymbol{f} = \operatorname{div}(x, y, z) = 3.$$

(The picture suggests, too, that the divergence is everywhere positive.) Thus the volume integral is

$$\iiint_V 3\, dV = 3 \times \text{volume of } V.$$

Details are left to the exercises.

This integral is simplest in cylindrical coordinates; we get

$$\iiint_V 3\, dV = \int_{\theta=0}^{\theta=2\pi} \int_{r=1}^{r=1} \int_{z=r^2}^{z=1} 3\, r\, dz\, dr\, d\theta = \frac{3\pi}{2}.$$

Now for the right side of Equation 5.7.1. The flux integral has two parts, one for each part of the surface. We've already handled the surface S_1 several times, most recently in Example 1, page 295. There, we parametrized S_1 by $\boldsymbol{X}(u, v) = (u, v, u^2 + v^2)$, with (u, v) in the unit disk D. Then the vector $\boldsymbol{X}_u \times \boldsymbol{X}_v = (-2u, -2v, 1)$ is normal to S_1. Notice, however, that this vector has positive z-coordinate, so it points into V. For an *outward* unit normal, therefore, we can take the reversed vector

Look closely at the picture to convince yourself.

$$\boldsymbol{n} = \frac{(2u, 2v, -1)}{\sqrt{1 + 4u^2 + 4v^2}}.$$

Some details are left to the exercises.

With this parametrization, therefore, the flux integral over S_1 becomes

$$\iint_{S_1} \boldsymbol{f} \cdot \boldsymbol{n}\, dS = \iint_D (u, v, u^2 + v^2) \cdot (2u, 2v, -1)\, du\, dv = \frac{\pi}{2}.$$

The surface S_2 is even easier to parametrize, using $\boldsymbol{X}(u, v) = (u, v, 1)$, with (u, v) in the unit disk D. Because S_2 is parallel to the xy-plane, moreover, the upward-pointing vector $\boldsymbol{k}$ is a suitable outward unit normal. Thus the flux integral over S_2 becomes

$$\iint_{S_2} \boldsymbol{f} \cdot \boldsymbol{n}\, dS = \iint_D (u, v, 1) \cdot (0, 0, 1)\, du\, dv = \iint_D 1\, du\, dv = \pi.$$

Therefore, the total flux integral is

$$\begin{aligned} \iint_S \boldsymbol{f} \cdot \boldsymbol{n}\, dS &= \iint_{S_1} \boldsymbol{f} \cdot \boldsymbol{n}\, dS + \iint_{S_2} \boldsymbol{f} \cdot \boldsymbol{n}\, dS \\ &= \frac{\pi}{2} + \pi = \frac{3\pi}{2}. \end{aligned}$$

In this case, therefore, the two sides of Equation 5.7.1 are equal, as the divergence theorem predicts.

Proving the divergence theorem: the idea. The divergence theorem is proved similarly to Green's theorem. Notice first that if $\boldsymbol{f} = (P, Q, R)$ and the outward unit normal to S has the form $\boldsymbol{n} = (n_1, n_2, n_3)$, then the divergence theorem says that

$$\iiint_V (P_x + Q_y + R_z)\, dV = \iint_S (P\, n_1 + Q\, n_2 + R\, n_3)\, dS.$$

Thus it's enough to prove, separately, three simpler identities:

$$\iiint_V P_x\, dV = \iint_S P\, n_1\, dS; \quad \iiint_V Q_y\, dV = \iint_S Q\, n_2\, dS;$$

$$\iiint_V R_z\, dV = \iint_S R\, n_3\, dS.$$

We'll prove the third identity, assuming that the boundary S of V has two parts: a lower surface S_1 and and an upper surface S_2 (as in the preceding example), and that both S_1 and S_2 are graphs of functions $z = g(x, y)$ and $z = h(x, y)$ respectively, for (x, y) in a region D in the xy-plane. Then, as we've seen, the vectors $(-g_x, -g_y, -1)$ and $(h_x, h_y, 1)$ are normal to the surfaces S_1 and S_2. (We used the downward-pointing normal for the lower surface S_1.)

Finally, we write out both sides of the third identity and compare results. On the right, the parametrizations above imply that

$$\iint_{S_1} R\, n_3\, dS = -\iint_D R(x, y, g(x, y))\, dA$$

and

$$\iint_{S_2} R\, n_3\, dS = \iint_D R(x, y, h(x, y))\, dA.$$

(The two integrals have opposite signs because of their different normal directions.) Adding these results gives the total surface integral:

$$\iint_S R\, n_3\, dS = \iint_D \left(R(x, y, h(x, y)) - R(x, y, g(x, y)) \right) dA.$$

On the left, the triple integral can be written in iterated form, as follows:

$$\iiint_V R_z\, dV = \iint_D \left(\int_{z=g(x,y)}^{z=h(x,y)} R_z\, dz \right) dA.$$

Applying the ordinary fundamental theorem to the inner integral (i.e., antidifferentiating in z) gives

$$\iiint_V R_z\,dV = \iint_D \left(R(x, y, h(x, y)) - R(x, y, g(x, y)) \right) dA.$$

Thus the surface and volume integrals are equal. Similar arguments apply to the integrals involving P and Q; they complete the idea of proof.

Exercises

1. In each part following, use Stokes' theorem to find the value of the surface integral

$$\iint_S (\text{curl } \boldsymbol{f}) \cdot \boldsymbol{n}\,dS.$$

Do so by transforming the surface integral into an equivalent line integral. Then calculate the line integral, rather than the surface integral.

   (a) Let S be the upper half of the sphere $x^2 + y^2 + z^2 = 1$, and let $\boldsymbol{f} = (x, y, z)$.

   (b) Let S be the upper half of the sphere $x^2 + y^2 + z^2 = 1$, and let $\boldsymbol{f} = (-y, x, z)$.

   (c) Let S be the part of the paraboloid $z = x^2 + y^2$ with $z \leq 1$, and let $\boldsymbol{f} = (y, z, x)$.

2. Carefully work out the volume and surface integrals in Example 2.

3. Consider the situation in Example 2, but use the vector field $\boldsymbol{f} = (x, 0, 0)$.

   (a) Calculate the volume integral $\iiint_V \text{div } \boldsymbol{f}\,dV$.

   (b) Calculate the surface integral $\iint_S \boldsymbol{f} \cdot \boldsymbol{n}\,dS$.

4. Let V be the cube in $\mathbb{R}^3$ defined by $0 \leq x \leq 1, 0 \leq y \leq 1, 0 \leq z \leq 1$. Let $\boldsymbol{f}$ be the vector field $\boldsymbol{f} = (x, y, z)$. Let S be the boundary of V; note that S has six faces.

   (a) What does the divergence theorem say in this case?

   (b) Calculate the triple integral $\iiint_V \text{div } \boldsymbol{f}\,dV$. (Hint: Very little calculation is involved!)

   (c) Calculate the flux integral $\iint_S \boldsymbol{f} \cdot \boldsymbol{n}\,dS$. (Hints: There are six parts, but all are quite simple. Notice that for each face of S, $\boldsymbol{i}$, $\boldsymbol{j}$, or $\boldsymbol{k}$ can be used as a unit normal. Be careful with signs!)

5. Let V be a solid in $\mathbb{R}^3$ with smooth boundary S. Let $\boldsymbol{f}$ be a smooth vector field (P, Q, R) in $\mathbb{R}^3$. is a gradient field, i.e., that Use the divergence theorem to show that

$$\iint_S (\text{curl } \boldsymbol{f}) \cdot \boldsymbol{n}\,dS = 0.$$

6. In each part following, use the divergence theorem to evaluate the flux integral $\iint_S \boldsymbol{f} \cdot \boldsymbol{n}\, dS$. In all parts, S is the unit sphere $x^2 + y^2 + z^2 = 1$, with outward unit normal.

   (a) $\boldsymbol{f} = (x, 2y, 3z)$

   (b) $\boldsymbol{f} = (x, y^2, 0)$

   (c) $\boldsymbol{f} = (0, y^2, 0)$

7. Let γ be the unit circle, oriented counterclockwise, in the xy-plane, and consider the vector field $\boldsymbol{f} = (-y, x, z)$. (Think of γ as a curve in 3-space.)

   (a) Calculate the line integral $\int_\gamma \boldsymbol{f} \cdot d\boldsymbol{X}$.

   (b) Let the surface S be the upper half of the unit sphere. Then γ is the boundary of S. What does Stokes' theorem say in this case? Is it true? (Hint: No calculation is needed; use the fact that the unit sphere has surface area 4π.)

   (c) Now let the surface S be the part of the paraboloid $z = 1 - x^2 - y^2$ that lies above the xy-plane. Again, γ is the boundary of S. What does Stokes' theorem say now? Is it true? (Hint: This time a surface integral calculation is needed, but it's relatively easy.)

8. Let $\boldsymbol{f}(x, y, z) = (z - x, x - y, y - z)$ and let T be the sphere $x^2 + y^2 + z^2 = 4$.

   (a) Evaluate curl $\boldsymbol{f}$.

   (b) Evaluate div $\boldsymbol{f}$.

   (c) Is $\displaystyle\iint_T \boldsymbol{f} \bullet \boldsymbol{n}\, dS$ positive, negative, or zero? Justify your answer.

9. Let S be the portion of the surface $z = x^2 + y^2$ that lies above the region R: $1 \le x \le 2, 2 \le y \le 3$.

   (a) Find a vector normal to S at the point $(1, 2, 5)$.

   (b) Find an equation for the plane tangent to S at the point $(1, 2, 5)$.

   (c) Find the area of S.

   (d) Find the area of the portion of the tangent plane in part (b) that lies above R.

10. Let $\boldsymbol{f}$ be a vector field in $\mathbb{R}^3$ such that on the surface of the unit sphere $\boldsymbol{f}(x, y, z) = (0, 0, z^3)$. If div $\boldsymbol{f} = a$ is a constant in all of $\mathbb{R}^3$, what is the value of a?

11. Suppose that $\boldsymbol{f}$ is a vector field such that curl $\boldsymbol{f} = (1, 2, 5)$ at every point in $\mathbb{R}^3$. Find an equation of a plane through the origin with the property that $\displaystyle\oint_\gamma \boldsymbol{f} \bullet d\boldsymbol{X} = 0$ for any closed curve γ lying in the plane.

CHAPTER

11

Infinite Series

11.1 Sequences and Their Limits

We'll study them in some detail.

This section, on infinite *sequences*, prepares the ground for the next topic—infinite *series*. Convergent series ◄ are defined in terms of the simpler, more basic idea of convergent sequences. We start with a quick introduction to sequences—what they are, what it means for them to converge or diverge, and how to find their limits.

The Idea and Language of Sequences; Simple Examples

A **sequence** is an infinite list of numbers, of the general form

$$a_1,\ a_2,\ a_3,\ a_4,\ \ldots,\ a_k,\ a_{k+1},\ \ldots.$$

Read "a sub three" and "a sub k."

Individual entries are called the **terms** of the sequence; a_3 and a_k, ◄ for instance, are the third term and the kth term, respectively. The full sequence is, technically speaking, an **ordered set**; the standard notation

$$\{a_k\}_{k=1}^{\infty}$$

(or simply $\{a_k\}$) uses set brackets to emphasize this view.

Our main interest in sequences is in their **limits**. For the simplest sequences, limits are evident at a glance. The next three examples are of this type; we include them mainly to illustrate ideas, notations, and terminology.

EXAMPLE 1 Discuss the sequence $\{a_k\}_{k=1}^{\infty}$ defined by the formula $a_k = 1/k$, $k = 1, 2, 3, \ldots$. Does this sequence have a limit?

Solution Sampling the first few terms,

$$\frac{1}{1},\ \frac{1}{2},\ \frac{1}{3},\ \ldots,\ \frac{1}{10},\ \frac{1}{11},\ \ldots,\ \frac{1}{100},\ \frac{1}{101},\ \ldots,$$

shows (to nobody's surprise) that the sequence converges to zero: As k increases, the terms approach zero arbitrarily closely. In symbols,

$$\lim_{k\to\infty} a_k = \lim_{k\to\infty} \frac{1}{k} = 0. \quad \blacksquare$$

EXAMPLE 2 Suppose that $\{b_j\}_{j=1}^{\infty}$ is defined ▶ by

$$b_j = \frac{(-1)^j}{j}, \qquad j = 1, 2, 3, \ldots.$$

This time we used j rather than k; the name of the "index variable" doesn't matter.

What's $\lim_{j\to\infty} b_j$?

Solution Writing out terms shows a similar pattern:

$$-\frac{1}{1}, \frac{1}{2}, -\frac{1}{3}, \ldots, \frac{1}{10}, -\frac{1}{11}, \ldots, \frac{1}{100}, -\frac{1}{101}, \ldots.$$

Although the terms oscillate in *sign*, they approach zero more and more closely. Eventually, all the terms—positive or negative—remain within any specified distance from zero. ▶ Hence, the sequence $\{b_j\}$ also converges to zero:

All terms past b_{1000}, for instance, are within 0.001 of zero.

$$\lim_{j\to\infty} b_j = \lim_{j\to\infty} \frac{(-1)^j}{j} = 0.$$ ■

EXAMPLE 3 Does the sequence $\{c_k\}_{k=0}^{\infty}$ with general term $c_k = (-1)^k$ converge?

Solution No; it diverges. Successive terms have the pattern

$$1, -1, 1, -1, 1, \ldots,$$

never settling on a single limit. ■

Fine Points

Sequences have their own notational quirks and conventions. Here are several to watch for.

Where To Start? The sequence in the preceding example began with c_0, not c_1. Other starting points, such as a_2 or even b_{-3}, occasionally arise. In practice, fortunately, the difference is usually unimportant. What usually matters for sequences is their long-run behavior, not the presence or absence of a few initial terms.

Index Names Don't Matter We can define the squaring function by writing either $f(t) = t^2$ or $f(x) = x^2$; the *variable name* makes no difference. In the same way, a sequence's *index name* is arbitrary: $\{a_k\}_{k=1}^{\infty}$ and $\{a_j\}_{j=1}^{\infty}$ mean exactly the same thing.

Reindexing The sequence

$$\frac{1}{1}, \frac{1}{2}, \frac{1}{3}, \ldots, \frac{1}{10}, \frac{1}{11}, \ldots$$

of Example 1 can be described symbolically in various different-looking, but still equivalent, ways. Here are two:

$$a_k = \frac{1}{k}, \quad k = 1, 2, \ldots \qquad \text{or} \qquad a_j = \frac{1}{j+1}, \quad j = 0, 1, 2, \ldots.$$

Depending on the situation, one description or the other may be preferable.

Sequences as Functions

Sequences are closely related to functions, as expressions such as

$$a_k = \frac{\sin k}{k} \qquad \text{and} \qquad f(x) = \frac{\sin x}{x}$$

illustrate. The formal definition of sequence makes this connection precise.

> **Definition** An **infinite sequence** is a real-valued function that is defined for *positive integer inputs.*

Alternative Views of Sequences. With the definition and the preceding examples, we have several useful ways to think of a sequence.

As a List As an infinite list of numbers: $a_1, a_2, a_3, \ldots$

As a Function As a function $a(n)$, where n takes only positive integer values. ◄ Thus, $a(1) = a_1$, $a(2) = a_2$, $a(3) = a_3$, and so on.

The function a may or may not make sense for other inputs.

As a Discrete Sample As a discrete sample of values of an ordinary function $f(x)$, defined for real $x \geq 1$. ◄

Most of the preceding examples are of this type.

Graphs of Sequences, Graphs of Functions

The graph of a function f consists of all points $\big(x, f(x)\big)$, as x runs through the domain of f. The graph of a sequence $\{a_k\}$ is therefore the set of points (k, a_k), as k runs through positive integer values.

EXAMPLE 4 Let a function f and a sequence $\{a_k\}$ be defined by

$$f(x) = \frac{\sin x}{x}; \qquad a_k = \frac{\sin k}{k}.$$

Plot graphs of both. What do the graphs say about limits?

Solution Here are the graphs:

Graphs of $f(x) = \sin(x)/x$ and $a_k = \sin(k)/k$

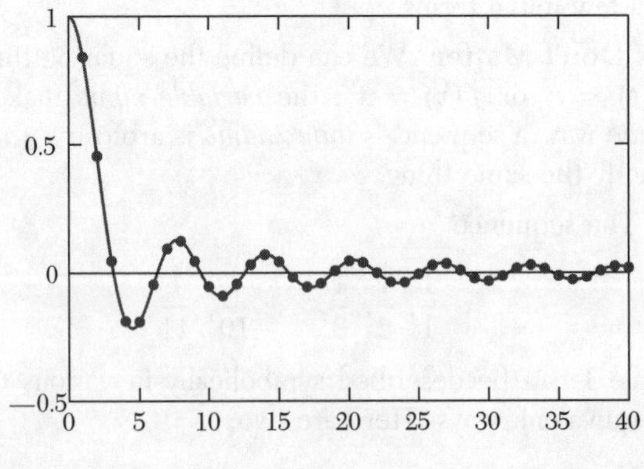

They show how the sequence $\{a_k\}$ is a "discrete sample" of the continuous function f. They show, too, that as x (or k) tends to infinity, this sequence and function, although oscillating in sign, tend to zero. ■

Not every sequence comes naturally from a familiar function. Nevertheless, graphs or tables often suggest limits.

EXAMPLE 5 A sequence $\{b_j\}$ has general term

$$b_j = 1 \cdot 2 \cdot 3 \cdot 4 \cdots (j-1) \cdot j = j!$$

(In words, b_j is j **factorial**.) Tabulate $\{b_j\}$; find its limit, if any.

Solution As the table suggests, $\{b_j\}$ diverges—quickly—to infinity:

As $k \to \infty$, $k! \to \infty$: Explosive Numerical Evidence							
k	1	2	4	8	16	32	64
$k!$	1	2	24	40,320	2.092×10^{13}	2.631×10^{35}	1.269×10^{89}

Can any doubt remain? ▶ ■

Graphs don't work well for this sequence. Do you see why not?

Limits of Sequences, Limits of Functions

Sequences are special sorts of functions; limits of sequences, therefore, are mild variants on limits at infinity. An informal definition will suffice. ▶

A formal definition describes more precisely what "approaches" means.

> **Definition** (Limit of a Sequence) Let $\{a_k\}$ be a sequence and L a real number. If a_k approaches L to within any desired tolerance as k increases without bound, then the sequence converges to L. In symbols,
>
> $$\lim_{k \to \infty} a_k = L.$$
>
> Otherwise, the sequence diverges.

Notice:

Divergence to Infinity If either $a_k \to \infty$ or $a_k \to -\infty$ as k increases without bound, then the sequence diverges to (positive or negative) infinity. We write, for instance,

$$\lim_{k \to \infty} k! = \infty.$$

Other Divergence Behavior Example 3 shows that a divergent sequence need not "blow up"; other patterns of "wandering" behavior (or no pattern at all) are possible.

Asymptotes A sequence, like a function, converges to a finite limit L if and only if its graph has a horizontal asymptote at $y = L$.

Sequences, Functions, and Limits. Many sequences, as we've said, are "discrete samples" of familiar functions. For such sequences we can often use our knowledge of the underlying function.

> **Fact** Let f be a function defined for $x \geq 1$. If $\lim_{x\to\infty} f(x) = L$, and $a_k = f(k)$ for all $k \geq 1$, then $\lim_{k\to\infty} a_k = L$.

EXAMPLE 6 Does the sequence with general term $a_k = \sin k$ converge or diverge?

Solution Here's the surprising graph.

The surprising graph of $a_k = \sin k$: structure but no convergence

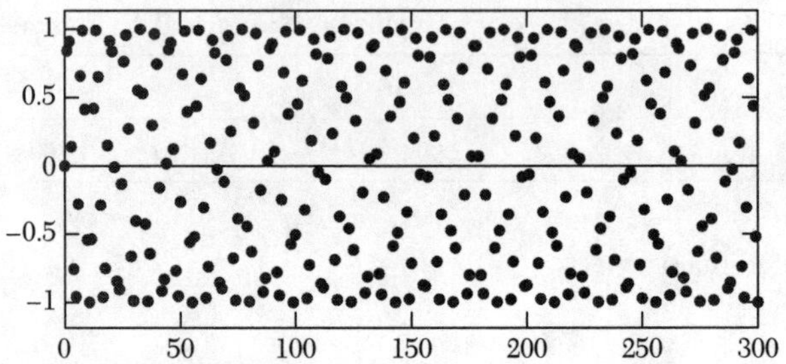

Although full of fascinating shapes, the graph never settles on a single limit, so the sequence diverges. ◀ ■

This example is based on an idea of Gilbert Strang.

Limits of Sequences and l'Hôpital's Rule

If a sequence happens to be given by a symbolic formula, l'Hôpital's rule can often be applied to the formula to find the limit.

EXAMPLE 7 Numerical evidence suggests that if $a_k = 2^k/k^2$, then $\{a_k\}$ diverges to infinity. Use l'Hôpital's rule to show this result symbolically.

Solution l'Hôpital's rule, applied to the continuous function $h(x) = 2^x/x^2$, says that ◀

Differentiate for yourself!

$$\lim_{x\to\infty} \frac{2^x}{x^2} = \lim_{x\to\infty} \frac{2^x \ln 2}{2x} = \lim_{x\to\infty} \frac{2^x \ln 2 \cdot \ln 2}{2} = \infty.$$

(The last equation holds because the numerator blows up as the denominator remains fixed.) Because $h(x) \to \infty$, $a_k \to \infty$ too. ■

EXAMPLE 8 Show that $\lim_{n\to\infty} n^{1/n} = 1$.

Solution l'Hôpital's rule doesn't apply directly, because no fraction is involved. Applying the natural logarithm produces a fraction:

$$a_n = n^{1/n} \Longrightarrow \ln(a_n) = \frac{\ln n}{n}.$$

Now l'Hôpital's rule *does* apply:

$$\lim_{n\to\infty} \frac{\ln n}{n} = \lim_{n\to\infty} \frac{1/n}{1} = 0.$$

We've shown that as $n \to \infty$, $\ln(a_n) \to 0$; it follows that $a_n \to 1$. ■

New Sequence Limits from Old

Limits of sequences, like limits of functions, can be combined in various ways to give *new* limits.

Standard Examples

Before combining limits, we'll need some known limits to combine. The following limits can be thought of as basic building blocks.

$$\lim_{n\to\infty} n^{1/n} = 1$$

$$\lim_{n\to\infty} x^{1/n} = 1 \quad (\text{for all } x > 0)$$

$$\lim_{n\to\infty} \frac{1}{n^k} = 0 \quad (\text{for all } k > 0)$$

$$\lim_{n\to\infty} r^n = 0 \quad (\text{if } -1 < r < 1)$$

(We derived the first of these limits in Example 8, using l'Hôpital's rule. The other limits are easier.)

Algebraic Combinations

Plausible-seeming calculations such as this one—

$$\lim_{k\to\infty}\left(\frac{1}{k} + \frac{3k}{k+1}\right) = \lim_{k\to\infty}\frac{1}{k} + 3\lim_{k\to\infty}\frac{k}{k+1} = 0 + 3 = 3,$$

rely implicitly on the following theorem. We've already seen it, in virtually identical form, for functions.

Theorem 1 **(Algebra with Limits)** Suppose that as $k \to \infty$,

$$a_k \to L \quad \text{and} \quad b_k \to M,$$

where L and M are finite numbers. Let c be any real constant. Then

$$(ca_k) \to cL, \qquad (a_k \pm b_k) \to L \pm M, \qquad \text{and} \qquad a_k b_k \to LM.$$

If $M \neq 0$, then $\dfrac{a_k}{b_k} \to \dfrac{L}{M}$.

The Squeeze Principle

For sequences as for functions, new, unknown limits can sometimes be found by "squeezing" them between old, known limits.

Theorem 2 **(The Squeeze Principle)** Suppose that for all $k > 0$,

$$a_k \le b_k \le c_k \quad \text{and} \quad \lim_{k\to\infty} a_k = \lim_{k\to\infty} c_k = L.$$

Then $\lim_{k\to\infty} b_k = L$.

EXAMPLE 9 Explain legalistically why $\lim_{k\to\infty} \frac{\sin k}{k} = 0$.

Solution The "squeeze inequality"

$$-\frac{1}{k} \le \frac{\sin k}{k} \le \frac{1}{k}$$

holds for all integers $k > 0$. As $k \to \infty$, both the left and the right sides tend to 0; so, therefore, must the middle expression. ■

When Must a Sequence Converge? An Existence Theorem

The ideal way to show that a sequence converges is to find a limit. Alas, sometimes that's difficult.

EXAMPLE 10 Does the sequence with general term

$$a_n = \left(1 + \frac{1}{n}\right)^n$$

converge or diverge? Why?

Solution The answer isn't obvious at a glance, so let's tabulate some values. ◀

A graph would show the same thing.

As $n \to \infty$, $(1 + 1/n)^n$ Appears to Converge

n	16	32	64	128	256	512	1024
$(1 + 1/n)^n$	2.63793	2.67699	2.69734	2.70774	2.71299	2.71563	2.71696

As $n \to \infty$, a_n seems to increase but not blow up. Therefore, apparently, the sequence converges to some limit. ◀ ■

The number $e \approx 2.71828$ is a reasonable guess.

A Convergence Theorem. The sequence in Example 10 seems to converge for two reasons: (1) it's increasing (i.e., $a_1 \le a_2 \le a_3 \ldots$), and (2) it's bounded above. Common sense suggests that any increasing sequence must either converge to a limit—as the preceding sequence appears to do—or blow up to ∞. Similarly, common sense suggests that a decreasing sequence (i.e., a sequence for which $a_1 \ge a_2 \ge a_3 \ldots$) either converges to a limit or diverges to $-\infty$. ◀

Think it through; do you agree?

The following theorem corroborates these commonsense impressions. It applies to any **monotone sequence**, i.e., any sequence that is *either* increasing or decreasing.

Theorem 3 Suppose that the sequence $\{a_k\}$ is *increasing* and *bounded above* by a number A. In other words,

$$a_1 \le a_2 \le a_3 \le \ldots a_k \le a_{k+1} \le \ldots \le A.$$

Then $\{a_k\}$ converges to some finite limit a, with $a \le A$. Similarly, if $\{b_k\}$ is *decreasing* and *bounded below* by a number B, then $\{b_k\}$ converges to a finite limit b, with $b \ge B$.

EXAMPLE 11 Consider the increasing sequence $\{a_k\}$ defined by

$$a_1 = 0; \qquad a_{k+1} = \sqrt{6 + a_k} \qquad \text{if } k \geq 1.$$

Show that $\{a_k\}$ converges and that the limit is less than or equal to 3.

Solution The first few terms suggest (and we take it for granted) that the sequence is increasing:

$$a_1 = 0; \qquad a_2 = \sqrt{6} \approx 2.45; \qquad a_3 \approx 2.91; \qquad a_4 \approx 2.98; \qquad a_5 \approx 2.997.$$

The sequence also seems to be bounded above by 3. This is easy to show. If $a_k < 3$, then, by definition,

$$a_{k+1} = \sqrt{6 + a_k} < \sqrt{6 + 3} = 3.$$

Thus, *no* term can exceed 3. It follows from the theorem, therefore, that the sequence converges to a limit no greater than 3. (The limit *is* 3, in fact, but that requires further proof.) ■

BASICS

In Exercises 1–4, find a symbolic expression for the general term a_k of the sequence.

1. $1, -\frac{1}{3}, \frac{1}{9}, -\frac{1}{27}, \frac{1}{81}, \ldots$
2. $3, 6, 9, 12, 15, \ldots$
3. $\frac{1}{2}, \frac{2}{4}, \frac{3}{8}, \frac{4}{16}, \frac{5}{32}, \ldots$
4. $\frac{1}{2}, \frac{1}{5}, \frac{1}{10}, \frac{1}{17}, \frac{1}{26}, \frac{1}{37}, \ldots$

In Exercises 5–20, first calculate a_1, a_2, a_5, and a_{10}, then determine $\lim_{k\to\infty} a_k$ (if the limit exists). (Round your answers to four decimal places.)

5. $a_k = (-3/2)^k$
6. $a_k = (-0.8)^k$
7. $a_k = (1.1)^k$
8. $a_k = \left(\sqrt{26}/17\right)^k$
9. $a_k = \left(\dfrac{\pi}{e}\right)^k$
10. $a_k = e^{-k}$
11. $a_k = (\arcsin 1)^k$
12. $a_k = (1/k)^k$
13. $a_k = \sin k$
14. $a_k = \arctan k$
15. $a_k = \cos(1/k)$
16. $a_k = \sin(k\pi)$
17. $a_k = \cos(k\pi)$
18. $a_k = \dfrac{k^2}{k^2 + k + 3}$
19. $a_k = \sqrt{\dfrac{2k}{k+3}}$
20. $a_k = \dfrac{k}{\sqrt{k} + 10}$

FURTHER EXERCISES

In Exercises 21–26, calculate a_1, a_{10}, and a_{100} (to four decimal places), then find the limit of the sequence. [**HINT:** l'Hôpital's rule may be useful.]

21. $a_n = \dfrac{n+2}{n^3 + 4}$
22. $a_m = m^2/e^m$
23. $b_j = \ln j/\sqrt[3]{j}$
24. $c_k = \dfrac{\ln(1 + k^2)}{\ln(4 + 3k)}$
25. $d_n = n\sin(1/n)$
26. $a_k = (2^k + 3^k)^{1/k}$
27. Show that $\lim_{n\to\infty} x^{1/n} = 1$ for all $x > 0$.
28. Let $a_k = \left(1 + \dfrac{x}{k}\right)^k$, where x is a real number.
    (a) Show that $\lim_{k\to\infty} \ln(a_k) = x$.
    (b) Use part (a) to evaluate $\lim_{k\to\infty} a_k$.
29. Evaluate $\displaystyle\lim_{n\to\infty}\left(1 - \frac{1}{2n}\right)^n$.

In Exercises 30–38, first calculate a_1, a_2, a_5, and a_{10}, then determine $\lim_{k\to\infty} a_k$ (if the limit exists). (Round your answers to four decimal places.)

30. $a_k = \dfrac{k!}{(k+1)!}$

31. $a_k = \ln\left(\dfrac{k}{k+1}\right)$

32. $a_k = \displaystyle\int_0^k e^{-x}\,dx$

33. $a_k = \displaystyle\int_k^\infty \frac{dx}{1+x^2}$

34. $a_k = \dfrac{\cos k}{\ln(k+1)}$

35. $a_k = e^{-k}\sin k$

36. $a_k = \sqrt{k^2+1} - k$

37. $a_k = \sqrt{k^2+k} - k$

38. $a_k = 3^{1/k}$

39. Give an example of a sequence that is
(a) convergent but not monotone.
(b) bounded but not monotone.
(c) monotone but not convergent.
(d) monotone decreasing and unbounded.
(e) monotone decreasing and convergent.
(f) unbounded but not monotone.

40. Let $a_n = \displaystyle\sum_{k=1}^{n} \frac{k}{n^2}$.
(a) Evaluate a_{10}.
(b) Explain why $\lim_{n\to\infty} a_n = \int_0^1 x\,dx = 1/2$.
[**HINT:** $k/n^2 = (k/n)\cdot(1/n)$.]

41. Define a sequence $\{a_n\}$ by $a_n = \displaystyle\sum_{k=1}^{n} \frac{1}{n+k}$.
(a) Show that this is a monotonically increasing sequence (i.e., $a_{n+1} > a_n$).
(b) Show that $a_n \le \dfrac{n}{n+1} < 1$.
(c) What do parts (a) and (b) imply about $\lim_{n\to\infty} a_n$?
(d) Explain why $\lim_{n\to\infty} a_n > 1/2$. [**HINT:** $a_1 = 1/2$.]
(e) Show that $\lim_{n\to\infty} a_n = \ln 2$. [**HINT:** $\dfrac{1}{n+k} = \dfrac{1}{1+k/n}\cdot\dfrac{1}{n}$, so a_n is a Riemann-sum approximation to an integral.]

42. Let $a_n = \sum_{k=1}^{n} k^2/n^3$. Evaluate $\lim_{n\to\infty} a_n$. [**HINT:** Think of a_n as a Riemann-sum approximation to an integral.]

43. Let $a_n = \dfrac{\sqrt[n]{n!}}{n}$.
(a) Show that $\ln a_n = \dfrac{1}{n}\displaystyle\sum_{k=1}^{n}\ln k - \frac{1}{n}\sum_{k=1}^{n}\ln n$.
(b) Use part (a) to show that $\ln a_n$ is a right-sum approximation to $\int_0^1 \ln x\,dx$.
(c) Use part (b) to show that $\lim_{n\to\infty} a_n = e^{-1}$.

44. Let $a_n = 4^n/n!$.
(a) Find a number N such that $a_{n+1} \le a_n$ for all $n \ge N$.
(b) Use part (a) to explain why $\lim_{n\to\infty} a_n$ exists.
(c) Evaluate $\lim_{n\to\infty} a_n$.

45. Suppose that $\{a_n\}$ is a sequence with the property $|a_{n+1}/a_n| \le (n+3)/(2n+1)$ for all $n \ge 1$. Show that $\lim_{n\to\infty} a_n = 0$. [**HINT:** Start by showing that $|a_{n+1}/a_3| \le (6/7)^{n-2}$ for all $n \ge 3$.]

46. (a) For which values of x does $\lim_{n\to\infty} x^n$ diverge?
(b) Find all values of x for which $\lim_{n\to\infty} x^n$ converges to 0.
(c) Are there any values of x for which $\lim_{n\to\infty} x^n$ converges to a number other than 0? For each such x, find the limit.

In Exercises 47–50, determine the values of x for which the sequence converges as $k\to\infty$. Evaluate $\lim_{k\to\infty} a_k$ for these values of x.

47. $a_k = e^{kx}$

48. $a_k = (\ln x)^k$

49. $a_k = (\arcsin x)^k$

50. $a_k = 2^{-k}(\arctan x)^k$

51. Let $a_n = \cos 1\cdot\cos 2\cdot\cos 3\cdot\cos 4\cdots\cos n$.
(a) Evaluate a_1, a_2, a_3, and a_5. (Round your answers to four decimal places.)
(b) Is the sequence $\{a_n\}$ bounded? Justify your answer.
(c) Is the sequence $\{a_n\}$ monotone? Justify your answer.
(d) Is the sequence $\{|a_n|\}$ bounded? Justify your answer.
(e) Is the sequence $\{|a_n|\}$ monotone? Justify your answer.

52. Use Theorem 3 to show that the sequence 0.7, 0.77, 0.777, 0.7777, 0.77777, 0.777777, 0.7777777, ... has a limit.

53. Let $a_n = \dfrac{1\cdot 3\cdot 5\cdots(2n-1)}{2\cdot 4\cdot 6\cdots(2n)}$.
(a) Calculate a_1, a_2, and a_5.
(b) Use Theorem 3 to show that $\lim_{n\to\infty} a_n$ is a finite number.

54. Show that $\displaystyle\lim_{n\to\infty} \sin\left(\frac{\pi}{2^2}\right)\cdot\sin\left(\frac{\pi}{3^2}\right)\cdots\sin\left(\frac{\pi}{n^2}\right) = 0$.
[**HINT:** $0 < \sin x < x$ when $0 < x < 1$.]

55. Does the sequence defined by $a_1 = 1$, $a_{n+1} = 1 - a_n$ converge? Justify your answer.

56. Does the sequence defined by $a_1 = 1$, $a_{n+1} = a_n/2$ converge? Justify your answer.

57. Consider the sequence defined by

$$a_1 = 1, \qquad a_{n+1} = \left(\frac{n}{n+1}\right)a_n.$$

(a) Show that the sequence converges.
(b) Find the limit. [**HINT:** Write out the first few terms and look for a pattern.]

58. Suppose that $\lim_{k\to\infty} a_k = L$, where L is a finite number, and that the terms of the sequence $\{b_k\}$ are defined by $b_k = L - a_k$. Explain why $\lim_{k\to\infty} b_k = 0$.

59. For which values of $x \ge 0$ does the sequence defined by $a_1 = x$, $a_{n+1} = \sqrt{a_n}$ converge? Justify your answer.

60. (a) Suppose that f is a differentiable function and that $f(0) = 0$. Show that $\lim_{n\to\infty} nf(1/n) = f'(0)$.
(b) Use the result in part (a), not l'Hôpital's rule, to evaluate $\lim_{n\to\infty} n \arctan(1/n)$.

61. Let $F_0 = 1$, $F_1 = 1$, and define $F_{n+1} = F_n + F_{n-1}$ for all $n \geq 1$. (The terms of this sequence are the Fibonacci numbers.) Assume that there is a number L such that $\lim_{n\to\infty} F_{n+1}/F_n = L$.
(a) Show that L is a solution of the equation $x^2 = x+1$.
(b) Suppose that x satisfies the equation $x^2 = x+1$. Use mathematical induction to show that $x^n = xF_n + F_{n-1}$ for all $n \geq 1$.
(c) Let $r_1 = (1+\sqrt{5})/2$ and $r_2 = (1-\sqrt{5})/2$. Use part (c) to show that $F_n = (r_1^n - r_2^n)/\sqrt{5}$.
(d) Use part (d) to evaluate L.

11.2 Infinite Series, Convergence, and Divergence

Introduction

An **infinite series**, or **series** for short, is an infinite sum, i.e., an expression of the form ▶

On the left side we used ***sigma notation*** *for brevity.*

$$\sum_{k=1}^{\infty} a_k = a_1 + a_2 + a_3 + a_4 + \cdots + a_k + a_{k+1} + \cdots.$$

Thus, a series results from *adding* the terms of a sequence $a_1, a_2, a_3, \ldots$. If, say, $a_k = 1/k^2$, then

$$\sum_{k=1}^{\infty} a_k = \sum_{k=1}^{\infty} \frac{1}{k^2} = \frac{1}{1} + \frac{1}{4} + \frac{1}{9} + \frac{1}{16} + \cdots.$$

If $a_k = k$, then

$$\sum_{k=1}^{\infty} a_k = \sum_{k=1}^{\infty} k = 1 + 2 + 3 + \cdots.$$

The idea of an infinite sum raises natural questions:

- What does it *mean* to add infinitely many numbers?
- Which series add up to a finite number? Which blow up?
- If a series has a finite sum, how can we find (or estimate) it?

The rest of this chapter addresses these questions.

Improper Sums, Improper Integrals

Standard examples of infinite series include

$$\sum_{k=1}^{\infty} \frac{1}{k^2}, \quad \sum_{k=1}^{\infty} \frac{1}{k}, \quad \sum_{k=1}^{\infty} \frac{1}{\sqrt{k}}, \quad \text{and} \quad \sum_{k=1}^{\infty} \frac{1}{2^k}.$$

The typographical resemblance to the improper integrals

$$\int_1^{\infty} \frac{dx}{x^2}, \quad \int_1^{\infty} \frac{dx}{x}, \quad \int_1^{\infty} \frac{dx}{\sqrt{x}}, \quad \text{and} \quad \int_1^{\infty} \frac{dx}{2^x}$$

is no accident; infinite series and improper integrals are similar in many respects. We'll see, in fact, that the first and fourth of the preceding series converge, while the second and third diverge—exactly as for the corresponding integrals. We'll often return to the important analogy between integrals and series.

Why Series Matter: A Preview

Understanding series takes some legwork. As a first glimpse of why the work is worth doing, consider the fact ◀ that for any real number x,

We'll see at the end of this chapter why it's true.

$$\cos x = 1 - \frac{x^2}{2!} + \frac{x^4}{4!} - \frac{x^6}{6!} + \frac{x^8}{8!} - \cdots. \tag{11.1}$$

If, say, $x = 1$, then ◀

The "dots" in each equation mean that the pattern continues.

$$\cos 1 = 1 - \frac{1}{2!} + \frac{1}{4!} - \frac{1}{6!} + \frac{1}{8!} - \cdots. \tag{11.2}$$

Why should we care about such equations? Why write something familiar—the cosine function—in terms of something exotic—a series?

One good answer is that the cosine function, although crucial in applications, ◀ has no *algebraic* formula. Without a concrete formula, finding accurate numerical values of $\cos x$ for given inputs x is a genuine problem. ◀

That's why every scientific calculator has a COS key.

For a few special values of x, such as $x = 0$, $x = \pi/6$, and $x = \pi/4$, finding $\cos x$ is easy. For most inputs x, it's genuinely hard.

The preceding equations help solve this problem. Equation 11.1, although not a formula in the ordinary sense, ◀ gives a concrete, computable recipe for approximating $\cos x$: Given an input x, calculate as far out as practically possible in the "infinite polynomial"

Ordinary formulas don't have dots in them.

$$1 - \frac{x^2}{2!} + \frac{x^4}{4!} - \frac{x^6}{6!} + \frac{x^8}{8!} - \cdots.$$

With luck, the result should closely approximate the true value of $\cos x$.

Equation 11.2 shows how to approximate $\cos 1$. It's easy to check ◀ that, to seven decimals,

Try it with your calculator.

$$1 - \frac{1}{2!} = 0.5000000;$$

$$1 - \frac{1}{2!} + \frac{1}{4!} = 0.5416667;$$

$$1 - \frac{1}{2!} + \frac{1}{4!} - \frac{1}{6!} = 0.5402778;$$

$$1 - \frac{1}{2!} + \frac{1}{4!} - \frac{1}{6!} + \frac{1}{8!} = 0.5403026.$$

The results converge with gratifying speed to the "right" answer—the true value of $\cos 1$ (≈ 0.5403023).

Open Questions. These calculations raise as many questions as they answer.

> *Where did Equation 11.1 come from? What do all the "dots" really mean? Are similar equations available for other functions—sine, arctangent, logarithm, and so on? How many terms are needed to guarantee accuracy to, say, five decimals?*

Questions such as these guide our study of series.

Convergence: Definitions and Terminology

Working successfully with series requires some up-front investment in definitions and technical language. ◀ After stating terms and definitions, we show by example why they're reasonable.

The payoff comes later, in clarity.

Series Language: Terms, Partial Sums, Tails, Convergence, Limit

Let $\sum_{k=1}^{\infty} a_k = a_1 + a_2 + a_3 + \cdots + a_k + a_{k+1} + \cdots$ be an infinite series. ▶ Recall that the summand a_k is called the kth term of the series. The nth **partial sum**, denoted by S_n, is the (finite) sum of all terms *through index* n:

To keep notation simple, k usually runs from 1 to ∞. Sometimes it's convenient to let k start at 0, 2, or elsewhere.

$$S_n = a_1 + a_2 + a_3 + \cdots + a_{n-1} + a_n = \sum_{k=1}^{n} a_k.$$

The nth tail, denoted by R_n, is the (infinite) sum of all terms *beyond index* n:

$$R_n = a_{n+1} + a_{n+2} + a_{n+3} + \cdots = \sum_{k=n+1}^{\infty} a_k.$$

As the notation R_n suggests, the nth tail is the remainder—what's left after the terms through index n are added. In symbols,

$$\sum_{k=1}^{\infty} a_k = (a_1 + a_2 + \cdots + a_n) + (a_{n+1} + a_{n+2} + a_{n+3} + \cdots) = S_n + R_n.$$

The crucial definition ▶ of convergence involves partial sums:

The notation is as before.

Definition If $\lim_{n\to\infty} S_n = S$, for some finite number S, then the series $\sum_{k=1}^{\infty} a_k$ converges to the limit S. Otherwise, the series diverges.

Notice the following aspects of the definition.

Divergent Series A divergent series is one for which the sequence of partial sums does *not* converge to a finite limit S. One possibility is that the partial sums S_n blow up to infinity. Another possibility is that the partial sums remain bounded, but never settle on a specific limit.

Improper Integrals, Improper Sums The preceding definition says, in symbols, that

$$\sum_{k=1}^{\infty} a_k = \lim_{n\to\infty} \sum_{k=1}^{n} a_k$$

if the limit exists. Convergence for improper integrals means much the same thing:

$$\int_{x=1}^{\infty} f(x)\,dx = \lim_{n\to\infty} \int_{x=1}^{n} f(x)\,dx$$

if *this* limit exists. ▶ An infinite series is an improper sum in exactly the sense that an integral may be improper.

In each case we take the limit of something proper.

What Convergence Means for Partial Sums and Tails To say that $\sum a_k$ converges to the limit (or **sum**) S means that $S_n \to S$ as $n \to \infty$. Since for all n, $S = S_n + R_n$, it follows that $R_n \to 0$ as $n \to \infty$. In particular, to decide whether a series converges, we need to examine the sequence of partial sums.

Two Sequences: Keep Them Straight Every series $\sum a_k$ involves two sequences: the sequence $\{a_k\}$ of *terms* and the sequence $\{S_n\}$ of *partial*

Many, maybe most, student difficulties arise from confusing them.

sums. Keeping these related but different sequences separate is essential. ◀ We'll take special care to do so in the next examples.

EXAMPLE 1 **(A Geometric Series)** Does the series

$$\sum_{k=0}^{\infty} \frac{1}{2^k} = 1 + \frac{1}{2} + \frac{1}{4} + \frac{1}{8} + \cdots$$

converge? If so, to what limit?

Solution In this series, the index k began at 0, not 1. So, therefore, does the sequence of partial sums. Direct calculation yields

$$S_0 = 1 = 2 - 1;$$
$$S_1 = 1 + \frac{1}{2} = \frac{3}{2} = 2 - \frac{1}{2};$$
$$S_2 = 1 + \frac{1}{2} + \frac{1}{4} = \frac{7}{4} = 2 - \frac{1}{4};$$
$$S_3 = 1 + \frac{1}{2} + \frac{1}{4} + \frac{1}{8} = \frac{15}{8} = 2 - \frac{1}{8};$$
$$S_4 = 1 + \frac{1}{2} + \frac{1}{4} + \frac{1}{8} + \frac{1}{16} = \frac{31}{16} = 2 - \frac{1}{16}.$$

The pattern is easy to see: For any $n \geq 1$,

$$S_n = 2 - \frac{1}{2^n}.$$

From this explicit formula for S_n, the conclusion follows: Because $S_n \to 2$ as $n \to \infty$, the series converges to 2. ■

A table of numerical values gives the same impression, and also illustrates the behavior of tails. Note that, for each n,

$$S_n = \sum_{k=0}^{n} \frac{1}{2^k}; \qquad R_n = \sum_{k=n+1}^{\infty} \frac{1}{2^k} = 2 - \sum_{k=0}^{n} \frac{1}{2^k}.$$

Check some for yourself.

Here are the numbers: ◀

Partial Sums and Tails of 1 + 1/2 + 1/4 + ⋯

n	0	1	2	3	4	5	6	7	8	9	10
S_n	1	1.5	1.75	1.875	1.938	1.969	1.984	1.992	1.996	1.998	1.999
R_n	1	0.5	0.25	0.125	0.063	0.031	0.016	0.008	0.004	0.002	0.001

EXAMPLE 2 Does $\sum_{k=1}^{\infty}(-1)^k = -1 + 1 - 1 + 1 - 1 + \cdots$ converge?

Solution No. Successive partial sums are of the form $-1, 0, -1, 0, \ldots$. Since the sequence $\{S_n\}$ diverges, so does the series. ■

EXAMPLE 3 Does the **harmonic series** $\sum_{k=1}^{\infty} \frac{1}{k}$ converge?

Solution The answer depends, as always, on the partial sums S_n. By definition,

$$S_n = 1 + \frac{1}{2} + \frac{1}{3} + \frac{1}{4} + \frac{1}{5} + \frac{1}{6} + \cdots + \frac{1}{n}.$$

This time, unfortunately, no simple formula for S_n in terms of n comes to mind, so we'll investigate the sequence $\{S_n\}$ numerically. A computer makes quick work of the calculations. Here are some results, to three decimal places:

$$S_1 = 1; \quad S_{20} = 3.598; \quad S_{50} = 4.499; \quad S_{100} = 5.187; \quad S_{1000} = 7.485.$$

The numerical evidence is ambiguous; the S_n's seem to keep growing, although slowly. Whether the S_n's converge to some finite number or diverge to infinity is not yet clear. ▶ ■

In fact, the series diverges. We'll show this fact soon by comparing the series to the integral $\int_1^{\infty} 1/x\,dx$.

When Does a Series (or an Integral) Converge?

Deciding whether a given infinite series converges or diverges is a delicate question; so was the analogous question for improper integrals. Such series as

$$\sum_{k=1}^{\infty} \frac{1}{k} \quad \text{and} \quad \sum_{k=1}^{\infty} \frac{1}{k^2}$$

pose the typical dilemma: Although successive terms of both series tend to zero, the *number* of terms is infinite. Convergence or divergence hinges on which of these conflicting tendencies "wins" in the long run.

The same dilemma arose ▶ for the improper integrals

Remember? See Section 10.1.

$$\int_1^{\infty} \frac{1}{x}\,dx \quad \text{and} \quad \int_1^{\infty} \frac{1}{x^2}\,dx.$$

Although the integrands tend to zero as x grows without bound, the total area bounded by their graphs may or may not converge to a finite limit. Indeed, we discovered earlier that these two integrals have opposite outcomes: The first integral diverges to infinity, and the second converges to one.

Convergence vs. Divergence, Graphically

Deciding whether the preceding two series converge or diverge is a bit harder than for the corresponding integrals, because (as we saw in Example 3) no convenient formulas for S_n are available.

For a series $\sum a_k$, plotting both $\{a_k\}$ and $\{S_n\}$ ▶ on the same axes illustrates the connection between the two—and sometimes suggests whether the series converge

The terms and the partial sums, respectively.

or diverges. The harmonic series $\sum 1/k$ generates this picture:

Terms and partial sums for $\sum 1/k$

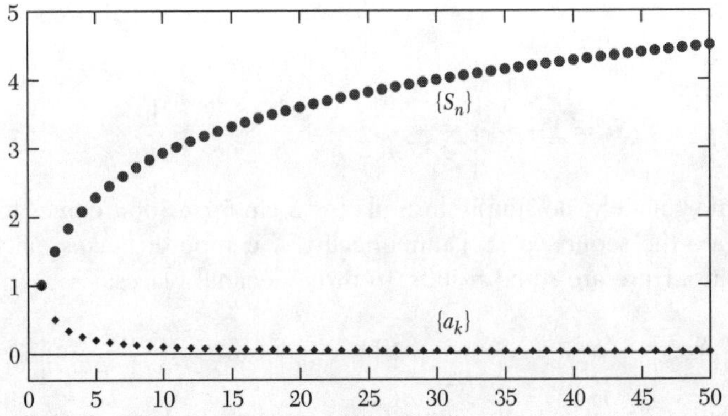

The picture for the series $\sum 1/k^2$ gives a different impression:

Terms and partial sums for $\sum 1/k^2$

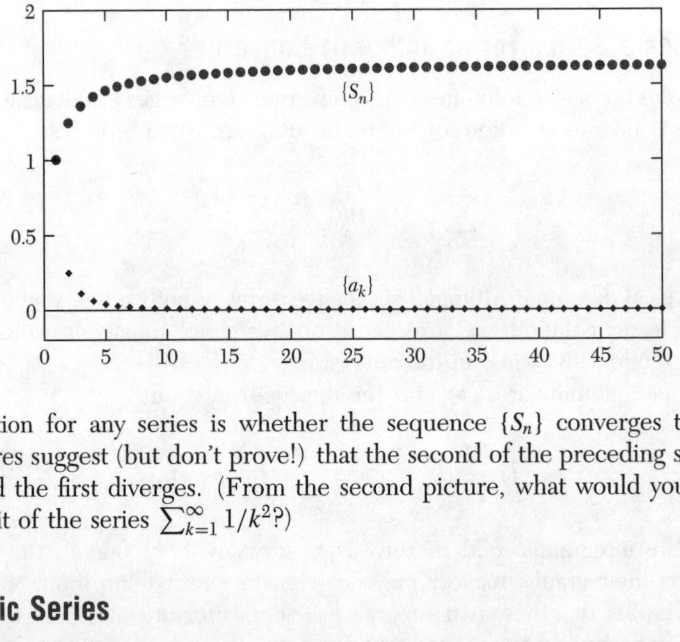

The question for any series is whether the sequence $\{S_n\}$ converges to a limit. The pictures suggest (but don't prove!) that the second of the preceding series converges and the first diverges. (From the second picture, what would you estimate as the limit of the series $\sum_{k=1}^{\infty} 1/k^2$?)

Geometric Series

Geometric series form the simplest and most important class of infinite series. A **geometric series** is one of the form

$$a + ar + ar^2 + ar^3 + ar^4 + \cdots = \sum_{k=0}^{\infty} ar^k;$$

Each term is r times the previous term.

a is called the **leading term**, and r is the **ratio**. ◀ The series

$$1 + \frac{1}{2} + \frac{1}{4} + \frac{1}{8} + \cdots = \sum_{k=0}^{\infty} \left(\frac{1}{2}\right)^k$$

of Example 1 is geometric, with $a = 1$ and $r = 1/2$.

Partial Sums of Geometric Series

Geometric series have one great advantage: It's easy to decide whether they converge or diverge and, if they converge, to find their limits. The reason it's easy is that—unlike for many other series—there's a convenient, explicit formula for an arbitrary partial sum S_n. The formula depends on the fact that if $r \neq 1$ and $n \geq 1$, then

$$1 + r + r^2 + r^3 + \cdots + r^n = \frac{1 - r^{n+1}}{1 - r}.$$

(Multiply both sides by $(1 - r)$ to convince yourself that the formula holds.)

Here, then, is our formula for S_n. If $r \neq 1$ and $n \geq 1$, then

$$S_n = a + ar + ar^2 + ar^3 + \cdots + ar^n = a\frac{1 - r^{n+1}}{1 - r}. \qquad (11.3)$$

From this formula follows the whole story of convergence and divergence for geometric series. As for any series, the present question is how the sequence $\{S_n\}$ of partial sums behaves as n tends to infinity. From the facts that

$$\begin{aligned} &\lim_{n\to\infty} r^n = 0 && \text{if} \quad |r| < 1, \\ &\lim_{n\to\infty} r^n = 1 && \text{if} \quad r = 1, \\ &\lim_{n\to\infty} r^n = \infty && \text{if} \quad r > 1, \\ &\lim_{n\to\infty} r^n \text{ does not exist} && \text{if} \quad r \leq -1, \end{aligned}$$

the conclusion follows. It's worth emphasizing:

> **Fact** If $|r| < 1$, the geometric series
>
> $$\sum_{k=0}^{\infty} ar^k = a + ar + ar^2 + ar^3 + \cdots$$
>
> converges to $\dfrac{a}{1-r}$. If $a \neq 0$ and $|r| \geq 1$, the series diverges.

EXAMPLE 4 The series $\dfrac{1}{3} - \dfrac{1}{6} + \dfrac{1}{12} - \dfrac{1}{24} + \cdots$ converges. To what limit?

Solution The series is geometric, with $a = 1/3$ and $r = -1/2$. It therefore converges to

$$\frac{a}{1-r} = \frac{1/3}{1+1/2} = \frac{2}{9} \approx 0.2222222.$$

A look at partial sums and tails supports the computation.

Partial Sums and Tails of $1/3 - 1/6 + 1/12 - 1/24 + \cdots$

n	0	1	2	3	4	5	6	7	8	9	10
S_n	0.3333	0.1667	0.2500	0.2083	0.2292	0.2188	0.2240	0.2214	0.2227	0.2220	0.2223
R_n	−0.1111	0.0555	−0.0278	0.0139	−0.0069	0.0035	−0.0017	0.0009	−0.0004	0.0002	−0.0001

■

Telescoping Series

Geometric series are the most important variety for which partial sums can be found explicitly. Telescoping series offer the same possibility. The next example illustrates how they work and explains the name "telescoping."

EXAMPLE 5 Show that $\sum_{k=1}^{\infty} \frac{1}{k(k+1)}$ converges, and find its limit.

To convince yourself that the two sides are equal, find a common denominator on the right side.

Solution A bit of algebra lets us rewrite the series in a more helpful form: ◀

$$\sum_{k=1}^{\infty} \frac{1}{k(k+1)} = \sum_{k=1}^{\infty} \left(\frac{1}{k} - \frac{1}{k+1} \right).$$

Writing out a few terms shows the "telescoping" pattern:

$$\begin{aligned} S_n &= \sum_{k=1}^{n} \left(\frac{1}{k} - \frac{1}{k+1} \right) \\ &= \left(\frac{1}{1} - \frac{1}{2} \right) + \left(\frac{1}{2} - \frac{1}{3} \right) + \left(\frac{1}{3} - \frac{1}{4} \right) + \cdots + \left(\frac{1}{n} - \frac{1}{n+1} \right) \\ &= 1 - \frac{1}{n+1}. \end{aligned}$$

Now it's clear that, as $n \to \infty$, $S_n \to 1$. That's the limit.

Watch the tails go to zero.

Again the numbers agree: ◀

Partial Sums and Tails of $\sum_{k=1}^{\infty} 1/(k^2+k)$

n	1	2	3	4	5	6	7	8	9	10	11
S_n	0.5000	0.6667	0.7500	0.8000	0.8333	0.8571	0.8750	0.8889	0.9000	0.9091	0.9167
R_n	0.5000	0.3333	0.2500	0.2000	0.1667	0.1429	0.1250	0.1111	0.1000	0.0909	0.0833

■

Algebra with Series

As with functions and sequences, combining series algebraically produces new series. Combining *convergent* series produces new *convergent* series, with limits related in the expected way.

Theorem 4 Suppose that $\sum_{k=1}^{\infty} a_k$ converges to S and that $\sum_{k=1}^{\infty} b_k$ converges to T. Let c be any constant. Then

$$\sum_{k=1}^{\infty} (a_k \pm b_k) \qquad \text{converges to } S \pm T;$$

$$\sum_{k=1}^{\infty} c a_k \qquad \text{converges to } cS.$$

These reasonable-looking properties of convergent *series* follow directly from the analogous properties of convergent *sequences*. ▶

After all, the limit of a series is defined as the limit of the sequence of partial sums.

A little series algebra, cleverly applied, can immensely simplify finding the limits of certain series.

EXAMPLE 6 Evaluate $\sum_{k=0}^{\infty} \frac{4+2^k}{3^k}$.

Solution The given series is the sum of two convergent geometric series. ▶ The theorem says, therefore, that

We can find limits of geometric series.

$$\sum_{k=0}^{\infty} \frac{4+2^k}{3^k} = \sum_{k=0}^{\infty} \frac{4}{3^k} + \sum_{k=0}^{\infty} \left(\frac{2}{3}\right)^k = 6+3 = 9.$$ ■

EXAMPLE 7 For the geometric series $\sum_{k=0}^{\infty} \frac{3}{2^k}$, calculate the tail R_{10}.

Solution A little algebra is all we need:

$$\begin{aligned} R_{10} = \sum_{k=11}^{\infty} \frac{3}{2^k} &= \frac{3}{2^{11}} + \frac{3}{2^{12}} + \frac{3}{2^{13}} + \cdots \\ &= \frac{3}{2^{11}}\left(1 + \frac{1}{2} + \frac{1}{2^2} + \frac{1}{2^3} + \cdots\right) \\ &= \frac{3}{2^{11}} \cdot 2 = \frac{3}{2^{10}} = \frac{3}{1024}. \end{aligned}$$

(The geometric series in the middle line sums to 2; hence the final answer.) ■

Detecting Divergent Series

For us, *convergent* series and their limits are mainly of interest. The preceding theorem, for instance, applies safely only to convergent series. It's important, therefore, to recognize *divergence* when it occurs. The **nth term test** sometimes helps.

A series $\sum a_k$ converges if and only if its partial sums S_n converge to a limit. For this to occur, the difference $S_n - S_{n-1}$ between successive partial sums must tend to zero. ▶ This difference is simply the nth term:

Otherwise the partial sum sequence wouldn't "level off."

$$S_n - S_{n-1} = (a_1 + a_2 + \cdots + a_{n-1} + a_n) - (a_1 + a_2 + \cdots + a_{n-1}) = a_n.$$

Thus, for a series to converge, its *terms* must tend to zero. We restate this fact in its most useful form:

> **Theorem 5** (The nth Term Test for Divergence) If $\lim_{n\to\infty} a_n \neq 0$, then $\sum a_n$ diverges.

What the Theorem Doesn't Say. It's important to notice that Theorem 5 does *not* guarantee that $\sum a_n$ converges if $a_n \to 0$. The harmonic series $\sum 1/k$ illustrates that the terms of a divergent series may tend to zero. The rather blunt nth term test may detect *divergence*, but never convergence. Sharper instruments are needed to detect convergence. We develop several in the next section.

EXAMPLE 8 Does $\displaystyle\sum_{k=1}^{\infty} \frac{k}{k+1000}$ converge?

Solution The nth term test says no. Because

$$a_n = \frac{n}{n+1000} \to 1 \quad \text{as} \quad n \to \infty,$$

the series diverges. ■

EXAMPLE 9 Given that $\displaystyle\sum_{k=0}^{\infty} \frac{3^k}{k!}$ converges, find $\displaystyle\lim_{k\to\infty} \frac{3^k}{k!}$.

Solution The limit is zero. By the preceding theorem, the terms of any convergent series must tend to zero. (Tabulating values ◀ of $3^n/n!$ supports this conclusion.) ■

Try some.

BASICS

1. This exercise is about partial sums of the geometric series $\sum_{k=0}^{\infty} ar^k$. By definition, a partial sum S_n of this series is $S_n = \sum_{k=0}^{n} ar^k = a + ar + ar^2 + ar^3 + \cdots + ar^n$. According to Equation 11.3, page 317, there's a simpler, more explicit formula for S_n:
$$S_n = a\,\frac{1-r^{n+1}}{1-r}.$$
   (a) Let $a = 1$, $r = 2$, and $n = 5$. Calculate both sides of the expression. Are they equal?
   (b) Repeat part (a) using $a = 1$, $r = 0.01$, and $n = 5$.
   (c) Suppose that $r = 1$. Explain why the formula does not hold in this case. Then find a formula for S_n when $r = 1$.
   (d) Calculate and simplify the quantity $S_n - rS_n$.
   (e) Use your result from part (d) to show that the formula holds if $r \neq 1$. [**HINT:** $S_n - rS_n = (1-r)S_n$.]
   (f) Find $3 + 6 + 12 + 24 + 48 + 96 + \cdots + 3072$. [**HINT:** Use the formula for S_n.]

2. The series $\displaystyle\sum_{k=0}^{\infty} a_k = \sum_{k=0}^{\infty} \frac{1}{k!}$ converges to $e \approx 2.718282$.
   (a) Evaluate a_1, a_2, a_5, and a_{10}.
   (b) Evaluate S_1, S_2, S_5, and S_{10}.
   (c) Explain why $\{S_n\}$ is an increasing sequence.
   (d) Evaluate R_1, R_2, R_5, and R_{10}.
   (e) Show that $R_n > 0$ for all $n \geq 0$.
   (f) Explain why $\{R_n\}$ is a decreasing sequence.
   (g) Using a calculator or a computer, determine a value of n for which S_n differs from e by less than 0.001.
   (h) Using a calculator or a computer, determine a value of n for which S_n differs from e by less than 10^{-5}.
   (i) Use parts (f) and (h) to show that $R_{50} < 10^{-5}$.
   (j) Explain why $\lim_{n\to\infty} R_n = 0$.

3. The series $\sum_{k=1}^{\infty} a_k = \sum_{k=1}^{\infty} \frac{1}{k^2}$ converges to $\frac{\pi^2}{6} \approx 1.64493$.
   (a) Evaluate a_1, a_2, a_5, and a_{10}.
   (b) Evaluate S_1, S_2, S_5, and S_{10}.
   (c) Is $\{S_n\}$ an increasing sequence? Justify your answer.
   (d) Evaluate R_1, R_2, R_5, and R_{10}.
   (e) Is $\{R_n\}$ a decreasing sequence? Justify your answer.
   (f) Show that $0 < R_n < 0.05$ for all $n \geq 20$.
   (g) Evaluate $\lim_{n\to\infty} R_n$.

4. Consider the series $\sum_{k=0}^{\infty} a_k = \sum_{k=0}^{\infty} \frac{1}{5^k}$.
   (a) Evaluate a_1, a_2, a_5, and a_{10}.
   (b) Evaluate S_1, S_2, S_5, and S_{10}.
   (c) Show that the sequence $\{S_n\}$ is increasing and bounded above. What does this imply about the sequence of partial sums?
   (d) Find the sum of the series (i.e., $\lim_{n\to\infty} S_n$).
   (e) Evaluate R_1, R_2, R_5, and R_{10}.
   (f) Show that $\{R_n\}$ is decreasing and bounded below.
   (g) Evaluate $\lim_{n\to\infty} R_n$.

5. Consider the series $\sum_{k=0}^{\infty} a_k = \sum_{k=0}^{\infty} (-0.8)^k$.
   (a) Evaluate a_1, a_2, a_5, and a_{10}.
   (b) Evaluate S_1, S_2, S_5, and S_{10}.
   (c) Find the sum of the series (i.e., $\lim_{n\to\infty} S_n$).
   (d) Evaluate R_1, R_2, R_5, and R_{10}.
   (e) Is the sequence $\{S_n\}$ increasing? Justify your answer.
   (f) Show that the sequence $\{R_n\}$ is neither increasing nor decreasing.
   (g) Show that the sequence $\{|R_n|\}$ is decreasing.
   (h) Evaluate $\lim_{n\to\infty} R_n$.

6. Consider the series $\sum_{k=0}^{\infty} a_k = \sum_{k=0}^{\infty} \frac{1}{k+2^k}$.
   (a) Evaluate S_1, S_2, S_5, and S_{10}.
   (b) Show that the sequence $\{S_n\}$ is increasing.
   (c) Explain why $a_k \leq 2^{-k}$ for all $k \geq 0$.
   (d) Use part (c) to show that $S_n \leq 2 - 2^{-n} < 2$. [**HINT:** $\sum_{k=0}^{n} 2^{-k}$ is a geometric series.]
   (e) Use parts (b) and (d) to show that the series $\sum_{k=0}^{\infty} a_k$ converges (i.e., that $\lim_{n\to\infty} S_n$ exists).
   (f) Show that $\lim_{n\to\infty} R_n = 0$.

7. Consider the series $\sum_{j=0}^{\infty} a_j = \sum_{j=0}^{\infty} \frac{1}{2+3^j}$.
   (a) Evaluate S_1, S_2, S_5, and S_{10}.
   (b) Show that the sequence $\{S_n\}$ is increasing and bounded above.
   (c) Does $\sum_{j=0}^{\infty} a_j$ converge? Justify your answer.

8. It is known that $\sum_{m=1}^{\infty} \frac{1}{m^4} = \frac{\pi^4}{90}$. Use this fact to evaluate
   (a) $\sum_{i=0}^{\infty} \frac{1}{(i+1)^4}$
   (b) $\sum_{k=3}^{\infty} \frac{1}{k^4}$

In Exercises 9–16, find the limit of the series.

9. $\frac{1}{16} + \frac{1}{32} + \frac{1}{64} + \frac{1}{128} + \cdots + \frac{1}{2^{i+4}} + \cdots$

10. $2 - 5 + 9 + \frac{1}{3} + \frac{1}{9} + \frac{1}{27} + \frac{1}{81} + \cdots + \frac{1}{3^n} + \cdots$

11. $\sum_{n=0}^{\infty} e^{-n}$

12. $\sum_{k=3}^{\infty} \left(\frac{e}{\pi}\right)^k$

13. $\sum_{m=1}^{\infty} (\arctan 1)^m$

14. $\sum_{i=10}^{\infty} \left(\frac{2}{3}\right)^i$

15. $\sum_{j=5}^{\infty} \left(-\frac{1}{2}\right)^j$

16. $\sum_{j=0}^{\infty} \frac{3^j + 4^j}{5^j}$

17. Show that the series $\sum_{k=1}^{\infty} \frac{1}{2+\sin k}$ diverges.

18. Use partial sums to explain why the series $\sum_{k=0}^{\infty} (-1)^k$ diverges.

In Exercises 19–26, find an expression for the partial sum S_n of the series. Use this expression to determine whether the series converges and, if so, to find its limit.

19. $\sum_{k=0}^{\infty} \big(\arctan(k+1) - \arctan k\big)$

20. $\sum_{i=2}^{\infty} \frac{1}{i(i-1)} = \sum_{i=2}^{\infty} \left(\frac{1}{i-1} - \frac{1}{i}\right)$

21. $\sum_{k=1}^{\infty} \frac{2}{k^2+k}$

22. $\sum_{j=1}^{\infty} \frac{j}{(j+1)!}$

23. $\sum_{m=1}^{\infty} \left(\frac{1}{\sqrt{m}} - \frac{1}{\sqrt{m+2}}\right)$

24. $\sum_{j=1}^{\infty} \ln\left(1 + \frac{1}{j}\right) = \sum_{j=1}^{\infty} \big(\ln(j+1) - \ln j\big)$

25. $\sum_{k=0}^{\infty} \cos(k\pi)$

26. $\sum_{j=2}^{\infty} (-1)^j j$

FURTHER EXERCISES

27. It is known that

$$\sum_{i=1}^{\infty} \frac{1}{i^2} = 1 + \frac{1}{4} + \frac{1}{9} + \frac{1}{16} + \frac{1}{25} + \cdots = \frac{\pi^2}{6}.$$

(a) Use this fact to evaluate

$$\sum_{j=1}^{\infty} \frac{1}{(2j)^2} = \frac{1}{4} + \frac{1}{16} + \frac{1}{36} + \frac{1}{64} + \frac{1}{100} + \cdots.$$

(b) Use part (a) to show that

$$\sum_{k=0}^{\infty} \frac{1}{(2k+1)^2} = 1 + \frac{1}{9} + \frac{1}{25} + \frac{1}{49} + \cdots = \frac{\pi^2}{8}.$$

(c) Evaluate

$$\sum_{m=1}^{\infty} \frac{(-1)^{m+1}}{m^2} = 1 - \frac{1}{4} + \frac{1}{9} - \frac{1}{16} + \frac{1}{25} - \frac{1}{36} + \cdots.$$

28. Express $\sum_{m=3}^{\infty} \frac{2^{m+4}}{5^m}$ as a rational number.

For each of the series in Exercises 29–37, find all values of x for which the series converges, then state the limit as a simple expression involving x. (Assume that $x^0 = 1$ for all x.)

29. $\sum_{k=0}^{\infty} x^k$

30. $\sum_{m=2}^{\infty} \left(\frac{x}{5}\right)^m$

31. $\sum_{j=5}^{\infty} x^{2j}$

32. $\sum_{k=1}^{\infty} x^{-k}$

33. $\sum_{n=3}^{\infty} (1+x)^n$

34. $\sum_{j=4}^{\infty} \frac{1}{(1-x)^j}$

35. $\sum_{k=11}^{\infty} \left(\frac{\sin x}{2}\right)^k$

36. $\sum_{m=2}^{\infty} (\ln x)^m$

37. $\sum_{n=0}^{\infty} (\arctan x)^n$

38. Find the limit of the sequence defined by $S_1 = 1$, $S_{n+1} = S_n + 1/3^n$. [**HINT:** Write out the first few terms to see the pattern.]

39. Find the limit of the sequence defined by $a_1 = 4$, $a_{n+1} = a_n - 1/2^n$.

In Exercises 40–55, determine whether the series converges or diverges. If a series converges, find its limit. Justify your answers.

40. $2 - 2 + 2 - 2 + 2 - 2 + \cdots$

41. $\frac{3}{10} - \frac{3}{20} + \frac{3}{40} - \frac{3}{80} + \frac{3}{160} - \frac{3}{320} + \cdots$

42. $1 - \frac{1}{2} + \frac{1}{2} - \frac{1}{3} + \frac{1}{3} - \cdots$

43. $1 - 1 + 2 - 1 - 1 + 3 - 1 - 1 - 1 + 4 - 1 - 1 - 1 - 1 + \cdots$

44. $\frac{4}{7^{10}} + \frac{4}{7^{12}} + \frac{4}{7^{14}} + \frac{4}{7^{16}} + \frac{4}{7^{18}} + \cdots$

45. $\sum_{n=0}^{\infty} \frac{n+1}{2n+1}$

46. $\sum_{j=0}^{\infty} (\ln 2)^j$

47. $\sum_{n=2}^{\infty} \frac{2}{n^2 - 1}$

48. $\sum_{k=1}^{\infty} \frac{k^\pi}{k^e}$

49. $\sum_{m=2}^{\infty} \frac{1}{(\ln 3)^m}$

50. $\sum_{j=0}^{\infty} \left(\frac{1}{2^j} + \frac{1}{3^j}\right)^2$

51. $\sum_{k=1}^{\infty} \left(\int_k^{k+1} \frac{dx}{x^2}\right)$

52. $\sum_{m=0}^{\infty} \left(\int_0^m e^{-x^2}\, dx\right)$

53. $\sum_{n=1}^{\infty} \left(1 + \frac{1}{n}\right)^n$

54. $\sum_{j=1}^{\infty} \sqrt[j]{\pi}$

55. $\sum_{n=1}^{\infty} \frac{\ln n}{\ln(3 + n^2)}$

56. A rubber ball rebounds to two-thirds the height from which it falls. If it is dropped from a height of 4 feet and is allowed to continue bouncing indefinitely, what is the total distance it travels?

57. Let $S_n = \sum_{k=1}^{n} \frac{1}{\sqrt{k}}$.

(a) Evaluate $\lim_{k \to \infty} \frac{1}{\sqrt{k}}$.

(b) Show that $S_n \geq \frac{n}{\sqrt{n}} = \sqrt{n}$ for all $n \geq 1$.
[**HINT:** If $k \leq n$, then $1/k \geq 1/n$.]

(c) Use part (b) to show that $\sum_{k=1}^{\infty} \frac{1}{\sqrt{k}}$ diverges.

58. Let $\{a_k\}$ be an increasing sequence such that $a_1 > 0$ and $a_k \leq 100$ for all $k \geq 1$.
(a) Does $\lim_{k\to\infty} a_k$ exist? Justify your answer.
(b) Show that $\sum_{k=1}^{\infty} a_k$ diverges.

59. Let $\{a_k\}$ be a sequence of positive terms such that $\sum_{k=1}^{n} a_k \leq 100$ for all $n \geq 1$. Explain why $\lim_{k\to\infty} a_k = 0$ must be true.

60. A certain series $\sum_{k=1}^{\infty} a_k$ has partial sums

$$S_n = \sum_{k=1}^{n} a_k = 5 - \frac{3}{n}.$$

(a) Evaluate $S_{100} = \sum_{k=1}^{100} a_k$.
(b) Evaluate $\sum_{k=1}^{\infty} a_k$.
(c) Evaluate $\lim_{k\to\infty} a_k$.
(d) Show that $a_k > 0$ for all $k \geq 1$.
[**HINT:** $a_{n+1} = S_{n+1} - S_n$.]

61. A certain series $\sum_{j=1}^{\infty} b_j$ has partial sums

$$S_n = \sum_{j=1}^{n} b_j = \ln\left(\frac{2n+3}{n+1}\right).$$

(a) Evaluate $\lim_{n\to\infty} S_n$.
(b) Does the series converge? Justify your answer.
(c) Show that $b_j < 0$ for all $j \geq 1$.

62. Suppose that the partial sums of the series $\sum_{k=1}^{\infty} a_k$ satisfy the inequality

$$\frac{6\ln n}{\ln(n^2+1)} < S_n < 3 + ne^{-n}$$

for all $n \geq 100$.
(a) Does the series converge? If so, to what limit? Justify your answers.
(b) What, if anything, can be said about $\lim_{k\to\infty} a_k$? Explain.

63. Let $S_n = \sum_{k=1}^{n} a_k$, and suppose that $0 \leq S_n \leq 100$ for all $n \geq 1$.
(a) Give an example of a sequence $\{a_k\}$ that satisfies the given conditions but $\sum_{k=1}^{\infty} a_k$ diverges.
(b) Show that if $a_k > 0$ for all $k \geq 1$, then $\sum_{k=1}^{\infty} a_k$ converges.
(c) Show that if $a_k > 0$ for all $k \geq 10^6$, then $\sum_{k=1}^{\infty} a_k$ converges.

64. Suppose that $\sum_{k=1}^{\infty} a_k$ diverges.
(a) Explain why $a_k > 0$ for all $k \geq 1$ implies that $\lim_{n\to\infty} S_n = \infty$.
(b) Give an example of a divergent series for which $\lim_{n\to\infty} S_n$ does not exist.

65. What is wrong with the following argument?

Let $S = 1 + 2 + 4 + 8 + \cdots$.

Then $2S = 2 + 4 + 8 + \cdots = S - 1$, so $S = -1$.

66. (a) Use the trigonometric identity $\tan x = \cot x - 2\cot(2x)$ to show that

$$\sum_{k=1}^{n} \frac{1}{2^k}\tan\left(\frac{x}{2^k}\right) = \frac{1}{2^n}\cot\left(\frac{x}{2^n}\right) - \cot x.$$

(b) Use part (a) to show that

$$\sum_{k=1}^{\infty} \frac{1}{2^k}\tan\left(\frac{x}{2^k}\right) = \frac{1}{x} - \cot x.$$

67. Let S_n denote the nth partial sum of the harmonic series

$$\sum_{k=1}^{\infty} \frac{1}{k} = 1 + \frac{1}{2} + \frac{1}{3} + \frac{1}{4} + \cdots$$

(i.e., $S_n = \sum_{k=1}^{n} 1/k$).
(a) Complete the following table; report answers to three decimal places. In the third row, enter *differences* between successive entries in the second row. A few entries are given.

n	10	20	30	40	50
S_n	2.929	3.598			
ΔS_n	—	0.671			
n	60	70	80	90	100
S_n					5.187
ΔS_n					

(b) Do the numbers in the table in part (a) suggest clearly whether the harmonic series converges or diverges? Why or why not?
(c) Complete the following table; report answers to three decimal places. In the third row, enter *differences* between successive entries in the second row. A few entries are given.

n	2	4	8	16	32
S_n	1.500	2.083			
ΔS_n	—	0.583			
n	64	128	256	512	1024
S_n					7.509
ΔS_n					0.693

(d) Do the numbers in the table in part (c) suggest clearly whether the harmonic series converges or diverges? Why or why not?
(e) The bottom row of the table in (c) shows that doubling n causes S_n to increase by about 0.693. Use this fact to guess values for S_{2048}, S_{4096}, and S_{8192}. (Don't try to calculate these numbers directly!)

68. Let $H_n = \sum_{k=1}^{n} \frac{1}{k}$, and let $I_n = \int_1^{n+1} \frac{dx}{x}$.
(a) Let L_n be the left Riemann-sum approximation, with n equal subdivisions, to I_n. Show that $L_n = H_n$.
(b) Use part (a) to show that the harmonic series diverges. [**HINT:** Start by comparing L_n and I_n.]

69. Let $H_n = \sum_{k=1}^{n} \frac{1}{k}$, and let $a_m = \sum_{j=1}^{2^{m-1}} \frac{1}{2^{m-1}+j}$. Then $H_{2^n} = 1 + \sum_{m=1}^{n} a_m$. [**NOTE:** a_m is the sum of a "block" of 2^{m-1} consecutive terms of the harmonic series—those from $n = 2^{m-1}+1$ through $n = 2^m$.]
(a) Show that $a_1 = 1/2$, $a_2 = 7/12$, and $a_3 = 533/840$.
(b) Show that $H_8 = 1 + a_1 + a_2 + a_3 = 761/280$.
(c) Show that $a_k \geq 1/2$ for all $k \geq 1$.
[**HINT:**
$$\frac{1}{2^{m-1}+j} \leq \frac{1}{2^{m-1}+2^{m-1}}$$
if $1 \leq j \leq 2^{m-1}$.]
(d) Use part (b) to show that $\lim_{n\to\infty} H_n = \infty$ (i.e., the harmonic series diverges).

70. Let $H_n = \sum_{k=1}^{n} \frac{1}{k}$, and let $I = \int_1^2 \frac{dx}{x}$.
(a) Let L_n be the left Riemann-sum approximation to I with n equal subdivisions. Show that
$$L_n = \sum_{k=0}^{n-1} \frac{1}{n+k} = \frac{1}{n} + \frac{1}{n+1} + \frac{1}{n+2} + \cdots + \frac{1}{2n-1}.$$
(b) Explain why $L_n > \ln 2$ for all $n \geq 1$.
(c) Show that $H_{2n} - H_n = L_n - \frac{1}{2n}$ for all $n \geq 1$.
(d) Show that $\lim_{n\to\infty} (H_{2n} - H_n) = \ln 2$.
(e) Explain why part (d) implies that the harmonic series diverges.
(f) Explain why $H_{2n} - H_n > \ln 2 - \frac{1}{2}$ for all $n \geq 1$. [**HINT:** Use parts (b) and (c).]
(g) Show that $H_2 > 1 + \ln 2 - \frac{1}{2}$.
(h) Show that that $H_{2^m} > 1 + m\left(\ln 2 - \frac{1}{2}\right)$ if $m \geq 1$. [**HINT:** Use parts (f) and (g).]

71. (a) Show that $\sin x \leq x \leq \tan x$ if $0 \leq x < \pi/2$.
(b) Use part (a) to show that $\cot^2 x \leq 1/x^2 \leq 1 + \cot^2 x$ if $0 \leq x < \pi/2$.
(c) Use part (b) to show that
$$\sum_{k=1}^{n} \cot^2\left(\frac{k\pi}{2n+1}\right) \leq \frac{(2n+1)^2}{\pi^2} \sum_{k=1}^{n} \frac{1}{k^2} \leq n + \sum_{k=1}^{n} \cot^2\left(\frac{k\pi}{2n+1}\right).$$
(d) It can be shown that
$$\sum_{k=1}^{n} \cot^2\left(\frac{k\pi}{2n+1}\right) = \frac{n(2n-1)}{3}.$$
Thus, part (c) is equivalent to
$$\frac{n(2n-1)}{3} \leq \frac{(2n+1)^2}{\pi^2} \sum_{k=1}^{n} \frac{1}{k^2} \leq n + \frac{n(2n-1)}{3}.$$
Use these inequalities to show that $\sum_{k=1}^{\infty} \frac{1}{k^2} = \frac{\pi^2}{6}$.

72. Let $I_n = \int_0^{\pi/4} \tan^n x\, dx$ for integer $n \geq 0$.
(a) Show that $\{I_n\}$ is a monotone decreasing sequence (i.e., $I_{n+1} \leq I_n$).
(b) Show that $I_n + I_{n-2} = \frac{1}{n-1}$.
(c) Use parts (a) and (b) to show that
$$\frac{1}{2(n+1)} \leq I_n \leq \frac{1}{2(n-1)}.$$
[**HINT:** Part (b) implies that $I_{n+2} + I_n = 1/(n+1)$.]
(d) Show that, if $n \geq 2$,
$$I_n = \frac{1}{n-1} - \int_0^{\pi/4} \tan^{n-2} x\, dx.$$
(e) Use part (d) to show that
$$I_{2n} = (-1)^n \left(\frac{\pi}{4} + \sum_{k=1}^{n} \frac{(-1)^k}{2k-1}\right)$$
for all integers $n \geq 1$.
(f) Use parts (c) and (e) to show that $\sum_{k=0}^{\infty} \frac{(-1)^k}{2k+1} = \frac{\pi}{4}$.
(g) Use part (d) to show that
$$I_{2n+1} = (-1)^n \left(\frac{1}{2}\ln 2 + \sum_{k=1}^{n} \frac{(-1)^k}{2k}\right)$$
for all integers $n \geq 1$.
(h) Use parts (c) and (g) to show that $\sum_{k=1}^{\infty} \frac{(-1)^{k+1}}{k} = \ln 2$.

73. Consider the series $\sum_{k=1}^{\infty} \frac{1}{k^p}$ with $p > 1$. This exercise outlines a proof that this series converges.
(a) Let $S_n = \sum_{k=1}^{n} \frac{1}{k^p}$. Show that the sequence of partial sums $\{S_n\}$ is increasing.
(b) Show that $S_{2m+1} = 1 + \sum_{k=1}^{m} \frac{1}{(2k)^p} + \sum_{k=1}^{m} \frac{1}{(2k+1)^p}$.
(c) Explain why $S_{2m+1} < 1 + 2\sum_{k=1}^{m} \frac{1}{(2k)^p}$.
[**HINT:** $1/(x+1) < 1/x$ for all $x > 0$.]
(d) Show that $S_{2m+1} < 1 + 2^{1-p} S_{2m+1}$. [**HINT:** First show that $S_{2m+1} < 1 + 2^{1-p} S_m$.]
(e) Show that $\{S_n\}$ is bounded above.

11.3 Testing for Convergence; Estimating Limits

Converge or Diverge: What's the Question?

In theory, the question of convergence is simple:

The series $\sum a_k$ converges to S if the sequence $\{S_n\}$ of partial sums tends to S.

The trouble, in practice, is that a simple, explicit formula for S_n is often unavailable. ▶ It might seem, then, that testing for convergence—let alone finding a limit—would be difficult or impossible. Surprisingly, that isn't so. All the convergence tests of this section and the next (comparison test, integral test, ratio test, and so on) offer clever, indirect ways of testing whether $\{S_n\}$ converges—even without knowing each S_n exactly.

The harmonic series poses this problem.

Nonnegative Series

This apparent sleight-of-hand depends on a surprisingly simple observation: ▶

Convince yourself of this simple but important fact.

If $a_k \geq 0$ for all k, then the sequence $\{S_n\}$ of partial sums of $\sum a_k$ is nondecreasing:

$$S_1 \leq S_2 \leq S_3 \leq \cdots S_n \leq S_{n+1} \cdots.$$

A series with this convenient property is called **nonnegative**. By Theorem 3 of Section 11.1, a nondecreasing sequence must either converge or diverge to infinity. Therefore, to decide whether a nonnegative series $\sum a_k$ converges, it's enough to answer this key question:

As $n \to \infty$, do the partial sums $\{S_n\}$ blow up or remain bounded?

If a simple formula for S_n is available (as it is, for instance, for geometric series), the answer may be obvious. If not, our best recourse is to try to compare ▶ the given series to something that is better understood (another series, an integral—anything that works). Sometimes an obvious comparison suggests itself.

Watch for frequent reappearances of the words "compare" and "comparison."

EXAMPLE 1 Does $\displaystyle\sum_{k=0}^{\infty} \frac{1}{2^k+1}$ converge?

Solution We saw in the preceding section ▶ that the similar series $\sum_{k=0}^{\infty} 1/2^k$ converges to 2. For all k,

See Example 1, page 314.

$$\frac{1}{2^k+1} < \frac{1}{2^k},$$

so each partial sum of $\sum_{k=0}^{\infty} 1/(2^k+1)$ is *less than* the corresponding partial sum of $\sum_{k=0}^{\infty} 1/2^k$. Because the latter sums tend to 2, the former must tend to a limit less than 2. Briefly, in symbols,

$$\frac{1}{2^k+1} < \frac{1}{2^k} \Longrightarrow \sum_{k=0}^{\infty} \frac{1}{2^k+1} < \sum_{k=0}^{\infty} \frac{1}{2^k} = 2.$$

Easily found by calculator or computer.

Numerical evidence ◀ agrees. For the original series,

$$S_{10} \approx 1.263523536; \qquad S_{20} \approx 1.264498827; \qquad S_{100} \approx 1.264499780. \quad \blacksquare$$

The Comparison Test: One Series vs. Another

In Example 1 we showed that one series converges by comparing it to another series that is *known* to converge. The following theorem makes this idea precise, duly noting all necessary hypotheses.

> **Theorem 6** **(Comparison Test for Nonnegative Series)** Suppose that for all $k \geq 1$, $0 \leq a_k \leq b_k$. Consider the two series $\sum_{k=1}^{\infty} a_k$ and $\sum_{k=1}^{\infty} b_k$.
>
> - If $\sum_{k=1}^{\infty} b_k$ *converges*, so does $\sum_{k=1}^{\infty} a_k$, and
>
> $$\sum_{k=1}^{\infty} a_k \leq \sum_{k=1}^{\infty} b_k.$$
>
> - If $\sum_{k=1}^{\infty} a_k$ *diverges*, so does $\sum_{k=1}^{\infty} b_k$.

Observe:

What We'd Expect The theorem's assertion is reasonable, and is not difficult to prove. A formal proof depends on the fact that, like the terms themselves, the partial sums of $\sum a_k$ are *all* less than the corresponding partial sums of $\sum b_k$.

Successful Comparisons In a "successful" comparison, either both series converge or both diverge. To use the comparison test, therefore, it's necessary first to *guess* whether the series in question converges or diverges. ◀

We did so implicitly in Example 1.

Comparing Tails We assumed earlier that the comparison inequality $0 \leq a_k \leq b_k$ holds for *all* possible k. This assumption isn't really necessary; if N is any positive integer, and $0 \leq a_k \leq b_k$ for all $k \geq N$, then

$$\sum_{k=N}^{\infty} a_k \leq \sum_{k=N}^{\infty} b_k.$$

We'll use this fact in the next example to estimate a limit numerically.

Comparing Integrals, Comparing Series An almost identical comparison theorem applies to improper integrals. ◀ Theorem 1, Section 10.2, says that if $0 \leq a(x) \leq b(x)$ for all $x \geq 1$, then

Sequences are "discrete" versions of functions; series are "discrete" versions of integrals.

$$\int_1^{\infty} a(x)\,dx \leq \int_1^{\infty} b(x)\,dx.$$

EXAMPLE 2 As we saw, $\sum_{k=0}^{\infty} 1/(2^k + 1)$ converges. How closely does $S_{100} \approx 1.264499781$ approximate the true limit S?

Solution For any series, the error in using S_{100} to estimate the limit comes from ignoring the tail, R_{100}. In symbols,

$$S = \sum_{k=0}^{\infty} a_k = \sum_{k=0}^{100} a_k + \sum_{k=101}^{\infty} a_k = S_{100} + R_{100}.$$

Because, for our series, the inequality

$$\frac{1}{2^k + 1} < \frac{1}{2^k}$$

holds for all $k \geq 101$, ▶ it follows that

The inequality actually holds for all *k; here we care only about* $k \geq 101$.

$$\sum_{k=101}^{\infty} \frac{1}{2^k + 1} < \sum_{k=101}^{\infty} \frac{1}{2^k}.$$

We can calculate the right side directly:

$$\sum_{k=101}^{\infty} \frac{1}{2^k} = \frac{1}{2^{101}} + \frac{1}{2^{102}} + \frac{1}{2^{103}} + \cdots = \frac{1}{2^{101}}\left(1 + \frac{1}{2} + \frac{1}{2^2} + \cdots\right) = \frac{1}{2^{100}}.$$

Thus, as expected, we commit *very* little error by ignoring the tail R_{100}; the estimate S_{100} differs from S by less than $1/2^{100} \approx 8 \times 10^{-31}$. ■

Compared to What?

The *idea* of comparison is easy. Harder, in practice, is deciding what to compare a series *to*. Our only reliable "benchmark" series, so far, are geometric series. The **integral test** enlarges our stock of known series considerably.

The Integral Test: Series vs. Integrals

We know that

$$\int_1^{\infty} \frac{1}{x}\,dx \quad \text{diverges,} \qquad \text{but} \qquad \int_1^{\infty} \frac{1}{x^2}\,dx \quad \text{converges.}$$

One might guess, therefore, that

$$\sum_{k=1}^{\infty} \frac{1}{k} \quad \text{diverges,} \qquad \text{but} \qquad \sum_{k=1}^{\infty} \frac{1}{k^2} \quad \text{converges.}$$

These guesses are correct, as graphical evidence has already suggested.

Integrals and Series: Comparing Areas

Thinking of the terms of a positive series $\sum_{k=1}^{\infty} a_k$ as rectangular areas ▶ clarifies the connection with the integral $\int_1^{\infty} a(x)\,dx$. The first picture shows one way of

Each has base 1.

doing so:

If $\int_1^\infty a(x)\,dx$ diverges, so must $\sum_{k=1}^{\infty} a_k$

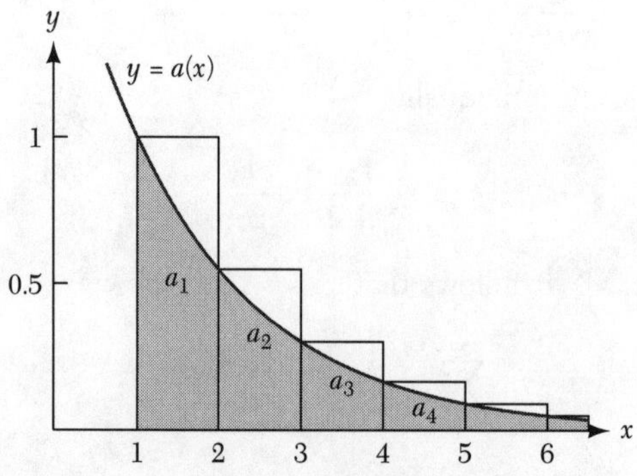

Look closely at the picture.

Total Areas The successive rectangles have *heights* $a(1) = a_1$, $a(2) = a_2$, $a(3) = a_3$, and so on; each *base* is 1. The respective *areas*, therefore, are a_1, a_2, a_3, and so on. Here's the key idea:

The series $a_1 + a_2 + a_3 + a_4 + \cdots$ represents the total "left-rule" rectangular area from 1 to ∞.

An Important Inequality The shaded area—less than the rectangular area—represents the *integral* $\int_1^\infty a(x)\,dx$. Thus, if the integral diverges, so must the series. ◄ Here's the message of the picture, in inequality form:

We collect our results in the next theorem.

$$a_1 + a_2 + a_3 + \cdots \geq \int_1^\infty a(x)\,dx.$$

If the right side diverges to infinity, so must the left.

A Decreasing Integrand The reasoning that led to the preceding inequality requires that the integrand a be *decreasing* for $x \geq 1$, as shown in the picture. We collect such technical hypotheses carefully in the next theorem.

In the next picture, an integral bounds a series from *above*.

If $\int_1^\infty a(x)\,dx$ converges, so must $\sum_{k=1}^{\infty} a_k$

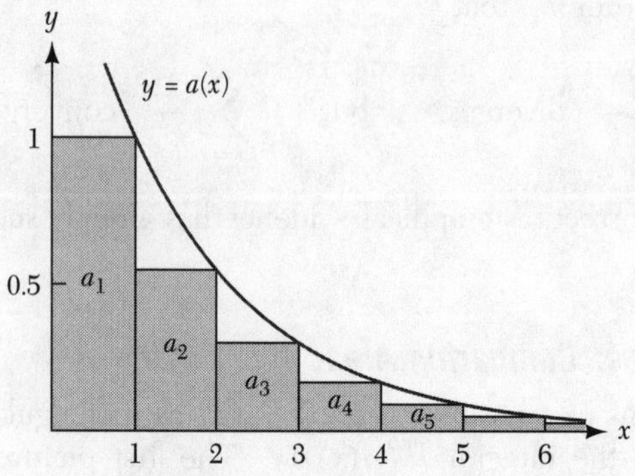

Again, successive rectangles have areas a_1, a_2, a_3, and so on. This time, comparing areas shows that ▶

Convince yourself; notice how the first term is "broken off."

$$a_1 + a_2 + a_3 + \cdots \leq a_1 + \int_1^\infty a(x)\,dx.$$

If the right side converges, so must the left.

Combining both of the preceding inequalities gives upper *and* lower bounds for the series:

$$\int_1^\infty a(x)\,dx \leq \sum_{k=1}^{\infty} a_k \leq a_1 + \int_1^\infty a(x)\,dx.$$

In particular:

The integral $\int_1^\infty a(x)\,dx$ and the series $\sum_{k=1}^\infty a_k$ either both converge or both diverge.

The final picture relates the tails of an integral and of a series.

Comparing tails: why $\sum_{k=n+1}^{\infty} a_k \leq \int_n^\infty a(x)\,dx$

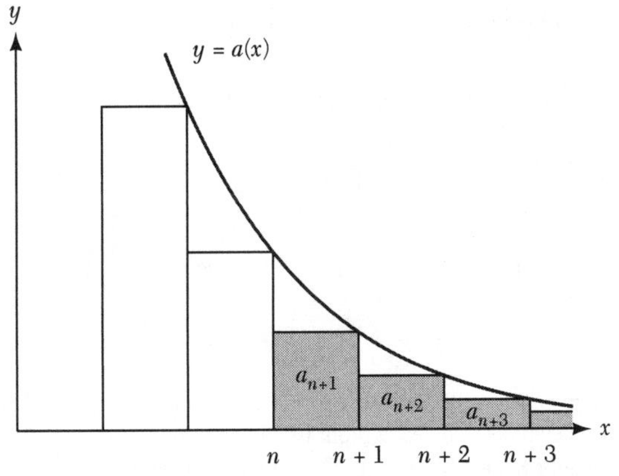

It shows, again by comparing areas, that for any n, the tail R_n satisfies

$$R_n = \sum_{k=n+1}^{\infty} a_k \leq \int_n^\infty a(x)\,dx.$$

The preceding pictures show a particular function $a(x)$, but the conclusions we drew are generic—they hold for *any* function $a(x)$ that is both positive and decreasing. ▶

What could go wrong if $a(x)$ isn't decreasing?

It's time to collect all the foregoing observations in a theorem. As before, $a_k = a(k)$ for all integers $k \geq 1$.

Theorem 7 **(Integral Test for Positive Series)** Suppose that, for all $x \geq 1$, the function $a(x)$ is continuous, positive, and decreasing. Consider the series $\sum_{k=1}^{\infty} a_k$ and the integral $\int_1^{\infty} a(x)\,dx$.

- If either diverges, so does the other.
- If either converges, so does the other. In this case,

$$\int_1^{\infty} a(x)\,dx \leq \sum_{k=1}^{\infty} a_k \leq a_1 + \int_1^{\infty} a(x)\,dx.$$

- If the series converges, then

$$R_n = \sum_{k=n+1}^{\infty} a_k \leq \int_n^{\infty} a(x)\,dx.$$

P-Series, Convergent and Divergent

Series of the form $\sum_{k=1}^{\infty} \frac{1}{k^p}$ are called ***p*-series**. They form an important family of examples, with behavior depending on the value of p.

EXAMPLE 3 Which p-series converge?

Solution We found in Chapter 10 that the integral $\int_1^{\infty} dx/x^p$ converges if and only if $p > 1$. Therefore, using the integral test, ◄

Are all the hypotheses really satisfied?

The p-series $\sum_{k=1}^{\infty} \frac{1}{k^p}$ *converges if and only if* $p > 1$.

In particular, the harmonic series $\sum 1/k$ diverges. ■

Harmonic Divergence. The fact that the harmonic series

$$1 + \frac{1}{2} + \frac{1}{3} + \frac{1}{4} + \frac{1}{5} + \cdots$$

diverges to infinity—even though the terms themselves tend to zero—has fascinated mathematicians for many centuries. One early proof (not the one given here) is attributed to Nicole Oresme, a 14th-century French bishop, scientist, and mathematician.

EXAMPLE 4 The p-series $\sum_{k=1}^{\infty} 1/k^3$ converges, by the integral test, to some limit S. How large must n be to ensure that S_n differs from S by less than 0.0001?

Solution We choose n so that the tail R_n is less than 0.0001. Theorem 7 shows how. By the last part, ◄

See the "Comparing tails" picture.

$$R_n \leq \int_n^{\infty} \frac{1}{x^3}\,dx.$$

The right side is easily calculated; we get $1/2n^2$. ◄ This quantity (and hence also R_n) are less than 0.0001 if $n \geq 71$. Hence, $S_{71} \approx 1.20196$ differs from the true limit by less than 0.0001. ■

See for yourself.

EXAMPLE 5 Does $\sum_{k=1}^{\infty} \frac{1}{10k+1}$ converge?

Solution The integral test says no.

$$\int_1^{\infty} \frac{1}{10x+1}\,dx = \lim_{n\to\infty} \int_1^{n} \frac{1}{10x+1}\,dx = \lim_{n\to\infty} \left.\frac{\ln(10x+1)}{10}\right]_1^n = \infty.$$

The comparison test *also* says no. Because for all $k \geq 1$,

$$\frac{1}{10k+1} \geq \frac{1}{11k},$$

it follows that

$$\sum_{k=1}^{\infty} \frac{1}{10k+1} \geq \sum_{k=1}^{\infty} \frac{1}{11k} = \frac{1}{11} \sum_{k=1}^{\infty} \frac{1}{k}.$$

Since the last series diverges to infinity, so must the first. ■

The Ratio Test: Comparison with a Geometric Series

In a geometric series

$$a + ar + ar^2 + ar^3 + ar^4 + \cdots$$

the ratio of successive terms is r; the series converges if and only if $|r| < 1$. ▶

Why? See the Fact on page 317.

The **ratio test** is based on the same principle. In the end, it amounts to a lightly disguised form of comparison with a geometric series.

> **Theorem 8** (Ratio Test for Positive Series) Suppose that $\sum_{k=1}^{\infty} a_k$ is a positive series and that
>
> $$\lim_{k\to\infty} \frac{a_{k+1}}{a_k} = L.$$
>
> - If $L < 1$, then $\sum a_k$ *converges.*
> - If $L > 1$, then $\sum a_k$ *diverges.*
> - If $L = 1$, either convergence or divergence is possible. The test is inconclusive.

Notice these possibilities:

When the Ratio Tends to 1 For many series, unfortunately, $a_{k+1}/a_k \to 1$ as $k \to \infty$. In such cases, the ratio test says *nothing*. This happens for every p-series $\sum 1/k^p$, for instance.

When the Ratio Test Works Best The ratio test works best for such series as $\sum 1/k!$, $\sum r^k$, and $\sum 1/(2^k + 3)$, in which the index k appears in an exponent or a factorial.

The Idea of a Proof

To illustrate the connection between the ratio test and geometric series, and to give the idea of a proof, let's suppose that

$$\lim_{k\to\infty} \frac{a_{k+1}}{a_k} = \frac{1}{2}.$$

Why must $\sum a_k$ converge?

The idea is that, for large k, $a_{k+1} \approx a_k/2$, so the series behaves similarly to a geometric series. Suppose, for instance, that $a_{k+1} < 0.6a_k$ for all $k \geq 1000$. ◄ Then

The inequality must hold for all large k because of the limit above.

$$a_{1001} < (0.6)a_{1000}, \ldots; \qquad a_{1002} < (0.6)a_{1001} < (0.6)^2 a_{1000},$$
$$a_{1003} < (0.6)^3 a_{1000},$$

so

$$\sum_{k=1000}^{\infty} a_k = a_{1000} + a_{1001} + a_{1002} + \cdots < a_{1000}\left(1 + (0.6) + (0.6)^2 + (0.6)^3 + \cdots\right).$$

The last inequality is the point; it shows that $\sum a_k$ converges *by comparison with the geometric series* $\sum a_{1000}(0.6)^k$.

The divergence statement can be proved in a similar way.

EXAMPLE 6 Show that $\sum_{k=0}^{\infty} \frac{1}{k!}$ converges. Guess a limit.

Solution The ratio test works nicely. ◄ Since

Check details carefully.

$$\lim_{k\to\infty} \frac{a_{k+1}}{a_k} = \lim_{k\to\infty} \frac{k!}{(k+1)!} = \lim_{k\to\infty} \frac{1}{k+1} = 0,$$

the series converges. In fact, it converges very, very fast. Here are some representative partial sums:

$$S_5 \approx 2.716666667;$$
$$S_{10} \approx 2.718281801;$$
$$S_{30} \approx 2.718281828459045235360287471 35.$$

Is e involved somehow? To 30 decimals,

$$e = 2.718281828459045235360287471 35.$$

In fact, this series can be shown rigorously to converge to e. We explore this phenomenon further in later sections. ■

EXAMPLE 7 Does $\sum_{k=0}^{\infty} \frac{100^k}{k!}$ converge?

Solution Yes, by the ratio test:

$$\lim_{k\to\infty} \frac{a_{k+1}}{a_k} = \lim_{k\to\infty} \frac{100^{k+1}}{(k+1)!} \cdot \frac{k!}{100^k} = \lim_{k\to\infty} \frac{100}{k+1} = 0.$$

Notice what the result means: Even though 100^k grows very fast, $k!$ grows faster still. ■

BASICS

1. Consider the series $\sum_{k=0}^{\infty} a_k$, where $a_k = \dfrac{1}{k+2^k}$.
   (a) Use the comparison test to show that the series converges. [**HINT:** $a_k \le 2^{-k}$ for all $k \ge 0$.]
   (b) Show that $0 \le R_{10} \le 2^{-10}$.
   (c) Compute an estimate of the limit of the series that is guaranteed to be within 0.001 of the exact value.
   (d) Is your estimate in part (c) an overestimate or an underestimate? Justify your answer.

2. Consider the series $\sum_{j=0}^{\infty} \dfrac{1}{2+3^j}$.
   (a) Show that the series converges.
   (b) Estimate the value of the limit of the series within 0.01.
   (c) Is your estimate in part (b) an overestimate or an underestimate? Justify your answer.

In Exercises 3–7, suppose that $a(x)$ is continuous, positive, and decreasing for all $x \ge 1$ and that $a_k = a(k)$ for all integers $k \ge 1$.

3. Rank the values $\int_1^n a(x)\,dx$, $\sum_{k=1}^{n-1} a_k$, and $\sum_{k=2}^{n} a_k$ in increasing order. [**HINT:** Draw a picture.]
4. Rank the values $\int_n^\infty a(x)\,dx$, $\sum_{k=n+1}^{\infty} a_k$, and $\int_{n+1}^\infty a(x)\,dx$ in increasing order.
5. Draw a carefully annotated picture that shows that $\int_1^{n+1} a(x)\,dx \le \sum_{k=1}^{n} a_k$.
6. Draw a carefully annotated picture that shows that $\sum_{k=2}^{n} a_k \le \int_1^n a(x)\,dx$.
7. Draw a carefully annotated picture that shows that

$$\sum_{k=n+1}^{\infty} a_k \le a_{n+1} + \int_{n+1}^{\infty} a(x)\,dx \le \int_n^\infty a(x)\,dx.$$

In Exercises 8–10, use the integral test to find upper and lower bounds on the limit of the series.

8. $\sum_{k=0}^{\infty} \dfrac{1}{k^2+1}$

9. $\sum_{k=1}^{\infty} \dfrac{1}{k\sqrt{k}}$

10. $\sum_{j=1}^{\infty} je^{-j}$

11. Show that $\sum_{n=2}^{\infty} \dfrac{1}{(\ln n)^2}$ diverges.
    [**HINT:** $\ln x \le \sqrt{x}$ for all $x \ge 1$.]

12. Show that $\sum_{k=3}^{\infty} \dfrac{1}{(\ln k)^k}$ converges.
    [**HINT:** $\ln k \ge \ln 3 \approx 1.0986$ for all $x \ge 3$.]

13. Consider the series $\sum_{k=1}^{\infty} \dfrac{2+\sin k}{k^2}$.
    (a) Explain why Theorem 7 cannot be used to prove that this series converges.
    (b) Show that this series converges.

In Exercises 14–17, use the ratio test to show that the series converges.

14. $\sum_{j=0}^{\infty} \dfrac{j^2}{j!}$

15. $\sum_{k=1}^{\infty} \dfrac{2^k}{k!}$

16. $\sum_{n=1}^{\infty} \dfrac{n^2}{2^n}$

17. $\sum_{m=1}^{\infty} \dfrac{m!}{(2m)!}$

FURTHER EXERCISES

18. Suppose that for all $k \ge 1$, $0 \le a_k \le b_k$. Let $S_n = \sum_{k=1}^{n} a_k$ and $T_n = \sum_{k=1}^{n} b_k$.
    (a) Suppose that $\sum_{k=1}^{\infty} b_k$ converges. Explain why there is a number M such that $S_n \le T_n \le M$ for all $n \ge 1$.
    (b) Explain why $\{S_n\}$ is an increasing sequence.
    (c) Explain why parts (a) and (b) together imply that $\sum_{k=1}^{\infty} a_k$ converges.
    (d) Suppose that $\sum_{k=1}^{\infty} a_k$ diverges. Explain why $\lim_{n\to\infty} S_n = \infty$.
    (e) Suppose that $\sum_{k=1}^{\infty} a_k$ diverges. Use part (d) to show that $\sum_{k=1}^{\infty} b_k$ diverges.

19. Suppose that $a(x)$ is continuous, positive, and decreasing for all $x \ge 1$, that $a_k = a(k)$ for all integers $k \ge 1$, and that $\int_1^\infty a(x)\,dx$ converges.
    (a) Explain why the sequence of partial sums $\{S_n\}$ is an increasing sequence.
    (b) Explain why $\int_1^n a(x)\,dx \le \int_1^\infty a(x)\,dx$.
    (c) Use parts (a) and (b) to show that the sequence of partial sums $\{S_n\}$ converges.

20. Suppose that $a_n \ge 0$ for all $n \ge 1$ and that $\sum_{n=1}^{\infty} a_n$ converges. Show that $\sum_{n=1}^{\infty} \sin(a_n)$ converges.
    [**HINT:** $|\sin x| \le |x|$ for all x.]

21. Does the series

$$1+\frac{1}{1\cdot 3}+\frac{1}{1\cdot 3\cdot 5}+\frac{1}{1\cdot 3\cdot 5\cdot 7}+\cdots+\frac{1}{1\cdot 3\cdot 5\cdot 7\cdots(2k+1)}+\cdots$$

converge? Justify your answer.

22. Show that $\sum_{k=3}^{\infty}\frac{1}{(\ln k)^{\ln k}}$ converges.
[**HINT:** $\ln k > e^2$ if $k > 1619$.]

23. Let $H_n = \sum_{k=1}^{n}\frac{1}{k}$ and let $S_n = \sum_{k=0}^{n}\frac{1}{2k+1}$.
   (a) Explain why $\lim_{n\to\infty} H_n = \infty$.
   (b) Show that $S_n \geq \frac{1}{2}H_n$.
   (c) What do the results in parts (a) and (b) imply about $\sum_{k=0}^{\infty} 1/(2k+1)$? Explain.

24. (a) Estimate a lower bound for $n!$ by comparing $\ln(n!)$ and $\int_1^n \ln x\,dx$.
   (b) Let a be a positive number. Use part (a) to find an integer N such that $a^N/N! < 1/2$.

25. For which values of p does the series

$$\sum_{n=3}^{\infty}\frac{1}{n(\ln n)^p}$$

converge? Justify your answer.

26. Consider the series $\sum_{k=1}^{\infty} a_k = \sum_{k=1}^{\infty}\frac{\ln k}{k}$.
   (a) Use the integral test to show that the series diverges. [**HINT:** The function $\ln x/x$ is monotone on $[3,\infty)$. Start by showing that $\sum_{k=3}^{\infty} a_k$ diverges.]
   (b) Use the comparison test to show that the series diverges. [**HINT:** $1 - x^{-1} \leq \ln x$ for all $x > 0$.]
   (c) Can the ratio test be used to show that the series diverges? Explain.

27. Consider the series

$$\sum_{k=1}^{\infty} a_k = \frac{1}{2}+\frac{1}{3}+\frac{1}{2^2}+\frac{1}{3^2}+\frac{1}{2^3}+\frac{1}{3^3}+\cdots.$$

   (a) Explain why $\lim_{k\to\infty} a_{k+1}/a_k$ does not exist.
   (b) What does the ratio test say about the convergence of the series $\sum_{k=1}^{\infty} a_k$?
   (c) Show that the series converges, and evaluate its limit. [**HINT:** Rewrite the given series as the sum of two series.]

28. (a) What does the ratio test say about the convergence of the series

$$\frac{1}{2}+\frac{1}{2}+\frac{1}{4}+\frac{1}{4}+\frac{1}{8}+\frac{1}{8}+\cdots?$$

   (b) Does the series in part (a) converge or diverge? Explain.

29. Give an example of a divergent series $\sum a_k$ such that $a_k > 0$ and $a_{k+1}/a_k < 1$ for all $k \geq 1$.

30. Use the ratio test to show that $\sum_{n=1}^{\infty} n^{-n}$ converges.

31. Use the ratio test to show that the series $\sum_{n=1}^{\infty}\frac{n^n}{n!}$ diverges.

In Exercises 32–35, use the comparison test to show that the series converges. Then find an upper bound on the limit of the series.

32. $\sum_{n=1}^{\infty}\frac{1}{n^2+\sqrt{n}}$

33. $\sum_{j=0}^{\infty}\frac{1}{j+e^j}$

34. $\sum_{m=1}^{\infty}\frac{1}{m\sqrt{1+m^2}}$

35. $\sum_{k=1}^{\infty}\frac{k}{\left(k^2+1\right)^2}$

In Exercises 36–56, determine whether the series converges or diverges. If the series converges, find an upper bound on its limit. Justify your answers.

36. $\frac{1}{100}+\frac{1}{200}+\frac{1}{300}+\cdots$

37. $\sum_{n=1}^{\infty}\frac{\arctan n}{1+n^2}$

38. $\sum_{m=1}^{\infty}\frac{m^3}{m^5+3}$

39. $\sum_{j=1}^{\infty}\frac{1}{100+5j}$

40. $\sum_{k=2}^{\infty}\frac{1}{k\ln k}$

41. $\sum_{n=1}^{\infty}\frac{1}{n\,3^n}$

42. $\sum_{k=1}^{\infty}\frac{1}{\ln\left(10^k\right)}$

43. $\sum_{j=1}^{\infty}\frac{j}{5^j}$

44. $\sum_{k=0}^{\infty}\frac{k^2}{5k^2+3}$

45. $1-\frac{1}{2}-\frac{1}{3}-\frac{1}{4}-\frac{1}{5}-\cdots$

46. $\sum_{j=1}^{\infty}\frac{j}{j^4+j^2-1}$

47. $\sum_{n=2}^{\infty}\frac{1}{\sqrt[3]{n^2-1}}$

48. $\sum_{k=1}^{\infty}\frac{\sqrt{k}}{k^2+k+1}$

49. $\sum_{m=0}^{\infty} e^{-m^2}$

50. $\sum_{j=0}^{\infty}\frac{j!}{(j+2)!}$

51. $\sum_{n=0}^{\infty}\frac{n!}{(2n)!}$

52. $\sum_{m=1}^{\infty}\frac{m^3}{m^4-7}$

53. $\sum_{k=1}^{\infty}\frac{k!}{(k+1)!-1}$

54. $\sum_{j=2}^{\infty}\frac{\ln j}{j^2}$

55. $\sum_{n=1}^{\infty}\left(\sum_{k=1}^{n} k^{-1}\right)$

56. $1-\frac{1}{2}+\frac{1}{2}-\frac{1}{4}+\frac{1}{3}-\frac{1}{6}+\frac{1}{4}-\frac{1}{8}+\frac{1}{5}-\frac{1}{10}+\cdots$

In Exercises 57–63, determine whether the series converges or diverges. If the series converges, find a number N such that $n \geq N$ implies that the partial sum S_n approximates the sum of the series within 0.001. If the series diverges, find a number N such that $n \geq N$ implies that $S_n \geq 1000$, or explain why there is no such N.

57. $\displaystyle\sum_{k=0}^{\infty} \frac{1}{k^2+3}$

58. $\displaystyle\sum_{m=1}^{\infty} \frac{\arctan m}{m}$

59. $\displaystyle\sum_{j=2}^{\infty} \frac{3^j}{4^{j+1}}$

60. $\displaystyle\sum_{k=0}^{\infty} \frac{1}{2+\cos k}$

61. $\displaystyle\sum_{m=2}^{\infty} \frac{\ln m}{m^3}$

62. $\displaystyle\sum_{k=0}^{\infty} \frac{k}{k^6+17}$

63. $\displaystyle\sum_{k=2}^{\infty} \frac{1}{k\,(\ln k)^5}$

64. (a) Where in the proof of the integral test (Theorem 7) is the assumption that $a(x)$ is a decreasing function used?
    (b) Suppose that the requirement that $a(x)$ be decreasing for all $x \geq 1$ is replaced by the "weaker" requirement that $a(x)$ be decreasing for all $x \geq 10$. How does this change in assumptions affect the conclusions of Theorem 7?

65. Consider the series $\displaystyle\sum_{k=1}^{\infty} \frac{1}{k!}$.
    (a) Explain why $\dfrac{1}{k!} \leq \dfrac{1}{2^{k-1}}$ for all $k \geq 1$.
    [**HINT:** $n \geq 2 \Longrightarrow 1/n \leq 1/2$.]
    (b) Show that $\dfrac{1}{k!} \leq \dfrac{1}{10!\,10^{k-10}}$ for all $k \geq 10$.
    (c) Explain why S_{10} underestimates the limit of the series, and find a bound on the magnitude of the approximation error. [**HINT:** Use part (b) to bound R_{10}.]

66. Let $n \geq 2$ be an integer.
    (a) Explain why
    $$\frac{1}{n!} + \frac{1}{(n+1)!} + \frac{1}{(n+2)!} + \cdots < \frac{1}{n!} + \frac{1}{n \cdot n!} + \frac{1}{n^2 \cdot n!} + \cdots.$$
    (b) Use part (a) to show that
    $$\sum_{k=n}^{\infty} \frac{1}{k!} < \frac{1}{n!}\,\frac{1}{1-\frac{1}{n}}$$
    (c) Use part (b) to show that $\displaystyle\sum_{k=1}^{\infty} \frac{1}{k!}$ converges.
    (d) Find an integer N such that the partial sum S_N approximates the sum of the series in part (c) within 0.00001.

67. Consider the series $\displaystyle\sum_{k=1}^{\infty} \left(\frac{1}{k!}\right)^2$.
    (a) Show that this series converges.
    (b) Find an integer N such that the partial sum S_N approximates the sum of this series within 5×10^{-6}.

68. Consider the series $\displaystyle\sum_{k=1}^{\infty} \frac{e^k}{k!}$. Find an integer N such that the partial sum S_N approximates the sum of this series within 5×10^{-6}.

69. Let x be a positive real number, and let N be an integer such that $N \geq x$.
    (a) Show that $\dfrac{x^k}{k!} \leq \dfrac{x^N}{N!}\left(\dfrac{x}{N+1}\right)^{k-N}$ for all $k \geq N$.
    (b) Use part (a) to show that
    $$\sum_{k=N}^{\infty} \frac{x^k}{k!} \leq \frac{x^N}{N!} \cdot \frac{1}{1-x/(N+1)}.$$

70. Let r be a positive number less than 1. Suppose that $\{a_k\}$ is a sequence of positive terms and that $a_{k+1}/a_k \leq r$ for all $k \geq 1$.
    (a) Show that $a_2 \leq a_1 r$ and that $a_3 \leq a_1 r^2$. (A similar argument shows that $a_{k+1} \leq a_1 r^k$.)
    (b) Use the result mentioned parenthetically in part (a) to show that $\sum_{k=1}^{\infty} a_k$ converges. [**HINT:** Use the comparison test and the formula for the sum of geometric series.]
    (c) Show that $R_n = \displaystyle\sum_{k=n+1}^{\infty} a_k \leq \frac{a_{n+1}}{1-r}$.

71. Let $a_k = 1/k$. Then $a_{k+1}/a_k < 1$ for all $k \geq 1$. Why can't the ideas outlined in Exercise 70 be used to "prove" that the harmonic series converges?

72. Use Exercise 70 to find an integer N such that the partial sum S_N approximates the sum of the series $\sum_{n=1}^{\infty} n^2/2^n$ within 0.0005.

73. Use Exercise 70 to find an integer N such that the partial sum S_N approximates the sum of the series $\sum_{n=1}^{\infty} (n!)^2/(2n)!$ within 0.0005.

74. Let $H_n = \displaystyle\sum_{k=1}^{n} \frac{1}{k}$ be the nth partial sum of the harmonic series.
    (a) Show that $\ln(n+1) < H_n < 1 + \ln n$. [**HINT:** Use the integral test.]
    (b) Use part (a) to show that $H_N > 10$ implies that $N > 8000$.
    (c) Show that $\displaystyle\lim_{n\to\infty} \frac{H_n}{\ln n} = 1$.
    (d) Show that the sequence defined by $a_n = H_n - \ln n$ is decreasing. [**HINT:** Explain why $\ln(n+1) - \ln n = \int_n^{n+1} x^{-1}\,dx > (n+1)^{-1}$.]
    (e) Use part (d) to show that $\lim_{n\to\infty} a_n$ exists. (This limit, denoted by γ, is called **Euler's constant**; $\gamma \approx 0.57722$.)

75. Let a_n and γ be as in Exercise 74, and let

$$f(x) = \ln(x+1) - \ln x - \frac{1}{x+1}.$$

(a) Show that $f(x) = -\int_x^\infty f'(t)\,dt$.

(b) Use part (a) to show that

$$f(x) > \frac{1}{2(x+1)^2}.$$

[**HINT:** If $x > 0$, then

$$-f'(x) = \frac{1}{x(x+1)^2} > \frac{1}{(x+1)^3}.]$$

(c) Show that $a_n - \gamma = \sum_{k=n}^\infty (a_k - a_{k+1})$.

(d) Use parts (b) and (c) to show that

$$\frac{1}{2(n+1)} < a_n - \gamma < \frac{1}{2n}.$$

[**HINT:** $f(k) = a_k - a_{k+1}$.]

76. **Cauchy condensation theorem.** Let $\sum_{n=1}^\infty a_n$ be a series of positive terms such that $a_{n+1} \le a_n$ for all n.

(a) Let $m \ge 1$ be an integer. Explain why

$$2^{m-1}a_{2^m} \le a_{2^{m-1}+1} + a_{2^{m-1}+2} + \cdots + a_{2^m} \le 2^{m-1}a_{2^{m-1}}.$$

(b) Use part (a) to show that $\displaystyle\frac{1}{2}\sum_{k=1}^{m} 2^k a_{2^k} \le \sum_{k=2}^{2^m} a_k$.

(c) Use part (b) to show that if $\sum_{k=1}^\infty 2^k a_{2^k}$ diverges, then $\sum_{k=1}^\infty a_k$ diverges.

(d) Let m and n be integers such that $n \le 2^m$. Use part (a) to show that $\sum_{k=2}^n a_k \le \sum_{k=1}^m 2^{k-1}a_{2^{k-1}}$.

(e) Use part (d) to show that if $\sum_{k=0}^\infty 2^k a_{2^k}$ converges, then $\sum_{k=1}^\infty a_k$ converges.

77. Use Exercise 76 to prove that the harmonic series diverges.

78. (a) Use the properties of geometric series and Exercise 76 to prove that $\sum_{k=1}^\infty 1/k^p$ converges if $2^{1-p} < 1$ (i.e., $p > 1$) and diverges otherwise.

(b) Use Exercise 76 and part (a) to prove that the series

$$\sum_{k=2}^\infty \frac{1}{k(\ln k)^p}$$

converges if $p > 1$ and diverges otherwise.

11.4 Absolute Convergence; Alternating Series

Not-Necessarily-Positive Series

The integral, comparison, and ratio tests, as stated in the last section, apply only to *nonnegative* series. Some interesting series, however, have both positive and negative terms.

EXAMPLE 1 From numerical and graphical evidence, does the **alternating harmonic series**

$$\sum_{k=1}^\infty \frac{(-1)^{k+1}}{k} = 1 - \frac{1}{2} + \frac{1}{3} - \frac{1}{4} + \frac{1}{5} - \cdots$$

converge or diverge? To what limit?

Solution As for any series, the question is how partial sums behave. Tabulating some of them ◀ shows a pattern:

Computed by machine, of course.

Partial Sums of $1 - 1/2 + 1/3 - 1/4 + 1/5 - \cdots$

n	1	2	3	4	5	6	7	8	9	10	...
S_n	1	0.500	0.833	0.583	0.783	0.617	0.760	0.635	0.746	0.646	...
n	51	52	53	54	55	56	57	58	59	60	...
S_n	0.703	0.684	0.702	0.684	0.702	0.684	0.702	0.685	0.702	0.685	...

Successive partial sums seem to hop back and forth across some limiting value. Plots of partial sums (together with terms) exhibit the same pattern:

Terms and partial sums for 1 – 1/2 + 1/3 – 1/4 + 1/5 – …

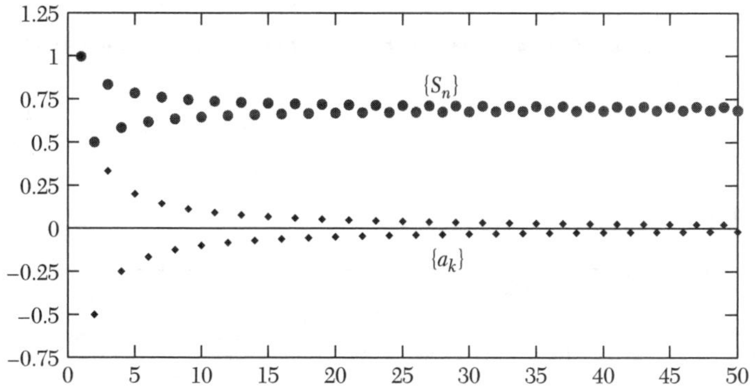

Because the terms alternate in sign, the partial sums successively rise and fall, alternately overshooting and undershooting the limiting value, which is apparently around 0.69. ▶ ■

It can be shown (with considerable effort) that the exact limit is $\ln 2 \approx 0.69315$.

In this section we develop tools for handling such not-necessarily-positive series.

Absolute vs. Conditional Convergence

The alternating harmonic series illustrates the phenomenon of **conditional convergence**. Although

$$1 - \frac{1}{2} + \frac{1}{3} - \frac{1}{4} + \frac{1}{5} - \cdots$$

itself converges, as the preceding example leads one to expect, the *ordinary* harmonic series ▶

Obtained from the previous series by taking the absolute value of each term.

$$1 + \frac{1}{2} + \frac{1}{3} + \frac{1}{4} + \frac{1}{5} + \cdots$$

diverges, as we saw from the integral test.

EXAMPLE 2 Does $\sum_{k=1}^{\infty} \frac{\sin k}{k^2}$ converge? Does $\sum_{k=1}^{\infty} \frac{|\sin k|}{k^2}$? Estimate limits.

Solution The first series, like the alternating harmonic series, has both negative and positive terms—although in no regular order this time. Plotting terms and partial sums suggests (but doesn't prove) that this series, too, converges:

Terms and partial sums for $\sum_{k=1}^{\infty} \sin(k)/k^2$

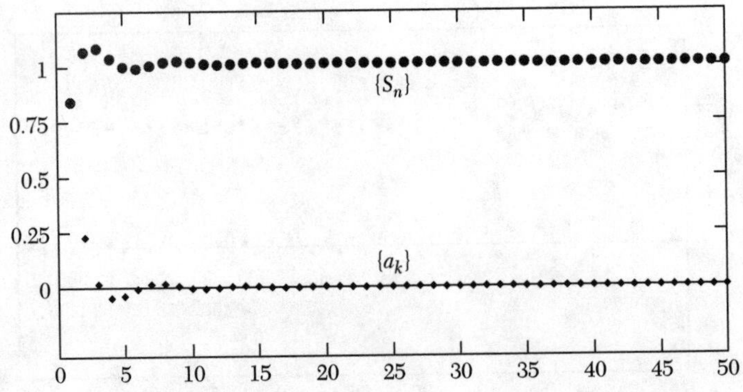

The partial sums wander up and down slightly but still appear to approach a horizontal asymptote, perhaps near $y = 1$. We'll show that this impression is correct.

The second series also seems to converge, this time to some limit near 1.25:

Terms and partial sums for $\sum_{k=1}^{\infty} |\sin(k)|/k^2$

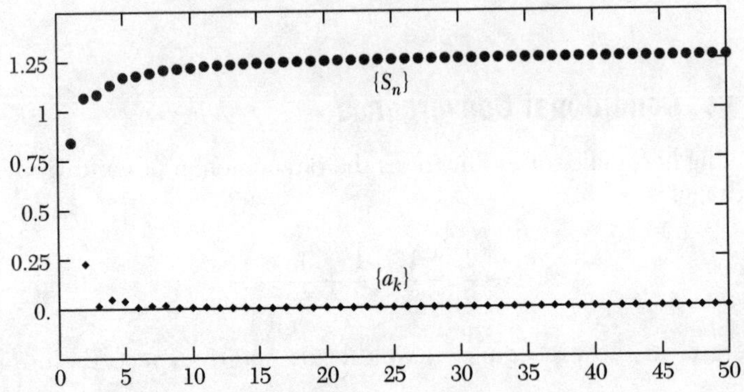

If $\sum a_k$ is the first series, $\sum |a_k|$ is the second.

The second series comes from the first by taking the absolute value of each term. ◄ The new series is nonnegative, so the comparison test applies. Since

$$0 \le \frac{|\sin k|}{k^2} \le \frac{1}{k^2}$$

for all $k \ge 1$ and $\sum_{k=1}^{\infty} 1/k^2$ converges, so must $\sum_{k=1}^{\infty} |\sin k|/k^2$. ■

This example illustrates the phenomenon of **absolute convergence**: Not only does the original series $\sum_{k=1}^{\infty} a_k$ converge, but so does its "absolute version" $\sum_{k=1}^{\infty} |a_k|$.

Definition Let $\sum_{k=1}^{\infty} a_k$ be any series.

- If $\sum_{k=1}^{\infty} |a_k|$ diverges but $\sum_{k=1}^{\infty} a_k$ converges, then $\sum_{k=1}^{\infty} a_k$ *converges conditionally*.
- If $\sum_{k=1}^{\infty} |a_k|$ converges, then $\sum_{k=1}^{\infty} a_k$ *converges absolutely*.

The Wacky World of Conditional Convergence. Conditionally convergent series have some surprising properties. Here is one of the oddest:

> *Let $\sum a_k$ be conditionally convergent, and let L be any real number. Then the terms of $\sum a_k$ can be reordered in such a way that the resulting series converges to L.*

(For more details, see your instructor.) Notice how drastically this peculiar property of conditionally convergent series upsets the naive hope that addition is commutative.

Pluses and Minuses of Pluses and Minuses

Let $\sum_{k=1}^{\infty} a_k$ be any series. If, perchance, $a_k \geq 0$ for all k, the advantage is simplicity: The partial sums are nondecreasing. The disadvantage, as the harmonic series shows, is that the partial sums may tend to infinity.

Mixing positive and negative terms may cost something in simplicity, but it's an advantage for convergence. As the alternating harmonic series shows, positive and negative terms can offset each other, thus helping in the cause of convergence.

Absolute Convergence Implies Ordinary Convergence

We saw in Chapter 10 for *integrals* that, if $\int_1^{\infty} |f(x)|\,dx$ converges, then so must $\int_1^{\infty} f(x)\,dx$, and

$$\left|\int_1^{\infty} f(x)\,dx\right| \leq \int_1^{\infty} |f(x)|\,dx.$$

The same principle applies to infinite series.

Theorem 9 If $\sum_{k=1}^{\infty} |a_k|$ converges, then so does $\sum_{k=1}^{\infty} a_k$, and

$$\left|\sum_{k=1}^{\infty} a_k\right| \leq \sum_{k=1}^{\infty} |a_k|.$$

The idea of a rigorous proof is to write the original series as a sum of two new series, one entirely positive and the other entirely negative. Using the comparison test, one can show that each of the new series converges.

The theorem shows that, as the picture suggested, the series $\sum_{k=1}^{\infty} \sin(k)/k^2$ of Example 2 does indeed converge, because $\sum_{k=1}^{\infty} |\sin k|/k^2$ does. Our limit estimates are also consistent with the theorem:

$$1 \approx \left|\sum_{k=1}^{\infty} \frac{\sin k}{k^2}\right| \leq \sum_{k=1}^{\infty} \frac{|\sin k|}{k^2} \approx 1.25.$$

EXAMPLE 3 For which values of x does the power series

$$\sum_{k=1}^{\infty} kx^k = x + 2x^2 + 3x^3 + 4x^4 + \cdots$$

converge? ◂

A power series is something like an "infinite polynomial." We discuss power series in the next section.

Solution First use the ratio test to check for absolute convergence. ◂

Watch the algebra.

$$\lim_{k\to\infty} \left| \frac{(k+1)x^{k+1}}{kx^k} \right| = \lim_{k\to\infty} \left| \frac{k+1}{k} \right| \cdot |x| = 1 \cdot |x| = |x|.$$

If $|x| < 1$, the original series converges absolutely. (Therefore, by Theorem 9, it also converges *without* absolute value signs.)

See for yourself.

It's easy to see ◂ that, for $|x| \geq 1$, the series diverges, by the nth term test. ■

Using the Theorem to Estimate Limits

Estimating a limit for any series—nonnegative or not—depends upon keeping the upper tail small. Theorem 9, combined with earlier estimates, can help.

We used Maple.

EXAMPLE 4 For the series $\sum_{k=1}^{\infty}(-1)^{k+1}k^{-3}$, we find ◂ that $S_{100} \approx 0.901542$. How closely does S_{100} approximate S, the true limit of the series?

Solution Since

$$S = \sum_{k=1}^{\infty} \frac{(-1)^{k+1}}{k^3} = \sum_{k=1}^{100} \frac{(-1)^{k+1}}{k^3} + \sum_{k=101}^{\infty} \frac{(-1)^{k+1}}{k^3} = S_{100} + R_{100},$$

we need only estimate R_{100}. ◂

The first inequality uses Theorem 9; the second uses the integral test.

$$|R_{100}| = \left| \sum_{k=101}^{\infty} \frac{(-1)^{k+1}}{k^3} \right| \leq \sum_{k=101}^{\infty} \frac{1}{k^3} < \int_{100}^{\infty} \frac{1}{x^3}\,dx.$$

Its lower limit is 100, not 101. Theorem 7, page 330, says why.

The last integral ◂ is easy to calculate:

$$\int_{100}^{\infty} \frac{1}{x^3}\,dx = \left. \frac{-1}{2x^2} \right]_{100}^{\infty} = \frac{1}{20{,}000} = 0.00005.$$

In other words, the estimate $S \approx S_{100} \approx 0.901542$ is good to at least four decimal places. ■

Alternating Series: Convergence and Estimation

For most series with both positive and negative terms, testing for absolute convergence is usually the best option. In the special (but surprisingly useful) case that the terms alternate *strictly* in sign, we can sometimes do better.

Definition An **alternating series** is one whose terms alternate strictly in sign, i.e., a series of the form

$$c_1 - c_2 + c_3 - c_4 + c_5 - c_6 + \cdots,$$

where each c_i is positive.

The alternating harmonic series ▶

See Example 1, especially the picture.

$$1 - \frac{1}{2} + \frac{1}{3} - \frac{1}{4} + \frac{1}{5} - \cdots$$

illustrates the best possibility. Because successive terms alternate in sign and decrease in size, successive partial sums straddle smaller and smaller intervals. If the terms also tend to zero, then the partial sums narrow down on a limit. ▶ The following theorem makes these observations formal and gives a convenient error bound.

Here, more than ever, the picture is key.

Theorem 10 **(Alternating Series Test)** Consider the series

$$\sum_{k=1}^{\infty}(-1)^{k+1}c_k = c_1 - c_2 + c_3 - c_4 + \cdots,$$

where

- $c_1 \ge c_2 \ge c_3 \ge \cdots \ge 0$;
- $\lim_{k\to\infty} c_k = 0$.

Then the series converges, and its limit S lies between any two successive partial sums; that is, for each $n \ge 1$, either $S_n \le S \le S_{n+1}$ or $S_{n+1} \le S \le S_n$. In particular,

$$|S - S_n| < c_{n+1}.$$

A formal proof is slightly tricky, but the underlying idea is simple. Because the limit S lies between successive partial sums, adding another term to any partial sum always "overshoots" the limit—hence the final inequality.

Although the hypotheses of this theorem seem restrictive, a surprising number of interesting series turn out to satisfy them.

Using the Theorems: Miscellaneous Examples

Combining Theorems 9 and 10 with results from earlier sections, we can handle many not-necessarily-positive series, detecting convergence or divergence and, when possible, estimating limits. The following examples illustrate some useful tricks of this trade.

EXAMPLE 5 **(An Alternating *p*-Series: Another Look)** What does Theorem 10 say about $\sum_{k=1}^{\infty}(-1)^{k+1}/k^3$ and its 100th partial sum, $S_{100} \approx 0.9015422$?

Solution In this context, $c_k = 1/k^3$. Now Theorem 10 says not only that the series converges—which we already knew—but also that

$$|S - S_{100}| < c_{101} = \frac{1}{101^3} \approx 0.000001.$$

Thus, $S_{100} \approx 0.9015422$ ▶ lies within 0.000001 of the true limit S. Equivalently, S lies between $S_{100} \approx 0.9015422$ and $S_{101} \approx 0.9015432$. ■

Does S_{100} overshoot or undershoot? Why?

EXAMPLE 6 **(The *n*th Term Test: Always Available)**

Does $\sum_{j=1}^{\infty}(-1)^j \frac{j}{j+1}$ converge or diverge? Why?

Solution The alternating series test looks tempting at first glance, but it doesn't apply. The given series *is* alternating, but another important hypothesis isn't satisfied:

$$\lim_{j\to\infty} \frac{j}{j+1} = \lim_{j\to\infty} \frac{1}{1+1/j} = 1,$$

not zero, as Theorem 10 requires. The *n*th term test ◄ does apply, however. Since $j/(j+1) \to 1$ as $j \to \infty$, it follows that $(-1)^j j/(j+1)$ has no limit as $j \to \infty$. ◄ It follows that the given series diverges. ■

Maybe we should call it the "jth term test" here. In any case, the index name is immaterial.

Successive terms are alternately near 1 and near −1.

EXAMPLE 7 Does the series

$$1+2+3+4+5-\frac{1}{6}+\frac{1}{7}-\frac{1}{8}+\frac{1}{9}-\cdots$$

converge? If so, find or estimate the limit.

Solution The alternating series test doesn't apply right out of the box, because the first five terms break the desired pattern. The problem isn't fatal, however. Basic series algebra ◄ lets us group our terms into two blocks as follows:

We studied it in Section 11.2.

$$(1+2+3+4+5) - \left(\frac{1}{6}-\frac{1}{7}+\frac{1}{8}-\frac{1}{9}+\cdots\right).$$

The first block is finite, so convergence isn't an issue; its sum is 15. The second block clearly satisfies all hypotheses of the alternating series test and so converges to some limit L. Any partial sum of the second block, moreover, differs from L by less than the magnitude of the next term (by the the last line of Theorem 10).

The entire series therefore converges to $S = 15 - L$, and any partial sum differs from S by no more than the next term. The partial sum $S_9 = 1+2+\cdots+1/9 \approx 14.962$, for instance, overshoots the true limit by less than 1/10. In other words, the exact limit S satisfies $14.862 \le S \le 14.962$. ◄ ■

Adding more terms would approximate S more closely.

EXAMPLE 8 Does $\sum_{n=1}^{\infty} \frac{\sin n}{n^3+n^2+n+1+\cos n}$ converge or diverge? Why?

Solution The problem is easier than it looks. Hoping for absolute convergence, we start by taking absolute values:

$$\left|\frac{\sin n}{n^3+n^2+n+1+\cos n}\right| = \frac{|\sin n|}{n^3+n^2+n+1+\cos n}.$$

From the general appearance of numerator and denominator, comparison suggests itself. Some simple inequalities make the job much easier: ◄

Convince yourself of each one.

$$\frac{|\sin n|}{n^3+n^2+n+1+\cos n} \le \frac{1}{n^3+n^2+n} \le \frac{1}{n^3}.$$

We know that $\sum 1/n^3$ converges; ◄ so, by comparison, must the absolute-value version of the given series. By Theorem 9, the original series must converge, too. ■

It's a p-series with $p = 3$.

BASICS

1. We showed in Example 7, page 342, that the series
$$1+2+3+4+5-\frac{1}{6}+\frac{1}{7}-\frac{1}{8}+\frac{1}{9}-\cdots$$
converges to some limit S.
   (a) Does the series converge conditionally or absolutely? Why?
   (b) Calculate S_{15} for this series. Does S_{15} overestimate or underestimate S? How do you know?
   (c) According to *Maple*, $S_{60} \approx 14.902$. Use this result to find good upper and lower bounds for S. Explain your answer.
   (d) We said in this section that the alternating harmonic series can be shown to converge to $\ln 2$. Use this fact to find the limit of the series exactly.

2. In Example 8, page 342, we showed that the series
$$\sum_{n=1}^{\infty} \frac{\sin n}{n^3+n^2+n+1+\cos n}$$
converges to some limit S, but we didn't find or estimate S.
   (a) Compute S_{50} (not by hand!).
   (b) Explain why $|R_{50}| \le \int_{50}^{\infty} \frac{dx}{x^3}$.
   (c) Use part (b) to give good upper and lower bounds for S.

3. (a) Suppose that the series $\sum_{k=1}^{\infty} a_k$ converges absolutely. Show that the series $\sum_{k=1}^{\infty} \frac{a_k}{k}$ converges.
   (b) Suppose that the series $\sum_{k=1}^{\infty} \frac{a_k}{k}$ converges. Must the series $\sum_{k=1}^{\infty} a_k$ also converge? Justify your answer.

4. Show that $\sum_{k=2}^{\infty} (-1)^k \frac{k}{k^2-1}$ converges conditionally.

In Exercises 5–10, show that the series converges. Then compute an estimate of the limit that is guaranteed to be in error by no more than 0.005.

5. $\sum_{k=1}^{\infty} \frac{(-1)^k}{k^4}$

6. $\sum_{k=1}^{\infty} \frac{(-1)^k}{k^2+2^k}$

7. $\sum_{k=0}^{\infty} \frac{(-2)^k}{7^k+k}$

8. $\sum_{k=0}^{\infty} \frac{(-3)^k}{(k^2)!}$

9. $\sum_{k=5}^{\infty} (-1)^k \frac{k^{10}}{10^k}$

10. $\sum_{k=0}^{\infty} \frac{(-1)^k}{(k+1)\,2^k}$

FURTHER EXERCISES

11. Does the series $\sum_{n=2}^{\infty} (-1)^n \frac{n}{2n-1}$ converge? Justify your answer.

12. Let $a_k = \int_k^{\infty} \frac{dx}{2x^2-1}$.
   (a) Evaluate $\lim_{k\to\infty} a_k$.
   [**HINT:** $1/2x^2 \le 1/(2x^2-1) \le 1/x^2$ if $x \ge 1$.]
   (b) Does $\sum_{k=1}^{\infty} (-1)^{k+1} a_k$ converge absolutely?
   (c) Does $\sum_{k=1}^{\infty} (-1)^{k+1} a_k$ converge?

13. (a) For which values of p does the series $\sum_{k=2}^{\infty} \frac{\ln k}{k^p}$ converge?
   (b) For which values of p does the series $\sum_{k=2}^{\infty} (-1)^k \frac{\ln k}{k^p}$ converge?
   (c) For which values of p does the series $\sum_{k=2}^{\infty} (-1)^k \frac{\ln k}{k^p}$ converge absolutely?
   (d) For which values of p does the series $\sum_{k=2}^{\infty} (-1)^k \frac{\ln k}{k^p}$ converge conditionally?

In Exercises 14–24, determine whether the series converges absolutely, converges conditionally, or diverges. If it converges, find upper and lower bounds on its limit. Justify your answers.

14. $\sum_{k=1}^{\infty} \frac{(-1)^k}{\sqrt{k}}$

15. $\sum_{j=1}^{\infty} \frac{(-1)^{j+1}}{j^2}$

16. $\sum_{n=1}^{\infty} \frac{(-3)^n}{n^3}$

17. $\sum_{k=4}^{\infty} (-1)^k \frac{\ln k}{k}$

18. $\sum_{m=8}^{\infty} \frac{\sin m}{m^3}$

19. $\sum_{n=1}^{\infty} \frac{\cos(n\pi)}{n}$

20. $\sum_{k=0}^{\infty} (-1)^k \frac{k}{2k+1}$

21. $\sum_{m=0}^{\infty} (-1)^m \frac{m^3}{2^m}$

22. $\sum_{k=0}^{\infty} \frac{(-2)^k}{3^k+k}$

23. $\sum_{j=0}^{\infty} (-1)^j \frac{j!}{(j^2)!}$

24. $\sum_{n=1}^{\infty} (-1)^{n+1} \frac{\arctan n}{n}$

25. Consider the series $\sum_{k=1}^{\infty}(-1)^{k+1}a_k$. Suppose that the terms of the sequence $\{a_k\}$ are positive and decreasing for all $k \geq 10^9$ and $\lim_{k\to\infty} a_k = 0$ but that $a_{10^9} > a_1$. Explain why the series converges. [**HINT:** Theorem 10 doesn't apply directly.]

26. Does the infinite series

$$1 - \frac{1}{2^3} + \frac{1}{3^2} - \frac{1}{4^3} + \frac{1}{5^2} - \frac{1}{6^3} + \frac{1}{7^2} - \frac{1}{8^3} + \cdots$$

converge or diverge? Justify your answer.

27. Suppose that $\sum_{j=1}^{\infty} b_j$ converges to a number S and that $b_j \geq 0$ for all $j \geq 1$.
   (a) Show that $\sum_{j=1}^{\infty}(-1)^{j+1}b_j$ converges.
   (b) Suppose that $0 \leq b_{j+1} \leq b_j$ for all $j \geq 1$ and that the partial sum $\sum_{j=1}^{100} b_j$ approximates S within 0.005. Explain why

$$0 \leq \sum_{j=1}^{\infty}(-1)^{j+1}b_j - \sum_{j=1}^{100}(-1)^{j+1}b_j < 0.005.$$

28. Give an example of a convergent series $\sum_{k=1}^{\infty} a_k$ with the property that $\sum_{k=1}^{\infty} (a_k)^2$ diverges.

29. Suppose that $a_k \geq 0$ for all $k \geq 1$. Is it possible that $\sum_{k=1}^{\infty} a_k$ converges conditionally? Explain.

30. Suppose that $\sum_{k=1}^{\infty} a_k$ diverges. Is it possible that $\sum_{k=1}^{\infty} |a_k|$ converges? Justify your answer.

31. (a) Show that $\displaystyle\sum_{k=n+1}^{\infty} \frac{(-1)^{k+1}}{k^2} \leq \sum_{k=n+1}^{\infty} \frac{1}{k^2}$ for any $n \geq 1$.
   (b) Explain why the result in part (a) implies that the series $\displaystyle\sum_{k=1}^{\infty} \frac{(-1)^{k+1}}{k^2}$ converges faster than the series $\displaystyle\sum_{k=1}^{\infty} \frac{1}{k^2}$.

32. This exercise outlines a proof of the alternating series test.

Let $S_n = \displaystyle\sum_{k=1}^{n}(-1)^{k+1}c_k$ denote the partial sum of the first n terms of a series satisfying the hypotheses of the alternating series test (Theorem 10).
   (a) Show that the sequence of even partial sums, $S_2, S_4, S_6, S_8, \ldots$, is monotone increasing.
   (b) Show that the sequence of odd partial sums, $S_1, S_3, S_5, S_7, \ldots$, is monotone decreasing.
   (c) Show that $S_{2m} \leq S_{2m-1}$ for any integer $m \geq 1$.
   (d) Use part (c) to show that the sequence of even partial sums and the sequence of odd partial sums both converge. [**NOTE:** Although both sequences converge, we must still show that they converge to the same limit.]
   (e) Show that $\lim_{m\to\infty}(S_{2m+1} - S_{2m}) = 0$. From this it follows that there is a real number S such that $\lim_{n\to\infty} S_n = S$.
   (f) Explain why $0 < S - S_{2m} < c_{2m+1}$ and $0 < S_{2m+1} - S < c_{2m+2}$.

33. Let $a_k = \dfrac{|\sin k|}{k^2}$.
   (a) Show that $\dfrac{a_{k+1}}{a_k} = |\cos 1 + \sin 1 \cot k| \cdot \dfrac{k^2}{(k+1)^2}$.
   (b) Explain why the result in part (a) implies that $\{a_k\}$ never becomes a decreasing sequence.
   (c) Show that $\displaystyle\sum_{k=1}^{\infty}(-1)^{k+1}a_k$ converges.

11.5 Power Series

Basic Ideas and Examples

A **power series** is a series of the form

$$a_0 + a_1x + a_2x^2 + a_3x^3 + a_4x^4 + \cdots + a_nx^n + \cdots = \sum_{k=0}^{\infty} a_kx^k.$$

Each a_k is a constant, called the **coefficient of x^k**. The symbol x denotes a variable. A power series may converge for some values of x and diverge for others.

EXAMPLE 1 **(A Geometric Power Series)** Among the simplest and most useful power series is

$$S(x) = 1 + x + x^2 + x^3 + x^4 + x^5 + \cdots = \sum_{k=0}^{\infty} x^k.$$

(For all $k \geq 0$, $a_k = 1$.) For which real numbers x does $S(x)$ converge?

Solution Setting $x = 1$ gives the divergent ▶ series

Why is it divergent?

$$S(1) = 1 + 1 + 1^2 + 1^3 + 1^4 + \cdots = 1 + 1 + 1 + 1 + 1 + \cdots.$$

If, say, $x = 1/2$, the series converges to 2: ▶

Check the arithmetic.

$$S(1/2) = 1 + \frac{1}{2} + \frac{1}{4} + \frac{1}{8} + \cdots + \frac{1}{2^n} + \cdots = 2.$$

Indeed, for *any* value of x, $S(x)$ is a geometric series in x, so it converges if and only if $|x| < 1$. We even know the limit: ▶

Recall these properties of geometric series? Section 11.2 has details.

$$\text{If } |x| < 1, \text{ then } 1 + x + x^2 + x^3 + x^4 + \cdots \text{ converges to } \frac{1}{(1-x)}.$$ ■

EXAMPLE 2 It's a fact ▶ that, for any number x,

In the next section we'll see why.

$$e^x = 1 + x + \frac{x^2}{2!} + \frac{x^3}{3!} + \frac{x^4}{4!} + \cdots.$$

Interpret the right side in the language of power series.

Solution Writing the series in the form

$$1 + x + \frac{x^2}{2!} + \frac{x^3}{3!} + \cdots = \sum_{k=0}^{\infty} \frac{1}{k!} x^k$$

shows the pattern of coefficients.

For what values of x does the series converge?

We'll use the ratio test to decide. Since the ratio test works only for positive series, we'll use it to check for absolute convergence. For any input x,

$$\lim_{k\to\infty} \frac{|a_{k+1}x^{k+1}|}{|a_k x^k|} = \lim_{k\to\infty} \frac{|x|^{k+1}}{(k+1)!} \cdot \frac{k!}{|x|^k} = \lim_{k\to\infty} \frac{|x|}{k+1} = 0.$$

Because $0 < 1$, the ratio test guarantees that this series converges absolutely—and therefore in the ordinary sense—for *all* values of x. ▶ It can also be shown, though we haven't shown it yet, that the series converges to e^x for any x. If, say, $x = 1$, then the series converges to e, as partial sums suggest numerically:

Recall: If a series converges with absolute value signs, then it converges without. See Theorem 9, page 339.

$S_{10} \approx 2.718281801$; $S_{20} \approx 2.718281828$; $S_{30} \approx 2.71828182845905$.

The last number agrees with e in all 15 decimal places. ■

Power Series and Polynomials

Power series are, roughly speaking, "unending" or "infinite-degree" polynomials. More precisely:

Terms Are Power Functions For both polynomials and power series, each summand is of the form $a_k x^k$, with k a nonnegative integer.

Partial Sums Are Ordinary Polynomials Every partial sum S_n of the power series $\sum_{k=0}^{\infty} a_k x^k$ has the form

$$S_n = a_0 + a_1 x + a_2 x^2 + a_3 x^3 + \cdots + a_n x^n,$$

i.e., an ordinary polynomial of degree n.

Easy to Use Polynomials are easy to differentiate and integrate, term by term. With due care taken for convergence, ◄ so are power series. We'll soon return to this theme and to its practical importance.

An important proviso!

Choosing Base Points

The polynomial expressions

$$\begin{aligned} p(x) &= x^2 - 2x + 2, \\ q(x) &= (x-1)^2 + 1, \\ r(x) &= (x-2)^2 + 2(x-2) + 2 \end{aligned}$$

all represent the same function. ◄ The differences have to do with different choices of **base point**. Version q, for instance, is said to be **expanded about** the base point $x = 1$, because q is written in powers of $(x - 1)$. Versions p and r are expanded about the base points $x = 0$ and $x = 2$, respectively.

Are you convinced? If not, multiply out q and r.

Which version is "best" depends on the problem at hand. Version q, for instance, focuses attention most clearly on the graph's vertex, which occurs at the base point $x = 1$. Finding values and derivatives of the function at the base point is especially easy.

The same choice of base point applies to power series. For example, the two power series ◄

They're different functions this time.

$$\sum_{k=0}^{\infty} 2^k x^k \qquad \text{and} \qquad \sum_{k=0}^{\infty} 2^k (x-1)^k$$

are written in powers of x and powers of $(x - 1)$, respectively; their respective base points are $x = 0$ and $x = 1$. Mathematically, the difference is small: The second form amounts only to a "shift" of one unit to the right. We'll usually, but not always, treat power series based at $x = 0$.

Power Series as Functions

Any power series

$$S(x) = \sum_{k=0}^{\infty} a_k x^k = a_0 + a_1 x + a_2 x^2 + a_3 x^3 + \cdots$$

defines, in a natural way, a *function* of x. For a given input x, $S(x)$ is the limit—if one exists—of the power series.

Domains of Power Series

Any function given by a "formula" in x has a natural domain: the set of x for which the formula makes sense. ◄ Power series are no different. The domain of a function $S(x)$ given by a power series is the set of inputs x for which the series converges—also known as the **interval of convergence**. We saw, for instance, that $\sum_{k=0}^{\infty} x^k$ converges for x in $(-1, 1)$; $\sum_{k=0}^{\infty} x^k/k!$ converges for x in $(-\infty, \infty)$.

For example, the natural domain of $\sqrt{x}$ is the set of nonnegative numbers.

EXAMPLE 3 A function $S(x)$ is defined by the power series

$$S(x) = \sum_{k=0}^{\infty} 2^k x^k = 1 + 2x + 4x^2 + 8x^3 + \cdots.$$

What's the domain of S? Is a simpler formula available?

Solution Think of $S(x)$ as a geometric series $1 + r + r^2 + r^3 + \cdots$, with $r = 2x$: ▶

In other words, $S(x)$ is "geometric in $2x$."

$$S(x) = 1 + 2x + (2x)^2 + (2x)^3 + \cdots.$$

A geometric series converges if $|r| < 1$; *this* one converges, therefore, if $|2x| < 1$, i.e., if $|x| < 1/2$. In that case, the limit is $1/(1-r) = 1/(1-2x)$. To summarize: The power series $S(x)$ converges for x in $(-1/2, 1/2)$; on that domain,

$$S(x) = 1 + 2x + (2x)^2 + (2x)^3 + \cdots = \frac{1}{1-2x}.$$ ■

Finding the Interval of Convergence

The first task, given a power series, is to find the interval of convergence. For many series, the ratio test is all that's needed. We illustrate with several important examples.

EXAMPLE 4 Show that the series

$$1 + 2x + 3x^2 + 4x^3 + 5x^4 + \cdots = \sum_{k=1}^{\infty} kx^{k-1}$$

converges only for x in $(-1, 1)$. Guess a limit.

Solution We'll use the ratio test to check for *absolute* convergence. The ratio of successive terms is

$$\frac{\left|a_{k+1}x^{k+1}\right|}{\left|a_k x^k\right|} = \frac{(k+1)|x|^k}{k|x|^{k-1}} = |x|\frac{k+1}{k}.$$

As $k \to \infty$, this ratio tends to $|x|$. Therefore, the series converges—with or without absolute-value signs—if $|x| < 1$.

If $|x| = 1$, the ratio test is inconclusive. However, it's easy to see that, if $x = \pm 1$,

$$\left|kx^{k-1}\right| = |k| \to \infty \qquad \text{as} \qquad k \to \infty.$$

Thus, by the nth term test, the series diverges if $x = \pm 1$. For the same reason, the series diverges if $|x| > 1$. The convergence interval is therefore $(-1, 1)$, as claimed.

Let's guess a limit. We saw in Example 1 that the equation

$$S(x) = 1 + x + x^2 + x^3 + x^4 + \cdots = \frac{1}{1-x}$$

holds for $|x| < 1$. Differentiating all three quantities suggests a limit for the original series. If $|x| < 1$, then it's reasonable to expect that

$$S'(x) = 1 + 2x + 3x^2 + 4x^3 + \cdots = \frac{1}{(1-x)^2}.$$

If, say, $x = 1/2$, then

$$\sum_{k=1}^{\infty} \frac{k}{2^{k-1}} = 1 + \frac{2}{2} + \frac{3}{4} + \frac{4}{8} + \cdots = \frac{1}{(1-1/2)^2} = 4.$$

Numerical evidence suggests that we're right. For the preceding series, $S_{20} \approx 3.999958038$. ■

Our guess is reasonable and correct, but it raises important questions:

- Is it legitimate to differentiate a series term by term?
- On what interval does the resulting series converge?

We address these questions in the next section.

EXAMPLE 5 Antidifferentiating the geometric series $S(x) = 1 + x + x^2 + x^3 + \cdots$ term by term gives the new series

$$T(x) = x + \frac{x^2}{2} + \frac{x^3}{3} + \frac{x^4}{4} + \frac{x^5}{5} + \cdots = \sum_{k=1}^{\infty} \frac{x^k}{k}.$$

Where does $T(x)$ converge? Guess a limit.

Solution Because $S(x)$ converges (absolutely) for $|x| < 1$, it's reasonable to expect the same of $T(x)$. The ratio test agrees: ◀

Check the algebraic details.

$$\lim_{k\to\infty} \frac{\left|a_{k+1}x^{k+1}\right|}{\left|a_k x^k\right|} = \lim_{k\to\infty} |x| \cdot \frac{k}{k+1} = |x|.$$

Therefore, as expected, $T(x)$ converges absolutely on the interval $(-1, 1)$.

What happens at the endpoints, $x = \pm 1$? Setting $x = \pm 1$ in $T(x)$ produces two by-now-familiar series: ◀

The harmonic series and the alternating harmonic series.

$$T(1) = \sum_{k=1}^{\infty} \frac{1}{k}; \qquad T(-1) = \sum_{k=1}^{\infty} \frac{(-1)^k}{k}.$$

As we saw earlier, the first series diverges; the second converges conditionally (by the alternating-series theorem). Thus, T converges for x in $[-1, 1)$.

Since $T(x)$ came from $S(x)$ by antidifferentiation, it's reasonable to guess a similar relationship for limits:

$$1 + x + x^2 + x^3 + \cdots = \frac{1}{1-x} \Longrightarrow x + \frac{x^2}{2} + \frac{x^3}{3} + \cdots = -\ln(1-x).$$

Numerical evidence suggests that we're right. If $x = 1/2$, the series gives $S_{20} \approx 0.69314714$; that's not far from $-\ln 1/2 \approx 0.69314718$. ■

Familiar questions arise again.

- Is it legitimate to antidifferentiate a series term by term?
- On what interval does the resulting series converge?

We address these questions in the next section.

EXAMPLE 6 Where does the power series $\sum_{k=0}^{\infty} k!\,x^k$ converge?

Solution Like any power series, this one converges at its base point, $x = 0$. But if $x \neq 0$, the ratio test (applied to absolute values) gives

$$\lim_{k\to\infty} \frac{\left|a_{k+1}x^{k+1}\right|}{\left|a_k x^k\right|} = \lim_{k\to\infty} \frac{(k+1)!\,|x|^{k+1}}{k!\,|x|^k} = \lim_{k\to\infty} (k+1)\cdot|x| = \infty.$$

This result amounts to a dramatic violation of the nth term test. Far from tending to zero, successive terms grow larger and larger. This power series has, in a sense, the smallest possible domain: It converges only if $x = 0$. ■

Power Series Convergence: Lessons from the Examples

The preceding examples illustrate several useful properties of power series and their convergence sets.

The Radius of Convergence. *Every* power series $\sum_{k=0}^{\infty} a_k x^k$ converges for $x = 0$. The real question is this:

How far from zero can x be without destroying convergence?

In every example so far, the convergence set turned out to be an *interval centered at* 0. ▶ This was no accident. The convergence domain for *any* power series is an interval centered at zero; ▶ its radius is called the **radius of convergence** of the power series. The following theorem guarantees all these claims.

In the last example, the power series converged only for $x = 0$. Stretching a point slightly, we'll call the set $\{0\}$ an interval of radius 0.

In particular, the convergence interval is symmetric about the "base point."

> **Theorem 11** Let $S(x) = \sum_{k=0}^{\infty} a_k x^k$ be a power series, and let C be any real number. If $S(x)$ converges for $x = C$, then $S(x)$ also converges for $|x| < |C|$.

The Idea of Proof for $C = 1$. Suppose that $\sum_{k=0}^{\infty} a_k x^k$ converges for $x = 1$. Then $\sum_{k=0}^{\infty} a_k$ converges. By the nth term test, $a_k \to 0$ as $k \to \infty$. Since the a_k's cannot blow up, they must be bounded in absolute value. ▶ In other words, there is a number $M > 0$ such that, for all k,

If the a_k's weren't bounded, they couldn't tend to zero.

$$|a_k| \le M\,.$$

Therefore

$$\left|a_k x^k\right| \le M|x|^k$$

holds for all k. Now the comparison test applies: For $|x| < 1$, the geometric series $\sum_{k=0}^{\infty} M|x|^k$ converges, so $\sum_{k=0}^{\infty} \left|a_k x^k\right|$ must converge, too. That's what we wanted to show. ▶

A general proof for any value of C isn't much harder.

At Endpoints, Anything Can Happen. In several of the preceding examples, the series converged on the *open* interval $(-1, 1)$; in another example, it converged on $[-1, 1)$. ▶ A series' interval of convergence may include either, both, or neither of its endpoints—any combination is possible. ▶ In practice, what really matters is a series' *radius* of convergence; what happens at the endpoints, although sometimes interesting, is usually less important.

In both cases, the radius of convergence is 1.

See the exercises at the end of this section.

Any Radius of Convergence Is Possible. In several examples, power series turned out to converge on $(-1, 1)$. Actually, any (positive) radius of convergence is possible. Indeed, for any positive constant R, the series $\sum_{k=0}^{\infty} x^k/R^k$ has radius of convergence precisely R.

Power Series Convergence, Graphically

For any $n \geq 0$, the nth partial sum of the power series $S(x) = \sum_{k=0}^{\infty} a_k x^k$ is the polynomial

$$p_n(x) = a_0 + a_1 x + a_2 x^2 + a_3 x^3 + \cdots + a_n x^n.$$

To say that the power series converges for x in $(-R, R)$ means that, for any x in that interval, there's a number $S(x)$ such that $p_n(x) \to S(x)$ as $n \to \infty$.

Sorting out precisely what this means is a worthy challenge. The following picture gives a graphical sense of the situation for the geometric power series $S(x) = \sum_{k=0}^{\infty} x^k$:

On $(-1, 1)$, $\sum_{k=0}^{\infty} x^k$ converges to $1/(1-x)$

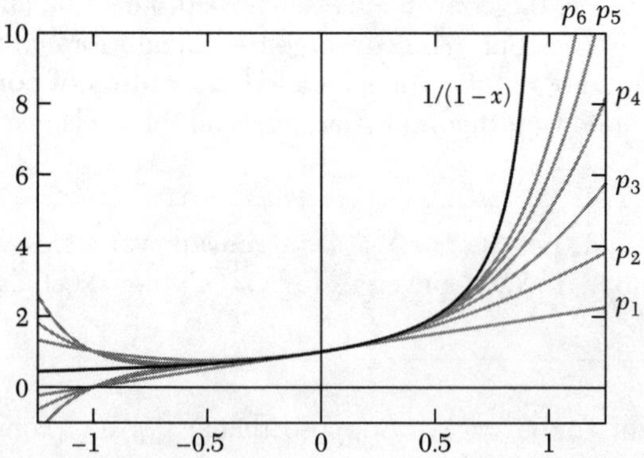

Labeled p_1 through p_6.

In the picture, the polynomial graphs ◀ represent the first six partial sums. Over the interval $(-1, 1)$, they appear to approach the graph of the limiting function more and more closely. Outside that interval, the polynomial graphs appear to diverge, rather than to approach any common limiting function.

BASICS

1. The power series $\sum_{k=1}^{\infty} x^k/k$ has radius of convergence 1. Plot the partial sum polynomials of degree 1, 2, 4, 6, 8, and 10 over the interval $[-2, 2]$. Is the interval of convergence of the series apparent? Explain.

In Exercises 2–7, find the radius of convergence of the power series.

2. $\displaystyle\sum_{j=1}^{\infty} \left(\frac{x}{2}\right)^j$

3. $\displaystyle\sum_{k=1}^{\infty} \frac{x^k}{k\,2^k}$

4. $\displaystyle\sum_{k=1}^{\infty} \frac{x^k}{\sqrt{k}}$

5. $\displaystyle\sum_{m=1}^{\infty} \frac{x^m}{m^2+1}$

6. $\displaystyle\sum_{n=1}^{\infty} n^n x^n$

7. $\displaystyle\sum_{n=0}^{\infty} \frac{x^n}{n! + n}$

In Exercises 8–15, find the radius and the interval of convergence of the power series.

8. $\sum_{k=1}^{\infty} (3x)^k$

9. $\sum_{m=0}^{\infty} \frac{(3x)^m}{m!}$

10. $\sum_{n=1}^{\infty} \frac{(3x)^n}{n}$

11. $\sum_{j=1}^{\infty} \frac{(3x)^j}{j^2}$

12. $\sum_{n=0}^{\infty} (x-2)^n$

13. $\sum_{n=2}^{\infty} \frac{(x-3)^{2n}}{n^4}$

14. $\sum_{n=2}^{\infty} \frac{(x-5)^n}{n \ln n}$

15. $\sum_{n=1}^{\infty} \frac{(x+1)^n}{n}$

FURTHER EXERCISES

16. Let $R > 0$ be any positive constant.
    (a) Show that $\sum_{k=0}^{\infty} \frac{x^k}{R^k}$ converges on $(-R, R)$.
    (b) Show that $\sum_{k=1}^{\infty} \frac{x^k}{kR^k}$ converges on $[-R, R)$.
    (c) Show that $\sum_{k=1}^{\infty} \frac{x^k}{k^2 R^k}$ converges on $[-R, R]$.
    (d) Concoct a power series that converges on $(-R, R]$.
17. For each of the following intervals, give an example of a power series that has the given interval as its interval of convergence. [**HINT:** See Exercise 16.]
    (a) $[-4, 4)$
    (b) $[-1, 5]$
    (c) $(-4, 0)$
    (d) $(8, 16]$
    (e) $[-11, -3)$
18. Suppose that $\sum_{n=1}^{\infty} a_n$ converges and that $|x| < 1$. Show that $\sum_{n=1}^{\infty} a_n x^n$ converges absolutely.
19. Suppose that the power series $\sum_{k=0}^{\infty} a_k x^k$ converges only if $-2 < x \le 2$.
    (a) Explain why the radius of convergence of this power series is 2.
    (b) Explain why the power series $\sum_{k=0}^{\infty} a_k (x-1)^k$ has radius of convergence 2.
    (c) Show that the interval of convergence of the power series $\sum_{k=0}^{\infty} a_k (x-3)^k$ is $(1, 5]$.
    (d) Find the interval of convergence of the power series $\sum_{k=0}^{\infty} a_k (x+1)^k$.
20. Suppose that the power series $\sum_{k=0}^{\infty} a_k (x-b)^k$ converges only if $-11 \le x < 17$.
    (a) What is the radius of convergence of the power series?
    (b) Determine the value of b.

In Exercises 21–26, find the interval of convergence (endpoint behavior too!) of the power series.

21. $\sum_{m=0}^{\infty} \left(\frac{x-3}{2}\right)^m$

22. $\sum_{j=0}^{\infty} \frac{(x-2)^j}{j!}$

23. $\sum_{k=1}^{\infty} \frac{(x-1)^k}{k4^k}$

24. $\sum_{n=1}^{\infty} \frac{(x-1)^n}{\sqrt{n}}$

25. $\sum_{i=1}^{\infty} \frac{(x+5)^i}{i(i+1)}$

26. $\sum_{m=1}^{\infty} \frac{2^m (x-1)^m}{m}$

27. Suppose that the power series $\sum_{k=0}^{\infty} a_k x^k$ converges if $x = -3$ and diverges if $x = 7$. Indicate which of the following statements *must* be true, which *may* be true, and which *cannot* be true. Justify your answers.
    (a) The power series converges if $x = -10$.
    (b) The power series diverges if $x = 3$.
    (c) The power series converges if $x = 6$.
    (d) The power series diverges if $x = 2$.
    (e) The power series diverges if $x = -7$.
    (f) The power series converges if $x = -4$.
28. Suppose that the power series $\sum_{k=0}^{\infty} a_k (x+2)^k$ converges if $x = -7$ and diverges if $x = 7$. Indicate which of the following statements *must* be true, which *may* be true, and which *cannot* be true. Justify your answers.
    (a) The power series converges if $x = -8$.
    (b) The power series converges if $x = 1$.
    (c) The power series converges if $x = 3$.
    (d) The power series diverges if $x = -11$.
    (e) The power series diverges if $x = -10$.
    (f) The power series diverges if $x = 5$.
    (g) The power series diverges if $x = -5$.
29. Consider a power series of the form $\sum_{k=0}^{\infty} a_k (x-1)^k$. Indicate whether each of the following statements *must* be true, *may* be true, or *cannot* be true. Justify your answers.
    (a) The power series converges only if $|x| > 2$.
    (b) The power series converges for all values of x.
    (c) If the radius of convergence of the power series is 3, the power series converges if $-2 < x < 4$.
    (d) The interval of convergence of the power series is $[-5, 5]$.
    (e) If the interval of convergence of the power series is $(-7, 9)$, the radius of convergence is 7.

30. The power series

$$\sum_{k=0}^{\infty} \frac{1}{k!}x^k = 1 + x + \frac{x^2}{2!} + \frac{x^3}{3!} + \cdots$$

converges, for any x, to e^x.

(a) If $x = -1$, the series converges to $1/e$. For this case, find S_{10}. By how much does S_{10} differ from the *actual* value of $1/e$? (Use a calculator or a computer.)

(b) If $x = -1$, the series is alternating. What does this say about the maximum possible error committed by S_{10} in estimating $1/e$? For which n must S_n estimate $1/e$ with error less than 10^{-10}?

31. Let $f(x) = \sum_{n=0}^{\infty} \frac{2x^n}{3^n + 5}$.

(a) Show that $f(10)$ is undefined (i.e., the series that defines $f(x)$ diverges when $x = 10$).

(b) Which of the numbers 0.5, 1.5, 3, and 6 are in the domain of f?

(c) Estimate $f(1)$ within 0.01 of its exact value.

32. Let $h(x) = \sum_{k=0}^{\infty} \frac{(x-2)^k}{k! + k^3}$.

(a) What is the domain of h (i.e., the set of x for which the series converges)?

(b) Estimate $h(0)$ within 0.005 of its exact value.

(c) Estimate $h(3)$ within 0.005 of its exact value.

33. Let $g(x) = \sum_{n=1}^{\infty} \frac{(x+4)^n}{n^3 5^n}$.

(a) What is the domain of g?

(b) Estimate $g(0)$ within 0.005 of its exact value.

(c) Estimate $g(-5)$ within 0.005 of its exact value.

34. (a) Evaluate $\lim_{x\to 1^-} \sum_{k=0}^{\infty} (-1)^k x^k$.

(b) Explain why the result in part (a) does *not* mean that $\sum_{k=0}^{\infty} (-1)^k$ converges.

11.6 Power Series as Functions

Any power series

$$S(x) = \sum_{k=0}^{\infty} a_k x^k = a_0 + a_1 x + a_2 x^2 + a_3 x^3 + \cdots$$

can be thought of as a function of x; its domain is the series' interval of convergence. As we saw in the last section, this domain is always an interval centered at 0. ◀ In this section we explore the remarkable—and useful—properties of functions defined by power series.

The convergence interval may or may not contain its endpoints; for most purposes, it doesn't matter much either way.

Calculus with Power Series

Given *any* power series $S(x)$, convergent or divergent, it's easy to differentiate or antidifferentiate term by term to produce new series $D(x)$ and $A(x)$:

$$S(x) = \sum_{k=0}^{\infty} a_k x^k = a_0 + a_1 x + a_2 x^2 + a_3 x^3 + \cdots;$$

$$D(x) = \sum_{k=1}^{\infty} k a_k x^{k-1} = a_1 + 2a_2 x + 3a_3 x^2 + \cdots;$$

$$A(x) = \sum_{k=0}^{\infty} a_k \frac{x^{k+1}}{k+1} = a_0 x + a_1 \frac{x^2}{2} + a_2 \frac{x^3}{3} + a_3 \frac{x^4}{4} + \cdots.$$

Prudence dictates a measure of caution; important questions remain to be answered.

- If the series S has radius of convergence r, is the same true of D and A?
- Even if we assume that S, D, and A all converge on $(-r, r)$, must D be the derivative of S, in the ordinary calculus sense? Is A necessarily an antiderivative of S?

The following convenient theorem answers all these questions in the affirmative. Happily, everything works just as we'd hope.

Theorem 12 (Derivatives and Antiderivatives) Let $S(x)$ be a power series with radius of convergence $r > 0$. Let $D(x)$ and $A(x)$ be defined as before. Then:

- Both D and A have radius of convergence r.
- For $|x| < r$, $D(x) = S'(x)$.
- For $|x| < r$, $A(x) = \int_0^x S(t)\,dt$.

The theorem says, among other things, that a function S given by a power series is differentiable and that its derivative is another power series, S', with the same radius of convergence. The same theorem applies to S', to S'', and so on, to show that S has *infinitely many* derivatives—all available by repeated term-by-term differentiation.

EXAMPLE 1 For any x, $e^x = 1 + x + \frac{x^2}{2!} + \frac{x^3}{3!} + \cdots$. Explain why.

Solution Let $S(x)$ represent the preceding series; we saw in the last section that $S(x)$ converges for *all* x. ▶ By Theorem 12, S' can be found by differentiating S term by term. However:

This S has infinite radius of convergence.

Differentiating S term by term leaves S unchanged.

But *every* differentiable function S for which $S' = S$ has the form $S(x) = Ce^x$. Since $S(0) = 1$, it follows that $C = 1$; hence, $S(x) = e^x$, as claimed. ■

Writing Known Functions as Power Series

In preceding examples, including the last one, we've written various functions—$1/(1-x)$, e^x, $\ln(1-x)$, and so on—in power series form. It's natural to ask whether *other* functions can be "represented" as power series and, if so, how. Theorem 12 suggests several techniques for doing so. First, let's address an even more basic question.

Why Bother? Examples with the Sine Function

Why write a function as a power series? What good, for instance, is the equation

$$\sin x = x - \frac{x^3}{3!} + \frac{x^5}{5!} - \frac{x^7}{7!} + \frac{x^9}{9!} - \cdots,$$

which holds for all real numbers x? ▶

It is an equation; assume so for now. We'll soon give convincing reasons.

Good Family Values. Transcendental functions—trigonometric functions, exponential functions, and so on—have no finite algebraic formulas. For such functions, power series are the next best thing. Using them, we can find very accurate, albeit approximate, values of many transcendental functions.

EXAMPLE 2 Use the preceding sine series to approximate sin 1 accurately.

Solution Substituting $x = 1$ into the sine series gives

$$\sin 1 = 1 - \frac{1}{3!} + \frac{1}{5!} - \frac{1}{7!} + \frac{1}{9!} - \frac{1}{11!} + \cdots,$$

Rapidly!

an alternating series with terms decreasing ◀ to zero. By the alternating series theorem, any partial sum of such a series differs from the limit by no more than the next term. In particular, the partial sum $1 - 1/3! + 1/5! - \cdots - 1/11! \approx 0.8414709846$ differs from sin 1 by no more than $1/13! \approx 2\times 10^{-10}$. In other words, our estimate is good to at least nine decimal places. This checks out numerically: $\sin 1 \approx 0.8414709848$. Not bad for so little work! ■

Hard Integrals Made Easy. As we've seen repeatedly, many integrals, even simple-looking ones, cannot be calculated in "closed form," i.e., by elementary antidifferentiation. Numerical methods—such as the midpoint rule—offer one recourse. Infinite series, being easy to integrate, offer another way to transcend such difficulties.

EXAMPLE 3 Find, in series form, an antiderivative for $\sin(x^2)$. Use it to estimate $I = \int_0^1 \sin(x^2)\,dx$. (For comparison, the midpoint rule applied to I gives $M_{50} \approx 0.31025$.)

Solution Even though the function $\sin(x^2)$ has no *elementary* antiderivative, it's easy to find an antiderivative in series form. Here's how.

Replacing x with x^2 in the sine series gives the new series

$$\sin(x^2) = x^2 - \frac{x^6}{3!} + \frac{x^{10}}{5!} - \frac{x^{14}}{7!} + \cdots.$$

(Because the original series converges for all x, so does this one.) Antidifferentiating term by term gives a new power series:

$$\int_0^x \sin(t^2)\,dt = \frac{x^3}{3} - \frac{x^7}{7\cdot 3!} + \frac{x^{11}}{11\cdot 5!} - \frac{x^{15}}{15\cdot 7!} + \cdots.$$

It converges for all x, by Theorem 12.

The new series ◀ is not an elementary function, but it's a perfectly honest antiderivative for $\sin(x^2)$. We can therefore find our definite integral in the obvious way:

$$\int_0^1 \sin(x^2)\,dx = \left.\frac{x^3}{3} - \frac{x^7}{7\cdot 3!} + \frac{x^{11}}{11\cdot 5!} - \frac{x^{15}}{15\cdot 7!} + \cdots\right]_0^1$$
$$= \frac{1}{3} - \frac{1}{7\cdot 3!} + \frac{1}{11\cdot 5!} - \frac{1}{15\cdot 7!} + \cdots.$$

The alternating series theorem applies to the last series, so the estimate

$$\int_0^1 \sin(x^2)\,dx \approx \frac{1}{3} - \frac{1}{7\cdot 3!} + \frac{1}{11\cdot 5!} - \frac{1}{15\cdot 7!} \approx 0.3102681578$$

is in error by no more than $1/19\cdot 9! \approx 1.5\times 10^{-7}$ (the size of the next term). The result agrees with the midpoint rule estimate through four decimal places. ■

New Series from Old: Help from Algebra and Calculus

For many familiar functions, power series can be found by simple algebra or calculus operations, starting from a few standard known series. Differentiating the sine series, for instance, gives (thanks to Theorem 12)

$$\cos x = 1 - \frac{x^2}{2!} + \frac{x^4}{4!} - \frac{x^6}{6!} + \cdots.$$

The new series, like the old, converges for all x.

We can also start with another famous series,

$$\frac{1}{1-x} = 1 + x + x^2 + x^3 + x^4 + x^5 + \cdots,$$

which converges for $|x| < 1$. With a little algebraic ingenuity we can produce many other useful series, all converging on the same set. Replacing x with $-x$, for instance, gives the alternating series

$$\frac{1}{1+x} = 1 - x + x^2 - x^3 + x^4 - x^5 + \cdots.$$

Replacing x with x^2 gives *another* alternating series:

$$\frac{1}{1+x^2} = 1 - x^2 + x^4 - x^6 + x^8 - x^{10} + \cdots.$$

Integrating *this* series term by term gives an even more striking result:

$$\arctan x = x - \frac{x^3}{3} + \frac{x^5}{5} - \frac{x^7}{7} + \frac{x^9}{9} - \cdots.$$

Setting $x = 1$ in this last series yields a remarkable and beautiful result, one that has been discovered and rediscovered by some of history's greatest mathematicians:

$$\frac{\pi}{4} = 1 - \frac{1}{3} + \frac{1}{5} - \frac{1}{7} + \frac{1}{9} - \cdots.$$

(A little caution is needed, however. These simple arguments show only that the series for $\arctan x$ is valid if $-1 < x < 1$. Showing carefully that the series converges to $\pi/4$ when $x = 1$ requires further argument.) ▶

But it's true!

Multiplying Power Series. Convergent power series can be multiplied together, something like polynomials, to form new convergent series. As always with series, convergence is a question. Here's the answer: The product of two power series converges wherever both factors converge. We illustrate with an example.

EXAMPLE 4 We already showed that

$$\frac{1}{1-x} = 1 + x + x^2 + x^3 + \cdots$$

and

$$\frac{1}{1+x} = 1 - x + x^2 - x^3 + \cdots.$$

Multiply these series. Where does the new series converge? What familiar function does the result represent?

Solution Symbolically, the problem looks like this:

$$(1 + x + x^2 + x^3 + \cdots) \cdot (1 - x + x^2 - x^3 + \cdots) = a_0 + a_1 x + a_2 x^2 + a_3 x^3 + \cdots.$$

We want numerical values for the constants on the right.

Both factors have infinitely many summands, so ordinary expansion quickly gets out of hand. To avoid this, we collect like powers right from the start. It's clear, for instance, that $a_0 = 1 \cdot 1 = 1$; no other combination of factors yields a constant result. Similarly, tracking the first and second powers of x gives

$$a_1 = 1 \cdot (-1) + 1 \cdot 1 = 0; \qquad a_2 = 1 \cdot 1 + 1 \cdot (-1) + 1 \cdot 1 = 1.$$

Try the next term or two for yourself. There's no substitute for experience.

Continuing this process ◀ quickly produces a simple pattern:

$$(1 + x + x^2 + x^3 + \cdots) \cdot (1 - x + x^2 - x^3 + \cdots) = 1 + x^2 + x^4 + x^6 + \cdots.$$

The result is a geometric series in powers of x^2; it converges for $|x^2| < 1$, i.e., if $-1 < x < 1$.

What function does the product series represent? Since the two factors represent the functions $1/(1 - x)$ and $1/(1 + x)$, it follows that the product *series* represents the product *function* $1/(1 - x^2)$. ■

The next example concerns yet another useful descendant of the geometric series.

EXAMPLE 5 Find a power series for $\ln(1 + x)$; use it to estimate $\ln 1.5$ with error less than 0.0001.

Solution Integrating the geometric series gives

$$\int \frac{1}{1+x}\,dx = \int \left(1 - x + x^2 - x^3 + \cdots\right) dx$$
$$= x - \frac{x^2}{2} + \frac{x^3}{3} - \frac{x^4}{4} + \cdots = \ln(1 + x).$$

(We used $C = 0$ as the constant of integration, because our "target" function $\ln(1+x)$ has the value 0 when $x = 0$.) The new series, like the old, converges for x in $(-1, 1)$.

To estimate $\ln 1.5$, we plug $x = 0.5$ into our series expression:

$$\ln 1.5 = 0.5 - \frac{0.5^2}{2} + \frac{0.5^3}{3} - \frac{0.5^4}{4} + \cdots = \sum_{k=1}^{\infty} (-1)^{k+1} \frac{0.5^k}{k}.$$

Now the alternating series theorem applies. To achieve our target accuracy, any partial sum S_n for which

$$\frac{0.5^{n+1}}{n+1} < 0.0001$$

will do. It's easy to see that $n = 10$ works, with room to spare. In fact, $S_{10} \approx 0.405435$; this compares favorably with the "exact" value $\ln 1.5 \approx 0.405465$. ■

A Brief Atlas of Power Series

For ease of reference, here is a short list of "standard" power series for basic calculus functions. Each series appears with a suitable interval of convergence. In the next section we develop additional tools for showing rigorously that the limits are as stated.

A Power Series Sampler

Function	Series	Convergence Interval
$\sin x$	$x-\frac{x^3}{3!}+\frac{x^5}{5!}-\frac{x^7}{7!}+\frac{x^9}{9!}-\cdots$	$(-\infty,\infty)$
$\cos x$	$1-\frac{x^2}{2!}+\frac{x^4}{4!}-\frac{x^6}{6!}+\frac{x^8}{8!}-\cdots$	$(-\infty,\infty)$
$\exp x$	$1+x+\frac{x^2}{2!}+\frac{x^3}{3!}+\frac{x^4}{4!}+\frac{x^5}{5!}+\cdots$	$(-\infty,\infty)$
$\frac{1}{1-x}$	$1+x+x^2+x^3+x^4+x^5+\cdots$	$(-1,1)$
$\frac{1}{1+x}$	$1-x+x^2-x^3+x^4-x^5+\cdots$	$(-1,1)$
$\frac{1}{1+x^2}$	$1-x^2+x^4-x^6+x^8-x^{10}+\cdots$	$(-1,1)$
$\arctan x$	$x-\frac{x^3}{3}+\frac{x^5}{5}-\frac{x^7}{7}+\frac{x^9}{9}-\cdots$	$[-1,1]$

What's Next? A Power Series for Any Function

As we've seen, knowing a power series expression for one function can lead, via various manipulations, to power series versions of related functions. A good question remains:

Given any function f, how can we find a power series "from scratch," without knowing a related series to begin with?

We answer this question in the next section.

BASICS

1. Let $f(x)=\sum_{k=0}^{\infty}\left(\frac{x}{2}\right)^k$.
   (a) What is the radius of convergence of the power series for f?
   (b) According to Theorem 12, $f'(x)=\sum_{k=1}^{\infty}\frac{kx^{k-1}}{2^k}$. What is the radius of convergence of the series for f'?
   (c) According to Theorem 12, $F(x)=\sum_{k=0}^{\infty}\frac{x^{k+1}}{(k+1)2^k}$ is an antiderivative of f. What is the radius of convergence of the series for F?

In Exercises 2–5, use the power series representation of the $(1-x)^{-1}$ to produce a power series representation of the function f.

2. $f(x)=\dfrac{x^2}{1+x}$
3. $f(x)=\dfrac{1}{1-x^2}$
4. $f(x)=\dfrac{1}{(1+x)^2}$
5. $f(x)=\dfrac{x}{1-x^4}$

In Exercises 6–9, find a power series representation of the function and the radius of convergence of this power series. Then plot the function and the fifth-order polynomial that is a partial sum of the power series on the same axes. [**HINT:** Write out the first few terms of the series before trying to find the form of the general term.]

6. $f(x)=\arctan(2x)$
7. $f(x)=\cos(x^2)$
8. $f(x)=x^2\sin x$
9. $f(x)=\ln\left(1+\sqrt[3]{x}\right)$

10. Let $f(x) = \ln(1+x)$. Show that the power series for f and f' have the same radius of convergence but not the same interval of convergence.
11. Use the partial sum of a series to estimate $1/\sqrt{e}$ with an error less than 0.005.
12. Use the partial sum of a series to estimate $\int_0^{0.2} xe^{-x^3}\,dx$ with an error less than 10^{-5}.
13. Evaluate $\sum_{n=1}^{\infty} n/2^n$ exactly. [**HINT:** If $f(x) = \sum_{n=0}^{\infty} x^n$, then $f'(x) = \sum_{n=1}^{\infty} nx^{n-1}$.]
14. Use power series to show that $\lim_{x\to 0} \dfrac{(\sin x - x)^3}{x(1-\cos x)^4} = -\dfrac{2}{27}$.
15. Show that $\lim_{x\to 0^+} \dfrac{x - \sin x}{(x \sin x)^{3/2}} = \dfrac{1}{6}$.

FURTHER EXERCISES

In Exercises 16–25, use power series to evaluate the limit. Check your answer using l'Hôpital's rule.

16. $\lim_{x\to 0} \dfrac{\sin x}{x}$
17. $\lim_{x\to 0} \dfrac{e^x - 1}{x}$
18. $\lim_{x\to 0} \dfrac{1-\cos x}{x}$
19. $\lim_{x\to 0} \dfrac{1-\cos x}{x^2}$
20. $\lim_{x\to 0} \dfrac{\arctan x}{x}$
21. $\lim_{x\to 0} \dfrac{e^x - e^{-x}}{x}$
22. $\lim_{x\to 0} \dfrac{\ln(1+x) - x}{x^2}$
23. $\lim_{x\to 0} \dfrac{x - \arctan x}{x^3}$
24. $\lim_{x\to 1} \dfrac{\ln x}{x-1}$
25. $\lim_{x\to 0} \dfrac{1-\cos^2 x}{x}$ [**HINT:** $1 - \cos^2 x = \big(1 - \cos(2x)\big)/2$.]

In Exercises 26–35, find a power-series representation of the function and the radius of convergence of this power series.

26. $f(x) = \dfrac{1}{2+x}$
27. $f(x) = \sin\left(\sqrt{x}\right)$
28. $f(x) = \sin x + \cos x$
29. $f(x) = 2^x = e^{x\ln 2}$
30. $f(x) = \ln\left(1+x^2\right)$
31. $f(x) = (x^2-1)\sin x$
32. $f(x) = \ln\left(\dfrac{1+x}{1-x}\right)$
33. $f(x) = \cos^2 x = \frac{1}{2}\big(1+\cos(2x)\big)$
34. $f(x) = \dfrac{5+x}{x^2+x-2} = \dfrac{2}{x-1} - \dfrac{1}{x+2}$
35. $f(x) = \sin^3(x) = \frac{1}{4}\big(3\sin x - \sin(3x)\big)$
36. Show that $\dfrac{1}{x-1} = \displaystyle\sum_{k=1}^{\infty} \dfrac{1}{x^k}$ if $|x| > 1$.
[**HINT:** $1/(x-1) = x/(x-1) - 1$.]
37. (a) Use the formula for the sum of a geometric series to show that
$$\sum_{k=1}^{\infty} \frac{x^k}{k} = -\ln|1-x|.$$
(b) What is the interval of convergence of the series in part (a)?
(c) Show that if $N \geq 1$, then
$$0 < \ln 2 - \sum_{k=1}^{N} \frac{1}{k\,2^k} \leq \frac{1}{(N+1)2^N}.$$
[**HINT:** $-\ln(1/2) = \ln 2$.]
38. Use a power series to show that $x - \frac{1}{2}x^2 < \ln(1+x) < x$ for all x in the interval $(0, 1)$.
39. Use power series to show that $1 - \cos x < \ln(1+x) < \sin x$ if $0 < x < 1$.
40. (a) Does the series $\sum_{n=1}^{\infty} \sin(1/n)$ converge? Justify your answer.
(b) Does the series $\displaystyle\sum_{n=1}^{\infty} \frac{1}{n}\sin(1/n)$ converge? Justify your answer.
41. (a) Does the series $\sum_{n=1}^{\infty} e^{-1/n}$ converge? Justify your answer.
(b) Does the series $\sum_{n=1}^{\infty} \left(1 - e^{-1/n}\right)$ converge? Justify your answer.
42. Use the fact that $\int_0^{\infty} t^n e^{-t}\,dt = n!$ to show that
$$\int_0^{\infty} e^{-t}\sin(xt)\,dt = \frac{x}{1+x^2} \quad \text{if } |x| < 1.$$
43. Let $I = \displaystyle\int_0^{\infty} \frac{xe^{-x}}{1-e^{-x}}\,dx$.
(a) Show that I is a convergent improper integral.
(b) Use the subsitution $u = 1 - e^{-x}$ to show that
$$I = -\int_0^1 \frac{\ln(1-u)}{u}\,du.$$
(c) Use part (b) and the series representation of $\ln(1-u)$ to show that $I = \pi^2/6$. [**HINT:** $\sum_{k=1}^{\infty} k^{-2} = \pi^2/6$.]
44. Let $f(x) = \dfrac{1}{1+x^4}$.
(a) Find a power series representation of f.
(b) What is the interval of convergence of the series in part (a)?
(c) Use the series found in part (a) to evaluate $\int_0^{0.5} f(x)\,dx$ with an error no greater than 0.001.

45. (a) Find the power series representation of an antiderivative of e^{-x^2}.
(b) Use the result from part (a) to estimate $\int_0^1 e^{-x^2}\,dx$ within 0.005 of its exact value.

46. Estimate $\int_0^1 \cos\left(x^2\right)\,dx$ with an error no greater than 0.005.

47. Estimate $\int_0^1 \sqrt{x}\sin x\,dx$ with an error no greater than 0.001.

48. Estimate $\displaystyle\int_0^{1000} \frac{\exp^{-10x}\sin x}{x}\,dx$ with an error no greater than 5×10^{-5}.

In Exercises 49–54, find the first four nonzero terms in the power series representation of the function.

49. $f(x) = e^{2x}\ln(1+x^3)$

50. $f(x) = \arctan x \sin(4x)$

51. $f(x) = \dfrac{e^x}{1-x}$

52. $f(x) = \tan x = \dfrac{\sin x}{\cos x}$

53. $f(x) = e^{\sin x}$

54. $f(x) = \ln(\cos x)$

55. Use the fact that $\cos x = -\cos(x-\pi)$ to write $\cos x$ as a series in powers of $x-\pi$.

56. Determine the coefficients a_k such that

$$\frac{1}{1-x} = \sum_{k=0}^{\infty} a_k(x-2)^k.$$

[**HINT:** $\dfrac{1}{1-x} = -\dfrac{1}{1+(x-2)}$.]

In Exercises 57–60, find the elementary function represented by the power series by manipulating a more familiar power series (e.g., the series for $\cos x$, $\sin x$, $(1-x)^{-1}$).

57. $\displaystyle\sum_{k=1}^{\infty} kx^{k-1}$

58. $\displaystyle\sum_{k=0}^{\infty} \frac{x^k}{(k+1)!}$

59. $\displaystyle\sum_{k=1}^{\infty} (-1)^{k+1}x^k$

60. $\displaystyle\sum_{k=1}^{\infty} \frac{(2x)^k}{k}$

61. **A proof that e is irrational.** Assume that $e = m/n$, where m and n are positive integers.
(a) Explain why

$$m!\left|\frac{1}{e} - \sum_{k=0}^{m}\frac{(-1)^k}{k!}\right| \le \frac{m!}{(m+1)!} = \frac{1}{m+1}.$$

(b) Explain why $m!/e$ is an integer.
(c) Explain why $m!\displaystyle\sum_{k=0}^{m}\frac{(-1)^k}{k!}$ is an integer.
[**HINT:** Start by explaining why $m!/k!$ is an integer if k is an integer and $0 \le k \le m$.]
(d) Parts (a)–(c) imply that

$$N = m!\left|\frac{1}{e} - \sum_{k=0}^{m}\frac{(-1)^k}{k!}\right|$$

is an integer that is less than or equal to $1/(m+1)$. Explain why it follows that $N = 0$.
(e) Explain why the conclusion of part (d) is impossible and therefore e cannot be a rational number. [**HINT:** $\sum_{k=m+1}^{\infty}(-1)^k/k! \neq 0$.]

62. Use power series to show that $y = e^x$ is a solution of the differential equation $y' = y$.

63. Use power series to show that $y = 2e^x$ is the solution of the initial value problem $y' = y$, $y(0) = 2$.

64. Use power series to show that $y = e^{3x}$ is the solution of the initial value problem $y' = 3y$, $y(0) = 1$.

65. Use power series to show that $y = \sin x$ is a solution of the differential equation $y'' = -y$.

66. Use power series to show that $y = (1-x)^{-1}$ is the solution of the initial value problem $y' = y^2$, $y(0) = 1$.

67. (a) Show that $f(x) = \tan x$ is the solution of the initial value problem $f'(x) = 1 + \left(f(x)\right)^2$, $f(0) = 0$.
(b) Use part (a) to find the first four nonzero terms in the power series representation of $\tan x$.

68. Let r be a fixed number, and define the function f by

$$f(x) = 1 + \sum_{n=1}^{\infty}\frac{r(r-1)(r-2)\cdots(r-n+1)}{n!}x^n.$$

(a) Show that the series defining f converges if $|x| < 1$.
(b) Show that $(1+x)f'(x) = rf(x)$.
(c) Let $g(x) = (1+x)^{-r}f(x)$. Show that $g'(x) = 0$.
(d) Show that part (c) implies that $f(x) = (1+x)^r$.
[**NOTE:** The power series for f is known as the **binomial series.**]

69. Use the binomial series defined in Exercise 68 to show that

$$\sqrt{1+x} \approx 1 + \frac{1}{2}x - \frac{1}{8}x^2 + \frac{1}{16}x^3 - \frac{5}{128}x^4 + \frac{7}{256}x^5 \mp \cdots.$$

In Exercises 70–73, use the series in Exercise 68 to find the first four nonzero terms of a power series representation of the given function.

70. $f(x) = \left(1+x^4\right)^3$

71. $f(x) = \sqrt[3]{1-x^2}$

72. $f(x) = \left(1+x^2\right)^{-3/2}$

73. $f(x) = \arcsin x$.

74. Use the partial sum of a series to estimate $\int_0^{0.4}\sqrt{1+x^3}\,dx$ with an error less than 5×10^{-4}.

11.7 Maclaurin and Taylor Series

In the preceding section we saw some of the practical advantages of writing a function as a power series. We also saw how to use a power series for one function to derive power series for related functions, using algebra, calculus, and other devices. In this section we show, given a suitable function, how to find its power series "from scratch," i.e., without starting from a related series. In the process we draw important connections among a function, its derivatives, and its series. First, however, a note of caution.

Does Every Function Have a Power Series?

Mathematical life would be simpler if *every* function $f(x)$ could be written as a power series—ideally, a series that converges to $f(x)$ for all x. Alas, it isn't so. At least two things can go wrong:

Smaller Domains A power series may have a smaller domain than the function it represents. For instance, the series equation

$$\frac{1}{1+x^2} = 1 - x^2 + x^4 - x^6 + x^8 - x^{10} + \cdots$$

holds if—but only if—$|x| < 1$, even though $f(x) = 1/(1+x^2)$ is defined (and well behaved) for *all* real numbers x.

No Series at All A function may have no series at all. Theorem 12 in the preceding section showed that every power series can be differentiated again and again on its interval of convergence. ◄ Thus, any function that *has* a power series must itself be repeatedly differentiable at $x = 0$. This fact rules out functions such as $f(x) = |x|$, which is continuous everywhere but not differentiable at $x = 0$.

Each derivative is another power series.

Despite these cautions, many important functions *can* be written as power series.

Coefficients and Derivatives at Zero

A simple equation relates the coefficients and the derivatives at zero of a power series. If

$$S(x) = a_0 + a_1 x + a_2 x^2 + a_3 x^3 + a_4 x^4 + \cdots,$$

then, simplest of all, $S(0) = a_0$. Differentiating S repeatedly ◄ gives

We can differentiate repeatedly, by Theorem 12 of the preceding section.

$$\begin{aligned} S(x) &= a_0 + a_1 x + a_2 x^2 + a_3 x^3 + a_4 x^4 + \cdots; \\ S'(x) &= a_1 + 2a_2 x + 3a_3 x^2 + 4a_4 x^3 + \cdots; \\ S''(x) &= 2a_2 + 6a_3 x + 12a_4 x^2 + \cdots; \\ S'''(x) &= 6a_3 + 24a_4 x + \cdots \\ &\vdots \end{aligned}$$

These equations show that

$$S(0) = a_0; \quad S'(0) = a_1; \quad S''(0) = 2a_2; \quad S'''(0) = 6a_3; \quad \ldots .$$

In general, the following holds true.

Fact If $S(x) = \sum_{k=0}^{\infty} a_k x^k$, then $a_k = \dfrac{S^{(k)}(0)}{k!}$ for all $k \geq 0$.

This important fact connects coefficients and derivatives; knowing either, we can find the other. We illustrate one of the uses of the fact with an example.

EXAMPLE 1 Assuming (for now) that $f(x) = \sin x$ *has* a power series, *find* that power series.

Solution To find the desired series $f(x) = a_0 + a_1 x + a_2 x^2 + a_3 x^3 + \cdots$, we need "only" find the coefficients $a_0, a_1, a_2, a_3, \ldots$.

In principle, finding infinitely many of anything sounds difficult. In practice, it's often easy; for the sine function, as for many functions of interest, the coefficients follow a simple pattern. The first few derivatives of f reveal the pattern:

$$a_0 = \frac{f(0)}{0!} = \sin 0 = 0; \qquad a_1 = \frac{f'(0)}{1!} = \cos 0 = 1;$$

$$a_2 = \frac{f''(0)}{2!} = \frac{-\sin 0}{2} = 0; \qquad a_3 = \frac{f'''(0)}{3!} = \frac{-\cos 0}{6} = -\frac{1}{6};$$

$$a_4 = \frac{f^{(4)}(0)}{4!} = \frac{\sin 0}{4!} = 0; \qquad a_5 = \frac{f^{(5)}(0)}{5!} = \frac{\cos 0}{5!} = \frac{1}{5!}.$$

Therefore, written as a series,

$$\sin x = x - \frac{x^3}{3!} + \frac{x^5}{5!} - \frac{x^7}{7!} + \cdots .$$

It's not hard to show that this series converges for *all* x, as we'd hope. ▶

But it's beside the main point, here.

Better yet, the series converges to $\sin x$. We haven't proved this fact yet, but numerical evidence is readily available. If, say, $x = 1$, $S_5 \approx 0.8414710096$, while $\sin 1 \approx 0.8414709848$. ■

Maclaurin Series

Using the preceding fact as a recipe for the coefficients, we can write a power series for any function f that has repeated derivatives at $x = 0$. ▶ Here's the formal definition, named for the 17th-century Scottish mathematician Colin Maclaurin:

*The **zeroth derivative** of f is f itself.*

Definition (Maclaurin Series) Let f be any function with infinitely many derivatives at $x = 0$. The **Maclaurin series** for f is the series $\sum_{k=0}^{\infty} a_k x^k$, with coefficients given by

$$a_k = \frac{f^{(k)}(0)}{k!}, \qquad k = 0, 1, 2, \ldots .$$

Finding Maclaurin Series: Help from Technology

Computing the necessary derivatives by hand to find a Maclaurin series can be tedious and error-prone. Fortunately, `Maple`and other software programs can find any partial sum of the Maclaurin series quickly and easily. (Such sums are called either **Maclaurin polynomials** or **Taylor polynomials**.) Here's how the process works (using one version of *Maple*) for the sine function:

```
> taylorpoly( sin(x), x=0, 7 );

                  3          5            7
        x - 1/6 x  + 1/120 x  - 1/5040 x
```

The output is the *seventh-degree Taylor polynomial for* $\sin x$, based at $x = 0$.

Taylor Series: Expansion About $x = a$

A Maclaurin series

$$\sum_{k=0}^{\infty} \frac{f^{(k)}(0)}{k!} x^k$$

is said to be expanded about $x = 0$, because all derivatives of f are calculated there. This choice, although often convenient, isn't necessary. A similar series can be found by expanding about *any* point $x = a$. The **Taylor series** for f, expanded about $x = a$, has the form

$$\sum_{k=0}^{\infty} \frac{f^{(k)}(a)}{k!} (x - a)^k .$$

For some functions, expansion about a point other than zero is convenient.

We'll let Maple *help.*

EXAMPLE 2 The function $f(x) = \ln x$ isn't defined at $x = 0$. Expand f about $x = 1$. ◂

See for yourself.

Solution Derivatives of $f(x) = \ln x$ are easy for either a human or a machine to calculate. ◂ Here are several:

$$f'(x) = \frac{1}{x}; \qquad f''(x) = -\frac{1}{x^2}; \qquad f'''(x) = \frac{2}{x^3};$$
$$f^{(4)}(x) = -\frac{6}{x^4}; \qquad f^{(5)}(x) = \frac{24}{x^5}; \qquad f^{(6)}(x) = -\frac{120}{x^6}.$$

At $x = 1$, therefore,

$$f'(1) = 1; \qquad f''(1) = -1; \qquad f'''(1) = 2;$$
$$f^{(4)}(1) = -6; \qquad f^{(5)}(1) = 24; \qquad f^{(6)}(1) = -120.$$

The general pattern is now visible. The Taylor series for the function $\ln x$, expanded about $x = 1$, has the form

$$\sum_{k=1}^{\infty} \frac{(-1)^{k+1}}{k} (x-1)^k = (x-1) - \frac{(x-1)^2}{2} + \frac{(x-1)^3}{3} - \frac{(x-1)^4}{4} + \cdots .$$

For the record, the computer agrees (to seventh degree, anyway):

```
> taylorpoly(ln(x),x=1,7);
                           2                3                4
    x - 1 - 1/2 (x - 1)  + 1/3 (x - 1)  - 1/4 (x - 1)
                    5                6                7
        + 1/5 (x - 1) - 1/6 (x - 1)  + 1/7 (x - 1)
```

■

Converging to the Right Place: Taylor's Theorem

Any function f that's infinitely differentiable at $x = 0$ has a Maclaurin series—ideally, one with a large radius of convergence. One possible problem remains: The series might converge at x, but perhaps to a limit other than $f(x)$.

Taylor's theorem guarantees that this unfortunate event almost never occurs. In the bargain, it lets us predict in advance how closely a Maclaurin polynomial approximates the "target" function f.

Theorem 13 (Taylor's Theorem) Suppose that f is repeatedly differentiable on an interval I containing 0. Let $\sum_{k=0}^{\infty} a_k x^k$ be the Maclaurin series of f, and let

$$P_n(x) = \sum_{k=0}^{n} a_k x^k = a_0 + a_1 x + a_2 x^2 + \cdots + a_n x^n$$

be the nth degree Maclaurin polynomial. Suppose also that for all x in I,

$$\left| f^{(n+1)}(x) \right| \le K_{n+1}.$$

Then

$$|f(x) - P_n(x)| \le \frac{K_{n+1}}{(n+1)!} |x|^{n+1}.$$

Notice the following features.

- Taylor's theorem estimates the *error* committed by $P_n(x)$ in estimating $f(x)$. Unless K_{n+1} grows very quickly with n, this error tends to 0 as $n \to \infty$, so the series converges to $f(x)$.
- The theorem involves familiar ingredients. Like our numerical integral-error estimates, this one depends on bounding a higher derivative of the function in question.
- A rigorous proof of Taylor's theorem is beyond the scope of this book. The underlying ideas, however, are based firmly on the mean value theorem. Indeed, Taylor's theorem is sometimes thought of as a general form of the mean value theorem. We pursued this idea briefly at the end of Section 4.3, in deriving error bounds for linear approximation.

EXAMPLE 3 For $f(x) = \sin x$, show that the Maclaurin series converges to $\sin x$ for any value of x.

Solution For $f(x) = \sin x$, *all* derivatives are sines or cosines or their opposites. Thus, for any n, the inequality

$$\left|f^{(n+1)}(x)\right| \le 1$$

holds for all x. By Taylor's theorem,

$$|P_n(x) - \sin x| \le \frac{1 \cdot |x|^{n+1}}{(n+1)!}.$$

The last quantity tends to 0 as n tends to infinity, so the series converges, for any x, to $\sin x$. The following graphs show what it means, geometrically, for the Maclaurin polynomials to converge to $\sin x$:

Graphs of sin x, $P_1(x)$, $P_3(x)$, $P_5(x)$, $P_7(x)$

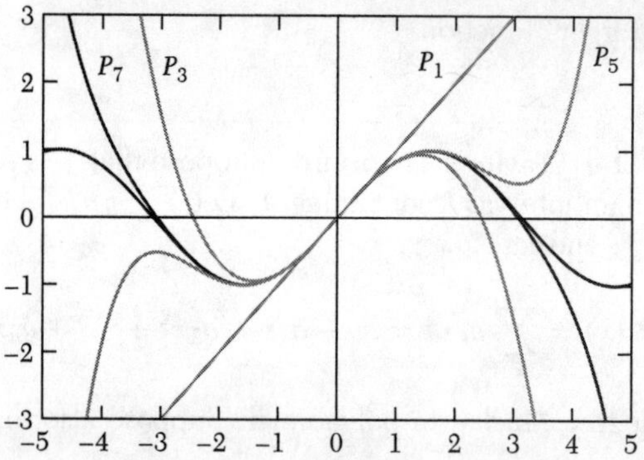

■

BASICS

1. Let $f(x) = x^4 - 12x^3 + 44x^2 + 2x + 1$.
   (a) Find the Maclaurin series representation of f.
   (b) Find the Taylor series representation of f expanded about $x = 3$.

2. Let $p(x) = (1+x)^n$, where n is a positive integer. Explain why the Maclaurin series for p is the polynomial p itself. [**HINT:** Use Theorem 13.]

3. Let $f(x) = \int_3^x \sqrt{t}e^{-t}\,dt$.
   (a) Show that if $x \approx 3$, then
   $$f(x) \approx \sqrt{3}e^{-3}(x-3) - \frac{5}{12}\sqrt{3}e^{-3}(x-3)^2 + \frac{23}{216}\sqrt{3}e^{-3}(x-3)^3.$$
   (b) Use Theorem 13 to bound the error made when $f(3.5)$ is estimated using the polynomial in part (a).

4. Let $f(x) = \sqrt{1+x}$.
   (a) Find the first three nonzero terms in the Maclaurin series for f.
   (b) Use Theorem 13 to bound the approximation error made if the Maclaurin polynomial from part (a) is used to estimate $f(1)$.

5. Explain why the power series
$$1 - x + \frac{x^2}{2} - \frac{x^4}{8} + \frac{x^5}{15} - \frac{x^6}{240} + \cdots$$
cannot be the Maclaurin series representation of the function f shown in the graph.

Graph of f

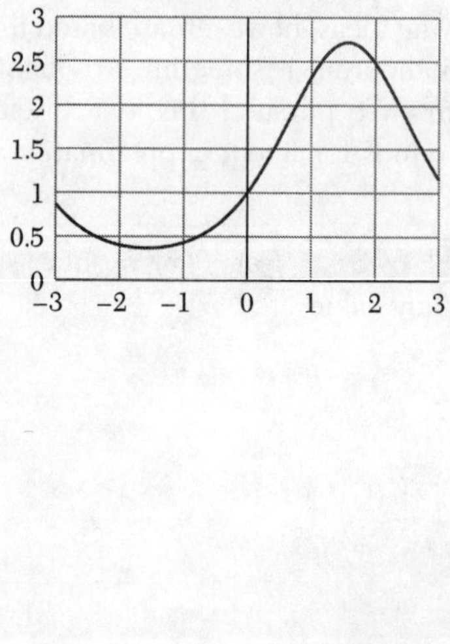

6. (a) Compute the quadratic Maclaurin polynomial for $f(x) = \sqrt{1+x}$.

(b) The electrical potential V at a distance r along the axis perpendicular to the center of a uniformly charged disk of radius a and charge density σ is

$$V = 2\pi\sigma\left(\sqrt{r^2+a^2} - r\right).$$

If r is large in comparison to a, then $a/r \approx 0$. Use part (a) to find an approximation for V.

FURTHER EXERCISES

7. Find numbers a_0, a_1, a_2, a_3, and a_4 such that

$$x^4 - 4x^3 + 5x = a_0 + a_1(x-1) + a_2(x-1)^2 + a_3(x-1)^3 + a_4(x-1)^4.$$

8. Suppose that f is a function such that $f(0) = 1$ and $f'(x) = 1 + \left(f(x)\right)^{10}$ for all x. Find the first four terms of the Maclaurin series for f.

9. Suppose that f is a function that is positive, increasing, and concave down on the interval $[-2, 2]$.
(a) Is the coefficient of x^2 in the Maclaurin series representation of f positive? Justify your answer.
(b) Let $g(x) = 1/\sqrt{1+f(x)}$. Is the coefficient of x^2 in the Maclaurin series representation of g positive? Justify your answer.

10. Let $f(x) = e^{2x}$. What is the coefficient of x^{100} in the Maclaurin series representation of f?

11. Let $f(x) = \dfrac{x}{1-x^3}$.
(a) Find the Maclaurin series for $f(x)$.
(b) What is the interval of convergence of the series in part (a)?
(c) Use part (a) to find a power series for $f''(x)$.
(d) Use part (a) to find the Maclaurin series for $\int_0^x f(t)\,dt$.

12. (a) Find the Maclaurin series representation of

$$f(x) = \frac{1}{2+x}.$$

[HINT: $\dfrac{1}{2+x} = \dfrac{1}{2} \cdot \dfrac{1}{1+(x/2)}$.]

(b) Find $f^{(259)}(0)$ exactly.

13. Let

$$f(x) = \begin{cases} \dfrac{1-\cos x}{x^2} & \text{if } x \neq 0, \\ \dfrac{1}{2} & \text{if } x = 0. \end{cases}$$

Evaluate $f^{(100)}(0)$.

14. Suppose that f is a function such that $f(1) = 1$, $f'(1) = 2$, and $f''(x) = (1+x^3)^{-1}$ for $x > -1$.
(a) Estimate $f(1.5)$ using a quadratic Taylor polynomial.
(b) Find an upper bound on the approximation error made in part (a).

15. Let $f(x) = \begin{cases} x^{-1}\sin x & \text{if } x \neq 0, \\ 1 & \text{if } x = 0. \end{cases}$
(a) Find the Maclaurin series representation of f.
(b) What is the interval of convergence of the power series found in part (a)?
(c) Use the series in part (a) to estimate $f'''(1)$ with an error no greater than 0.005.

16. During an encounter with a friendly extraterrestrial, it is revealed to you that the answer to life, the universe, and everything is $f(1)$ for some function f. It is also revealed that $f(0) = 26$, $f'(0) = 22$, $f''(0) = -16$, $f'''(0) = 12$, and $\left|f^{(4)}(x)\right| \leq 7x^4$ if $|x| \leq 2$.
(a) Find an upper bound on the value of $f(1)$.
(b) Find a lower bound on the value of $f(1)$.

17. Use Theorem 13 to show that the Maclaurin series for e^x converges to e^x for all x.

18. Use Theorem 13 to show that the Maclaurin series for $1/(1+x)$ converges to $1/(1+x)$ if $-1/2 < x < 1$. [HINT: Consider the cases $-1/2 < x < 0$ and $0 \leq x < 1$ separately.]

19. Suppose that f is a function such that $f^{(n)}$ exists for all $n \geq 1$.
(a) Explain why the Maclaurin series for f converges to f if $\left|f^{(n)}(x)\right| \leq n$ for all $n \geq 1$.
(b) Does Theorem 13 guarantee that the Maclaurin series for f converges to f if $\left|f^{(n)}(x)\right| \leq 2^n$ for all $n \geq 1$?

20. Let $f(x) = \begin{cases} e^{-1/x^2} & \text{if } x \neq 0, \\ 0 & \text{if } x = 0. \end{cases}$
(a) Use the definition of the derivative (and l'Hôpital's rule) to show that $f'(0) = 0$.
(b) Using methods similar to those in part (a), it can be shown that $f^{(k)}(0) = 0$ for all integers $k \geq 0$. Use this fact to find the Maclaurin series for f.
(c) What is the radius of convergence of the series in part (b)?
(d) For which values of x does the series in part (b) converge to $f(x)$?

21. Use the trigonometric identity $\cos(x + y) = \cos x \cos y - \sin x \sin y$ to find the Taylor series representation for $\cos x$ expanded about $a = \pi/3$.

22. Find the Taylor series representation for $\sin x$ expanded about $a = \pi/4$.

23. (a) Use the binomial series (see Exercise 68 in Section 11.6) to show that

$$\arcsin x = \int_0^x \frac{dt}{\sqrt{1-t^2}} = x + \sum_{n=1}^{\infty} \frac{1 \cdot 3 \cdot 5 \cdots (2n-1)}{2 \cdot 4 \cdot 6 \cdots (2n)} \frac{x^{2n+1}}{2n+1}.$$

(b) If $n \geq 1$ is an integer, then

$$\int_0^{\pi/2} \sin^{2n+1} x\, dx = \frac{2 \cdot 4 \cdot 6 \cdots (2n)}{3 \cdot 5 \cdot 7 \cdots (2n+1)}.$$

Use this fact to evaluate $\displaystyle\int_0^1 \frac{x^{2n+1}}{\sqrt{1-x^2}}\, dx$.

(c) Use parts (a) and (b) to show that

$$\int_0^1 \frac{\arcsin x}{\sqrt{1-x^2}}\, dx = \sum_{k=0}^{\infty} \frac{1}{(2k+1)^2}.$$

(d) Use the substitution $u = \arcsin x$ to show that

$$\int_0^1 \frac{\arcsin x}{\sqrt{1-x^2}}\, dx = \frac{\pi^2}{8}.$$

(e) Use parts (c) and (d) to show that

$$\sum_{k=1}^{\infty} \frac{1}{k^2} = \frac{\pi^2}{6}.$$

24. Let f be a function that has continuous derivatives on an interval containing a and x.

(a) Explain why $f(x) = f(a) + \int_a^x f'(t)\, dt$.

(b) Use part (a) and integration by parts to show that

$$f(x) = f(a) + f'(a)(x-a) - \int_a^x (t-x) f''(t)\, dt = f(a) + f'(a)(x-a) + \int_a^x (x-t) f''(t)\, dt.$$

[**HINT:** Let $dv = dt$ and $v = t - x$. (This is legitimate since x is not the integration variable.)]

(c) Use part (b) and integration by parts to show that

$$f(x) = f(a) + f'(a)(x-a) + \frac{1}{2} f''(a)(x-a)^2 + \frac{1}{2}\int_a^x (x-t)^2 f'''(t)\, dt.$$

(d) Use part (c) and integration by parts to show that

$$f(x) = f(a) + f'(a)(x-a) + \frac{1}{2} f''(a)(x-a)^2 + \frac{1}{3!} f'''(a)(x-a)^3 + \frac{1}{3!}\int_a^x (x-t)^3 f^{(4)}(t)\, dt.$$

(e) Show that repeated integration by parts leads to this alternative form of Taylor's theorem:

$$f(x) = f(a) + f'(a)(x-a) + \frac{1}{2} f''(a)(x-a)^2 + \frac{1}{3!} f'''(a)(x-a)^3 + \cdots + \frac{1}{n!} f^n(a)(x-a)^n + \frac{1}{n!}\int_a^x (x-t)^n f^{(n+1)}(t)\, dt.$$

25. Let $R_n(x) = \dfrac{1}{n!}\displaystyle\int_a^x (x-t)^n f^{(n+1)}(t)\, dt$. (This is the integral form of the remainder in Taylor's theorem derived Exercise 24.) Show that if $a = 0$ and $\left|f^{(n+1)}(t)\right| \leq K_{n+1}$ for all t, then

$$|f(x) - P_n(x)| = |R_n(x)| \leq \frac{K_{n+1}}{(n+1)!} |x|^{n+1}.$$

26. Let $f(x) = (1+x)^r$. In this exercise we prove that the binomial series (see Exercise 68 in Section 11.6) converges to f.

(a) Use part (e) of Exercise 24 to show that $R_n(x) = f(x) - P_n(x)$ is

$$R_n(x) = \frac{r \cdot (r-1) \cdot (r-2) \cdots (r-n)}{n!} \int_0^x \frac{(x-t)^n}{(1+t)^{n+1-r}}\, dt.$$

(b) Suppose that $0 \leq t \leq x < 1$. Show that $\dfrac{|x-t|}{1+t} \leq |x|$.

(c) Use part (b) to show that if $0 \leq x < 1$, then

$$|R_n(x)| \leq \frac{|(r-1) \cdot (r-2) \cdots (r-n)|}{n!} x^n \left|(1+x)^r - 1\right|.$$

(d) Use part (c) to show that the binomial series converges to f if $0 \leq x < 1$.

(e) Adapt the reasoning in parts (b)–(d) to show that the binomial series converges to f if $-1 < x < 0$.

11.8 Chapter Summary

In this long and relatively technical chapter, we studied the sophisticated topic of infinite series. Infinite series are formed by adding—in a special sense, and with due concern for convergence—infinitely many numbers or functions. Making sense of such infinite summation requires special care—hence the subtlety of the subject.

Convergence. An infinite series converges if its partial sums, formed from only finitely many terms, converge as a sequence. Unfortunately, the sequence of partial sums, being so defined, is often hard to understand or handle directly. For **geometric series**, however, partial sums and limits are easy to calculate.

Convergence Tests. Among the first questions to ask about any series is whether it converges or diverges. As with improper integrals, the answer may be far from obvious at a glance. To address this problem, several tests for convergence and divergence were discussed, including the **integral test**, the **comparison test**, and the **ratio test**. All these tests apply, in raw form, only to series of positive terms. Series that contain both positive and negative terms need special care. Given that care, they too can be tested for convergence.

Estimating Limits. Many infinite series, even ones known to converge, are difficult to evaluate exactly. The problem of estimating their limits therefore arises naturally. As with integration, technology proves helpful, both for calculating partial sums and for estimating the error committed in using partial sums to approximate series.

Power Series: Series as Functions. Power series, or "infinite polynomials," are among the most useful and convenient functions of calculus. Moreover, many calculus functions can be written in power series form. Finding such a power series and understanding its functional properties are the subjects of the last two sections of the chapter.

Taylor and Maclaurin Series; Taylor's Theorem. Taylor's theorem—one of the key theorems of calculus—guarantees that under appropriate conditions a function can be written in series form, usually as a Maclaurin series. The partial sums of a Maclaurin series, being polynomials, are easy to handle, and they offer useful approximations to less convenient functions.

Appendix A

Polar coordinates and polar curves

Any point P in the xy-plane has a familiar and natural "address": its **rectangular** (or **Cartesian**) coordinates. The point P below left, for instance, has rectangular coordinates $(4, 3)$:

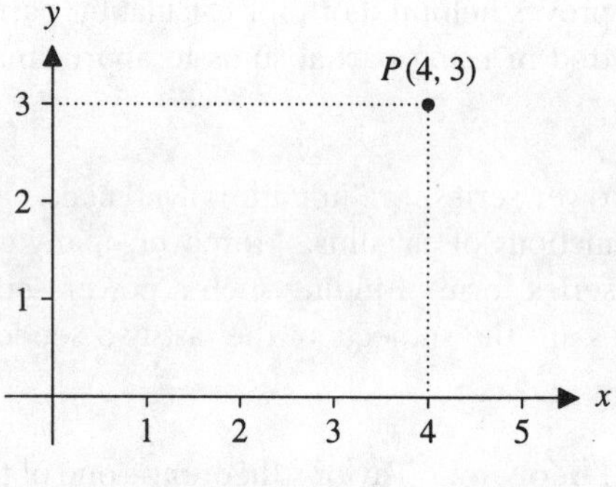

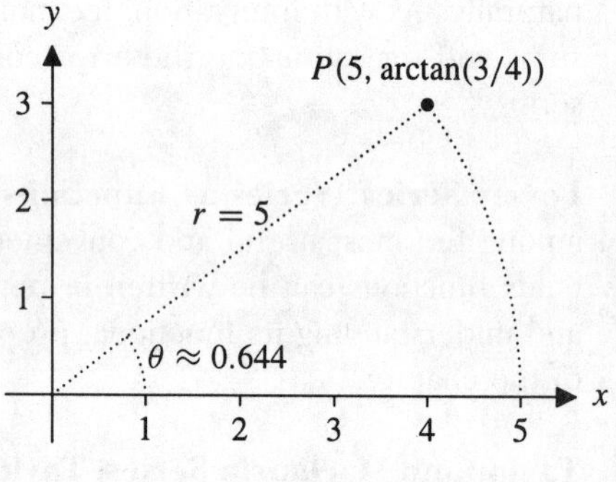

I.e., either positive or negative, depending on direction.

The x- and y-coordinates of P measure, respectively, directed distances from P to the two perpendicular coordinate axes. To reach $P(4, 3)$ from the origin, one moves 4 units *right* and 3 units *up*.

In radians!

Convince yourself that $\theta = \arctan(3/4)$.

Polar coordinates offer another way of "locating" a point in the plane. In the **polar coordinate system**, a point P has coordinates r and θ; they tell, respectively, the *distance from the origin O to P* and the *angle from the positive x-axis to the ray from O to P*. The picture above right shows that the point P with rectangular coordinates $(4, 3)$ has **polar coordinates** $(5, \arctan(3/4)) \approx (5, 0.644)$.

Polar coordinate systems

But not invariably.

A **rectangular coordinate system** in the Euclidean plane starts with an **origin** O and **two perpendicular coordinate axes**. Usually the x-axis is horizontal and the y-axis vertical; x-coordinates increase to the *right* and y-coordinates increase *upward*.

I.e., a half-line.

A **polar coordinate system** starts with different ingredients: an **origin** O, called the **pole**, and a ray, beginning at the origin, called the **polar axis**. The polar axis normally points to the right, along the positive x-axis.

With these ingredients, and a unit for measuring distance, we can assign polar coordinates (r, θ) to any point P:

r is the distance from O to P; θ is any angle from the polar axis to the segment $\overline{OP}$.

The picture below shows several points, with their polar coordinates, plotted on a **polar grid**:➻

Check carefully that each point's coordinates are "correct."

Points on a polar grid

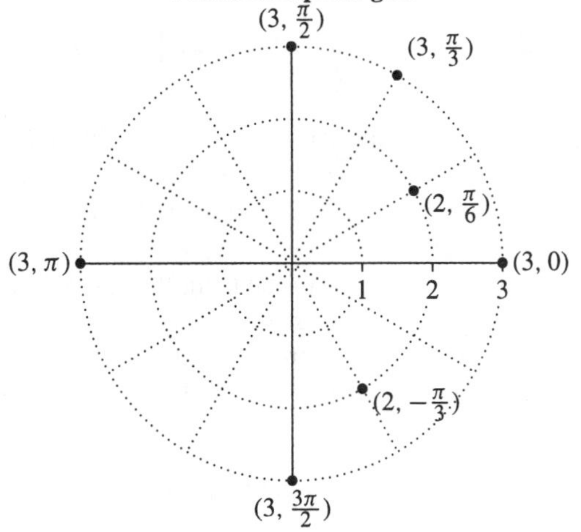

Polar vs. rectangular grids. A rectangular coordinate system leads naturally to a **rectangular grid**, with vertical lines $x = a$ and horizontal lines $y = b$. In a polar system, holding the coordinates r and θ constant produces, respectively, concentric circles and radial lines. The result is a web-like **polar grid**. Here are grids of both types:

A rectangular grid

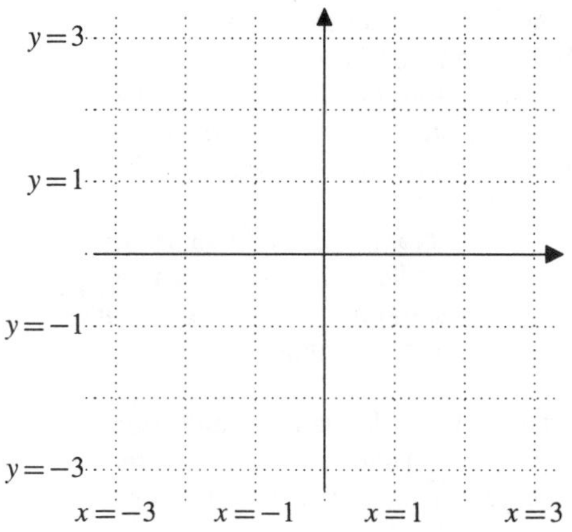

A polar grid

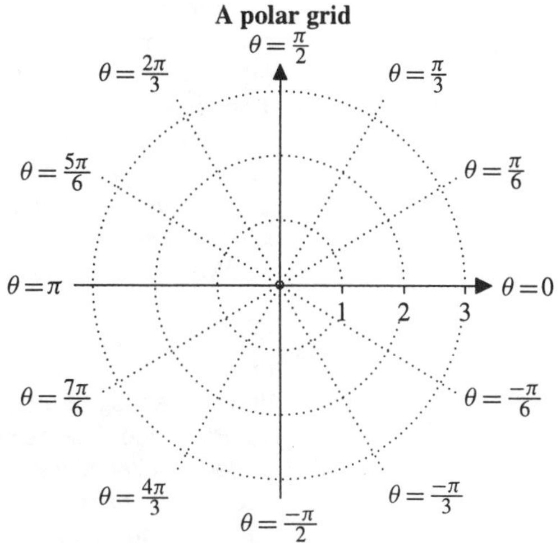

On the scales. In a rectangular coordinate system, the two axes may (and often do) carry different scales (i.e., units of measurement). On a graph, therefore, a vertical inch and a horizontal inch may represent different "distances." "Circles," in particular, may look far from round.

A polar coordinate system, by contrast, has just one axis—the polar axis. As a result, "distance" does *not* depend on direction. Circles look round.

Polar coordinates: not unique. A point in the plane—$P(4,3)$, for instance—has just *one* possible pair of rectangular coordinates. Different rectangular coordinate pairs (x_1, y_1) and (x_2, y_2) correspond to different points in the plane.

Polar coordinates, by contrast, are *not* unique. Every point in the plane has *many* possible pairs of polar coordinates. For example, all of the following polar coordinate pairs—

And many others.

$$\left(2, \frac{\pi}{4}\right), \left(2, \frac{9\pi}{4}\right), \left(2, -\frac{7\pi}{4}\right), \left(2, -\frac{15\pi}{4}\right), \left(-2, \frac{5\pi}{4}\right), \left(-2, -\frac{3\pi}{4}\right)$$

represent the *same* point: the one with rectangular coordinates $(\sqrt{2}, \sqrt{2})$. Notice especially the last two pairs: a *negative* r-coordinate means that to locate the point P, one moves r units *opposite* to the θ-direction. The point $(-2, 5\pi/4)$, for instance, lies two units from the origin on the ray $\theta = \pi/4$). The origin O allows even more freedom: it's represented by *any* pair of the form $(0, \theta)$—regardless of θ.

That's the ray opposite to $\theta = \pi/4$.

This "ambiguity" of polar coordinates arises for a simple reason. All angles that differ by integer multiples of 2π determine the same direction. In practice this ambiguity can present some annoyance, but seldom a serious problem. Two simple rules will help:

E.g., $\theta = \pi/4$, $\theta = 9\pi/4$, $\theta = -7\pi/4$, etc.

See the exercises for more on these rules.

Multiples of 2π: For any r and θ, the pairs (r, θ) and $(r, \theta + 2\pi)$ describe the same point.

Negative r: For any r and θ, the pairs (r, θ) and $(-r, \theta + \pi)$ describe the same point.

Polar coordinates on Earth. The polar grid somewhat resembles an overhead view of Earth, looking "down" at the North Pole. In cartographers' language, lines of **longitude** (aka **meridians**) converge at the pole; the concentric circles are lines of **latitude** (aka **parallels**). The **prime meridian** (or polar axis, in calculus language), for which $\theta = 0$, has been taken for hundreds of years to be the line of longitude that passes through the Greenwich Observatory, just east of London, England. For the same reason, Greenwich Mean Time (the time of day along the prime meridian) is used worldwide was a reference point.

What makes Greenwich "prime," rather than, say, India or Arabia, where navigation and time-keeping flourished even in antiquity? Nothing intrinsic: Greenwich just happened to be a center of attention when the terms were defined—an early (and quite literal!) instance of Eurocentrism.

Polar coordinates in the *plane*, it should also be said, aren't perfectly suited to measuring the (almost) spherical earth. In practice geographers use a related system, similar in some respects, called *spherical* coordinates.

Polar graphs

The ordinary graph of an equation in x and y is the set of points (x, y) whose coordinates satisfy the equation. The graph of $x^2 + y^2 = 1$, for instance, is the circle of radius 1 about the origin. The point $(2, 3)$ does *not* lie on this graph because $2^2 + 3^2 \neq 1$.

The idea of a **polar graph** is similar—but not quite identical. The graph of an equation in r and θ is the set of points whose *polar coordinates* r and θ satisfy the equation. For instance, the polar point $(3, 0)$ lies on the graph of $r = 2 + \cos\theta$ but the polar point $(2, \pi)$ does not.

Because $3 = 2 + \cos 0$.

A warning. The fact that a point in the plane has more than one pair of polar coordinates means that polar plotting requires extra care. At first glance, for instance, the point P with polar coordinates $(-3, \pi)$ seems *not* to satisfy the polar equation $r = 2 + \cos\theta$. A closer look, shows, however, that P can *also* be written with polar coordinates $(3, 0)$—which *do* satisfy the given equation. Here's the moral:

As we saw above.

> *A point P lies on the graph of a polar equation if P has* any *pair of polar coordinates that satisfy the equation.*

Drawing polar graphs

The simplest polar graphs come from *functions*, usually of the form $r = f(\theta)$. Given such a function and a specific θ-domain, it's a routine matter to tabulate points, and then plot them. We illustrate by example.

■ **Example 1.** Plot the equation $r = 2 + \cos\theta$, for $0 \le \theta \le 2\pi$.

Solution: Let $f(\theta) = 2 + \cos\theta$; we want the r-θ graph of f. First we'll tabulate some values:

Values of $r = f(\theta) = 2 + \cos\theta$

θ	0	$\frac{\pi}{6}$	$\frac{\pi}{3}$	$\frac{\pi}{2}$	$\frac{2\pi}{3}$	$\frac{5\pi}{6}$	π	$\frac{7\pi}{6}$	$\frac{4\pi}{3}$	$\frac{3\pi}{2}$	$\frac{5\pi}{3}$	$\frac{11\pi}{6}$	2π
r	3	2.87	2.5	2	1.5	1.14	1	1.14	1.5	2	2.5	2.87	3

Next we plot the data (polar "graph paper" makes the job easier) and fill in the gaps smoothly. Here's the result:

A polar graph: $r = 2 + \cos\theta$

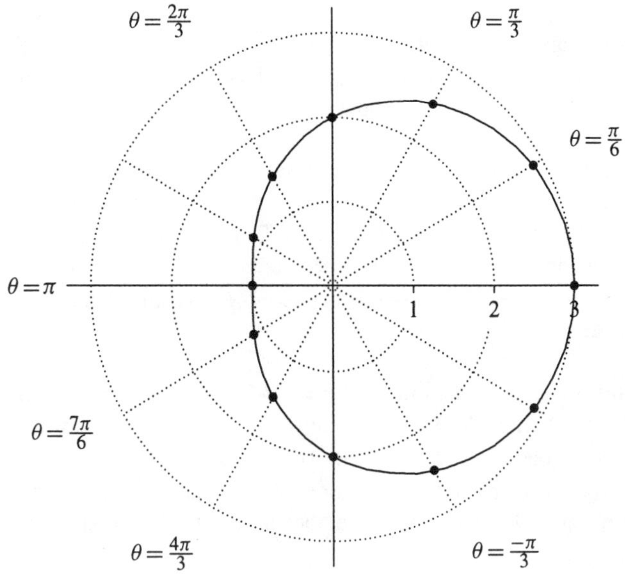

Polar graphs: a sampler

Several polar graphs are shown below. The simplest polar graphs of all—those of the equations $r = a$ and $\theta = b$—we've seen already.

Cardioids and limaçons. Graphs of the form $r = a \pm b\cos\theta$ and $r = a \pm b\sin\theta$, where a and b are positive numbers, are called **limaçons**; if $a = b$, the term **cardioid** is used. (The graph in the previous example is a limaçon.) The graphs below illustrate the variety of limaçons, and show the effects of the constants a and b.

I.e., "heart-like".

$r = 2 + 2\cos\theta$

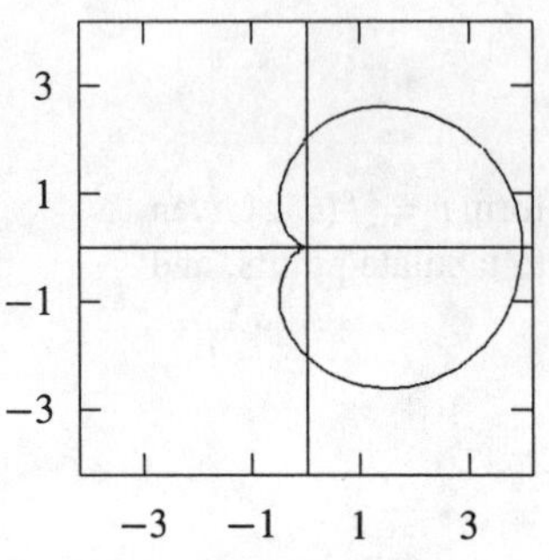

$r = 2 - \cos\theta$

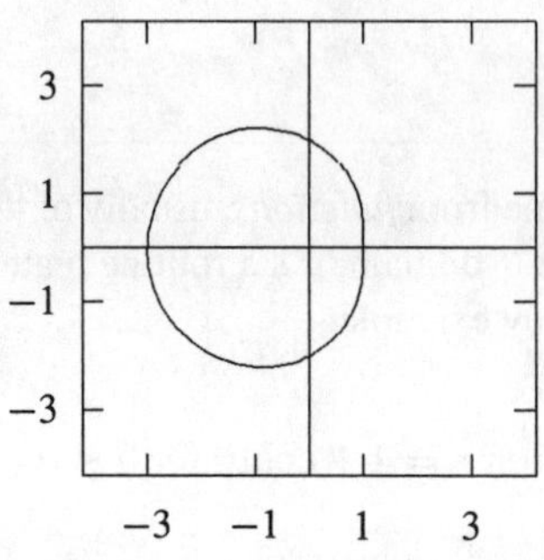

$r = 1 + 2\cos\theta$

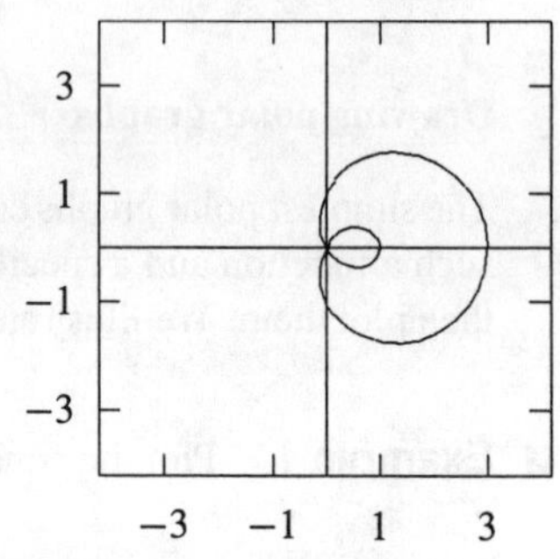

$r = 2 + 2\sin\theta$

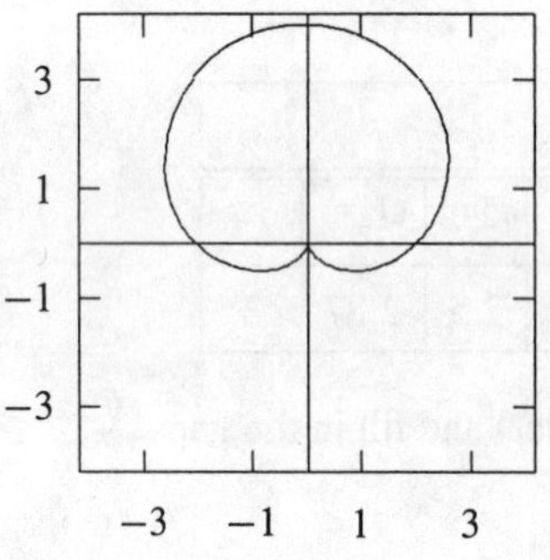

$r = 2 - \sin\theta$

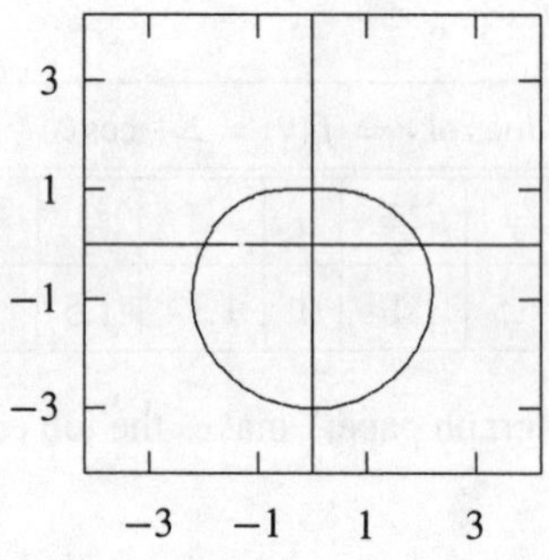

$r = 1 + 2\sin\theta$

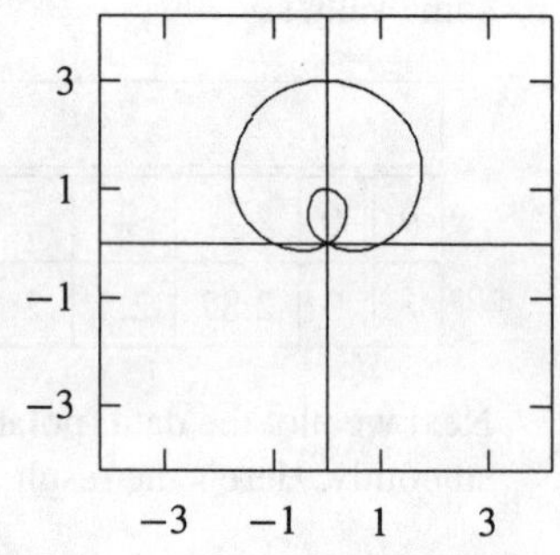

Notice:

What θ-range? The graphs above were drawn by letting θ vary through the interval $[0, 2\pi]$. Since all functions involved are 2π-periodic, any other interval of length 2π would produce the same result.

We might, instead have used the interval $-\pi \le t \le \pi$.

Symmetry. Three of the limaçons above—those that involve the cosine function—are symmetric about the x-axis, i.e., the line $\theta = 0$. (The other three are symmetric about the y-axis.) This symmetry occurs because the cosine function is *even*: for any θ, $\cos\theta = \cos(-\theta)$. The other graphs are symmetric about the y-axis because the sine function is odd. (For more on symmetry, see the exercises.)

Inner loops. Each of the limaçons $r = 1+2\cos\theta$ and $r = 1+2\sin\theta$ has an **inner loop**. A close look at the graphs and the formulas shows that these loops correspond to *negative* values of r. For $r = 1 + 2\cos\theta$, for instance, we have $r = 0$ when $\theta = 2\pi/3$ or $\theta = 4\pi/3$, and $r < 0$ for $2\pi/3 < \theta < 4\pi/3$. For these θ-values, therefore, the curve is drawn on the *opposite* side of the origin.

Roses. Equations of the form $r = a\cos(k\theta)$ and $r = a\sin(k\theta)$, where a is a constant and k is a positive integer, produce graphs called **roses**. The following pictures explain the name:

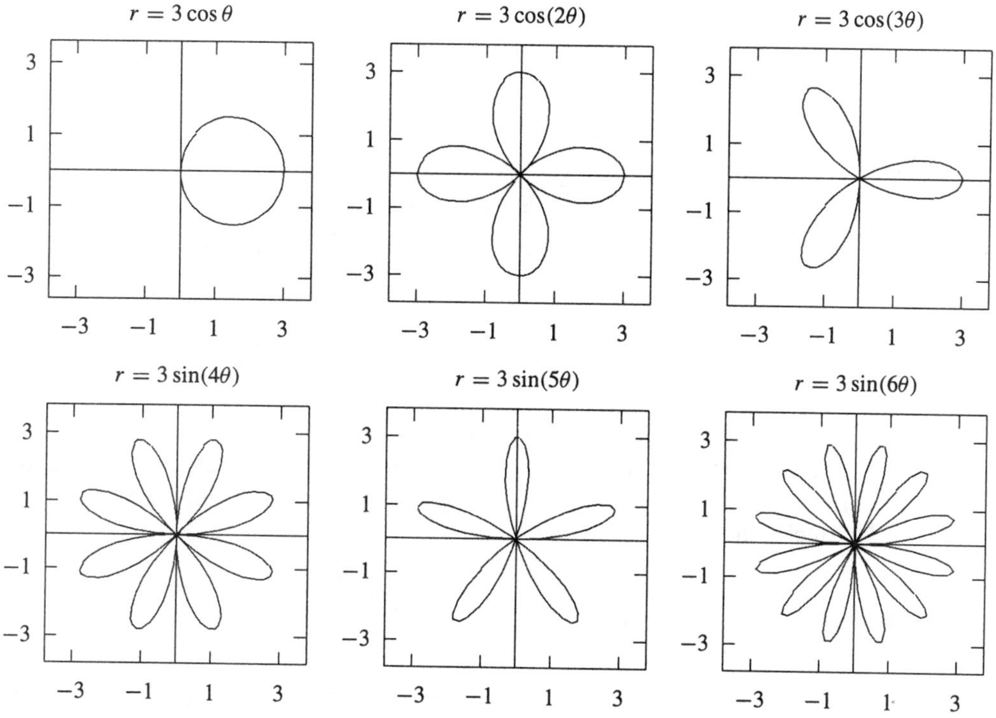

Notice:

Symmetry. Like limaçons (and for the same reason), all roses are symmetric about an axis—"cosine roses" about the x-axis and "sine roses" about the y-axis.

The rose's radius. The coefficient a in $r = a\cos(k\theta)$ and $r = a\sin(k\theta)$ determines the rose's "radius."

How many petals? The coefficient k in $r = a\cos(k\theta)$ and $r = a\sin(k\theta)$ determines the number of "petals": k if k is *odd*, $2k$ if k is even. But here's a subtlety, best revealed by plotting some roses by hand:

If k is odd, then each petal is traversed *twice* for $0 \le \theta \le 2\pi$.

In other words: for odd k, the rose $r = a\cos(k\theta)$ (or $r = a\sin(k\theta)$) has k *double petals.*

Trading polar and rectangular coordinates

How are the polar coordinates (r, θ) of a point P related to the rectangular coordinates (x, y) of the *same* point? How can either type of coordinates be found from the other? Here's a useful picture:

Relating polar and rectangular coordinates

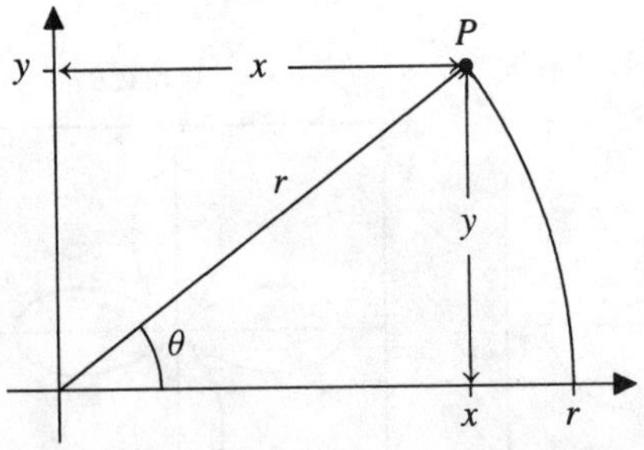

Convince yourself of each.

The picture illustrates many relations among x, y, r, and θ. Among the simplest:

$$x = r\cos\theta; \quad y = r\sin\theta; \quad r^2 = x^2 + y^2; \quad \tan\theta = \frac{y}{x}.$$

(The last equation holds only if $x \neq 0$, of course.)

These relations let us convert from one type of coordinates to the other. Equations in x and y, for instance, are easy to rewrite in terms of r and θ. The next two examples show the process in action.

■ **Example 2.** Find a polar equation for the straight line $y = mx + b$.

Solution: Substituting $x = r\cos\theta$ and $y = r\sin\theta$ into the equation for the line gives

$$y = mx + b \iff r\sin\theta = m(r\cos\theta) + b \iff r = \frac{b}{\sin\theta - m\cos\theta}.$$

(One moral: rectangular coordinates are better-suited to straight lines than polar coordinates!) □

Look again. Do you agree?

■ **Example 3.** The graph of $r = 3\cos\theta$ *looks* like a circle. Is it a circle? Which circle?

Solution: It *is* a circle. Changing to rectangular coordinates shows why:

$$r = 3\cos\theta \implies r^2 = 3r\cos\theta \implies x^2 + y^2 = 3x.$$

Check details.

The last equation does, as expected, define a circle. To decide *which* circle, complete the square:

$$x^2 + y^2 = 3x \implies x^2 - 3x + y^2 = 0 \implies \left(x - \frac{3}{2}\right)^2 + y^2 = \frac{9}{4}.$$

Do you agree? Look again at the picture.

The circle, therefore, has radius $3/2$ and center at $(3/2, 0)$—just as the picture suggests. □

Exercises

In Exercises 1–6, a point is given in *rectangular* coordinates. Plot the given point, then give three different pairs of *polar* coordinates for the same point. For at least one pair, r should be *negative*. [NOTE: Answers involve "familiar" angles θ, so give *exact* answers rather than decimal approximations.]

1. $(1, 1)$
2. $(-1, 1)$
3. $(1, \sqrt{3})$
4. $(-\sqrt{3}, 1)$
5. $(\pi, 0)$
6. $(0, \pi)$

In Exercises 7–12, a point is given in *rectangular* coordinates. Plot the given point, then give three different pairs of *polar* coordinates for the same point. For at least one pair, r should be *negative*. [NOTE: Angles are not necessarily "familiar," so use a calculator (in radian mode!). Round answers to 3 decimal places.]

7. $(1, 2)$
8. $(-1, 2)$
9. $(1, 4)$
10. $(1, 100)$
11. $(0.356, 0.478)$
12. $(-0.356, -0.478)$

In Exercises 13–18, a point is given in *polar* coordinates. Plot the given point, then give *rectangular* coordinates for the same point. [NOTE: Answers involve "familiar" angles θ, so give *exact* answers rather than decimal approximations.]

13. $(2, \pi/4)$
14. $(-2, 5\pi/4)$
15. $(1, 13\pi/6)$
16. $(42, 0)$
17. $(a, 0)$ (a any positive number)
18. $(-a, 0)$ (a any positive number)

In Exercises 19–24, a point is given in *polar* coordinates. Plot the given point, then give *rectangular* coordinates for the same point. [NOTE: Angles are not necessarily "familiar," so use a calculator (in radian mode!). Round answers to 3 decimal places.]

19. $(1, 1)$
20. $(-1, 1)$
21. $(2, 2)$
22. $(2, -2)$
23. $(1, \arctan 1)$
24. $(1, \arctan 2)$

25. The point $P(1, 0)$ has *identical* rectangular and polar coordinates. What other points have this property? Why?

26. Imitate Example 1, page 371, to plot the limaçon $r = 2 + \sin\theta$. Proceed as follows:

(a) Make a table of values like that in Example 1—let θ range from 0 to 2π in steps of $\pi/6$. Round r-values to 2 decimals.

(b) Copy the polar grid shown in Example 1. On it, plot the points calculated in part (a). Join them with a smooth curve.

(c) Discuss the symmetry of the resulting limaçon. What is its axis of symmetry?

(d) Look carefully at the table of values in Example 1; notice how the r-values are "symmetric" about $\theta = \pi$. (The graph in Example 1 is also symmetric about $\theta = \pi$.) What similar type of symmetry does your table from part (a) above show? Does your graph agree?

27. Imitate Example 1, page 371, to plot the cardioid $r = 1 + \cos\theta$. Proceed as follows:

(a) Make a table of values like that in Example 1—let θ range from 0 to 2π in steps of $\pi/6$. Round r-values to 2 decimals.

(b) Copy the polar grid shown in Example 1. On it, plot the points calculated in part (a). Join them with a smooth curve. Why does the name "cardioid" fit?

(c) Discuss the symmetry of this cardioid. What is its axis of symmetry? How does the table of values in part (a) reflect the cardioid's symmetry?

28. Plot and discuss the limaçon $r = 1 - 2\cos\theta$, as follows:

(a) Make a table of values like that in Example 1—let θ range from 0 to 2π in steps of $\pi/6$. Round r-values to 2 decimals. Plot the points, and join them with a smooth curve.

(b) For what values of θ is $r = 0$? How do these values appear on the graph?

(c) On what θ-interval is $r < 0$? How does this interval show up on the graph?

In Exercises 29–34, draw a plot of the given polar equation. To avoid tedious point plotting, use the models of cardioids and limaçons shown on page 372. [NOTE: A calculator or computer is OK but shouldn't be necessary. Be sure to label your graphs with appropriate units.]

29. $r = 3 + 3\cos\theta$

30. $r = 3 - \cos\theta$

31. $r = 1 + \sqrt{3}\cos\theta$

32. $r = 4 + 4\sin\theta$

33. $r = 4 - 2\sin\theta$

34. $r = 2 - 4\sin\theta$

In Exercises 41–46, sketch the given polar rose. To avoid tedious point plotting, use the models of roses shown on page 373. [NOTE: A calculator or computer is OK but shouldn't be necessary. Be sure to label your graphs with appropriate units.]

41. $r = 2\sin\theta$

42. $r = 2\sin(2\theta)$

43. $r = 2\sin(3\theta)$

44. $r = 2\cos(4\theta)$

45. $r = 2\cos(5\theta)$

46. $r = 2\cos(1001\theta)$
[NOTE: rough is OK!]

53. We claimed in this section that for any numbers r and θ, the pairs (r, θ), $(r, \theta + 2\pi)$, and $(-r, \theta + \pi)$ all describe the same point in the plane.

(a) What does the claim say if $r = 1$ and $\theta = 0$? Is it true? Why or why not? Explain in your own words.
(b) What does the claim say if $r = -1$ and $\theta = \pi/4$? Is it true? Why or why not? Explain in your own words.
(c) The point with rectangular coordinates $(1, 0)$ can be written in polar coordinates as $(1, 2k\pi)$, where k is any integer, or as $(-1, (2k-1)\pi)$, where k is any integer. In the same sense, describe all the possible polar coordinates of the point with rectangular coordinates $(1, 1)$.

In Exercises 54–57, change the given equation in r and θ to an equivalent equation in x and y. Then plot the result using whichever form seems simpler.

54. $r = 2\sec\theta$
55. $r = 4$
56. $\tan\theta = 1$
57. $r = 2\sin\theta$

In Exercises 58–61, change the given equation in x and y to an equivalent equation in r and θ. Then plot the result using whichever form seems simpler.

58. $x^2 + y^2 = 9$
59. $y = 4$
60. $y = 2x$
61. $(x-1)^2 + y^2 = 1$

62. (Use a graphing calculator or other polar plotting tool.) This exercise concerns limaçons of the form $r = 1 + a\cos\theta$, where a is any real constant. The following questions are open-ended. Answer them by experimenting with plots for various values of a: positive, negative, large, small, etc.

(a) For which *positive* values of a does the graph have an inner loop?
(b) What happens at a=1?
(c) What happens as $a \to 0$?
(d) How are the graphs for a and $-a$ (e.g., $r = 1 + 0.5\cos\theta$ vs. $r = 1 - 0.5\cos\theta$) related to each other?
(e) What happens as $a \to \infty$?

In Exercises 63–66, the graph is some sort of *spiral.* Plot each one, either by hand or by calculator (if the latter, be sure to adjust the θ-range appropriately).

63. $r = \theta$, for $0 \le \theta \le 2\pi$ (an ordinary spiral)
64. $r = \theta$, for $-2\pi \le \theta \le 2\pi$ (another ordinary spiral)
65. $r = \ln(\theta)$, for $1 \le \theta \le 4\pi$ (a logarithmic spiral)
66. $r = \exp(\theta)$, for $-\pi \le \theta \le \pi$ (an exponential spiral)

Appendix B

Calculus in polar coordinates

In the last section we explored the polar coordinate system, polar equations, and the geometry of polar curves. This section is about *calculus* on polar curves. As for rectangular curves, two main problems stand out for polar curves: finding the *slope* at a point and finding the area enclosed.

To illustrate these questions, consider these pictures:

Interesting points on a cardioid: $r = 1 + \cos\theta$

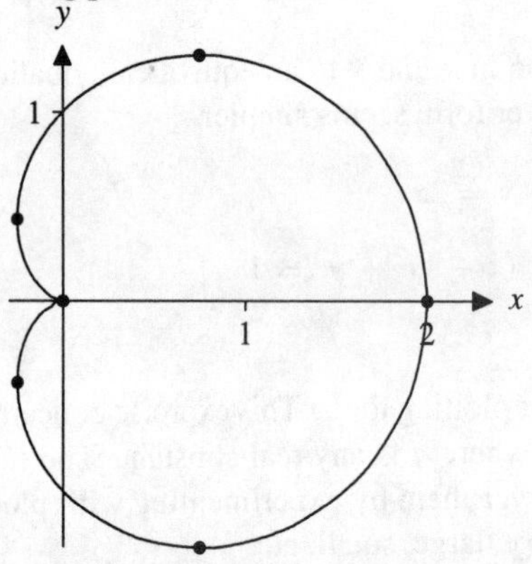

How large a rose? $r = \cos(2\theta)$

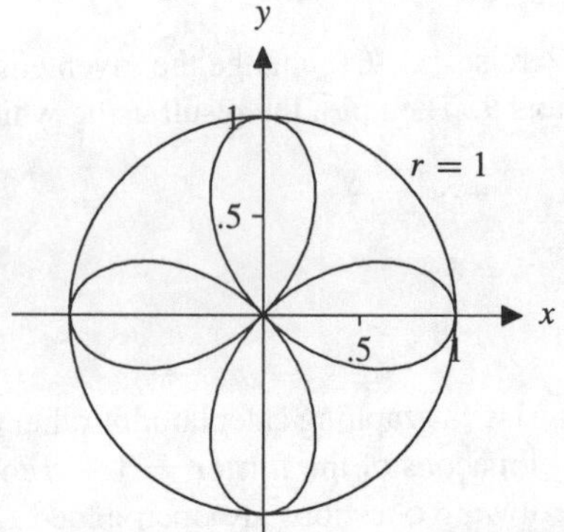

Among the questions we'll address:

Horizontal and vertical tangents. Where—exactly—does the cardioid have horizontal and vertical tangent lines? (The bulleted points on the cardioid seem to be of interest, but what are their coordinates?)

Make a guess. Record it here in the margin.

Slope at any point. What's the slope of either curve at *any* point?

Areas. How much area does one petal of the rose enclose? What fraction of the *circular* area does the rose enclose?

Make a guess. Record it here in the margin.

Slopes on polar curves

We'll consider curves of the form $r = f(\theta)$, where f is a differentiable function. For given θ_0, we want the slope at the polar point (r_0, θ_0).

Both curves shown above are of this type.

An obvious—but wrong—guess. Let's acknowledge, just for the record, that the desired slope is *not* $f'(\theta_0)$. A look at either graph above quickly confirms this. For $f(\theta) = 1 + \cos\theta$, for instance, $f'(\theta) = -\sin\theta$, which varies only between -1

and 1. Yet it's clear at a glance that slopes on the cardioid take *all* real values; at three points, moreover, the cardioid has *vertical* tangent lines. Thus $f'(\theta)$ is *not* the slope we seek, though it will play a role.

Polar curves and parametric equations

We showed in earlier work how to find the slope at a point on a curve defined by parametric equations. Here, for reference, is the result:

> **Fact: (Slope of a parametric curve.)** Let a smooth curve C be given by parametric equations $x = f(t)$, $y = g(t)$, with $a \le t \le b$. If $f'(t) \neq 0$, then the slope dy/dx of C at (x, y) is given by
>
> $$\frac{dy}{dx} = \frac{g'(t)}{f'(t)} = \frac{dy/dt}{dx/dt}.$$

To *use* the result, we'll need first to write a polar curve $r = f(\theta)$ in parametric form. This is surprisingly easy to do. The key facts are the "conversion" formulas from polar to rectangular coordinates:

We explained them in the last section.

$$x = r\cos\theta; \qquad y = r\sin\theta.$$

For points on our polar curve, $r = f(\theta)$; for such points, therefore,

$$x = r\cos\theta = f(\theta)\cos\theta; \qquad y = r\sin\theta = f(\theta)\sin\theta.$$

These are our desired parametric equations. We've rewritten our curve, originally given in r-θ form, as a pair of parametric equations, with θ as parameter. All that's left is to apply the slope formula above. We'll need $dx/d\theta$ and $dy/d\theta$ below; let's compute them now. By the product rule,

$$x = f(\theta)\cos\theta \implies \frac{dx}{d\theta} = f'(\theta)\cos\theta - f(\theta)\sin\theta;$$

$$y = f(\theta)\sin\theta \implies \frac{dy}{d\theta} = f'(\theta)\sin\theta + f(\theta)\cos\theta.$$

Everything is now in place. Here's the result, stated with appropriate technical hypotheses:

See the exercises for more on these hypotheses.

> **Fact: (Slope of a polar curve.)** Let a curve C be given in polar coordinates by a function $r = f(\theta)$, $\alpha \le \theta \le \beta$, where f and f' are continuous on (α, β), and not simultaneously zero. Then for θ in (α, β), the slope of C at $(r, \theta) = (f(\theta), \theta)$ is given by
>
> $$\frac{dy}{dx} = \frac{dy/d\theta}{dx/d\theta} = \frac{f'(\theta)\sin\theta + f(\theta)\cos\theta}{f'(\theta)\cos\theta - f(\theta)\sin\theta}$$
>
> wherever the denominator is not zero.

Using this Fact we can answer the questions on tangents raised at the beginning of this section. The next example shows how; it illustrates, too, some of the caution needed in working with polar coordinates.

■ **Example 1.** Consider the cardioid $r = 1 + \cos\theta$, shown on page 378. Where is the curve horizontal? Where is it vertical? What happens at the origin?

Take a close look.

Solution: Here, $f(\theta) = 1 + \cos\theta$ and $f'(\theta) = -\sin\theta$, so

$$\frac{dy/d\theta}{dx/d\theta} = \frac{f'(\theta)\sin\theta + f(\theta)\cos\theta}{f'(\theta)\cos\theta - f(\theta)\sin\theta} = \frac{(1 - 2\cos\theta)(1 + \cos\theta)}{(2\cos\theta + 1)\sin\theta}.$$

Check our work.

(We used the Fact above and a little algebra.)

The cardioid can have a **horizontal tangent line** only if the *numerator* $dy/d\theta$ is zero, i.e., if

$$(1 - 2\cos\theta)(1 + \cos\theta) = 0 \iff \cos\theta = \frac{1}{2} \quad \text{or} \quad \cos\theta = -1.$$

For θ in $[0, \pi]$, one or the other condition holds only if $\theta = \pi/3$ or $\theta = \pi$. (By symmetry, it's enough to look only on the *upper half* of the cardioid.)

The picture shows that, in fact, the upper half of the cardioid is horizontal *only* at $\theta = \pi/3$, i.e., at the point with polar coordinates $(3/2, \pi/3)$, and rectangular coordinates $(3/4, 3\sqrt{3}/4) \approx (0.75, 1.30)$. Symmetry dictates that the cardioid is also horizontal at $\theta = 5\pi/6$.

See for yourself!

What happens at $\theta = \pi$? Isn't the cardioid horizontal there, too? Not necessarily—at $\theta = \pi$, *both* $dx/d\theta$ and $dy/d\theta$ are zero and so the slope expression is undefined. (In fact, the cardioid has a "cusp" at the origin, and hence no well-defined slope there.)

A **vertical tangent line** can occur only where the *denominator* $dx/d\theta$ is zero, i.e., if

$$-\sin\theta\,(2\cos\theta + 1) = \iff \sin\theta = 0 \quad \text{or} \quad \cos\theta = -\frac{1}{2}.$$

What are their rectangular coordinates?

For θ in $[0, \pi]$, one or the other holds if $\theta = 0$ or $\theta = 2\pi/3$. As the picture shows, both $\theta = 0$ and $\theta = 2\pi/3$ correspond to points of vertical tangency on the cardioid. These points have polar coordinates $(2, 0)$ and $(1/2, 2\pi/3)$, respectively.

Superimposing the cardioid on a polar grid supports all the calculations above:

Horizontal and vertical points on a cardioid: $r = 1 + \cos\theta$

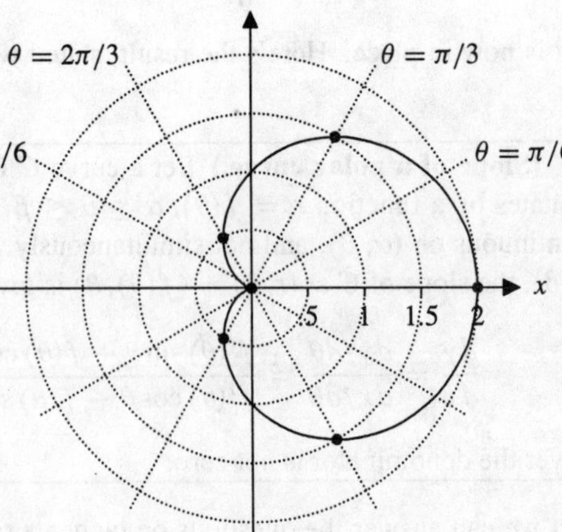

□

■ **Example 2.** Let $a > 0$ be any positive constant. What does the slope formula say about the circle $r = a$?

Solution: For the circle $r = a$, $f(\theta) = a$ and $f'(\theta) = 0$. Now the Fact says

$$\begin{aligned} \frac{dy}{dx} &= \frac{f'(\theta)\sin\theta + f(\theta)\cos\theta}{f'(\theta)\cos\theta - f(\theta)\sin\theta} \\ &= \frac{a\cos\theta}{-a\sin\theta} \\ &= -\cot\theta = -\frac{1}{\tan\theta}. \end{aligned}$$

Note what the last expression means: the tangent line to the circle at any point $P(a, \theta)$ is *perpendicular* to the ray from the origin to P. □

Why perpendicular? Because the slopes are negative reciprocals.

■ **Example 3.** Let $r = f(\theta)$ describe a polar curve C; suppose that $f(\theta_0) = 0$. What does the formula

$$\frac{dy}{dx} = \frac{f'(\theta)\sin\theta + f(\theta)\cos\theta}{f'(\theta)\cos\theta - f(\theta)\sin\theta}$$

say about the slope of C at the point $(0, \theta_0)$?

Solution: The slope formula becomes much simpler if $f(\theta_0) = 0$. In that case,

Remember: f and f' aren't simultaneously zero.

$$\frac{dy}{dx} = \frac{f'(\theta_0)\sin\theta_0}{f'(\theta_0)\cos\theta_0} = \tan\theta_0.$$

Note what this means: if a smooth polar curve passes through the origin at $\theta = \theta_0$, its tangent line is simply the ray $\theta = \theta_0$. For the rose $r = \cos(2\theta)$, for example, $r = 0$ when $\theta = \pi/4$, $\theta = 3\pi/4$, $\theta = 5\pi/4$, and $\theta = 7\pi/4$. As the following picture shows, the curve passes through the origin in just these directions:

The rose $r = \cos(2\theta)$: tangent lines at the origin

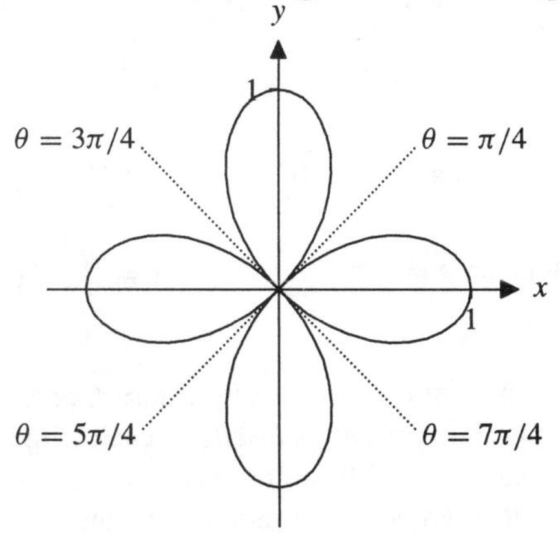

□

Finding area in polar coordinates

The standard area problem in *rectangular* coordinates concerns the area defined by an ordinary, $y = f(x)$-style graph, for $a \le x \le b$. In *polar* coordinates the standard problem is a little different: to find the area bounded by a *polar* curve $r = f(\theta)$, for $\alpha \le \theta \le \beta$. These pictures illustrate the "generic" situations; the areas in question are shown shaded:

Area in rectangular coordinates

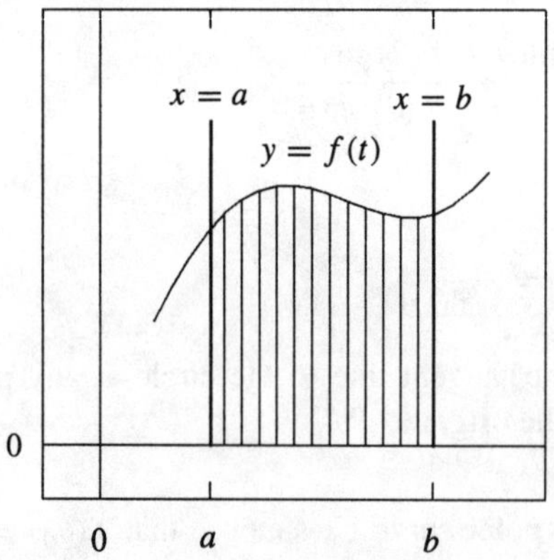

Area in polar coordinates

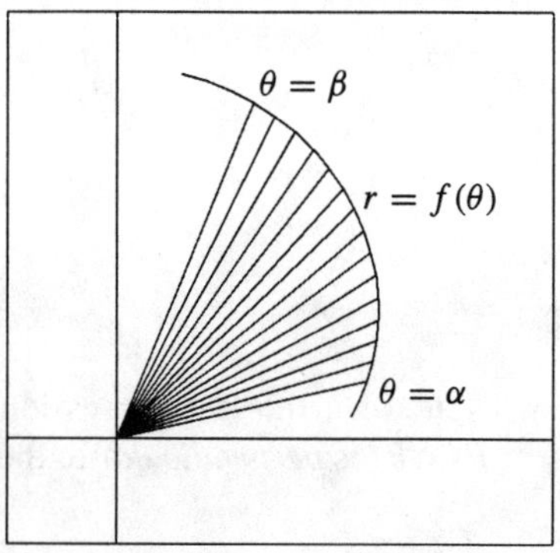

Areas by integration: rectangular vs. polar

The different styles of shading in the two figures above—one vertical, the other radial—were chosen intentionally. They reflect two slightly different approaches to finding areas. In order to emphasize both similarities and differences between the rectangular and polar situations, the next two paragraphs appear in "parallel."

Area, Cartesian style. In rectangular coordinates, area is approximated by subdividing the region into thin *vertical strips*, each one based on the x-axis. Each strip corresponds to a small subinterval, say $[x_i, x_i + \Delta x]$, obtained by partitioning the domain interval $a \le x \le b$ into n equal pieces.

We've done this often before, so we'll omit some details.

If Δx is small, then the strip is approximately a *rectangle*, with base Δx and height $f(x_i)$. Therefore:

$$\text{area of one strip} \approx f(x_i)\Delta x.$$

Adding all n strips gives

$$\text{total area} \approx \sum_{i=1}^{n} f(x_i)\Delta x.$$

Therefore, taking the limit as $\Delta x \to 0$,

$$\text{total area} = \lim_{n\to\infty} \sum_{i=1}^{n} f(x_i)\Delta x = \int_a^b f(x)\,dx.$$

Area, polar style. In polar coordinates, by contrast, area is approximated by subdividing the region into thin *pie-shaped wedges*, each with its vertex at the origin. Each wedge corresponds to a small subinterval, say $[\theta_i, \theta_i + \Delta\theta]$, obtained by partitioning the domain interval $\alpha \le \theta \le \beta$ into n equal pieces.

Look carefully at the polar picture: it shows 20 radial wedges.

If Δx is small, then each wedge is approximately a sector of a *circle*, with radius $f(\theta_i)$ and making the angle $\Delta\theta$ at the origin.

Remember: $r = f(\theta)$ gives the radius for given θ.

Now, a brief detour. Just ahead, we'll need to know the area of a **circular sector** as described above. The answer, though perhaps less familiar than the area of a rectangle, is easy to find. A *full* circle of radius R (for which θ runs from 0 to 2π) encloses total area πR^2. A wedge making angle $\Delta\theta$ at the origin represents the fraction $(\Delta\theta)/(2\pi)$ of the total circle, so

$$\text{area of wedge} = \pi R^2 \times \frac{\Delta\theta}{2\pi} = \frac{R^2}{2}\Delta\theta.$$

If, say, $R = f(\theta_i)$, then a circular wedge has area $f(\theta_i)^2 \Delta\theta/2$.

Back to the main road. Our detour showed that in the "generic" case pictured above,

$$\text{area of one wedge} \approx \frac{f(\theta_i)^2}{2} \Delta\theta.$$

Adding all n wedges gives

$$\text{total area} \approx \sum_{i=1}^{n} \frac{f(\theta_i)^2}{2} \Delta\theta.$$

Taking the limit as $\Delta\theta \to 0$ gives our area formula:

$$\text{total area} = \lim_{n\to\infty} \sum_{i=1}^{n} \frac{f(\theta_i)^2}{2} \Delta\theta = \int_{\alpha}^{\beta} \frac{f(\theta)^2}{2}\, d\theta.$$

The formula is worth remembering:

> **Fact: (Area in polar coordinates.)** Let f be a continuous function, and let R be the region in the xy-plane bounded by the polar curve $r = f(\theta)$ and the rays $\theta = \alpha$ and $\theta = \beta$. Then
>
> $$\text{area of } R = \int_{\alpha}^{\beta} \frac{f(\theta)^2}{2}\, d\theta.$$

■ **Example 4.** What's the area of one leaf of the polar rose $r = f(\theta) = \cos(2\theta)$? What fraction of the circle $r = 1$ does the entire rose cover?

Solution: As the picture on page 381 shows, the eastward-pointing leaf lies between $\theta = -\pi/4$ and $\theta = \pi/4$. To ease calculations we'll integrate from $\theta = 0$ to $\theta = \pi/4$ and double the result. Here is the integral calculation—a table of integrals helps along the way:

But check our work.

$$\int_0^{\pi/4} \frac{\cos^2(2\theta)}{2}\, d\theta = \frac{\theta}{4} + \frac{\sin(4\theta)}{16}\Big]_0^{\pi/4} = \frac{\pi}{16}.$$

Thus *one* leaf has area $\pi/8$; all four leaves have area $\pi/2$—exactly *half* the area of the circle $r = 1$. □

A rose with any other n . . . A more general (and more surprising) result than that of the last example is true:

See the exercises for more on this.

> *For any even n, the rose $r = \cos(n\theta)$ encloses total area $\pi/2$—half the area of the enclosing circle.*

■ **Example 5.** How much area does the cardioid $r = 1 + \cos\theta$ enclose?

Solution: With the integral formula, the answer is easy:

$$\begin{aligned}\text{area} &= \frac{1}{2}\int_0^{2\pi} f(\theta)^2\,d\theta \\ &= \frac{1}{2}\int_0^{2\pi}\left(1+2\cos\theta+\cos^2\theta\right)d\theta \\ &= \frac{1}{2}\left(\theta+2\sin\theta+\frac{\theta}{2}+\frac{\sin(2\theta)}{4}\right)\Big]_0^{2\pi} \\ &= \frac{3\pi}{2}\approx 4.71.\end{aligned}$$

□

Exercises

1. We claimed in this section that

$$x = f(\theta)\cos\theta \implies \frac{dx}{d\theta} = f'(\theta)\cos\theta - f(\theta)\sin\theta;$$
$$y = f(\theta)\sin\theta \implies \frac{dy}{d\theta} = f'(\theta)\sin\theta + f(\theta)\cos\theta.$$

Use the product rule to show that these claims are valid.

2. We claimed in Example 1 that on the cardioid $r = 1+\cos\theta$, the slope dy/dx at (r,θ) is

$$\frac{(1-2\cos\theta)(1+\cos\theta)}{(2\cos\theta+1)\sin\theta}.$$

Use the Fact on page 379 to verify this.

3. Consider the spiral $r=\theta$, for $0\le\theta\le 4\pi$.
   (a) Draw the spiral.
   (b) At what points does the spiral have horizontal tangent lines? Vertical tangent lines?
   (c) Write an equation in rectangular form for the line tangent to the spiral at the polar point $(1,1)$. Draw this tangent line on your graph.

4. Consider the limaçon $r = 1+2\sin\theta$.
   (a) Draw the limaçon.
   (b) Write an equation in rectangular form for the line tangent to the limaçon at the polar point $(1,0)$. Draw this tangent line on your graph.
   (c) Find the slope of the line tangent to the limaçon at the polar point $(0, 7\pi/6)$. Draw this tangent line on your graph.
   (d) Find the slope of the line tangent to the limaçon at the polar point $(0, 11\pi/6)$. Draw this tangent line on your graph.
   (e) At what points does the limaçon have horizontal tangent lines?

5. This open-ended problem is about the family of limaçons of the form $r = 1+a\cos\theta$.

(a) After plotting example limaçons for various values of a, determine which ones have a "dimple" (i.e., a pushed-in section on either the right or the left) and which ones don't. [HINT: Any limaçon with a dimple has *three* vertical tangent lines.]

(b) The answer to part (a) is that there's no dimple if $|a| \leq 1/2$. Show this using derivatives. [HINT: Which values of a lead to *three* vertical tangent lines.]

In Exercises 6–9, draw the region bounded by the given polar curves, and find its area.

6. $r = 1, \theta = 0, \theta = \pi$
7. $r = 3, \theta = 0, \theta = \pi$
8. $r = a, \theta = 0, \theta = \pi$
9. $r = 1, \theta = 0, \theta = \beta$

In Exercises 10–13, draw the given region and find its area.

10. The outer loop of the limaçon $r = 1 + 2\sin\theta$.
11. The inner loop of the limaçon $r = 1 + 2\sin\theta$.
12. The region bounded by $r = \sec\theta, \theta = 0, \theta = \pi/4$.
13. The region bounded by $r = \sec\theta, \theta = 0, \theta = \arctan m$.

14. Consider the $2n$-leafed rose $r = \cos(n\theta)$, where n is even.

    (a) Find the area of one leaf.

    (b) Find the area of all $2n$ leaves. What fraction of the circle $r = 1$ does the entire rose fill up?

15. Consider the n-leafed rose $r = \cos(n\theta)$, where n is odd.

    (a) Find the area of one leaf.

    (b) Find the area of n leaves. What fraction of the circle $r = 1$ does the entire rose fill up?

16. Draw and then calculate the area bounded by one "turn" of the spiral $r = \theta$, i.e., from $\theta = 0$ to $\theta = 2\pi$.

17. Draw and then calculate the area bounded by one "turn" of the exponential spiral $r = e^{\theta}$, i.e., from $\theta = 0$ to $\theta = 2\pi$.

18. Draw and then calculate (or estimate numerically) the area bounded by one "turn" of the logarithmic spiral $r = \ln(\theta)$, i.e., from $\theta = 2\pi$ to $\theta = 4\pi$.

19. Use polar coordinates to find the area of the region inside the circle $r = 1$ and to the right of $x = 1/2$.

20. Let $0 < a < 1$. Use polar coordinates to find the area of the region inside the circle $r = 1$ and to the right of $x = a$.

In this section we used the idea that a *polar curve* can also be thought of as a *parametric* curve. Specifically, the curve defined by the polar function $r = f(\theta)$, for $\alpha \le \theta \le \beta$, can *also* be defined by the parametric equations

$$x = f(\theta)\cos\theta; \qquad y = f(\theta)\sin\theta,$$

$\alpha \le \theta \le \beta$. Consider, for example, the unit circle $r = 1$, $0 \le \theta \le 2\pi$. Then $f(\theta) = 1$, so the parametric form is simply

$$x = f(\theta)\cos\theta = \cos\theta; \quad y = f(\theta)\sin\theta = \sin\theta; \quad 0 \le \theta \le 2\pi.$$

In Exercises 21–28, a curve is given in polar form. First rewrite the curve in parametric form, then plot it. (Use a graphing calculator or computer.) Do you see the "expected" results?

21. $r = 2; \quad 0 \le \theta \le 2\pi$

22. $r = 2; \quad 0 \le \theta \le \pi$

23. $r = \sec\theta; \quad -\pi/4 \le \theta \le \pi/4$

24. $r = \csc\theta; \quad \pi/4 \le \theta \le 3\pi/4$

25. $r = \theta; \quad 0 \le \theta \le 2\pi$

26. $r = \cos\theta; \quad 0 \le \theta \le \pi$

27. $r = \cos(2\theta); \quad 0 \le \theta \le 2\pi$

28. $r = 1 + \cos\theta; \quad 0 \le \theta \le 2\pi$

29. The formula for the slope of a polar curve requires that f and f' not be simultaneously zero. Show that if this condition holds, then $dy/d\theta$ and $dx/d\theta$ are not simultaneously zero. [HINT: Recall that $dy/d\theta = f'(\theta)\sin\theta + f(\theta)\cos\theta$ and $dx/d\theta = f'(\theta)\cos\theta - f(\theta)\sin\theta$. Look at $(dy/d\theta)^2 + (dx/d\theta)^2$.]

Appendix C

Matrices and matrix algebra: a crash course

This appendix assumes basic familiarity with vectors, including vector addition, scalar multiplication, and the dot product.

What is a matrix?

A **matrix** is a rectangular array of real numbers or symbols.➥ Here are some examples:

The symbols stand for real numbers.

$$A = \begin{bmatrix} 1 & 2 & 3 \end{bmatrix} \quad X = \begin{bmatrix} x \\ y \\ z \end{bmatrix} \quad B = \begin{bmatrix} a & b & c \\ d & e & f \\ g & h & i \end{bmatrix} \quad C = \begin{bmatrix} 1 & 2 & 3 \\ 4 & 5 & 6 \end{bmatrix} \quad I_2 = \begin{bmatrix} 1 & 0 \\ 0 & 1 \end{bmatrix}$$

As the examples suggest, matrices may have any number of rows and any number of columns. A little vocabulary and notation will help us navigate among the possibilities:

Dimensions. An $m \times n$ matrix has m *rows* and n *columns*. Among the matrices above, A is a 1×3 matrix, X is 3×1, and C is 2×3. An $m \times m$ matrix, such as B and I_2 above, is called **square**.

Rows, columns, and indices. Matrix notation uses *two* index variables,➥ one for rows and one for columns. Keeping clear which is which is essential. Fortunately, there's an unbreakable rule:

The letters i, j, and k are popular—but not sacred—choices.

First rows, then columns.

For instance, M_{23} is the entry in row 2, column 3, and a 4×7 matrix has 4 rows and 7 columns—not the other way around.

Matrices and vectors. Matrices and vectors are closely related. An $m \times 1$ matrix is sometimes called a **column vector**; a $1 \times m$ matrix is a **row vector**. (See the matrices X and A, above.) In fact, we can think of any matrix—regardless of its dimensions—as built up either from column vectors or from row vectors. Which point of view is more useful depends on the circumstances; we'll find uses for both.

Entries. The numbers or symbols in any matrix M are called **entries**; the i, jth entry, denoted M_{ij}, is the one in row i and column j. Above, for instance, $A_{12} = 2$, $X_{31} = z$, and $B_{23} = f$, while A_{21}, X_{13}, and B_{45} are not defined. In general, of course, M_{ij} and M_{ji} need not be equal, even if both quantities are defined.

Matrices B and C illustrate this.

Diagonals and identity matrices. The **main diagonal** of a matrix M consists of all entries of the form M_{ii}. In an $m \times m$ (square) matrix, the main diagonal starts at the upper left corner and ends at the lower right corner. The $m \times m$ matrix that has ones along the main diagonal and zeros everywhere else is called the $m \times m$ **identity matrix**; it's denoted I_m (I_2 appears above). (The name is appropriate because I_m behaves, as we'll see in a moment, as an "identity" for matrix multiplication.)

In a non-square matrix, the main diagonal does not end at the lower right corner.

What are matrices for?

Like tables, arrays, spreadsheets, and the like, matrices are used to store information concisely and efficiently—especially when the information has, like the matrix itself, a natural "two-dimensional" structure. For instance, the following table gives road mileage between several pairs of Texas cities:

	Austin	Dallas	Houston	San Antonio
Austin	0	195	161	78
Dallas	195	0	240	272
Houston	161	240	0	195
San Antonio	78	272	195	0

All the numerical information fits naturally into a 4×4 matrix, say T:

"T" is For Texas—the cities are now understood, not explicitly stated.

$$T = \begin{bmatrix} 0 & 195 & 161 & 78 \\ 195 & 0 & 240 & 272 \\ 161 & 240 & 0 & 195 \\ 78 & 272 & 195 & 0 \end{bmatrix}$$

How much does it cost, in dollars, to drive from one Texas city to another? If we multiply every entry in T by 0.31 (many companies reimburse employees for auto expenses at \$0.31 per mile), we get a new matrix

$$E = \begin{bmatrix} 0 & 60.45 & 49.91 & 24.18 \\ 60.45 & 0 & 74.40 & 84.32 \\ 49.91 & 74.40 & 0 & 61.69 \\ 24.18 & 84.32 & 61.69 & 0 \end{bmatrix};$$

each entry of E tells the mileage reimbursement associated with the given trip. Under these circumstances, it's reasonable to write, simply, $E = 0.31\,T$.

More below on other matrix operations.

The matrices T and E have some special properties, such as having all zeros along the main diagonal. (The distance from any city to itself is zero!) Notice, too, that each entry has an identical twin in the "mirror" position across the main diagonal. This property has a formal name:

> **Definition:** An $n \times n$ matrix M is **symmetric** if, for all i and j from 1 to n,
> $$M_{ij} = M_{ji}.$$

Not every useful matrix is symmetric, but enough are to have spawned a considerable mathematical theory. In multivariable calculus, symmetric matrices occur in the context of second derivatives of functions of several variables.

Matrix algebra

Addition, subtraction, and scalar multiplication. Matrices can be added, subtracted, and multiplied by scalars in *exactly* the same way as vectors, their close cousins. The following simple examples illustrate what this means; all the symbols stand for real numbers:

$$\begin{bmatrix} a & b \\ c & d \end{bmatrix} + \begin{bmatrix} e & f \\ g & h \end{bmatrix} = \begin{bmatrix} a+e & b+f \\ c+g & d+h \end{bmatrix}; \quad r\begin{bmatrix} a & b \\ c & d \end{bmatrix} = \begin{bmatrix} ra & rb \\ rc & rd \end{bmatrix}.$$

(The Texas mileage expense matrix is another example of scalar multiplication.) The examples show that matrices—just like vectors—can be added, subtracted, and multiplied by scalars in the simplest possible way, "entry-by-entry."➼ (In particular, $A \pm B$ can make sense only if A and B have the same dimensions.) The formal definitions say the same thing, but in full symbolic regalia:

Vectors are special types of matrices, so the similarity in behavior is no surprise.

> **Definition:** Let A and B be $m \times n$ matrices and let r be a scalar. Then $A + B$, $A - B$, and rA are new $m \times n$ matrices, defined respectively by
> $$(A + B)_{ij} = a_{ij} + b_{ij}; \quad (A - B)_{ij} = a_{ij} - b_{ij}; \quad (rA)_{ij} = ra_{ij}.$$

Like their vector counterparts (and for the same reasons), these matrix operations enjoy pleasant algebraic properties, such as **commutativity** of addition (i.e., $A+B = B + A$) and **distributivity** (i.e., $r(A + B) = rA + rB$).

■ **Example 1.** In multivariable calculus, matrices often have to do with linear equations and linear functions. Consider the following system of 3 linear equations in 3 unknowns:

$$\begin{aligned} 1x + 2y + 3z &= 7 \\ 2x + 3y + 1z &= 8 \\ 3x + 2y + 1z &= 9 \end{aligned}$$

Rewrite this system using matrices.

Solution: The system suggests three different matrices:

$$D = \begin{bmatrix} 1 & 2 & 3 \\ 2 & 3 & 1 \\ 3 & 2 & 1 \end{bmatrix}, \quad X = \begin{bmatrix} x \\ y \\ z \end{bmatrix}, \quad \text{and} \quad E = \begin{bmatrix} 7 \\ 8 \\ 9 \end{bmatrix}.$$

Here D is called the **coefficient matrix** of the system, B contains the "right side" of the system of equations, and X stores the variable names. (We'll see shortly why it's convenient for X to have the shape of a column, not a row.)

Solving linear systems of equations (including very large ones) is an important mathematical problem, but it can be tedious and error-prone. Storing information in matrices as one goes along helps avoid errors and unnecessary duplication. The system in this example can be written entirely in matrix notation as

$$\begin{bmatrix} 1 & 2 & 3 \\ 2 & 3 & 1 \\ 3 & 2 & 1 \end{bmatrix}\begin{bmatrix} x \\ y \\ z \end{bmatrix} = \begin{bmatrix} 7 \\ 8 \\ 9 \end{bmatrix};$$

written entirely in symbols, the equation is simply $DX = E$.

Notice especially the left side of the preceding equation: DX is a matrix *product*—an idea to which we now turn. □

Matrix multiplication. How can two matrices A and B be multiplied to form a third matrix, AB, that deserves to be called the matrix product? One reasonable guess, following the pattern for addition, is simply to multiply element-by-element. It turns out, however, that another definition of matrix multiplication is much more useful. We'll first state the definition formally and then explore what it means.

> **Definition:** (**Matrix multiplication**) Let A be an $m \times p$ matrix and B a $p \times n$ matrix. The product AB is an $m \times n$ matrix, with entries given by
>
> $$(AB)_{ij} = \sum_{k=1}^{p} A_{ik}B_{kj} = A_{i1}B_{1j} + A_{i2}B_{2j} + \cdots + A_{ip}B_{2p}.$$

This somewhat forbidding-looking definition is best unpacked through simple examples. Consider these carefully:

$$\begin{bmatrix} a & b \\ c & d \end{bmatrix}\begin{bmatrix} x \\ y \end{bmatrix} = \begin{bmatrix} ax+by \\ cx+dy \end{bmatrix} \qquad \begin{bmatrix} 2 & 3 \end{bmatrix}\begin{bmatrix} x \\ y \end{bmatrix} = \begin{bmatrix} 2x+3y \end{bmatrix}$$

$$\begin{bmatrix} 1 & 2 & 3 \\ 2 & 3 & 1 \\ 3 & 2 & 1 \end{bmatrix}\begin{bmatrix} x \\ y \\ z \end{bmatrix} = \begin{bmatrix} 1x+2y+3z \\ 2x+3y+1z \\ 3x+2y+1z \end{bmatrix} \qquad \begin{bmatrix} a & b \\ c & d \end{bmatrix}\begin{bmatrix} x & z \\ y & w \end{bmatrix} = \begin{bmatrix} ax+by & az+bw \\ cx+dy & cz+bw \end{bmatrix}$$

Here are some lessons the definition and examples teach:

Possible shapes. The product AB makes sense if (but only if) A has the same number of columns as B has rows. Equivalently, the rows of A must be exactly as long as the columns of B. In particular, A, B, and AB may *all* have different shapes. At the other extreme, if both A and B are (square) $n \times n$ matrices, then so is AB.

Order matters. Matrix multiplication is *not* commutative. More often than not, $AB \neq BA$—even if both products happen to make sense.◂ For example, reversing the factors in the last example above gives

Of which there is no guarantee.

$$\begin{bmatrix} x & z \\ y & w \end{bmatrix}\begin{bmatrix} a & b \\ c & d \end{bmatrix} = \begin{bmatrix} ax+cz & bx+dz \\ ay+cw & by+dw \end{bmatrix}.$$

Matrix multiplication and dot products. Matrix multiplication is closely linked to the dot product. Indeed, if A is a $1 \times n$ row vector and B is an $n \times 1$ column vector, then the matrix product AB has just one entry: the dot product of A and B, thought of as vectors. In fact, *every* matrix product AB is found by taking appropriate dot products. More precisely:➻

Read this carefully—it's a nice way to remember the recipe for matrix multiplication.

The ijth entry of AB is the dot product of the ith row of A and the jth column of B.

Zero and identity matrices. The $n \times n$ matrix O with *all* entries zero is called the **zero matrix**. It behaves as any self-respecting zero should: For every $n \times n$ matrix A, $AO = O = OA$ and $A + O = A = O + A$. In a similar vein, the $n \times n$ **identity matrix** I_n (with ones on the main diagonal and zeros elsewhere)➻ is a multiplicative identity: For every $n \times n$ matrix A, $AI_n = A = I_nA$.

See I_2 at the beginning of this appendix.

Matrix algebraic expressions. With various matrix operations understood, we can make sense of algebraic expressions and equations that involve matrices. In Example 1, for example, we expressed the system of linear equations as the *matrix* equation $DX = E$, where all the symbols represent matrices of appropriate shapes and sizes. Similarly, we can now understand more complicated matrix equations, such as

$$PAQ = B, \quad A(3B + C) = 3AB + AC, \quad \text{and} \quad AB = I_3.$$

For these expressions to make good sense requires, of course, that the various matrices have compatible dimensions.

Multiplicative inverses. Two real numbers a and b are called multiplicative inverses if $ab = 1$, i.e., if their product is the multiplicative identity for real numbers. Thus 3 and $1/3$ are inverses, as are $-17/12$ and $-12/17$; in fact, the real numbers a and $1/a$ are inverses for any $a \neq 0$.

Similar ideas hold for square matrices.➻ Two $n \times n$ matrices A and B are called **inverses** if $AB = I_n$, i.e., if their product is an identity matrix. In this case, we write $A = B^{-1}$ and $B = A^{-1}$. For instance, it's easy to check➻ that if

Square matrices work best since they can be multiplied in either order.

This would be a good time.

$$A = \begin{bmatrix} 3 & 2 \\ 2 & 1 \end{bmatrix} \quad \text{and} \quad B = \begin{bmatrix} -1 & 2 \\ 2 & -3 \end{bmatrix}, \quad \text{and} \quad I_2 = \begin{bmatrix} 1 & 0 \\ 0 & 1 \end{bmatrix},$$

then $AB = BA = I_2$, so $B = A^{-1}$ and $A = B^{-1}$.

(To be fully rigorous about inverses, we should require both $AB = I_n$ and $BA = I_n$. It turns out, however, that for square matrices the two conditions are equivalent. This can be shown using techniques from linear algebra.)

Not every square matrix *has* an inverse. For example, the $n \times n$ zero matrix O certainly doesn't, since for every $n \times n$ matrix A, $AO = O \neq I_n$. But O is not the only square matrix without an inverse. The next example—which applies to *any* 2×2 matrix—suggests why some square matrices have inverses and some don't.

■ **Example 2.** Consider the 2×2 matrices

$$A = \begin{bmatrix} a & b \\ c & d \end{bmatrix} \quad \text{and} \quad B = \begin{bmatrix} \frac{d}{ad-bc} & -\frac{b}{ad-bc} \\ -\frac{c}{ad-bc} & \frac{a}{ad-bc} \end{bmatrix}.$$

How are A and B related?

Solution: For B to make any sense, the common denominator $ad - bc$ must be non-zero. But if the denominator isn't zero, then straightforward calculation shows that $AB = I_2$, so A and B are inverses. □

Try it—the result is satisfying!

More to the story .... The general theory of matrices and their inverses and how (if possible) to find one from the other is studied in much more detail in linear algebra courses. For multivariable calculus we need only basic ideas and definitions.

Matrices through the ages. Matrices have a very long history. Ordered tables of numbers and recipes for manipulating them go back as far as the ancient Babylonian and Chinese mathematicians, who may have used matrices in solving practical problems that led to linear equations. The modern abstract theory of matrices (and the word "matrix") dates back to the mid-1800's. The English lawyer and mathematician Arthur Cayley first defined the matrix operations of addition, scalar multiplication, and inversion.

Determinants

The quantity $ad - bc$ in Example 2 is called the **determinant** of the matrix $A = \begin{bmatrix} a & b \\ c & d \end{bmatrix}$; it's denoted by $\det A$. The determinant turns out to ... well ... *determine* whether or not A has an inverse.

Defining determinants. Determinants can be defined for square matrices M of any size. In all cases, $\det M$ is a number calculated from the entries of M; it tells (among other things) whether M has an inverse. The general definition of $\det M$ is a bit complicated; fortunately, we'll need the idea only in dimensions two and three.

Not a vector or a matrix.

Definition: (Determinant) For a 2×2 matrix, the determinant is

$$\det \begin{bmatrix} a & b \\ c & d \end{bmatrix} = ad - bc.$$

For a 3×3 matrix, the determinant is

$$\det \begin{bmatrix} a & b & c \\ d & e & f \\ g & h & i \end{bmatrix} = aei - afh + bfg - bdi + cdh - ceg.$$

Observe that, in both cases, the determinant involves several summands, with alternating positive and negative signs. Each summand is the product of one factor from each row and each column. There are six such summands for a 3×3 matrix. One way to organize the summands and keep track of signs in the 3×3 case is to write the matrix in a "double" array:

This is easy to see for the 2×2 case; the 3×3 case needs a closer look.

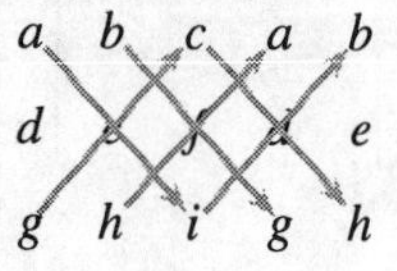

Then the determinant is the sum of six *threefold diagonal products*, with "southeast-pointing" diagonals counted positive and "northeast-pointing" diagonals negative.➻

Try it.

For larger matrices, determinants are normally calculated (if at all) with help from technology. However they're calculated, determinants have a useful relationship to inverses.

Fact: An $n \times n$ matrix M has an inverse if and only if $\det M \neq 0$.

Example 2 showed what the Fact means (and why it's true) for 2×2 matrices.➻ More general discussion can be found in any linear algebra text.

The exercises pursue the matter a little further, too.

Determinants geometrically: area and volume. What else can determinants tell us about matrices? For multivariable calculus purposes there are good geometric answers.

For a 2×2 matrix $M = \begin{bmatrix} a & b \\ c & d \end{bmatrix}$, the determinant measures the *area* of a certain parallelogram. To see why this is so, think of the rows of M as vectors (a, b) and (c, d) in the xy-plane; the parallelogram in question is "spanned" by these two vectors:

Two vectors span a parallelogram

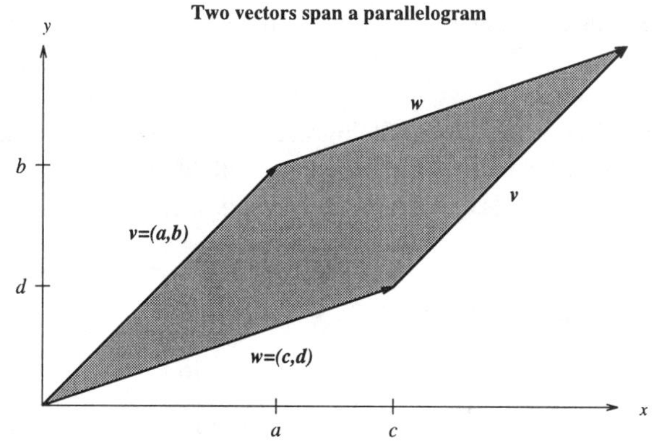

Similarly, for a 3×3 matrix we can regard the rows as vectors in xyz-space, and consider the three-dimensional solid (called a **parallelepiped**) that's spanned by these three vectors, as shown:

The solid spanned by three vectors

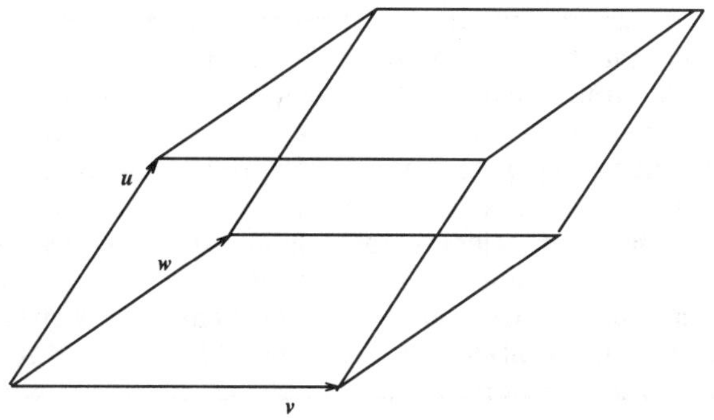

This time, the determinant measures the *volume*. Here are the precise statements:

> **Fact: (What the determinant tells.)** For a 2×2 matrix M,
>
> $$|\det M| = \text{area of parallelogram spanned by rows of } M.$$
>
> For a 3×3 matrix M,
>
> $$|\det M| = \text{volume of parallelepiped spanned by rows of } M.$$

We will not prove this Fact here; more discussion of its two claims appear in Sections 1.6 and 1.9, respectively. Observe:

Determinant zero. If $\det M = 0$, then the rows of M span zero area or zero volume. In the 2×2 case this means that the two row vectors of M are collinear—one is simply a multiple of the other. In the 3×3 case, spanning zero *volume* means that the three row vectors of M are coplanar. (In linear algebra jargon, these conditions say that the rows of M are "linearly dependent.")

Positive or negative? The Fact refers only to the magnitude $|\det M|$; determinants may be either positive or negative. For our purposes, the sign of the determinant is usually less important than its magnitude.

■ **Example 3.** Let P be the parallelogram with adjacent edges determined by the vectors $(1, 1)$ and $(2, 0)$. Let S be the solid parallelepiped with edges determined by the vectors $(1, 0, 0)$, $(0, 1, 0)$, and $(1, 1, 1)$. Find the area of P and the volume of S.

Solution: Consider the matrices

$$M = \begin{bmatrix} 1 & 1 \\ 2 & 0 \end{bmatrix} \quad \text{and} \quad N = \begin{bmatrix} 1 & 0 & 0 \\ 0 & 1 & 0 \\ 1 & 1 & 1 \end{bmatrix}.$$

Remember, it's the absolute value that counts.

Easy calculations show that $\det M = -2$ and $\det N = 1$. Therefore P has area 2 and Q has volume 1. □

Determinants—how important? Determinants were especially popular in the 18th and 19th centuries, when their theory was developed by such famous mathematicians as Leibniz, Maclaurin, Gauss, Lagrange, and Cauchy. But mathematical fashions change. More recently, the importance of determinants (as opposed to other mathematical tools for addressing similar questions) has been questioned. As evidence, consider the title of a recent journal article: *Down with determinants!*, by S. Axler, American Mathematical Monthly, February 1995, pp. 139–154.

Exercises

Note. Several problems below use the following matrices:

$$A = \begin{bmatrix} 1 & 2 \\ 3 & 5 \end{bmatrix}; \quad B = \begin{bmatrix} -5 & 2 \\ 3 & -1 \end{bmatrix}; \quad C = \begin{bmatrix} 0 & 1 & 2 \\ 0 & 0 & 1 \\ 1 & 0 & 2 \end{bmatrix};$$

$$D = \begin{bmatrix} 0 & 1 & 4 \\ 2 & 0 & 1 \end{bmatrix}; \quad E = \begin{bmatrix} x & u \\ y & v \\ z & w \end{bmatrix}; \quad X = \begin{bmatrix} x \\ y \end{bmatrix}; \quad U = \begin{bmatrix} u & v \end{bmatrix};$$

1. (a) Calculate AB, AD, AX, UA, and UX.
   (b) For each product MN in the preceding part, consider the reversed product NM. Compute those that make sense.
   (c) Are there any pairs of inverses among the matrices above?
2. (a) Calculate CE, C^2, DC, UD, EX.
   (b) For each product MN in the preceding part, consider the reversed product NM. Compute those that make sense.
3. (a) Find the determinants of A, B, AB, C, and C^2.
   (b) How is the determinant of AB related to the determinants of A and B?
   (c) How is the determinant of C^2 related to the determinant of C?
4. For real numbers a and b, $ab = 0$ implies that at least one of a and b is zero. Matrices do *not* have this property. To illustrate this, find two 2×2 matrices A and B, with all non-zero entries, such that $AB = \left[\begin{smallmatrix} 0 & 0 \\ 0 & 0 \end{smallmatrix}\right]$.
5. Consider the 2×2 matrices $M = \left[\begin{smallmatrix} a & b \\ c & d \end{smallmatrix}\right]$ and $N = \left[\begin{smallmatrix} e & f \\ g & h \end{smallmatrix}\right]$. Show that $\det(MN) = (\det M)(\det N)$. (Use "brute force"—i.e., just crank out both sides of the equation.)
6. This exercise is about Example 3, page 394.
   (a) Draw the parallelogram P. Convince yourself by elementary methods (e.g., areas of triangles) that P has area 2, as the determinant formula says.
   (b) Consider the matrix M in Example 3. Let M' be the same as M, but with the first and second rows reversed. Show that $\det M' = -\det M$? What does $\det M'$ mean geometrically?
   (c) Draw the parallelepiped Q. (Do this by hand.) Can you convince yourself by elementary methods that Q has volume 1, as the determinant formula says?
7. Consider the Texas mileage matrix T, on page 388; we observed that T is symmetric. What property of the real world guarantees this symmetry?
8. Recall that a square matrix M is *symmetric* (see page 389) if $M_{ij} = M_{ji}$ for all index values i and j. Similarly, a square matrix M is called **skew-symmetric** if $M_{ij} = -M_{ji}$ for all i and j. In each of the following parts, write out the 3×3 matrix M determined by the given rule, and decide whether the matrix is symmetric, skew-symmetric, or neither.

$$\begin{array}{c} {}^{1}\quad {}^{2}\quad {}^{3} \\ \begin{matrix} 1 \\ 2 \\ 3 \end{matrix} \begin{bmatrix} & & \\ & & \\ & & \end{bmatrix} \end{array}$$

(a) $M_{ij} = i + j$

(b) $M_{ij} = i - j$

(c) $M_{ij} = i^2 + j^2$

(d) $M_{ij} = i^2 - j$

9. Let M be a 3×3 matrix. (See the previous problem for the definition of a skew-symmetric matrix.)

   (a) Show that if M is skew-symmetric, then all the diagonal entries of M must be zeros.

   (b) Can M be both symmetric and skew-symmetric? If so, give an example. If not, why not?

10. Let A and B be matrices, both of dimension $m \times n$. We'll denote by $A \star B$ the new $m \times n$ matrix formed by multiplying element-by-element, i.e., $(A \star B)_{ij} = A_{ij} B_{ij}$. (The matrix $A \star B$ is called the **Hadamard** product of A and B, after the French mathematician Jacques Hadamard.)

    (a) Calculate the Hadamard products

$$\begin{bmatrix} 1 & 2 & 3 \\ 4 & 5 & 6 \end{bmatrix} \star \begin{bmatrix} 7 & 8 & 9 \\ 10 & 11 & 12 \end{bmatrix} \quad \text{and} \quad \begin{bmatrix} a & b \\ c & d \end{bmatrix} \star \begin{bmatrix} e & f \\ g & h \end{bmatrix}.$$

    (b) Ordinary matrix multiplication is not commutative. Is Hadamard multiplication commutative? Why or why not?

    (c) Among all 2×2 matrices, which matrix I deserves to be called an *identity* for Hadamard multiplication? Why?

    (d) Let A and B be 2×2 matrices. We'll say that A and B are **Hadamard inverses** if $A \star B = I$, where I is the matrix of the previous part. Which 2×2 matrices have Hadamard inverses? If A has a Hadamard inverse, how is it calculated?

11. Let M be a 2×2 matrix, and suppose that $MN = NM$ for all 2×2 matrices N. Show that M is a scalar multiple of the identity matrix.

Appendix D

Table of Derivatives and Integrals

Basic Forms

1. $\frac{d}{dx}c = (c)' = 0$

2. $\frac{d}{dx}x = (x)' = 1$

3. $\frac{d}{dx}x^n = \left(x^n\right)' = nx^{n-1}$

4. $\frac{d}{dx}e^x = \left(e^x\right)' = e^x$

5. $\frac{d}{dx}\ln x = (\ln x)' = \frac{1}{x}$

6. $\frac{d}{dx}\sin x = (\sin x)' = \cos x$

7. $\frac{d}{dx}\cos x = (\cos x)' = -\sin x$

8. $\frac{d}{dx}\tan x = (\tan x)' = \sec^2 x$

9. $\frac{d}{dx}\sec x = (\sec x)' = \sec x \tan x$

10. $\frac{d}{dx}\arcsin x = (\arcsin x)' = \frac{1}{\sqrt{1-x^2}}$

11. $\frac{d}{dx}\arctan x = (\arctan x)' = \frac{1}{1+x^2}$

12. $\int x^n\,dx = \frac{x^{n+1}}{n+1}, \quad n \neq -1$

13. $\int \frac{dx}{x} = \ln|x|$

14. $\int e^x\,dx = e^x$

15. $\int b^x\,dx = \frac{1}{\ln b} b^x$

16. $\int \sin x\,dx = -\cos x$

17. $\int \cos x\,dx = \sin x$

18. $\int \tan x\,dx = \ln|\sec x| = -\ln|\cos x|$

19. $\int \cot x\,dx = \ln|\sin x| = -\ln|\csc x|$

20. $\int \sec x\,dx = \ln|\sec x + \tan x| = \ln\left|\tan\left(\frac{x}{2} + \frac{\pi}{4}\right)\right|$

21. $\int \csc x\,dx = \ln|\csc x - \cot x| = \ln\left|\tan\left(\frac{x}{2}\right)\right|$

22. $\int \sec^2 x\,dx = \tan x$

23. $\int \csc^2 x\,dx = -\cot x$

24. $\int \sec x \tan x\,dx = \sec x$

25. $\int \csc x \cot x\,dx = -\csc x$

26. $\int \frac{dx}{x^2 + a^2} = \frac{1}{a}\arctan\left(\frac{x}{a}\right), \quad a \neq 0$

27. $\int \frac{dx}{x^2 - a^2} = \frac{1}{2a}\ln\left|\frac{x - a}{x + a}\right|$

28. $\int \frac{dx}{\sqrt{a^2 - x^2}} = \arcsin\left(\frac{x}{a}\right), \quad a > 0$

29. $\int \ln x\,dx = x(\ln x - 1)$

Expressions Containing $ax + b$

30. $\int (ax + b)^n\,dx = \frac{(ax + b)^{n+1}}{a(n + 1)}, \quad n \neq -1$

31. $\int \frac{dx}{ax + b} = \frac{1}{a}\ln|ax + b|$

32. $\int \frac{x}{ax + b}\,dx = \frac{x}{a} - \frac{b}{a^2}\ln|ax + b|$

33. $$\int \frac{x}{(ax+b)^2}\,dx = \frac{b}{a^2(ax+b)} + \frac{1}{a^2}\ln|ax+b|$$

34. $$\int \frac{dx}{x(ax+b)} = \frac{1}{b}\ln\left|\frac{x}{ax+b}\right|$$

35. $$\int \frac{dx}{x^2(ax+b)} = -\frac{1}{bx} + \frac{a}{b^2}\ln\left|\frac{ax+b}{x}\right|$$

36. $$\int \sqrt{ax+b}\,dx = \frac{2}{3a}\sqrt{(ax+b)^3}$$

37. $$\int x\sqrt{ax+b}\,dx = \frac{2(3ax-2b)}{15a^2}\sqrt{(ax+b)^3}$$

38. $$\int \frac{dx}{\sqrt{ax+b}} = \frac{2\sqrt{ax+b}}{a}$$

39. $$\int \frac{dx}{x\sqrt{ax+b}} = \frac{1}{\sqrt{b}}\ln\left|\frac{\sqrt{ax+b}-\sqrt{b}}{\sqrt{ax+b}+\sqrt{b}}\right|, \quad b>0$$

40. $$\int \frac{dx}{x\sqrt{ax-b}} = \frac{2}{\sqrt{b}}\arctan\sqrt{\frac{ax-b}{b}}, \quad b>0$$

41. $$\int x^n\sqrt{ax+b}\,dx = \frac{2}{a(2n+3)}\left(x^n\sqrt{(ax+b)^3} - nb\int x^{n-1}\sqrt{ax+b}\,dx\right)$$

42. $$\int \frac{dx}{x^n\sqrt{ax+b}} = -\frac{\sqrt{ax+b}}{(n-1)bx^{n-1}} - \frac{(2n-3)a}{(2n-2)b}\int \frac{dx}{x^{n-1}\sqrt{ax+b}}$$

Expressions Containing ax^2+c, $x^2 \pm p^2$, and p^2-x^2, $p>0$

43. $$\int \frac{dx}{p^2-x^2} = \frac{1}{2p}\ln\left|\frac{p+x}{p-x}\right|$$

44. $$\int \frac{dx}{ax^2+c} = \frac{1}{\sqrt{ac}}\arctan\left(x\sqrt{\frac{a}{c}}\right), \quad a>0,\ c>0$$

45. $$\int \frac{dx}{ax^2-c} = \frac{1}{2\sqrt{ac}}\ln\left|\frac{x\sqrt{a}-\sqrt{c}}{x\sqrt{a}+\sqrt{c}}\right|, \quad a>0,\ c>0$$

46. $$\int \frac{dx}{(ax^2+c)^n} = \frac{1}{2(n-1)c}\,\frac{x}{(ax^2+c)^{n-1}} + \frac{2n-3}{2(n-1)c}\int \frac{dx}{(ax^2+c)^{n-1}}, \quad n>1$$

47. $$\int x\left(ax^2+c\right)^n dx = \frac{1}{2a}\,\frac{(ax^2+c)^{n+1}}{n+1}, \quad n \neq -1$$

48. $$\int \frac{x}{ax^2+c}\,dx = \frac{1}{2a}\ln\left|ax^2+c\right|$$

49. $$\int \sqrt{x^2 \pm p^2}\,dx = \frac{1}{2}\left(x\sqrt{x^2 \pm p^2} \pm p^2\ln\left|x+\sqrt{x^2 \pm p^2}\right|\right)$$

50. $$\int \sqrt{p^2-x^2}\,dx = \frac{1}{2}\left(x\sqrt{p^2-x^2} + p^2\arcsin\left(\frac{x}{p}\right)\right), \quad p>0$$

51. $\displaystyle\int \frac{dx}{\sqrt{x^2 \pm p^2}} = \ln\left|x + \sqrt{x^2 \pm p^2}\right|$

Expressions Containing Trigonometric Functions

52. $\displaystyle\int \sin^2(ax)\,dx = \frac{x}{2} - \frac{\sin(2ax)}{4a}$

53. $\displaystyle\int \sin^3(ax)\,dx = -\frac{1}{a}\cos(ax) + \frac{1}{3a}\cos^3(ax)$

54. $\displaystyle\int \sin^n(ax)\,dx = -\frac{\sin^{n-1}(ax)\cos(ax)}{na} + \frac{n-1}{n}\int \sin^{n-2}(ax)\,dx, \quad n > 0$

55. $\displaystyle\int \cos^2(ax)\,dx = \frac{x}{2} + \frac{\sin(2ax)}{4a}$

56. $\displaystyle\int \cos^3(ax)\,dx = \frac{1}{a}\sin(ax) - \frac{1}{3a}\sin^3(ax)$

57. $\displaystyle\int \cos^n(ax)\,dx = \frac{\cos^{n-1}(ax)\sin(ax)}{na} + \frac{n-1}{n}\int \cos^{n-2}(ax)\,dx$

58. $\displaystyle\int \sin(ax)\cos(bx)\,dx = -\frac{\cos\big((a-b)x\big)}{2(a-b)} - \frac{\cos((a+b)x)}{2(a+b)}, \quad a^2 \neq b^2$

59. $\displaystyle\int \sin(ax)\sin(bx)\,dx = \frac{\sin\big((a-b)x\big)}{2(a-b)} - \frac{\sin\big((a+b)x\big)}{2(a+b)}, \quad a^2 \neq b^2$

60. $\displaystyle\int \cos(ax)\cos(bx)\,dx = \frac{\sin\big((a-b)x\big)}{2(a-b)} + \frac{\sin\big((a+b)x\big)}{2(a+b)}, \quad a^2 \neq b^2$

61. $\displaystyle\int x\sin(ax)\,dx = \frac{1}{a^2}\sin(ax) - \frac{x}{a}\cos(ax)$

62. $\displaystyle\int x\cos(ax)\,dx = \frac{1}{a^2}\cos(ax) + \frac{x}{a}\sin(ax)$

63. $\displaystyle\int x^n\sin(ax)\,dx = -\frac{x^n}{a}\cos(ax) + \frac{n}{a}\int x^{n-1}\cos(ax)\,dx, \quad n > 0$

64. $\displaystyle\int x^n\cos(ax)\,dx = \frac{x^n}{a}\sin(ax) - \frac{n}{a}\int x^{n-1}\sin(ax)\,dx, \quad n > 0$

65. $\displaystyle\int \tan^n(ax)\,dx = \frac{\tan^{n-1}(ax)}{a(n-1)} - \int \tan^{n-2}(ax)\,dx, \quad n \neq 1$

66. $\displaystyle\int \sec^n(ax)\,dx = \frac{\sec^{n-2}(ax)\tan(ax)}{a(n-1)} + \frac{n-2}{n-1}\int \sec^{n-2}(ax)\,dx, \quad n \neq 1$

Expressions Containing Exponential and Logarithm Functions

67. $\displaystyle\int xe^{ax}\,dx = \frac{e^{ax}}{a^2}(ax - 1)$

68. $\displaystyle\int x^n e^{ax}\,dx = \frac{1}{a}x^n e^{ax} - \frac{n}{a}\int x^{n-1}e^{ax}\,dx, \quad n > 0$

69. $\displaystyle\int e^{ax}\sin(bx)\,dx = \frac{e^{ax}}{a^2+b^2}\big(a\sin(bx) - b\cos(bx)\big)$

70. $\displaystyle\int e^{ax}\cos(bx)\,dx = \frac{e^{ax}}{a^2+b^2}\big(a\cos(bx) + b\sin(bx)\big)$

71. $\displaystyle\int x^n \ln(ax)\,dx = x^{n+1}\left(\frac{\ln(ax)}{n+1} - \frac{1}{(n+1)^2}\right), \quad n \neq -1$

72. $\displaystyle\int (\ln x)^n\,dx = x(\ln x)^n - n\int (\ln x)^{n-1}\,dx$

73. $\displaystyle\int \frac{dx}{a+be^{px}} = \frac{x}{a} - \frac{1}{ap}\ln\left|a+be^{px}\right|$

Expressions Containing Inverse Trigonometric Functions

74. $\displaystyle\int \arcsin(ax)\,dx = x\arcsin(ax) + \frac{1}{a}\sqrt{1-a^2x^2}$

75. $\displaystyle\int \arccos(ax)\,dx = x\arccos(ax) - \frac{1}{a}\sqrt{1-a^2x^2}$

76. $\displaystyle\int \operatorname{arccsc}(ax)\,dx = x\operatorname{arccsc}(ax) + \frac{1}{a}\ln\left|ax+\sqrt{a^2x^2-1}\right|$

77. $\displaystyle\int \operatorname{arcsec}(ax)\,dx = x\operatorname{arcsec}(ax) - \frac{1}{a}\ln\left|ax+\sqrt{a^2x^2-1}\right|$

78. $\displaystyle\int \arctan(ax)\,dx = x\arctan(ax) - \frac{1}{2a}\ln(1+a^2x^2)$

79. $\displaystyle\int \operatorname{arccot}(ax)\,dx = x\operatorname{arccot}(ax) + \frac{1}{2a}\ln(1+a^2x^2)$

Index